**PRENTICE-HALL SERIES
IN TECHNICAL MATHEMATICS**

Frank L. Juszli, *Editor*

RICHARD S. PAUL

Department of Mathematics
The Pennsylvania State University

M. LEONARD SHAEVEL

Department of Physics
The Pennsylvania State University

essentials of technical mathematics

PRENTICE-HALL, INC., *Englewood Cliffs, New Jersey*

Library of Congress Cataloging in Publication Data

PAUL, RICHARD S
   Essentials of technical mathematics.

   (Prentice-Hall technical mathematics series)
   1. Mathematics—1961– I. Shaevel, M.
Leonard, joint author. II. Title.
QA39.2.P39   512′.1   73–16322
ISBN 0–13–288084–9

10 9 8 7 6 5 4 3 2

Printed in the United States of America

Cover illustration courtesy of Dorine Lerner

PRENTICE-HALL INTERNATIONAL, INC., *London*
PRENTICE-HALL OF AUSTRALIA, PTY. LTD., *Sydney*
PRENTICE-HALL OF CANADA, LTD., *Toronto*
PRENTICE-HALL OF INDIA PRIVATE LIMITED, *New Delhi*
PRENTICE-HALL OF JAPAN, INC., *Tokyo*

contents

v

7
trigonometry
192

8
systems of
linear equations
243

9
exponents
and radicals
271

10
variation
296

11
complex numbers

308

12
quadratic equations

342

13
exponential and
logarithmic functions

370

14
graphs of the
trigonometric functions

406

15
trigonometric formulas and equations
434

16
oblique triangles and applications of angular measurement
458

17
inequalities
479

18
analytic geometry
504

19
sequences and series
542

preface

This textbook has been written for students in the various programs of engineering technology that are offered in community colleges, technical institutes, and junior colleges. We are in an age where success for such students requires an ability to meet the mathematical demands of technology. Our aim is to provide a firm understanding of those essential mathematical principles and to illustrate meaningful ways in which they are directly applicable to the engineering technologies.

Four of the most frequent criticisms that we receive from students concerning mathematics textbooks are: (1) they are difficult to read, (2) there are not enough examples and those given are difficult to follow, (3) the exercises do not relate to the material covered, and (4) the student does not see meaningful application or relevance in the subject matter.

To eliminate these difficulties, *Essentials of Technical Mathematics* has been written expressly *for* and *to* the student. Considerable effort has been made to present the material in a readable manner. The mathematical concepts and methods involved are thoroughly and carefully explained and motivated wherever possible. We have included more than 1,000 examples which are explained in step-by-step detail and more than 3,500 exercises which are structured to relate to the examples and material in the text. Each exercise set has been designed to include a reasonable number of problems which the student can master without difficulty in order to fix in his mind the basic concepts under study. As a result, the student will quickly gain some degree of confidence. The relevance of the subject matter to the engineering technologies is indicated by the illustrative applications in the examples, the material of the text, and the exercises. These applications require no prior knowledge on the part of the student.

Chapter 0, not necessarily intended as part of a formal course, provides a very brief review of some fundamental arithmetic operations and geometrical concepts; it has been included for the convenience of the student. With the exception of Chapter 0, each chapter contains a review section which includes a programmed-style review covering essential mathematical concepts, and numerous review exercises.

Answers to the odd-numbered problems, all review exercises, and all slide-rule problems appear at the end of the book.

By relegating the unusually thorough descriptions of slide-rule operations to the appendix, they may be presented at any convenient point in the mathematics program without interrupting the continuity of the text. They also form a convenient reference for the student.

Available from the publisher is an extensive instructor's manual which contains answers to all exercises and detailed solutions to a great many of them, including all applied problems.

We express our thanks to Keuffel and Esser Company for the photographs, material, and exercises on the slide rule which have been adapted to our needs and appear in the appendix; and to Cary Baker and Bob Duchacek and the entire Prentice-Hall staff for their cooperation and assistance.

Finally, we express our most grateful appreciation to Professor Frank Kocher of the Department of Mathematics of The Pennsylvania State University whose objective critique of the entire manuscript and important suggestions for its improvement are in evidence throughout the text.

RICHARD S. PAUL

M. LEONARD SHAEVEL

essentials
of
technical
mathematics

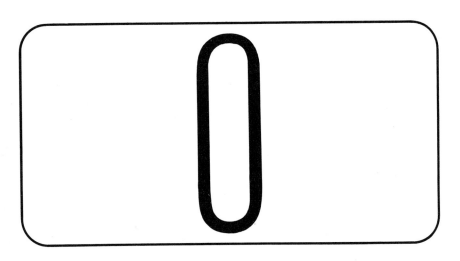

preliminary topics

0-1
INTRODUCTION

Although it is assumed that you have had some prior preparation in mathematics, a brief review of certain arithmetical operations and selected geometrical concepts may be useful. This chapter provides such a review for your convenience; it is not meant to be a summary of all prerequisites for a course of this type. The topics covered are those to which an immediate second exposure may be beneficial. We urge you to devote whatever time is necessary to those topics in which you need review.

0-2
BASIC TERMINOLOGY

Whole numbers are those numbers which are used in counting, such as 0, 1, 2, 13, 44, and 610; they are also called **integers**. Whole numbers are either **even** or **odd**, depending, respectively, on whether they are divisible by 2 or not. Thus, 2, 10, and 1564 are even integers and 1, 3, 29, and 97 are odd integers.

A **prime** number is a whole number greater than 1 that is divisible only by itself and 1. Thus, 2, 3, 5, 7, 11, 13, and 17 are prime numbers, but 15 is not, since 15 is divisible by 3 and 5 as well as 15 and 1.

Whole numbers are represented by combinations of the ten **digits** 0, 1, 2, 3, 4, 5, 6, 7, 8, and 9. An analysis of the numeral 234 would be that the digit 4 occupies the units place and represents four 1's (4), the digit 3 occupies the tens place and represents three groups of 10 (30), and the digit 2 occupies the hundreds place and represents two groups of 100 (200).

It seems reasonable to assume that you are totally familiar with addition, subtraction, multiplication, and division of whole numbers, and so we dispense with any further discussion of these operations. It is appropriate, however, to state two important properties that are valid for all types of numbers:

1. *The addition and multiplication of a group of numbers can be performed in any order.*

Thus, $2 + 3 + 4 = 4 + 2 + 3 = 4 + 3 + 2 = 9$. Similarly, $(3)(2)(4) = (2)(3)(4) = (4)(3)(2) = 24$. Since the numbers 2, 3, and 4 when multiplied together give 24, each of 2, 3, and 4 is called a **factor** of 24, and 24 is called the **product** of 2, 3, and 4.

2. *The operations of subtraction and division must be performed in the given order.*

Thus, $6 \div 3$ is not the same as $3 \div 6$. In $6 \div 3 = 2$, 6 is called the **dividend**, 3 is called the **divisor**, and 2 is the **quotient**.

In computations involving more than one operation, multiplications and divisions should be performed first, followed by the additions and subtractions. For example,

a. $6 + (6)(2) = 6 + 12 = 18$.

b. $12 - (2)(2) = 12 - 4 = 8$.

c. $(12)(2) + 6 \div 3 = 24 + 2 = 26$.

However, any series of operations within parentheses should be performed first. Thus,

d. $12 + (2 + 3) - (4)(6 - 3)$
$= 12 + 5 - (4)(3)$
$= 12 + 5 - 12$
$= 17 - 12$
$= 5$.

e. $16 + (12 \div 3)(4 - 3)$
    $= 16 + (4)(1)$
    $= 16 + 4$
    $= 20.$

To avoid any ambiguity in part c above, $6 \div 3$ can be written $\frac{6}{3}$, and it is preferable to write

$$(12)(2) + \tfrac{6}{3} \qquad \text{instead of} \qquad (12)(2) + 6 \div 3.$$

EXERCISE 0-2

Perform the indicated operations in Problems 1–20.

1. $10 + (2)(4) - (8 \div 2)$
2. $(12 - 4)(6) + 2(5 + 1)$
3. $1600 - (5)(210 \div 2)(2)$
4. $(2 \div 2) + 7(6)(3 - 1) - 17$
5. $(6 + 4 - 5) \div (2 + 3 - 4) + (7)(6 + 2)$
6. $(36 \div 6)(6 \div 2)(1 + 2) - 6$
7. $(2 + 16)(8 - 2) \div (5 - 2)(1)$
8. $(12 - 4)(3 + 1) + 2(16 \div 4) - 12$
9. $(223)(21)(84)$
10. $(1062)(2)(56)$
11. $8525 \div 341$
12. $10,209 \div 123$
13. $(621)(4)(16) \div 8$
14. $26,451 + 26,245 + 86,216 - 726 + 624 - 18,707$
15. $576 \div (2 + 16)$
16. $8622 \div (13 - 4)$
17. $(12,604)(865)$
18. $33,454 \div 86$
19. $1,864,200 \div 17,925$
20. $1,003,005 \div 6471$
21. The tens and units digits of the number 862 are reversed and the new number is subtracted from 862. What is the result?
22. What digit must be changed in 6257 to make it an even number?

23. If the dividend and quotient of a division are both 6, what is the divisor?

24. The product of three factors is 504. Two of the factors are 6 and 7. Find the third factor.

25. In the number 62,417, what place does the 6 occupy?

0-3

COMMON FRACTIONS

A common fraction such as $\frac{2}{5}$ consists of the **numerator** 2 and the **denominator** 5. It can be considered as denoting the division of the **dividend** 2 by the **divisor** 5 and, as a result, is often referred to as being the **quotient** of 2 divided by 5, or, simply, the quotient $\frac{2}{5}$.

Fractions are of two types, depending on the relative values of the numerator and denominator. If the numerator is less than the denominator, such as in $\frac{3}{8}$, the fraction is said to be a **proper** fraction and its value is less than 1. If, on the other hand, the numerator is equal to or greater than the denominator, the fraction, which has a value of at least 1, is said to be **improper**. Fractions such as $\frac{3}{3}$, $\frac{16}{7}$, and $\frac{214}{161}$ are examples of improper fractions. Any improper fraction can be expressed as the sum of an integer and a proper fraction, called a **mixed number**. For example,

$$\tfrac{23}{6} = 3 + \tfrac{5}{6} = 3\tfrac{5}{6}.$$

Here the improper fraction $\frac{23}{6}$ is equivalent to the mixed number $3\frac{5}{6}$.

To write an improper fraction as a mixed number, it is only necessary to divide the numerator by the denominator, thus obtaining the integer part, and add to this the fraction whose numerator is the remainder of the division and whose denominator is the denominator of the original fraction. Thus,

a. $\tfrac{16}{13} = 1 + \tfrac{3}{13} = 1\tfrac{3}{13}.$

b. $\tfrac{14}{3} = 4 + \tfrac{2}{3} = 4\tfrac{2}{3}.$

c. $\tfrac{27}{4} = 6 + \tfrac{3}{4} = 6\tfrac{3}{4}.$

A fundamental principle used in computations involving fractions is that *multiplying or dividing both the numerator and denominator of a fraction by the same nonzero number does not change the value of the fraction.* In effect, the fraction is being multiplied by 1. Thus,

a. $\dfrac{3}{8} = \dfrac{(3)(4)}{(8)(4)} = \dfrac{12}{32}.$

b. $12 = \dfrac{12}{1} = \dfrac{(12)(2)}{(1)(2)} = \dfrac{24}{2}.$

c. $\dfrac{12}{16} = \dfrac{12 \div 4}{16 \div 4} = \dfrac{3}{4}.$

d. $\dfrac{210}{26} = \dfrac{210 \div 2}{26 \div 2} = \dfrac{105}{13}.$

By multiplying both the numerator and the denominator of a fraction by the same number, we can express the fraction as an equivalent fraction having any desired denominator. Thus, if the fraction $\frac{5}{8}$ is to be written as an equivalent fraction whose denominator is 24, it should be clear that since the denominator must be multiplied by 3, so must the numerator.

$$\frac{5}{8} = \frac{(5)(3)}{(8)(3)} = \frac{15}{24}.$$

A fraction is said to be in **lowest terms** if the numerator and denominator have no common factor other than 1. Thus, $\frac{18}{12}$ is not in lowest terms since 18 and 12 have a common factor of 6, that is, $18 = (6)(3)$ and $12 = (6)(2)$. However, by dividing the numerator and denominator by 6 we get the equivalent fraction

$$\frac{18}{12} = \frac{18 \div 6}{12 \div 6} = \frac{3}{2},$$

which is in lowest terms. Moreover, if we write the fraction $\frac{18}{12}$ as

$$\frac{(6)(3)}{(6)(2)},$$

the division by 6 is seen to be nothing more than the familiar process of **cancellation**, whereby factors common to the numerator and denominator cancel each other, denoted by slashes ($/$):

$$\frac{\cancel{(6)}(3)}{\cancel{(6)}(2)} = \frac{3}{2}.$$

The process of repetitive cancellation can be used as follows, where in each step, a factor 2 is removed:

$$\frac{104}{224} = \frac{\overset{52}{\cancel{104}}}{\underset{112}{\cancel{224}}} = \frac{\overset{26}{\cancel{52}}}{\underset{56}{\cancel{112}}} = \frac{\overset{13}{\cancel{26}}}{\underset{28}{\cancel{56}}} = \frac{13}{28}.$$

In a similar fashion,

$$\frac{\overset{1}{\cancel{(12)}(6)\cancel{(3)}(14)}}{\underset{7}{(22)(14)\cancel{(21)}(18)}} = \frac{\overset{2}{(12)(6)\cancel{(14)}}}{\underset{1}{(22)(14)\cancel{(7)}(18)}} = \frac{\overset{1}{(12)(6)\cancel{(2)}}}{\underset{11}{\cancel{(22)}(14)(18)}}$$

$$= \frac{\overset{1}{(12)\cancel{(6)}}}{\underset{3}{(11)(14)\cancel{(18)}}} = \frac{\overset{4}{\cancel{(12)}}}{\underset{1}{(11)(14)\cancel{(3)}}}$$

$$= \frac{\overset{2}{\cancel{(4)}}}{\underset{7}{(11)\cancel{(14)}}} = \frac{2}{77}.$$

This can be written more compactly as

$$\frac{\overset{2}{\underset{1}{\cancel{4}}}\ \overset{}{1}\ \overset{}{1}\ \overset{1}{\cancel{2}}}{\cancel{(12)}\cancel{(6)}\cancel{(3)}\cancel{(14)}} \Big/ \underset{\underset{1\ \ 1}{11\ \ 7\ \ \cancel{7}\ \ \cancel{3}}}{\cancel{(22)}\cancel{(14)}\cancel{(21)}\cancel{(18)}} = \frac{2}{77}.$$

The sum (or difference) of two fractions having the same denominator is a fraction having the same denominator as the original fractions but whose numerator is the sum (or difference) of the numerators of the original fractions. For example,

a. $\dfrac{2}{3} + \dfrac{5}{3} = \dfrac{2+5}{3} = \dfrac{7}{3}.$

b. $\dfrac{8}{9} - \dfrac{6}{9} = \dfrac{8-6}{9} = \dfrac{2}{9}.$

c. $\dfrac{7}{18} + \dfrac{3}{18} - \dfrac{5}{18} = \dfrac{7+3-5}{18} = \dfrac{5}{18}.$

d. $\dfrac{12}{27} - \dfrac{6}{27} + \dfrac{4}{27} - \dfrac{1}{27} = \dfrac{12-6+4-1}{27} = \dfrac{9}{27} = \dfrac{1}{3}.$

To add fractions when the denominators are not alike, the fractions must first be expressed as equivalent fractions, all of which have the same denominator. The appropriate denominator to choose is the smallest number such that each of the denominators is a factor of it. Such a number is called the **least common denominator**, denoted L.C.D. For example, the denominators of the fractions $\frac{1}{5}$, $\frac{3}{8}$, and $\frac{7}{10}$ are each a factor of 40, and 40 is the **least** such number which has 5, 8, and 10 as factors. Thus, 40 is the L.C.D.

In general, to determine the L.C.D. of a group of fractions, first express each denominator as a product of prime factors. The L.C.D. is the product of the prime factors, each occurring in the product the greatest number of times it occurs in any one denominator. For example, to determine the L.C.D. of $\frac{1}{15}$, $\frac{3}{8}$, and $\frac{7}{10}$ we write the denominators in the following fashion:

$$15 = (5)(3)$$
$$8 = (2)(2)(2)$$
$$10 = (5)(2).$$

The factor 2 occurs at most three times and the factors 5 and 3 each occur at most once. Thus,

$$\text{L.C.D.} = (2)(2)(2)(5)(3) = 120.$$

Similarly, the L.C.D. of $\frac{1}{6}$, $\frac{2}{9}$, and $\frac{5}{12}$ would be determined as follows:

$$6 = (3)(2)$$
$$9 = (3)(3)$$
$$12 = (3)(2)(2)$$

and

$$\text{L.C.D.} = (2)(2)(3)(3) = 36.$$

To add the fractions $\frac{1}{3}$, $\frac{8}{9}$, and $\frac{5}{12}$ we first determine that the L.C.D. is 36 and write equivalent fractions as follows:

$$\frac{1}{3} = \frac{12}{36}, \qquad \frac{8}{9} = \frac{32}{36}, \qquad \frac{5}{12} = \frac{15}{36}.$$

Hence,

$$\frac{12}{36} + \frac{32}{36} + \frac{15}{36} = \frac{12 + 32 + 15}{36} = \frac{59}{36}.$$

Similarly,

$$\frac{2}{3} + \frac{6}{8} + \frac{4}{16} = \frac{32}{48} + \frac{36}{48} + \frac{12}{48} = \frac{80}{48} = \frac{5}{3}.$$

Subtraction of fractions is performed in a similar manner:

$$\frac{7}{4} - \frac{3}{2} + \frac{1}{5} = \frac{35}{20} - \frac{30}{20} + \frac{4}{20} = \frac{35 - 30 + 4}{20} = \frac{9}{20}.$$

We can write a mixed number as an improper fraction by expressing the integer part as a fraction and adding it to the fractional part. Thus,

a. $9\frac{3}{4} = \frac{9}{1} + \frac{3}{4} = \frac{36}{4} + \frac{3}{4} = \frac{39}{4}.$

b. $8\frac{3}{7} = \frac{8}{1} + \frac{3}{7} = \frac{56}{7} + \frac{3}{7} = \frac{59}{7}.$

To add mixed numbers, add the integers first and treat the fractions as before. Thus,

$$7 + 3\frac{3}{4} + 2\frac{5}{3} = (7 + 3 + 2) + (\tfrac{3}{4} + \tfrac{5}{3})$$
$$= 12 + (\tfrac{9}{12} + \tfrac{20}{12})$$
$$= 12 + \tfrac{29}{12} = 12 + 2\tfrac{5}{12} = 14\tfrac{5}{12}.$$

For subtraction of mixed numbers, a slight modification in the procedure may be necessary, as illustrated in the following example:

$$3\tfrac{3}{4} - 2\tfrac{7}{8}.$$

If we proceed as before, we have

$$(3 - 2) + (\tfrac{3}{4} - \tfrac{7}{8}) = 1 + (\tfrac{6}{8} - \tfrac{7}{8})$$
$$= 1 + \left(\frac{6 - 7}{8}\right).$$

Since we cannot yet subtract 7 from 6, we rewrite the problem as

$$3\tfrac{3}{4} - 2\tfrac{7}{8} = \tfrac{15}{4} - \tfrac{23}{8} = \tfrac{30}{8} - \tfrac{23}{8} = \tfrac{7}{8}.$$

Here we transformed each mixed number to a fraction immediately. Certainly, this method is applicable to either addition or subtraction; you should use the method with which you are most comfortable.

The product of two or more fractions is a fraction whose numerator is the product of the numerators of the fractions and whose denominator is the product of the denominators of the fractions. Thus,

$$\frac{3}{5} \cdot \frac{8}{4} \cdot \frac{2}{9} = \frac{\cancel{(3)}\cancel{(8)}^{2}\cancel{(2)}}{\cancel{(5)}\cancel{(4)}_{1}\cancel{(9)}_{3}} = \frac{4}{15}$$

and

$$(5\tfrac{2}{3})(2\tfrac{3}{4})(\tfrac{2}{17}) = \left(\frac{17}{3}\right)\left(\frac{11}{4}\right)\left(\frac{2}{17}\right) = \frac{\cancel{(17)}^{1}(11)\cancel{(2)}^{1}}{\cancel{(3)}\cancel{(4)}_{2}\cancel{(17)}_{1}} = \frac{11}{6}.$$

For division of fractions, we first introduce the notion of the **reciprocal** of a number. The reciprocal of a number is a number that, when multiplied by the original number, is 1. Thus, the reciprocal of 2 is $\frac{1}{2}$, since $(2)(\frac{1}{2}) = 1$. Similarly, the reciprocals of 3, $\frac{2}{3}$, and $\frac{5}{9}$ are $\frac{1}{3}$, $\frac{3}{2}$, and $\frac{9}{5}$, respectively. As a rule, we can state that the reciprocal of a nonzero number is 1 divided by that number. Thus, the reciprocal of $\frac{2}{3}$ is

$$\frac{1}{\frac{2}{3}}$$

since

$$\left(\frac{\frac{2}{3}}{1}\right)\left(\frac{1}{\frac{2}{3}}\right) = 1.$$

Also

$$\left(\frac{2}{3}\right)\left(\frac{3}{2}\right) = \frac{(2)(3)}{(3)(2)} = 1.$$

From an observation of the last example, we see that the reciprocal of a number can be formed by interchanging its numerator with its denominator, sometimes referred to as **inverting**.

To find the quotient

$$\frac{\frac{3}{8}}{\frac{4}{5}}$$

we multiply the numerator and denominator by the reciprocal of the denominator. Hence,

$$\frac{\frac{3}{8}}{\frac{4}{5}} = \frac{\left(\frac{3}{8}\right)\left(\frac{5}{4}\right)}{\left(\frac{4}{5}\right)\left(\frac{5}{4}\right)} = \frac{\left(\frac{3}{8}\right)\left(\frac{5}{4}\right)}{1} = \left(\frac{3}{8}\right)\left(\frac{5}{4}\right) = \frac{15}{32}.$$

Looking at the next-to-last step, however, we see that to divide $\frac{3}{8}$ by $\frac{4}{5}$ it was only necessary to multiply $\frac{3}{8}$ by the reciprocal of $\frac{4}{5}$—that is, invert the divisor and multiply, which is the rule with which you are probably familiar. One additional point that must be emphasized is that **division by zero has no meaning**; hence, the reciprocal of zero does not exist. It is completely **erroneous** to say that $\frac{1}{0} = 0$, and you should fix this fact firmly in mind. A few more examples will clarify these ideas.

a. $\frac{7}{8} \div \frac{5}{4} = \left(\frac{7}{8}\right)\left(\frac{4}{5}\right) = \frac{7}{10}.$

b. $\frac{3}{8} \div 4 = \left(\frac{3}{8}\right)\left(\frac{1}{4}\right) = \frac{3}{32}.$

c. $8\frac{1}{5} \div \frac{5}{3} = \left(\frac{41}{5}\right)\left(\frac{3}{5}\right) = \frac{123}{25}.$

To conclude this section three observations are appropriate:

1. *If two fractions are to be compared as to relative value, the fractions can be transformed to fractions having the same denominator and their numerators then compared.*

Thus, if comparing $\frac{7}{8}$ and $\frac{8}{9}$, we write

$$\frac{7}{8} = \frac{63}{72} \qquad \text{and} \qquad \frac{8}{9} = \frac{64}{72}$$

and conclude that $\frac{7}{8}$ is smaller than $\frac{8}{9}$ since 63 is less than 64.

2. *Dividing the denominator of a fraction by a number is equivalent to multiplying the numerator by the same number.*

For example, given the fraction $\frac{1}{3}$ and dividing the denominator by 2,

$$\frac{1}{\frac{3}{2}} = \frac{1(2)}{3} = \frac{2}{3}.$$

Similarly, given the fraction $\frac{8}{6}$ and dividing the denominator by 7,

$$\frac{8}{\frac{6}{7}} = \frac{(8)(7)}{6} = \frac{28}{3}.$$

3. *Dividing the numerator of a fraction by a number is equivalent to multiplying the denominator by the same number.*

Hence, given the fraction $\frac{3}{4}$ and dividing the numerator by 8,

$$\frac{\frac{3}{8}}{4} = \frac{3}{(4)(8)} = \frac{3}{32}.$$

EXERCISE 0-3

In Problems 1–18 perform the indicated operations.

1. $\frac{2}{3} + \frac{3}{4} + \frac{5}{8}$

2. $\frac{4}{5} + \frac{9}{8} + \frac{1}{12}$

3. $3\frac{2}{3} + 4\frac{5}{8} - 6\frac{7}{11}$

4. $4\frac{5}{7} - 2\frac{7}{5} + 1\frac{2}{35}$

5. $7\frac{2}{3} - 4\frac{3}{4}$

6. $8\frac{1}{12} - 6\frac{7}{8}$

7. $\left(\frac{3}{4}\right)\left(\frac{8}{5}\right)\left(\frac{4}{9}\right)$

8. $\left(\frac{3}{5}\right)\left(\frac{4}{11}\right)\left(\frac{7}{3}\right)\left(\frac{25}{4}\right)$

9. $\dfrac{\left(\frac{2}{3}\right)\left(\frac{4}{5}\right) + 1}{\left(\frac{6}{7}\right) + 2}$

10. $\left(\frac{8}{5} + \frac{3}{9}\right) \div \left(2 + \frac{4}{7}\right)$

11. $\dfrac{\frac{8}{3}}{\frac{5}{4}}$

12. $\dfrac{7}{\frac{\frac{1}{3}}{8}}$

13. $\dfrac{\dfrac{3\frac{1}{5}}{\frac{1}{3}}}{\frac{14}{15} + 1}{\frac{7}{15}}$

14. $\dfrac{7 - \frac{2}{3}}{15 - (6 \div \frac{3}{4})}$

15. $\dfrac{\dfrac{(7\frac{1}{2})(6\frac{2}{3})}{1}}{(6\frac{2}{3})(4) + 1}$

16. $\dfrac{1 + \frac{1}{2}}{1 + \dfrac{1}{1 + \frac{1}{2}}}$

17. $\dfrac{(8\frac{1}{3})\left(1 - \dfrac{\frac{1}{2}}{3}\right)}{(2 + \frac{1}{3})\left(6 - \dfrac{1}{\frac{1}{2}}\right)}$

18. $\dfrac{(\frac{1}{5} - \frac{1}{10})(6\frac{1}{2} \div \frac{4}{7}) + 1}{1 - (\frac{1}{4})(1 + \frac{2}{3})(1 - \frac{3}{4})}$

19. Arrange the following numbers in order from the largest to the smallest: $\frac{3}{5}, \frac{15}{28}, \frac{22}{10}, \frac{2}{3}, \frac{3}{4}$.

20. How many $(\frac{1}{8})$'s are there in $4\frac{1}{4}$?

0-4
OPERATIONS WITH DECIMALS

A **decimal fraction** is a fraction whose denominator is a power of 10, that is, 10, 100, 1000, etc. In writing a decimal fraction, such as $\frac{6}{10}$, $\frac{15}{100}$, $\frac{121}{1000}$, or $\frac{62}{1000}$, we may replace the denominator by placing a decimal point in the numerator, positioning the point so that the number of digits to its right is equal to the number of zeros in the denominator of the original fraction. Thus,

a. $\frac{6}{10} = .6,$ one digit to the right.

b. $\frac{15}{100} = .15,$ two digits to the right.

c. $\frac{121}{1000} = .121,$ three digits to the right.

d. $\frac{1624}{100} = 16.24,$ two digits to the right.

e. $\frac{62}{1000} = .062,$ three digits to the right.

Note that since $\frac{6}{10} = .6$ and $\frac{6}{10} = \frac{60}{100} = .60$, adding zeros to the right of the decimal point after the last digit of a decimal does not change its value. Note also that we can consider an integer to be a decimal number by placing a decimal point after the rightmost digit.

In order to add or subtract decimals, we must place each number so that the decimal points are aligned, one under another, and then add or subtract the numbers without regard to the decimal point, remembering to include in the result a decimal point that is aligned with the others. For example,

a. $.123 + 1.624 + .0621 + .1$

$$
\begin{array}{r}
.1230 \\
1.6240 \\
.0621 \\
.1000 \\
\hline
1.9091
\end{array}
$$

b. $.726 - .0246$

$$
\begin{array}{r}
.7260 \\
-.0246 \\
\hline
.7014
\end{array}
$$

In the multiplication of decimals, the decimal point can, as usual, be disregarded as far as the calculation is concerned. After obtaining the product, however, we place a decimal point at such a position so that there will be as many digits to its right as the sum of the number of digits to the right of the decimal point in each of the factors. If the number of digits in the product is not sufficient, additional zeros can be inserted to the left of the leftmost digit. Thus,

$$
\begin{array}{r}
6\,2.4 \\
1.2\,3 \\
\hline
1\,8\,7\,2 \\
1\,2\,4\,8 \\
6\,2\,4 \\
\hline
7\,6.7\,5\,2
\end{array}
$$

The factor 62.4 has one digit to the right of the decimal point and the factor 1.23 has two digits to the right. Their sum, namely 3, is the number of digits to the right of the decimal point in the product. As an additional example consider

$$
\begin{array}{r}
.1\,2\,3 \\
.0\,1\,2 \\
\hline
2\,4\,6 \\
1\,2\,3 \\
\hline
.0\,0\,1\,4\,7\,6
\end{array}
$$

Here, two zeros were inserted in the result to agree with the requirement that there be six digits to the right of the decimal point. The inclusion of the additional zeros is necessary to determine the correct value of the decimal.

It should be clear that to multiply any decimal by 10 we need only move the decimal point one position to the right. Similarly, we move the decimal point two, three, and four places to the right to multiply by 100, 1000, and 10,000, respectively. Hence,

a. $(12.12)(10) = 121.2$

b. $(12.12)(100) = 1212. = 1212$

c. $(12.12)(10000) = 121200. = 121200$

To divide a decimal by a decimal, both the numerator and denominator should first be multiplied by the power of 10 that will make the denominator an integer. Thus,

$$
\frac{62.314}{72.62} \quad \text{becomes} \quad \frac{6231.4}{7262.}
$$

or

$$
72.62\,\overline{)62.314} \quad \text{becomes} \quad 7262\,\overline{)6231.4}
$$

The division is then performed without regard to the decimal point, after which a decimal point must be inserted in the quotient in such a manner that it is aligned with the one in the dividend.

$$
\begin{array}{r}
.85808 \\
7262\overline{)6231.40000} \\
58096 \\
\hline
42180 \\
36310 \\
\hline
58700 \\
58096 \\
\hline
60400 \\
58096 \\
\hline
2304
\end{array}
$$

In this case the result is not an exact quotient, and it is necessary to round off the quotient by disregarding all digits beyond a desired place. If the first digit to the right of the last retained digit is less than a 5, the last retained digit remains unchanged. If the first digit to the right of the last retained digit is at least 5, the last retained digit is increased by 1—unless that digit, the one to the right of the last retained digit, is a 5 followed by zeros. In such cases we adopt the convention that the last retained digit will be left an even integer and, hence, will be left unchanged if it is even or increased by 1 if it is odd.

The digits used to represent the accuracy of a number are said to be **significant figures**. Thus, if we wanted to consider four, three, two, or one significant figures in the quotient above, by the given rules we would have

.8581	four significant figures
.858	three significant figures
.86	two significant figures
.9	one significant figure.

If a decimal is less than 1, the zeros to the right of the decimal point that occur **before** the first nonzero digit **are not** considered significant figures. Hence, .023, .00062141, and .0001 have two, five, and one significant figures, respectively. However, zeros to the right of the last nonzero digit **are** significant. Hence, .06100 has four significant figures and .000060 has two significant figures. In problems involving multiplication or division of approximate numbers, the result usually should not have more significant figures than the least number of significant figures occurring in any one of the original numbers.

To divide a decimal by a power of 10, merely move the decimal point to the left the same number of places as there are zeros in the power of ten. Hence,

a. $\dfrac{26.24}{10} = 2.624$

b. $\dfrac{.08624}{100} = .0008624$

c. $\dfrac{1183.421}{1000} = 1.183421$

As a final note on decimals, we can write every decimal as a common fraction by replacing the decimal point by the appropriate power of 10. Thus,

a. $62.5 = \frac{625}{10} = \frac{125}{2}$.

b. $2.432 = \frac{2432}{1000} = \frac{304}{125}$.

Similarly, we can change a fraction to a decimal by placing a decimal point in the numerator and dividing. Hence,

a. $\dfrac{5}{8} = \dfrac{5.}{8} = .625$

b. $\frac{43}{64} = .671875$

EXERCISE 0-4

1. Add the following numbers, assuming they are exact.
   a. $.021 + .1206 + 12.6 + 123$
   b. $1.006 + 1.0 + .629 + .4$
   c. $71.62 + 0.01 + .0006 + 4.1$
   d. $84.0264 + .621 + .006 + 71$

2. Subtract, assuming the numbers are exact.
   a. $.6241 - .5968$ b. $.0026 - .00094$
   c. $1.006 - .99$ d. $71.6241 - 12.345$

3. Multiply, giving the answer to the appropriate number of significant figures.
   a. $(72.64)(.023)$ b. $(861.4)(.6241)$
   c. $(.00601)(1.005)(.04)$ d. $(.621)(0.101)(.8)(1000)$

4. Divide, giving the answer to the appropriate number of significant figures.
   a. $24.530688 \div .312$ b. $246.21 \div 864.34$
   c. $.00241 \div 1.624$ d. $0.624 \div 1.002$

5. Change each of the given decimals to fractions.

    a. .6241   b. 1.0241

    c. .006   d. 22.01

    e. .62415   f. 11.61

6. Change each of the given fractions to decimals.

    a. $\frac{1}{8}$ b. $\frac{63}{64}$

    c. $\frac{3}{32}$ d. 160.7/2500

7. Simplify and write the answer as a fraction.

    a. $\dfrac{(34)(1.4)(.25)(8.1)}{(.16)(1.5)(17)(2.7)}$ b. $\dfrac{(.018)(3.2)(36)(.48)}{(.80)(.012)(4.2)(33)}$

8. If 18.621 and 7.21 are the measured dimensions in inches of a rectangular plate, the area of the plate is found by determining the product of these two numbers. Why is it incorrect to give the area as 134.25741 square inches?

0-5
GEOMETRICAL CONCEPTS AND FORMULAS

It will be necessary at times to make reference to some fundamental results of plane geometry. Although you should be familiar with these, the most important ones are summarized below.

    a. The *Pythagorean theorem* is a statement that the square of the hypotenuse of a right triangle is equal to the sum of the squares of the legs of the triangle. Symbolically,

$$c^2 = a^2 + b^2$$

and

$$c = \sqrt{a^2 + b^2}.$$

    b. The sum of the angles in any triangle is 180°.

    c. Two triangles are *similar* if their corresponding angles are equal. The triangles in the figure below are similar.

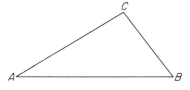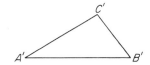

If two triangles are similar, the lengths of their corresponding sides are proportional. That is, for example,

$$\frac{AB}{A'B'} = \frac{CB}{C'B'}.$$

d. In an *isosceles* triangle (one in which at least two sides are equal) the angles opposite the equal sides are equal.

e. An *equilateral* triangle is one having three equal sides.

f. A *right angle* is an angle of 90°.

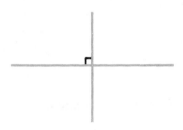

g. Two lines are *perpendicular* to each other if they intersect at right angles.

h. The *bisector* of an angle is the line which divides the angle into two equal angles.

i. If two parallel lines are cut by a transversal, the alternate interior angles (*a* and *b*) formed are equal.

j. The area of a triangle is equal to one-half the product of the base and height.

k. A *central angle* of a circle is an angle formed by two radii of the circle. In the diagram below, angle *A* is a central angle.

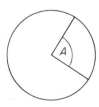

l.   Two triangles are *congruent* if they can be made to coincide, that is, if they have the same size and shape.

m.   In the following useful formulas, r denotes a radius, and h an altitude.

(1) Circle: area $= \pi r^2$, circumference $= 2\pi r$.

(2) Right circular cylinder:

$$\text{volume} = \pi r^2 h$$

$$\text{lateral surface area} = 2\pi r h$$

$$\text{total surface area} = 2\pi r h + 2\pi r^2.$$

(3) Right circular cone:

$$\text{volume} = \tfrac{1}{3}\pi r^2 h.$$

(4) Sphere:

$$\text{volume} = \tfrac{4}{3}\pi r^3$$

$$\text{surface area} = 4\pi r^2.$$

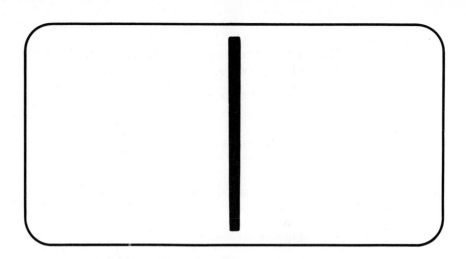

the real number system

SETS

In some respects, beginning the study of mathematics is similar to beginning the study of a language. However, the vocabulary of the language of mathematics, in addition to words and phrases which are explicitly defined, also includes a number of concepts and associated symbols. One of these concepts is that of a *set*. Our purpose here is not to present a detailed discussion of sets and set theory, but rather to introduce to you those concepts and symbols of set theory which will permit more meaningful and convenient discussions in the remainder of the text.

The notion of a **set** is fundamental in mathematics. In simplest terms, a set can be described as a well-defined collection of objects. By *well-defined* we mean that we can determine which objects do, and which do not, belong to the set. For example, we can speak of the set of even numbers between 5 and 11, namely the numbers 6, 8, and 10.

Every object in a set is called an **element** or **member** of the set. Usually, braces are used to indicate that a set is being specified, and we can give a set a name by means of a symbol, such as a letter. For example, the set A whose elements are the numbers 6, 8, and 10 can be written as

$$A = \{6, 8, 10\}.$$

The members of a set may be any objects and are not necessarily limited to being numbers. Thus we may speak of the set W whose members are the names of all the chemical elements contained in ordinary water (H_2O):

$$W = \{\text{Hydrogen, Oxygen}\}.$$

If x is an element of a set A, we can use the **membership symbol** $\in$ and write $x \in A$, read "x is a member of A" or "x belongs to A." If x is not a member of A, we accordingly write $x \notin A$. Hence, if $A = \{6, 8, 10\}$, then $6 \in A$, $8 \in A$, and $10 \in A$, but $7 \notin A$.

Two sets, A and B, are said to be *equal*, written $A = B$, if and only if they consist of the same elements. That is, $A = B$ when every element of A is an element of B and every element of B is an element of A. Thus, if $M = \{1, 2, 3\}$ and $N = \{3, 1, \frac{4}{2}\}$, then $M = N$. Note that *the order of listing the elements of a set is of no consequence.* Likewise, *repetition of an element in the listing of a set does not affect the set itself*; thus $\{4, 5\} = \{4, 5, 4\}$.

When a set is specified by listing all its elements, such a list is called a **roster**. Although this is not always feasible, especially if a set has many elements, there are, however, two other means by which a set may be specified: (1) a **verbal description** and (2) **set-builder notation**. Both methods are illustrated in Example 1.

EXAMPLE
1

a. $E = \{\text{names of all known chemical elements}\}$ is a set specified by a *verbal description* of its members.

b. $V = \{x \mid x \text{ is a whole number greater than } 5\}$ is a set specified by *set-builder* notation. The *literal symbol* x represents an arbitrary member of the set and the vertical bar means "such that." The complete statement is read: "V is the set of all x such that x is a whole number greater than 5." Hence, V is the set consisting of the numbers 6, 7, 8, etc.

Sets can be classified as either *finite* or *infinite,* depending on whether or not the *distinct* elements in the set are of a limited number. That is, when the process of counting distinct elements is unending, the set is said to be **infinite**; when the counting process terminates, the set is **finite**. In Example 1, set E is finite, whereas set V is infinite. We can also write V as

$$V = \{6, 7, 8, \ldots\},$$

where the three dots indicate that the listing of elements continues indefinitely. Note that although each element of V is not identified individually, it is clear as to what these elements are. At times, three dots are also used in connection with finite sets. For instance, we understand the set $T = \{2, 4, 6, \ldots, 20\}$ to be the set of all even numbers from 2 to 20.

Another important term in set theory is that of *subset*. A set A is said to be a **subset** of a set B, denoted $A \subset B$, if and only if every element of A is also an element of B. For example, if $A = \{1, 2\}$ and $B = \{1, 2, 3, 4\}$, then both elements which comprise A are also in B and, hence, $A \subset B$. However, since $3 \in B$ and $3 \notin A$, clearly B is not a subset of A. If $A \subset B$, we also say that A *is contained* in B or that B *contains* A. Since for every set A it is true that $x \in A$ when $x \in A$, it follows that $A \subset A$; that is, *every set is a subset of itself*. Convince yourself of this fact.

EXERCISE 1-1

In Problems 1–10, classify the given statement as either true or false.

1. $2 \in \{1, 3, 2\}$
2. $\{2\} \subset \{1, 2, 3\}$
3. $\frac{1}{7} \in \{1, \frac{1}{2}, \frac{1}{3}, \frac{1}{4}, \ldots\}$
4. $\{1, 2, 3, \ldots, 50\}$ is infinite
5. $\{1, 2, 3\} = \{2, 3, 1\}$
6. $\{a, 1, 2\} = \{1, a, 2, a\}$
7. $\{1, 2, 3\} \subset \{1, 2\}$
8. $\{1, 2, 3\} \subset \{1, 2, 3\}$
9. $\{1, 0\} = \{1\}$
10. $\{4\} \in \{2, 3, 4\}$

In Problems 11–17, state whether the given set is finite or infinite.

11. $A = \{$the electrons in the chair in which you are seated$\}$
12. $B = \{$the grains of sand in the Sahara Desert$\}$
13. $C = \{$the people on earth at 2:00 P.M. today$\}$
14. $D = \{$the man-hours necessary to produce a blueprint$\}$
15. $E = \{1, \frac{1}{2}, \frac{1}{3}, \frac{1}{4}, \ldots\}$
16. $F = \{1, 1, 1, 1, \ldots\}$
17. $G = \{0, 1, 0, 1, \ldots\}$

In Problems 18–20, specify the given set by means of a listing of its members.

18. $A = \{x \mid x \text{ is a whole number between } \frac{1}{2} \text{ and } \frac{17}{2}\}$
19. $B = \{x \mid x \text{ is an even number between 13 and 19}\}$
20. $C = \{x \mid x \text{ is a digit used to write numerals in the decimal number system}\}$

REAL NUMBERS AND THE REAL NUMBER LINE

The numbers 1, 2, 3, 4, 5, . . . are no doubt familiar to you as those numbers you first encountered as a child. More formally, they are called the **natural numbers** or the **positive integers**. When considered as a set, the natural numbers are usually referred to by the letter **N**:

$$\mathbf{N} = \{1, 2, 3, \ldots\}.$$

Unfortunately, the system of positive integers is, by itself, insufficient to completely describe many mathematical statements. For example, if we consider the equation $x + x = 0$, we are unable to find any positive integer which can serve as a replacement for x so as to make the equation true. The same comment applies to the equation $x + 2 = 1$. Since it is imperative that these and similar equations have solutions, it is necessary to extend our number system and introduce other numbers—**zero** (0) and the **negative integers** $-1, -2, -3, \ldots$. The infinite set consisting of the positive integers, the negative integers, and zero is called the set of **integers** and is denoted by **Z**:

$$\mathbf{Z} = \{\ldots, -4, -3, -2, -1, 0, 1, 2, 3, 4, \ldots\}.$$

Since every positive integer is also an integer, this means that $\mathbf{N} \subset \mathbf{Z}$.

If we now consider the equation $x + x = 1$, we find that even in the system of integers there is no suitable replacement for x. As before, to provide a solution we must extend our number system. We thus are led to the notion of a *rational number*. **Rational numbers** are those numbers, such as $\frac{1}{2}$ and $\frac{5}{3}$, which can be expressed as a ratio (or quotient) of two integers. That is, a rational number is one which can be expressed as p/q, where p and q are integers and q is not zero (since division by zero is not defined). Denoting the infinite set of rational numbers by **Q**, we have

$$\mathbf{Q} = \left\{ \frac{p}{q} \,\middle|\, p \in \mathbf{Z}, q \in \mathbf{Z}, \text{ and } q \neq 0 \right\}.$$

The symbol $\neq$ is read "is not equal to." Other examples of rational numbers are $\frac{-3}{7}, \frac{9}{5}, \frac{0}{3}, \sqrt{\frac{9}{4}}$, and $\frac{2}{4}$. The integer 6 is a rational number since $6 = \frac{6}{1}$. In fact, all integers are rational numbers; hence, $\mathbf{Z} \subset \mathbf{Q}$. Note that $\frac{1}{2}, \frac{2}{4}, \frac{3}{6}, \frac{20}{40}$, and .5 all represent the same rational number. To remind yourself of the meaning of rational number, it may be helpful to think of the word *rational* as *ratio-nal*.

Geometrically, rational numbers can be represented by points on a horizontal line. On such a line we select an initial point, referred to as the **origin,** to represent the number 0 (see Fig. 1-1). At some arbitrary distance to the right of the origin we label a second point +1. The distance chosen is called the *unit distance*. The unit distance is then successively marked off both to the right and to the left of the origin.

With each point on the line we can associate a directed distance or **signed number** which depends on the position of the point with respect to the origin. Positions to the right of the origin are considered *positive* (+), and positions to the left are considered *negative* (−). Thus, with the point $\frac{3}{2}$ units to the right of the origin there corresponds the signed number $+\frac{3}{2}$, positive three-halves (Fig. 1-2). With the point three units

FIGURE 1-1   **FIGURE 1-2**

to the left of the origin there corresponds the signed number −3, negative three. We shall adopt the custom of omitting the + sign before signed numbers unless the sign is to be emphasized. As a result, we shall feel free to speak of the numbers 7 and +7 interchangeably.

Certainly all of the rational numbers, and hence the integers, can be represented by points on the line. However, it can be shown that there are points whose directed distances from 0 *cannot* be represented by ratios of two integers. The numbers which represent the directed distances of this type are said to be **irrational**—that is, not rational. For example, $\sqrt{2}$ and π are known to be irrational numbers (remember that $\frac{22}{7}$ and 3.14 are only approximations to π). Irrational numbers provide solutions to equations such as $x \cdot x = 2$, which rational numbers would not.

Together, the rational numbers and the irrational numbers form the infinite set of **real numbers,** denoted **R.** In Fig. 1-3 various points and

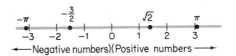

FIGURE 1-3

their associated signed numbers are identified. It should be clear that to each real number there corresponds a unique point on the line, and that to each point on the line there corresponds a unique real number. For this reason we say there is a *one-to-one correspondence* between real numbers and points on the line. In fact, we sometimes refer to points on the line as numbers. For example, we can speak of the *number* 7 and the *point* 7 interchangeably. Now it should be obvious why we refer to this line as the **real number line.**

If a and b are any two points on the real number line, there are three possibilities as to their relative positions. Either a and b coincide, or a lies to the left of b, or a lies to the right of b (Fig. 1-4). If a and b coin-

FIGURE 1-4

cide, then $a = b$. If a lies to the left of b, we say a is less than b and write $a < b$, the **inequality symbol** $<$ being read "is less than." On the other hand, if a lies to the right of b, we say a is greater than b and write $a > b$, the inequality symbol $>$ being read "is greater than." Note that writing $a < b$ is equivalent to writing $b > a$. Combining the notions of equality and inequality, we introduce two additional inequality symbols, $\leq$ and $\geq$. The symbol $\leq$ is read "is less than or equal to." Thus, $a \leq b$ means $a < b$ or $a = b$. Similarly, $a \geq b$ means a is greater than b or a is equal to b. In this case we say "a is greater than or equal to b."

Summarizing the meanings of our new symbols,

$a < b$	a is less than b
$a > b$	a is greater than b
$a \leq b$	a is less than or equal to b
$a \geq b$	a is greater than or equal to b.

Since we can use the terms *real number* and *point on the real number line* interchangeably, we can write $6 < 8$, which means 6 lies to the left of 8 on the real number line. From Fig. 1-5 it should be clear that $3 < 7$,

FIGURE 1-5

$7 > 3$, $-5 < -2$, $-2 < 1$, $-5 < 0$, $0 > -2$, $7 \leq 7$, and $3 \geq 0$. If $a > 0$, we say a is *positive*; if $a < 0$, then a is said to be *negative*.

There are many instances where inequality symbols play important roles in describing physical phenomena. For example, if two objects are in contact and are not in motion with respect to one another, the static frictional force f acting between the objects satisfies the inequality $f \leq \mu N$, where μ (Greek letter mu) is a fixed number and N is a force acting between the objects.

For the present, the real number system is sufficient for our demands. Later in our studies, however, to meet other conditions we shall find it necessary to introduce other types of numbers.

EXERCISE 1-2

In Problems 1–14, classify the given statement as either true or false.

1. $-3 \in \mathbf{N}$

2. $\frac{2}{3} \in \mathbf{Q}$

3. $0 \notin \mathbf{Q}$

4. $\mathbf{Q} \subset \mathbf{N}$

5. $\pi \in \mathbf{R}$

6. $-\frac{7}{3} \in \mathbf{Q}$

7. $\{\pi\} = \{\frac{22}{7}\}$

8. $0 \notin \mathbf{Z}$

9. $\frac{0}{7} \in \mathbf{Q}$

10. $\{-3\} \subset \mathbf{R}$

11. $\{.5\} = \{\frac{1}{2}\}$

12. $\{.33\} = \{\frac{1}{3}\}$

13. $-\frac{4}{2} \in \mathbf{Z}$

14. $\mathbf{Q} = \left\{\frac{p}{q} \,\middle|\, p, q \in \mathbf{Z}\right\}$

In Problems 15–20, state whether the given set is finite or infinite.

15. {all positive integers less than 7}

16. {all positive real numbers less than 7}

17. $\{x \mid x$ is a positive even integer$\}$

18. $\{7x \mid x \in \mathbf{N}\}$

19. {all real numbers greater than 1 and less than 5}

20. {all integers greater than -1 and less than 1}

In Problems 21–28, determine whether the given statement is true or false.

21. $2 > -3$

22. $0 > -1$

23. $7 \geq 7$

24. $6 < 16$

25. $-4 < -5$

26. $-2 < -3$

27. $0 > -4$

28. $-4 \leq -4$

1-3
INTERVALS

If a and b are real numbers and $a < b$, the set of all real numbers between a and b, *excluding* a and b, is called an **open interval**, denoted

$\{x | a < x < b\}$ or, more simply, (a, b). It is an infinite subset of **R** which we geometrically indicate on the real number line by the bold line segment in Fig. 1-6. The fact that a and b, called **endpoints**, are not to be considered part of the set is indicated by the use of parentheses. If we wish, however, to consider the case when a and b *are* to be included, then we have the set $\{x | a \leq x \leq b\}$, called a **closed interval**; it is denoted by $[a, b]$, where the brackets indicate that a and b are both elements of the set (see Fig. 1-7).

Open interval (a, b) Closed interval $[a, b]$

FIGURE 1-6   **FIGURE 1-7**

EXAMPLE
2

The set $\{x | 3 < x < 5\} = (3, 5)$ is an open interval and consists of all real numbers x strictly between 3 and 5. The numbers 3 and 5 are not in the set; hence, we use parentheses in Fig. 1-8. Note that to write $3 < x < 5$ means $3 < x$ and $x < 5$ *simultaneously*.

Consider the set $\{x | x \geq -2\}$. The endpoint, namely -2, is included. This set, which consists of all real numbers greater than or equal to -2, can be represented on the number line as shown in Fig. 1-9. The arrow

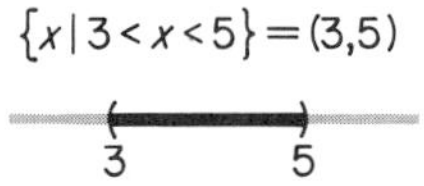

$\{x | 3 < x < 5\} = (3, 5)$ $\{x | x \geq -2\} = [-2, \infty)$

FIGURE 1-8 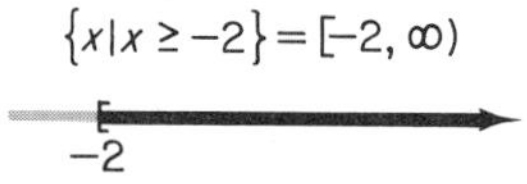   **FIGURE 1-9**

$(a, b] = \{x | a < x \leq b\}$

$[a, b) = \{x | a \leq x < b\}$

$[a, \infty) = \{x | x \geq a\}$

$(a, \infty) = \{x | x > a\}$

$(-\infty, a] = \{x | x \leq a\}$

$(-\infty, a) = \{x | x < a\}$

$(-\infty, \infty) = \{x | -\infty < x < \infty\}$
$= \{x | x \in \mathbf{R}\}$

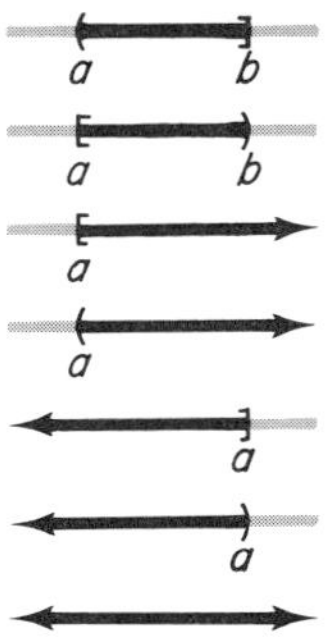

FIGURE 1-10

indicates that the bold portion of the line extends indefinitely to the right, and the bracket, of course, means that -2 is included in the set. We can also denote this set by writing $[-2, \infty)$, where the **infinity symbol** ∞ does not denote a number but is merely a convenience for indicating here that the interval extends indefinitely to the right.

Extending the concept of interval, we have the additional intervals shown on the preceding page in Fig. 1-10, where the meaning of $-\infty$ is analogous to that of ∞.

EXERCISE 1-3

In Problems 1–16, find an alternative way of expressing the given interval and indicate the interval geometrically on the real number line.

1. $[3, 6]$
2. $\{x \mid x > -\frac{3}{2}\}$
3. $(-2, 3)$
4. $[0, \infty)$
5. $(-\infty, 6)$
6. $[2, 6]$
7. $[-7, \infty)$
8. $(-\infty, -4]$
9. $\{x \mid x \geq 4\}$
10. $\{x \mid x > 3, x < 9\}$
11. $\{x \mid 2 \leq x \leq 5\}$
12. $\{x \mid -2 \leq x \leq 3\}$
13. $[3, 4)$
14. $(5, 6]$
15. $\{x \mid 5 < x \leq 12\}$
16. $\{x \mid 2 \leq x < 2.1\}$

1-4
AXIOMS FOR THE REAL NUMBERS

Certain fundamental principles, which we assume to be true, govern the system of real numbers. They are known as **axioms** (or *laws*) of the real number system. Most of the algebraic concepts that we shall consider follow, in one way or another, from these axioms. For this reason it is of utmost importance that you not only understand these axioms but also are able to make effective use of them by properly applying them.

In presenting the axioms, we shall symbolically represent real numbers (or quantities) by letters of the alphabet. In this context the letters are referred to as **literal numbers**. A symbol that represents only one number throughout a discussion is called a **constant**. For example, 2, π, and $\frac{7}{3}$ are constants. In fact, since their values *never* change, they are called **numerical constants**. On the other hand, when literal symbols such as a, b, c, x, y, etc., are used to represent unspecified numbers that remain fixed in a discussion, the symbol is referred to as an **arbitrary constant**.

If a symbol is thought of as representing any one of a set of possible numbers throughout a discussion, it is called a **variable**. For example, in the formula $A = \pi r^2$ for the area A of a given circle, r represents the radius of the given circle and is a constant. However, considering *all* circles, r is a variable and can assume an infinite number of values. Thus, in one case r denotes a constant and in the other case a variable. It is customary to use letters at the beginning of the alphabet, such as a, b, and c, to represent constants, and letters at the latter part, such as x, y, and z, to represent variables. Remember, however, that there are exceptions.

We shall begin our discussion of the axioms of the real numbers with the *axioms of equality*. Although these axioms may appear rather trivial, they are actually quite powerful and vital—for if we did not have them at our disposal, developing a system of mathematics could possibly become a futile effort. One more comment: We shall denote the product of a and b by $a \cdot b$, $(a)(b)$, ab, or $a(b)$, depending on convenience.

I.   Axioms of Equality

If a, b, and c are any real numbers, then

a.   **Reflexive Axiom:**

$$a = a.$$

This means that every quantity is equal to itself.

b.   **Symmetric Axiom:**

$$\text{If } a = b, \text{ then } b = a.$$

For example, if $x + 6 = y$, then $y = x + 6$.

c.   **Transitive Axiom:**

$$\text{If } a = b \text{ and } b = c, \text{ then } a = c.$$

Thus, two quantities both of which are equal to a third quantity are equal to each other. For example, if $x = y$ and $y = 4$, then $x = 4$.

d.   **Substitution Axioms:**

1.   *If $a = b$ and $a + c = d$, then $b + c = d$.*
2.   *If $a = b$ and $a \cdot c = d$, then $b \cdot c = d$.*

These axioms essentially assert that we can replace a quantity in an expression by an equal quantity. For example,

$$\text{if } x = 1 \text{ and } x + 7 = y, \text{ then } 1 + 7 = y;$$
$$\text{if } x = 2 \text{ and } xy = 12, \text{ then } 2y = 12.$$

e. **Addition and Multiplication Axioms:**

    1. *If $a = b$, then $a + c = b + c$.*

    2. *If $a = b$, then $ac = bc$.*

These axioms assert that if equal numbers are added to (or multiplied by) equal numbers, the resulting sums (or products) are equal. To illustrate,

$$\text{if } x = 2, \text{ then } x + 3 = 2 + 3;$$

$$\text{if } x = 2, \text{ then } x(3) = 2(3).$$

II. The Closure Axioms

If a and b are any real numbers, then both

$$a + b \qquad and \qquad ab$$

are unique real numbers.

For example, since 7 and π are real numbers, we are assured that their sum, $7 + \pi$, and their product, 7π, are also real numbers.

III. The Commutative Axioms

If a and b are any real numbers, then

$$a + b = b + a \qquad and \qquad ab = ba.$$

This means we may add or multiply any two real numbers in any order. For example, $3 + 4 = 4 + 3$ and $3 \cdot \pi = \pi \cdot 3$.

IV. The Associative Axioms

If a, b, and c are any real numbers, then

$$a + (b + c) = (a + b) + c \qquad and \qquad a(bc) = (ab)c.$$

This means that in addition and multiplication, numbers can be grouped in any order. For example, $7 + (2 + 3) = (7 + 2) + 3$, $6(\frac{1}{3} \cdot 5) = (6 \cdot \frac{1}{3}) \cdot 5$, and $2x + (y + 3z) = (2x + y) + 3z$.

Finding the product of several numbers can be accomplished only by considering products of numbers taken two at a time. For example, to find the product of x, y, and z, we could first multiply x by y and then multiply that product by z, or we could instead multiply x by the product of y and z. The associative axiom of multiplication asserts that both results are identical, regardless of how the numbers are grouped. That is, $xyz = (xy)z = x(yz)$. Thus, there is no ambiguity in writing the expression xyz. Of course this concept can be extended to more than three numbers and applies equally well to addition.

V. The Identity Elements

a. *There exists a unique real number* 0 *such that for every real number a we have*

$$a + 0 = a.$$

The number 0 *is called the* **identity for addition.**

For example, $9 + 0 = 9$. It is also true that if a is any real number, then $a \cdot 0 = 0$.

b. *There exists a unique real number* 1, *different from* 0, *such that for every real number a we have*

$$a \cdot 1 = a.$$

We call 1 *the* **identity for multiplication.**

For example, $8 \cdot 1 = 8$.

VI. The Inverse Elements

a. *For each real number a there exists a unique real number* $-a$ *such that*

$$a + (-a) = 0.$$

The number $-a$ *is called the* **additive inverse, or negative,** *of a.*

For example, the negative of 3 is -3 and $3 + (-3) = 0$. The additive inverse of a number is not necessarily a negative number. To illustrate, the additive inverse of -5 is 5 since $-5 + (5) = 0$; that is, the negative of -5 is 5.

b. *For each real number a which is different from* 0, *there exists a unique real number* $\dfrac{1}{a}$ *such that*

$$a\left(\frac{1}{a}\right) = 1.$$

The number $\dfrac{1}{a}$ *is called the* **multiplicative inverse, or reciprocal,** *of a.*

For example, the reciprocal of 3 is $\frac{1}{3}$ and $3(\frac{1}{3}) = 1$; also, 3 is the reciprocal of $\frac{1}{3}$ and $(\frac{1}{3})(3) = 1$. **We emphasize that the reciprocal of 0 is not defined.**

VII.   The Distributive Axioms (Axioms of Addition and Multiplication)

If a, b, and c are any real numbers, then

$$a(b + c) = ab + ac \quad \text{and} \quad (b + c)a = ba + ca.$$

For example,

$$6(7 + 2) = 6 \cdot 7 + 6 \cdot 2 = 42 + 12 = 54$$

$$(2 + 3) \cdot 4 = 2 \cdot 4 + 3 \cdot 4 = 8 + 12 = 20$$

$$x(yz + 4) = x(yz) + x(4) = xyz + 4x.$$

The distributive axiom can be extended to the form $a(b + c + d) = ab + ac + ad$. In fact, it can be extended to sums involving any other number of terms.

It is by the additive inverse axiom that we formally define the operation of *subtraction*:

Definition.   If a and b are any real numbers, then b subtracted from a, or the difference between a and b, denoted $a - b$, is defined by

$$a - b = a + (-b),$$

where $-b$ is the additive inverse of b.

Subtraction is therefore defined in terms of a sum (that is, addition). For example, $2 - 1$ means $2 + (-1)$.

In a similar fashion we can define *division*:

Definition.   If a and b are any real numbers and $b \neq 0$, then a divided by b, or the *quotient* of a by b, denoted a/b or $a \div b$, is defined by

$$\frac{a}{b} = a \cdot \frac{1}{b} \quad \text{(first form)},$$

where $1/b$ is the reciprocal of b.

Thus, $\frac{5}{9}$ means $5(\frac{1}{9})$ since $\frac{1}{9}$ is the reciprocal of 9. Note that division is defined in terms of a product (that is, multiplication) and also that **division by zero is not defined**. An alternative definition of division is as follows:

$$\frac{a}{b} = c \text{ means } a = bc \quad \text{(second form)}.$$

Thus, $\frac{6}{2} = 3$ since $6 = 2 \cdot 3$.

This is an ideal time to show you why division by zero is undefined. Suppose, for example, $\frac{4}{0}$ denoted some number a. Then $\frac{4}{0} = a$ means $4 = 0 \cdot a$. But from our knowledge of numbers, there is no number a such that $0 \cdot a = 4$; indeed, $0 \cdot a = 0$. Thus, $\frac{4}{0}$ cannot denote a number. Note, however, that although $\frac{4}{0}$ is not defined, the number $\frac{0}{4}$ *is* defined and is 0.

In branches of modern mathematics, axioms are used to logically deduce consequences, called **theorems**. The technique is similar to the processes performed in the following examples.

EXAMPLE
3

a. $(3 + x) - y = (3 + x) + (-y)$ (definition of subtraction)

$\qquad\qquad\quad = 3 + [x + (-y)]$ (associative law)

$\qquad\qquad\quad = 3 + (x - y)$ (definition of subtraction).

b. By the definition of subtraction,

$$6 - \pi = 6 + (-\pi).$$

However, by the commutative axiom,

$$6 + (-\pi) = -\pi + 6.$$

Thus, by the transitive axiom,

$$6 - \pi = -\pi + 6.$$

More concisely we can write

$$6 - \pi = 6 + (-\pi) = -\pi + 6.$$

c. By the first form of the definition of division,

$$\frac{2 \cdot 3}{5} = (2 \cdot 3) \cdot \frac{1}{5}.$$

But by the associative axiom,

$$(2 \cdot 3) \cdot \frac{1}{5} = 2 \cdot \left(3 \cdot \frac{1}{5}\right).$$

However, by the definition of division, $3 \cdot \frac{1}{5} = \frac{3}{5}$ and thus by the transitive axiom,

$$\frac{2 \cdot 3}{5} = 2\left(\frac{3}{5}\right).$$

If we were to generalize the result in Example 3c, we could say that if a, b, and c were any real numbers where $c \neq 0$, then

$$\frac{ab}{c} = a\left(\frac{b}{c}\right).$$

Similarly, the following properties are also true and deserve your attention.

Property	Example
$a(b-c) = ab - ac$	$2(2x - 3y) = 2(2x) - 2(3y)$
	$(-1)(x - y) = (-1)(x) - (-1)(y)$
$\dfrac{a}{1} = a$	$\dfrac{7}{1} = 7, \dfrac{0}{1} = 0$
$\dfrac{0}{a} = 0$ when $a \neq 0$	$\dfrac{0}{7} = 0$
$\dfrac{a}{a} = 1$ when $a \neq 0$	$\dfrac{2}{2} = 1$
$a\left(\dfrac{b}{a}\right) = b$	$2\left(\dfrac{7}{2}\right) = 7$
$\dfrac{ab}{c} = \left(\dfrac{a}{c}\right)(b) = a\left(\dfrac{b}{c}\right)$	$\dfrac{2 \cdot 7}{3} = \dfrac{2}{3} \cdot 7 = 2 \cdot \dfrac{7}{3}$
$\dfrac{a}{bc} = \left(\dfrac{a}{b}\right)\left(\dfrac{1}{c}\right) = \left(\dfrac{1}{b}\right)\left(\dfrac{a}{c}\right)$	$\dfrac{2}{3 \cdot 7} = \dfrac{2}{3} \cdot \dfrac{1}{7} = \dfrac{1}{3} \cdot \dfrac{2}{7}$
$\dfrac{a}{b} \cdot \dfrac{c}{d} = \dfrac{ac}{bd}$	$\dfrac{2}{3} \cdot \dfrac{4}{5} = \dfrac{2 \cdot 4}{3 \cdot 5} = \dfrac{8}{15}$
	$\dfrac{2}{3} \cdot \dfrac{1}{x} = \dfrac{2 \cdot 1}{3 \cdot x} = \dfrac{2}{3x}$

We point out that not only should you be aware of the manipulative aspects of the axioms which govern the real number system, but you should also be aware of and familiar with the terminology involved.

EXERCISE 1-4

1. State the axioms or definitions used to justify each of the following statements. Assume $x, y,$ and z are real numbers.
    a. If $2 + 3 = 5$, then $5 = 2 + 3$.
    b. $5(4 + 7) = (4 + 7)5$
    c. $(-1)[3 + (-2)] = (-1)(3) + (-1)(-2)$
    d. $5(2 \cdot 3) = 3(5 \cdot 2)$
    e. $2(x + y) = 2x + 2y$
    f. $(x + 5) + y = y + (x + 5)$
    g. $9 \cdot 8$ is a real number.
    h. $x(3y) = 3(xy)$
    i. $z(x - y) = (x - y)z$
    j. $926,832 + 0 = 926,832$
    k. $z \cdot 1 = z$
    l. $2[27 + (x + y)] = 2[(y + 27) + x]$

m. $3 - x = 3 + (-x)$

n. $\frac{2}{4} = 2 \cdot \frac{1}{4}$

2. Show that the following statements are true.

a. $(5a)(x + 3) = 5ax + 15a$

b. $(2 - x) + y = 2 + (y - x)$

c. $(x - y)2 = 2x - 2y$

3. Use the second form for the definition of division to determine the values, if any, of the following:

a. $\frac{0}{1}$ b. $\frac{1}{0}$ c. $\frac{0}{0}$

Hint: In part c, show that the symbol $\frac{0}{0}$ could represent *any* number and in this sense is an ambiguous symbol.

1-5
THE ARITHMETIC OF REAL NUMBERS

If a is any real number, then one and only one of the following statements is true: $a < 0$, $a = 0$, or $a > 0$. Thus, with every nonzero real number there is associated a positive ($+$) or negative ($-$) sign denoting direction. We have spoken before of these so-called signed numbers and we shall now consider how the operations of arithmetic apply to them. You should thoroughly study these properties, since facility in manipulating signed numbers is essential to your study of mathematics.

In formulating the arithmetic rules for signed numbers, we shall use the notion of *absolute value*. Stated simply, the **absolute value** of a real number x is its distance from the origin, disregarding direction—that is, its *undirected distance*. Symbolically, the absolute value of x is denoted $|x|$. For example, since 5 and -5 are both five units from the origin (Fig. 1-11), we can write $|5| = 5$ and $|-5| = 5$. Similarly, $|3| = 3$, $|-8| = 8$, $|+\frac{1}{2}| = \frac{1}{2}$, $|-\pi| = \pi$, $|\sqrt{2}| = \sqrt{2}$, and $|0| = 0$. Note that $|x|$ is never negative; it is always positive or zero, that is, *nonnegative*. Symbolically, $|x| \geq 0$. Two or more numbers which have equal absolute values are sometimes said to be *numerically equal* or *equal in magnitude*.

The operation of addition of two real numbers will be geometrically represented by using the real number line. In this respect, a positive number will be interpreted as a movement to the right on the line and a negative number as a movement to the left. For example, consider the

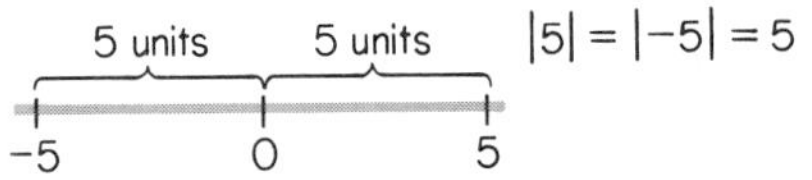

$5 \text{ units} \qquad 5 \text{ units} \qquad |5| = |-5| = 5$

$-5 \qquad 0 \qquad 5$

FIGURE 1-11

sum $(+4) + (+3)$. This means move four units to the right of the origin, and then move three more units to the right. In Fig. 1-12 these movements are illustrated by arrows. In the addition, $+4$ is represented by an arrow extending four units to the right of zero. The number $+3$ is represented by an arrow three units long, also extending to the right, but it begins (so to speak) where the first arrow ended. Since the second arrow terminates at the point $+7$, we conclude that

$$(+4) + (+3) = +7. \qquad (1)$$

The sum $(-2) + (-1)$ can be geometrically interpreted as that number obtained by moving two units to the left of 0 and then moving one more unit to the left (Fig. 1-13). Thus,

$$(-2) + (-1) = -3. \qquad (2)$$

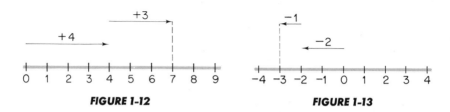

FIGURE 1-12   FIGURE 1-13

Reflection on results (1) and (2) leads to the following general rule:

The sum of two real numbers of like signs is the number obtained by adding their absolute values and prefixing their common sign to the result.

EXAMPLE
4

a. $(+2) + (+7) = +(2 + 7) = +9 = 9.$

b. $(-3) + (-5) = -(3 + 5) = -8.$

c. $(-7) + (-\pi) = -(7 + \pi).$

d. $(-12) + (-10) = -22.$

e. $0 + (-4) = -4$ since 0 is the additive identity.

The sum $(+7) + (-5)$ is that number obtained by moving seven units to the right of 0 and then from this position moving five units to the left (Fig. 1-14). That is,

$$(+7) + (-5) = +2. \qquad (3)$$

More simply we can write $7 + (-5) = 2$. Note, however, that

$$5 + (-7) = -2, \qquad (4)$$

which is shown geometrically in Fig. 1-15. From (3) and (4) we conclude:

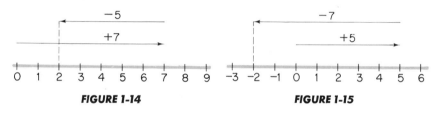

FIGURE 1-14   **FIGURE 1-15**

The sum of two real numbers of unlike signs is the number obtained by subtracting the smaller absolute value from the larger absolute value and prefixing to the result the sign of the number having the larger absolute value.

EXAMPLE 5

a. $(-4) + 9 = +5 = 5$.

b. $4 + (-15) = -11$.

c. $3 + (-3) = 0$ by the additive inverse axiom.

d. $0 + (-3) = -3$ by the additive identity axiom.

Recall from Sec. 1-4 that subtraction of real numbers is defined in terms of addition: $a - b = a + (-b)$. Hence,

To subtract a real number from another real number, change the sign of the number to be subtracted and proceed as in addition.

EXAMPLE 6

a. $2 - 7 = 2 + (-7) = -5$.

b. $-12 - 10 = -12 + (-10) = -22$.

c. $-6 - (-10) = -6 + (10) = 4$.

d. $2 - (-7) = 2 + (7) = 9$; more generally, $a - (-b) = a + b$.

e. $0 - 6 = 0 + (-6) = -6$.

f. $0 - (-4) = 0 + (4) = 4$.

g. By the associative axiom, the previous rules can be extended to more complicated situations:

$$11 - 5 + 4 = 11 + (-5) + 4$$
$$= 11 + (-5 + 4)$$
$$= 11 + (-1)$$
$$= 10.$$

The rules governing the multiplication of signed numbers will be presented next. Since division is defined in terms of multiplication $\left(\dfrac{a}{b} = a \cdot \dfrac{1}{b}\right)$, the same rules apply to division.

The product (or quotient) of two real numbers of like signs is the number obtained by multiplying (or dividing) their absolute values and, if desired, prefixing the result with a positive $(+)$ sign.

This means that the product or quotient of *two positives* or *two negatives* is positive.

EXAMPLE
7

a. $(+5)(+6) = +(5 \cdot 6) = 30.$

b. $(-3)(-4) = (3)(4) = 12.$

c. $\dfrac{+14}{+2} = +\dfrac{14}{2} = 7.$

d. $\dfrac{-3}{-5} = \dfrac{3}{5}.$

When multiplying or dividing two numbers of unlike signs, use the following rule:

The product (or quotient) of two real numbers of unlike signs is the number obtained by multiplying (or dividing) their absolute values and prefixing a negative $(-)$ sign to the result.

EXAMPLE
8

a. $(-1)(3) = -(1 \cdot 3) = -3.$ More generally, if a is any real number, then
$$(-1)(a) = -a.$$

b. $2\left(-\dfrac{1}{2}\right) = -\left(2 \cdot \dfrac{1}{2}\right) = -1.$

c. $\dfrac{-4}{2} = -\left(\dfrac{4}{2}\right) = -2.$

d. $\dfrac{6}{-7} = -\dfrac{6}{7}.$

e. $(-1)(x + y) = -(x + y).$

f. $(-1)(6x - 2y) = (-1)[(6x) + (-2y)]$ (definition of subtraction)
$ = (-1)(6x) + (-1)(-2y)$ (distributive axiom)
$ = -6x + 2y.$

Symbolically, the laws of signed numbers for multiplication and division presented above can be given as follows: If a and b are real numbers, then

$$(a)(b) = ab = (-a)(-b) \qquad \frac{-a}{-b} = \frac{a}{b}$$

$$a(-b) = -(ab) = (-a)(b) \qquad \frac{a}{-b} = -\frac{a}{b} = \frac{-a}{b}$$

Therefore, $6(-7) = -(6\cdot7) = (-6)(7)$, $(2)(3) = (-2)(-3)$, and $\dfrac{2}{-4} = -\dfrac{2}{4} = \dfrac{-2}{4}$. Also, from Example 8a,

$$\boxed{-a = (-1)(a)}$$

Thus $-6 = (-1)(6)$ and $-x = (-1)(x)$. Moreover, it is also true that

$$\boxed{-(-a) = a}$$

For example, $-(-6) = 6$ and $-(-x) = x$. We now turn to more examples which illustrate the rules given in this section.

EXAMPLE
9

a. $\dfrac{7(-4)}{3} = \dfrac{-28}{3} = -\dfrac{28}{3}$.

b. $(7 - 6) - 11 = 1 - 11 = 1 + (-11) = -10$.

c. $(-2)(-6 + 4) = (-2)(-2) = 4$.

d. $\dfrac{6(-2)}{-3} = \dfrac{-12}{-3} = \dfrac{12}{3} = 4$.

e. $\dfrac{2(-3)}{6} = \dfrac{-6}{6} = -\dfrac{6}{6} = -1$.

f. $\dfrac{-6 + (-2)}{-4} = \dfrac{-8}{-4} = \dfrac{8}{4} = 2$.

g. $\dfrac{2}{3(-5)} = \dfrac{2}{-(3\cdot5)} = \dfrac{2}{-15} = -\dfrac{2}{15}$.

h. $\dfrac{-6}{(-5)(-7)} = \dfrac{-6}{35} = -\dfrac{6}{35}$.

i. $(-6)(-2)(-1) = (-6\cdot-2)(-1) = (12)(-1) = -12$ by the associative axiom.

j. $\dfrac{1}{-3} = -\dfrac{1}{3} = \dfrac{-1}{3}$.

k. $2 - |-6| = 2 - (6) = -4$.

l. $-|6| = -(6) = -6$.

m. $-2 + |-3| = -2 + 3 = 1$.

n. $|4| - |-7| = (4) - (7) = -3$; $|4 - 7| = |-3| = 3$.

o. To subtract -2 from 17 we have $17 - (-2) = 17 + 2 = 19$.

p. $(-6)(x + y) = (-6)(x) + (-6)(y)$
$\qquad = (-6x) + (-6y)$
$\qquad = -6x - 6y$.

With expressions involving signed numbers where numerous operations are involved, the correct order of performing operations sometimes becomes a problem for a student. In general, we say that multiplication and division take precedence over addition and subtraction; that is, perform any multiplications and divisions before the required additions and subtractions. Thus,

$$1 + 5 - 3 \cdot 4 \div 6 = 1 + 5 - 12 \div 6$$
$$= 1 + 5 - 2$$
$$= 6 - 2$$
$$= 4.$$

In cases such as the one above, any ambiguity can be removed by rearrangement of terms and the use of symbols of grouping. Symbols of grouping will be discussed in Chapter 2. The example above should be written in the preferred form

$$1 + 5 - \frac{3 \cdot 4}{6}.$$

EXERCISE 1-5

Evaluate the expressions in Problems 1–34.

1. $|-6|$
2. $|-\pi|$
3. $|270|$
4. $|\frac{7}{2}|$
5. $|0|$
6. $|6(-2)|$
7. $-2 + (-4)$
8. $-6 + 2$
9. $6 + (-4)$
10. $7 - 2$
11. $7 - (-4)$
12. $-7 - (-4)$
13. $-8 - (-6)$
14. $(-2)(9)$
15. $7(-9)$
16. $(-2)(-12)$
17. $-(6)$
18. $-(-9)$
19. $-(-6 + x)$
20. $-7(x)$
21. $-12(x - y)$
22. $-[-6 + (-y)]$
23. $-2 \div 6$
24. $-2 \div -4$
25. $4 \div (-2)$
26. $2(-6 + 2)$
27. $3[-2(3) + 6(2)]$
28. $(-2)(-4)(-1)$
29. $(-5)(-5)$
30. $x(0)$

31. $|7| - |23|$

32. $-|23| - 7$

33. $\dfrac{2}{-3}$

34. $\dfrac{7}{(-2)(-14)}$

Problems 35–44 each indicate a pair of signed numbers. For each pair, (a) find their sum, (b) subtract the second number from the first, (c) subtract the first number from the second, (d) find the product of the numbers, and (e) divide the first number in each pair by the absolute value of the second number.

35. $-5, -3$

36. $1, -5$

37. $3, 7$

38. $-6, 2$

39. $0, 12$

40. $-5, 0$

41. $\dfrac{1}{2}, -4$

42. $\dfrac{1}{2}, 8$

43. $-\dfrac{1}{2}, -\dfrac{1}{4}$

44. $0, -\dfrac{1}{2}$

In Problems 45–54, evaluate the given expressions.

45. $(-3a) + |0|$

46. $(-2)(4)(-8) + (6)(-2)$

47. $\dfrac{(-1)(14)(5) - (-3)}{7}$

48. $2(-3)(-|5|)$

49. $-(-|-4| + 3)(-6) - |-4 - 2|$

50. $[921 - (-1032)] + 6$

51. $\dfrac{16(-3 - 5 + 6)(0 - 4)}{3}$

52. $\dfrac{(-4) - (-9)(2) + 6(-1)}{-[-8 + 4 - (-2)]}$

53. $16 - \dfrac{4(-2)}{2} - \dfrac{(32)(-4)}{-(-1)}$

54. $\dfrac{16(-3) - 7(-2)(-3)}{(-4) - (-4)}$

55. For an arrangement of three particles in a straight line, the location $\bar{x}$ (read "x bar") of the center of mass is defined to be

$$\bar{x} = \frac{m_1 x_1 + m_2 x_2 + m_3 x_3}{m_1 + m_2 + m_3}{}^{*}.$$

Find $\bar{x}$ if $m_1 = 2$, $m_2 = 3$, $m_3 = 4$, $x_1 = -2$, $x_2 = -|-3|$, and $x_3 = 8$.

56. Repeat Problem 55 if $x_1 = -x_2 = x_3 = -4$ and $m_1 = m_2 = m_3 = 2$.

1-6
EXPONENTS

When a number of quantities are multiplied together to form a product, each quantity is called a **factor** of the product. Thus, for the product $2ab$ we can say that 2, a, and b are factors.

*The numbers appearing to the right of the letters are called **subscripts** and x_1, for example, is read "x sub one."

The product

$$2 \cdot 2 \cdot 2 \cdot 2 \cdot 2 \cdot 2$$

is abbreviated 2^6, and in general, for n a positive integer, 2^n represents the product of n 2's. More specifically, we have the following definition:

Definition.   If n is a positive integer, x^n is the product of n factors each of which is x. The letter n is called the **exponent** while x is called the **base**. The symbol x^n is read "the nth power of x" or "x raised to the nth power."

Thus, for the symbol 2^6 the base is 2, the exponent is 6, and 2^6 is read "the sixth power of 2" or "2 raised to the sixth power." Since $2^6 = 64$, we say that 2^6 is an *exponential form* for 64. Can you think of another exponential form for 64?

EXAMPLE 10

a. $x^4 = x \cdot x \cdot x \cdot x$.

b. $6^2 = 6 \cdot 6$ (the second power of 6 or 6 squared).

c. $2^3 = 2 \cdot 2 \cdot 2$ (the third power of 2 or 2 cubed).

d. $(-5)^4 = (-5)(-5)(-5)(-5) = 625$. In general, **a negative number raised to an even power is positive.**

e. $(-6)^3 = (-6)(-6)(-6) = -216$. **A negative number raised to an odd power is negative.**

f. $4^1 = 4$.

g.   Since $25 = 5^2$, we say 5^2 is an exponential form for 25.

As shown in Example 10f, if an exponent is not written, it is understood to be 1. We caution you to keep in mind that *an exponent applies only to the quantity immediately to the left and below it*. Thus, in $(-3)^2$ the base is -3, but in -3^2 the base is 3. That is,

$$(-3)^2 = (-3)(-3) = 9 \qquad \text{but} \qquad -3^2 = -(3)(3) = -9.$$

We turn now to the fundamental rules of exponents.

To begin, suppose we consider the product of two numbers in exponential form, where both numbers have the same base:

$$x^m \cdot x^n = (\underbrace{xxxx \cdots xxxx}_{m \text{ factors}})(\underbrace{xxxxx \cdots xxxx}_{n \text{ factors}})$$

$$= (\underbrace{xxxxxxxxx \cdots x}_{m + n \text{ factors}})$$

$$= x^{m+n}.$$

Essentially, we have only to raise x to that power obtained by adding the exponents of the given numbers. We thus have the following rule:

Rule 1. $x^m \cdot x^n = x^{m+n}$ where m and n are positive integers.

That is, **the product of powers of the *same base* is a power of that base, the exponent being the sum of the individual exponents.**

EXAMPLE
11

a. $3^5 \cdot 3^9 = 3^{5+9} = 3^{14}$.

b. $2^2 \cdot 2 = 2^{2+1} = 2^3 = 8$.

c. $y^6 \cdot y^6 = y^{12}$.

d. $x^a \cdot x^2 = x^{a+2}$.

e. $(-2)(-2) = (-2)^2 = 4$.

f. $(-x)(-x)^6 = (-x)^7 = -x^7$.

g. $(-2)^4(-2^2) \neq (-2)^6$, but $(-2)^4(-2^2) = 16(-4) = -64$.

h. $(x-2)^5(x-2)^9 = (x-2)^{14}$.

Rule 1 can be extended to products of more than two factors. Thus,

$$x^2 \cdot x^5 \cdot x^4 = x^{2+5+4} = x^{11}.$$

Another rule of exponents concerns a "power of a power." Since

$$(x^3)^2 = x^3 \cdot x^3 = x^{3+3} = x^{3 \cdot 2}$$

and

$$(x^2)^4 = x^2 \cdot x^2 \cdot x^2 \cdot x^2 = x^{2+2+2+2} = x^{2 \cdot 4},$$

it should be clear that

$$(x^m)^n = \underbrace{x^m \cdot x^m \cdots x^m}_{n \text{ factors}}$$

$$= x^{\overbrace{m+m+\cdots+m}^{n\ m\text{'s}}}$$

$$= x^{mn}.$$

Rule 2. $(x^m)^n = x^{mn}$ where m and n are positive integers.

That is, **a power of a power of the base x is the base x raised to the power which is equal to the product of the exponents.**

EXAMPLE
12

a. $(10^2)^6 = 10^{2 \cdot 6} = 10^{12}$.

b. $(x^2)^3 = x^{2 \cdot 3} = x^6$.

c. $(y^8)^5 = y^{40}$.

d. $[(-x)^2]^3 = (-x)^6 = x^6$.

e. $[(x+7)^2]^{10} = (x+7)^{20}$.

Be sure you understand the distinction between Rules 1 and 2. Note that

$$2^3 \cdot 2^2 = 2^5 \qquad \text{but that} \qquad (2^3)^2 = 2^6.$$

Students sometimes confuse these rules.

In a manner similar to that used above, the following rules of exponents can also be derived:

Rule 3. $(xy)^n = x^n y^n$ where n is a positive integer.

Rule 4. $\left(\dfrac{x}{y}\right)^n = \dfrac{x^n}{y^n}$ where n is a positive integer and $y \neq 0$.

EXAMPLE
13

In the following, the number to the right refers to the rule used.

a. $(2 \cdot 5)^3 = 2^3 \cdot 5^3$ (3).

b. $(xy)^2 = x^2 y^2$ (3).

c. $(ab)^{2n} = a^{2n} b^{2n}$ (3).

d. $(3x)^3 = 3^3 x^3 = 27x^3$. Note that $3x^3 \neq (3x)^3$.   (3)

e. $2^4 \cdot 3^4 = (2 \cdot 3)^4 = 6^4$ (3).

f. $(w^2 r^4)^3 = (w^2)^3 (r^4)^3$ (3)
$\qquad\qquad = w^6 r^{12}$ (2).

g. $\left(\dfrac{x}{y}\right)^3 = \dfrac{x^3}{y^3}$ (4).

h. $\left(\dfrac{2z}{3}\right)^2 = \dfrac{(2z)^2}{3^2}$ (4)
$\qquad\qquad = \dfrac{2^2 z^2}{9} = \dfrac{4z^2}{9}$ (3).

i. $\left(\dfrac{1}{x}\right)^4 = \dfrac{1^4}{x^4} = \dfrac{1}{x^4}$ (4).

j. $\left(\dfrac{2}{3}\right)^3 = \dfrac{2^3}{3^3} = \dfrac{8}{27}$ (4).

k. $\dfrac{10^3}{5^3} = \left(\dfrac{10}{5}\right)^3 = 2^3 = 8$ (4).

l. $\left(\dfrac{a^2 b^3}{c^4}\right)^5 = \dfrac{(a^2 b^3)^5}{(c^4)^5}$ (4)
$\qquad\qquad = \dfrac{(a^2)^5 (b^3)^5}{(c^4)^5}$ (3)
$\qquad\qquad = \dfrac{a^{10} b^{15}}{c^{20}}$ (2).

m. $\left(\dfrac{ab^2}{c^2d}\right)^6 = \dfrac{(ab^2)^6}{(c^2d)^6} = \dfrac{a^6b^{12}}{c^{12}d^6}$ (4), (3).

n. $(-2x^2)^3 = [(-1)(2x^2)]^3 = (-1)^3(2x^2)^3 = (-1)(2^3x^6) = (-1)(8x^6)$
   $= -8x^6$ (3).

Better yet, we can write

$$(-2x^2)^3 = [(-2)(x^2)]^3 = (-2)^3(x^2)^3 = -8x^6.$$

o. $(xyz)^n = [(xy)z]^n = (xy)^n z^n = x^n y^n z^n$. Thus Rule 3 can be extended to the case of more than two factors.

Consider the quotient

$$\frac{4^5}{4^3}.$$

By the familiar process of cancellation,

$$\frac{4^5}{4^3} = \frac{4 \cdot 4 \cdot 4 \cdot 4 \cdot 4}{4 \cdot 4 \cdot 4} = 4^2.$$

The same result would have been obtained by raising 4 to that power obtained by subtracting the exponent 3 in the denominator from the exponent 5 in the numerator. That is,

$$\frac{4^5}{4^3} = 4^{5-3} = 4^2.$$

Similarly,

$$\frac{4^3}{4^5} = \frac{4 \cdot 4 \cdot 4}{4 \cdot 4 \cdot 4 \cdot 4 \cdot 4} = \frac{1}{4^2} = \frac{1}{4^{5-3}}.$$

A generalization of these results is given by Rules 5–7 where it is assumed that m and n are positive integers.

Rule 5. $\dfrac{x^m}{x^n} = x^{m-n}$ if $m > n$ and $x \neq 0$.

Rule 6. $\dfrac{x^m}{x^n} = \dfrac{1}{x^{n-m}}$ if $m < n$ and $x \neq 0$.

Rule 7. $\dfrac{x^n}{x^n} = 1$ if $x \neq 0$.

EXAMPLE
14

In each of the following, the number to the right refers to the rule used.

a. $\dfrac{2^6}{2^4} = 2^{6-4} = 2^2 = 4$ (5).

b. $\dfrac{x^5}{x^2} = x^{5-2} = x^3$ (5).

c. $\dfrac{x^7}{x^{10}} = \dfrac{1}{x^{10-7}} = \dfrac{1}{x^3}$ (6).

d. $\dfrac{2^8}{2^{12}} = \dfrac{1}{2^{12-8}} = \dfrac{1}{2^4} = \dfrac{1}{16}$ (6).

e. $\dfrac{y^{10}}{y^{10}} = 1$ if $y \neq 0$ (7).

f. $\dfrac{(x^6)^3}{(x^3)^4} = \dfrac{x^{18}}{x^{12}} = x^6$ (2) and (5).

g. $\dfrac{(x^2)(x^4)^3}{x^{90}} = \dfrac{(x^2)(x^{12})}{x^{90}} = \dfrac{x^{14}}{x^{90}} = \dfrac{1}{x^{76}}$ (2), (1), and (6).

h. $\left(\dfrac{1}{x^2}\right)(x^3) = \dfrac{x^3}{x^2} = x$ (5).

EXERCISE 1-6

Simplify each of the following.

1. $(2^3)(2^2)$

2. $(3^4)(3^2)$

3. $x^6 x^9$

4. $x^{20} x^{15}$

5. $w^4 w^8$

6. $y^9 y^{91}$

7. $x^6 x^4 x^3$

8. $x^2 x^3 x^4$

9. $\dfrac{x^2 x^6}{y^7 y^{10}}$

10. $\dfrac{x^{100} x^{200}}{z^9 z^{90}}$

11. $(x^{12})^4$

12. $(x^6)^6$

13. $\dfrac{(x^2)^5}{(y^5)^{10}}$

14. $(2x^3)^3$

15. $\left(\dfrac{x^2}{y^3}\right)^5$

16. $\left(\dfrac{2}{x}\right)^4$

17. $(2x^2 y^3)^3$

18. $(xy^2)^4$

19. $\left(\dfrac{w^2 s^3}{y^2}\right)^2$

20. $\dfrac{(xy^2)^2}{(z^2 w)^3}$

21. $\dfrac{x^8}{x^2}$

22. $\dfrac{y^4}{y^9}$

23. $\left(\dfrac{2x^2}{4x^4}\right)^3$

24. $\dfrac{2^6 \cdot 2^9}{(2^6)^3}$

25. $\dfrac{(x^3)^6}{x(x^3)}$

26. $\left(\dfrac{2}{3} x^2 y\right)^2$

27. $\dfrac{(x^2)^3 (x^3)^2}{(x^3)^4}$

28. $(-1)^5 (-1)^8$

29. $(x^a y^b)^c$

30. $\left(\dfrac{x^a}{y^b}\right)^a$

31. $(-3)^4(-3^2)$

32. $(-3)^2(-x)^2$

33. $(-x)^2(-x^2)$

34. $(-x)^2(-x)^3$

35. $(-x)^2(-x)^3(-x)^5$

36. $(2^3 x^3 y^4 z^5)^2(-1)^2$

37. $\left(\dfrac{1}{x^2 y^3}\right)^4$

38. $\dfrac{1}{(x^2)^3}(x^3)^2$

1-7
THE ZERO EXPONENT

We shall now attach a meaning to the symbol x^0. Since $x^m \cdot x^n$ is true for positive integral exponents, it would be desirable that we define x^0 so that

$$x^n \cdot x^0 = x^{n+0} = x^n.$$

But we know that

$$x^n \cdot 1 = x^n.$$

Since x^0 plays the same role in multiplication as 1 does, it seems reasonable to define x^0 to be 1. But suppose x were zero. Since $0 = 0^1 = 0^2 = 0^3$ etc., we cannot expect 0 raised to any power to be 1. The ambiguity attached to the symbol 0^0 renders it undefinable as a specific number in mathematics.

Definition. $x^0 = 1$ if $x \neq 0$. The symbol 0^0 is undefined.

EXAMPLE
15

a. $2^0 = 1$.

b. $100^0 = 1$.

c. $(x - 3)^0 = 1$ when $x \neq 3$.

d. $(-\pi)^0 = 1$.

1-8
NEGATIVE EXPONENTS

Now let us attach a meaning to a negative exponent such as in x^{-n}. For consistency with Rule 1 we certainly want

$$x^n \cdot x^{-n} = x^{n+(-n)} = x^0 = 1.$$

But we know that

$$x^n \cdot \left(\dfrac{1}{x^n}\right) = 1.$$

It follows, then, that x^{-n} plays the same role in multiplication as $1/x^n$ does. This causes us to make the following definition:

Definition. $x^{-n} = \dfrac{1}{x^n} = \dfrac{1}{\underbrace{x \cdot x \cdot x \cdots x}_{n \text{ factors}}}$ $x \neq 0.$

This definition holds for all integers n and is therefore true when n is replaced by $-n$. Hence

$$x^{-(-n)} = \frac{1}{x^{-n}}.$$

But

$$x^{-(-n)} = x^n.$$

Thus

$$\boxed{\dfrac{1}{x^{-n}} = x^n, \qquad x \neq 0.}$$

EXAMPLE
16

a. $3^{-4} = \dfrac{1}{3^4} = \dfrac{1}{81}.$

b. $x^{-5} = \dfrac{1}{x^5}.$

c. $\dfrac{x^3}{x^5} = \dfrac{1}{x^2} = x^{-2}.$

d. $\dfrac{1}{3^{-2}} = 3^2 = 9.$

e. $2x^{-3} = 2\left(\dfrac{1}{x^3}\right) = \dfrac{2}{x^3},$ but $(2x)^{-3} = \dfrac{1}{(2x)^3} = \dfrac{1}{2^3 x^3} = \dfrac{1}{8x^3}.$

f. $\dfrac{1}{1^{-1}} = 1^1 = 1.$

Clearly,

$$\frac{x^{-2}}{y^{-2}} = (x^{-2})\left(\frac{1}{y^{-2}}\right) = \left(\frac{1}{x^2}\right)(y^2) = \frac{y^2}{x^2}.$$

Thus a nonzero *factor* in the numerator (or denominator) of a fraction may be equivalently expressed as a factor in the denominator (or numerator) by changing the sign of its exponent. For example,

$$\frac{x^{-2}y^3}{2z^{-2}} = \frac{y^3 z^2}{2x^2}.$$

EXAMPLE 17

a. $7x^{-2} = \dfrac{7}{x^2}$, but $(7x)^{-2} = \dfrac{1}{(7x)^2} = \dfrac{1}{49x^2}$.

b. $x^{-1} + y^{-1} = \dfrac{1}{x} + \dfrac{1}{y}$. *Note:* $x^{-1} + y^{-1} \neq \dfrac{1}{x+y}$, but $x^{-1}y^{-1} = \dfrac{1}{xy}$.

c. $\dfrac{8x^{-1}}{2x^2} = \dfrac{8}{2} \cdot \dfrac{1}{x^2} \cdot \dfrac{1}{x} = \dfrac{4}{x^3}$.

d. $\dfrac{2+x}{y^{-2}} = (2+x)y^2$, but $\dfrac{2+x}{y^{-2}} \neq 2 + xy^2$ and $\dfrac{2+x}{y^{-2}} \neq 2 + x - y^2$.

Although the rules of exponents stated in this chapter assumed the exponents to be positive integers, **the rules of exponents are equally true for any exponents.** *Henceforth, we assume this fact.*

EXAMPLE 18

a. $x^{-2} \cdot x^{-3} = x^{(-2)+(-3)} = x^{-5}$.

b. $(2x^3y^{-6})^2 = 2^2(x^3)^2(y^{-6})^2 = 4x^6y^{-12} = \dfrac{4x^6}{y^{12}}$.

c. $(x^2y^{-4})^{-3} = (x^2)^{-3}(y^{-4})^{-3} = x^{-6}y^{12} = \dfrac{y^{12}}{x^6}$.

d. $10^{-6} \cdot 10^{-4} = 10^{-10} = \dfrac{1}{10^{10}}$.

e. $\dfrac{2^{10}}{2^{10}} = 2^{10-10} = 2^0 = 1$.

We remark that it is sometimes preferred, after a calculation or algebraic manipulation, to express an answer in terms of positive exponents only.

EXERCISE 1-8

In Problems 1–18 write the expressions in terms of positive exponents only. Simplify if possible.

1. x^{-5}

2. $x^{-2}x^{-3}$

3. xy^{-4}

4. $2^{-2}x$

5. $\dfrac{1}{x^{-7}}$

6. $\dfrac{1^0}{(xy)^{-4}}$

7. $\dfrac{1^0}{x^{-1}}$

8. $x^{-1}yz^{-2}$

9. $\dfrac{x^3y^{-2}}{z^2}$

10. $3x^{12}y^{-3}$

11. $2x^2b^{-4}$

12. $x + y^{-1}$

13. $2x^{-1}$

14. $(5xy^2)^{-2}$

15. $(3x)^{-2}$

16. a^2x^{-2}

17. $(x^{-2}y^{11})^{-3}$

18. $x + (x^2y^{-3}z^{-5})^{-6}$

Simplify the following and express all answers in terms of positive exponents.

19. $\dfrac{x^{-2}y^{-6}z^2}{xy^{-1}}$

20. $-\dfrac{2x^{-3}}{4x^{-5}}$

21. $\dfrac{x^{-2}yzw}{x^{-3}y^5}$

22. $\dfrac{x^2y^{-5}}{(x^{-8}y^6)^{-3}}$

23. $\dfrac{2(x^2y)^2}{3y^{13}z^{-2}}$

24. $\dfrac{(xyz^{-1})^{-2}}{(x^2)^{-4}}$

25. $\dfrac{3^0}{(2^{-3}x^{-4}y^6)^3} = \dfrac{2^9 x^{12}}{Y^{18}}$

26. $\left[\left(\dfrac{x}{y}\right)^{-2}\right]^{-4} = \left(\dfrac{x}{Y}\right)^8 = \dfrac{x^8}{Y^8}$

1-9
SCIENTIFIC NOTATION

In science and technology, it is not unusual for decimal numbers of extremely large or small values to occur. **Scientific notation** provides a convenient means of expressing such numbers by writing them as the product of a number between 1 and 10 and some integral power of 10. Some powers of 10 are

$$10 = 10^1 \qquad\qquad \frac{1}{10} = 10^{-1}$$

$$100 = (10)(10) = 10^2 \qquad\qquad \frac{1}{100} = \frac{1}{10^2} = 10^{-2}$$

$$1000 = (10)(10)(10) = 10^3 \qquad\qquad \frac{1}{1000} = \frac{1}{10^3} = 10^{-3}$$

$$10{,}000 = (10)(10)(10)(10) = 10^4 \qquad \frac{1}{10{,}000} = \frac{1}{10^4} = 10^{-4}$$

To write a number in scientific notation, first determine the factor between 1 and 10 by writing the digits in the original number with a decimal point to the right of the leftmost nonzero digit. The exponent of 10 for the second factor corresponds to the number of places the decimal point was moved from its original position. If moved to the left, the exponent is positive; if moved to the right, it is negative.

EXAMPLE
19

a. $16 = (1.6)(10^1) = 1.6 \times 10$.

b. $257 = (2.57)(10^2) = 2.57 \times 10^2$.

c. $6543 = (6.543)(10^3) = 6.543 \times 10^3$.

d. $.00624 = (6.24)(10^{-3}) = 6.24 \times 10^{-3}$.

e. $21{,}000{,}000{,}000 = (2.1)(10^{10}) = 2.1 \times 10^{10}$.

f. $.000000000040926 = (4.0926)(10^{-11}) = 4.0926 \times 10^{-11}$.

g. $23.789 = (2.3789)(10) = 2.3789 \times 10$.

h. $672.0 = 6.720 \times 10^{2}$.

Similarly,

i. $6.24 \times 10^{-4} = .000624$

j. $2.6142 \times 10^{8} = 261{,}420{,}000$

k. $7.0030 \times 10^{-6} = .0000070030$

EXAMPLE
20

Coulomb's law states that the force F (in newtons) acting between two small objects having electric charges q_1 and q_2 (in coulombs) is

$$F = \frac{(9 \times 10^{9})q_1 q_2}{r^{2}},$$

where r is the distance between the objects (in meters). Given that in the hydrogen atom the electron and proton each have a charge of 1.6×10^{-19} coulombs and are separated by a distance of $.53 \times 10^{-10}$ meters, compute the force between them.

Here we treat the numerical factors and powers of 10 separately:

$$F = \frac{(9 \times 10^{9})(1.6 \times 10^{-19})(1.6 \times 10^{-19})}{(.53 \times 10^{-10})^{2}}$$

$$= \frac{(9 \times 1.6 \times 1.6)(10^{9} \times 10^{-19} \times 10^{-19})}{(.53)^{2}(10^{-20})}$$

$$= 82 \times 10^{9-19-19+20}$$

$$= 82 \times 10^{-9} = 8.2 \times 10^{-8} \text{ newtons.}$$

EXERCISE 1-9

In Problems 1–10 express the given number in scientific notation.

1. $.0000060214$ 2. $213{,}146{,}100{,}000{,}000$

3. 10.4 4. $.006241001$

5. $26{,}245.1001$ 6. $2{,}600{,}000$

7. 142 8. $.0071$

9. $.76$ 10. $.1$

Write the numbers in Problems 11–18 in decimal notation.

11. 2.62×10^{8} 12. 1.234×10^{-8}

13. 6.24×10^{-10} 14. 1.006×10^{4}

15. 2.020×10^{-1} **16.** 6.0411×10^{12}

17. 7.611×10^5 **18.** 2.0×10

19. The mass of the earth, in kilograms, is usually taken to be

$$5,983,000,000,000,000,000,000,000.$$

Express this mass in scientific notation.

20. The speed of light in a vacuum is taken to be 186,000 miles per second. Express this speed in scientific notation.

21. The computational value of the gravitational constant (in meter-kilogram-second units) is

$$.0000000000667.$$

Express this constant in scientific notation.

22. A current I of 10^4 amperes is produced in a conductor, whose length L is 4 meters, at a point where the earth's magnetic field B is 5×10^{-5} webers per square meter at right angles to the conductor. The force F, in newtons, on the conductor is given by

$$F = IBL$$
$$= (10^4)(5 \times 10^{-5})(4).$$

Evaluate F.

23. While the universe is about 10^{10} years old, man has existed only for about 10^6 years. Determine the number of years the universe existed before man, and express that number in scientific notation.

24. If the mass of an electron is 9.11×10^{-28} grams, how many electrons would it take to make 1 gram?

25. The potential at a distance of 1.73×10^{-4} meters from a point charge of 4.62×10^{-2} coulombs is

$$\frac{(9 \times 10^9)(4.62 \times 10^{-2})}{1.73 \times 10^{-4}} \text{ volts.}$$

Evaluate this potential.

1-10
RADICALS

If $r^n = x$ where n is a positive integer, then r is said to be an **nth root of x**. For example, since $(-2)^3 = -8$, we call -2 a third (or *cube*) root of -8. Similarly, since $3^2 = 9$, 3 is called a square root of 9. Also, since $(-3)^2 = 9$, -3 is a square root of 9. Thus, both 3 and -3 are square roots of 9. In particular, the **principal nth root** of x is that nth root of x which is positive if x is positive and is negative if x is negative, assuming such a real root exists. Thus, although $+3$ and -3 are both square roots of 9, since 9 is a positive number it follows that $+3$ is the principal square root of 9. Likewise, -2 is the principal cube root of -8.

We denote the principal nth root of x by the symbol $\sqrt[n]{x}$, where n is called the **index**, x is the **radicand**, and $\sqrt{}$ is the **radical sign**. The symbol $\sqrt[n]{x}$ is itself called a **radical**. Thus, $-2 = \sqrt[3]{-8}$ and $3 = \sqrt{9}$, but $-3 \neq \sqrt{9}$. We can say, however, that $-3 = -\sqrt{9}$. The index 2 is usually omitted for square roots. An nth root of 0 is zero and we therefore define $\sqrt[n]{0} = 0$. For convenience we shall drop the word *principal* and simply say that 3 is *the* square root of 9, etc.

You will find the appearance of radicals in many of your other courses, for they are found in virtually every area of technology. To illustrate, the expressions

$$\frac{1}{2\pi\sqrt{LC}}, \qquad \sqrt{\frac{\beta}{d}}, \qquad \text{and} \qquad \sqrt{\frac{M}{2\rho_0 N_0}}$$

are, respectively, the resonant frequency of an alternating current circuit, the velocity of a compressional wave, and the distance between ions in a cubic crystal.

EXAMPLE
21

    a. $\sqrt{25} = 5$ (and 5 is *the* square root of 25).

    b. $\sqrt[4]{16} = 2$ (and 2 is *the* fourth root of 16).

    c. $\sqrt[3]{-1} = -1$ (and -1 is *the* cube root of -1).

    d. $\sqrt{6^2} = \sqrt{36} = 6$.

    e. $\sqrt{(-6)^2} = \sqrt{36} = 6$.

    f. $\sqrt{x^8} = x^4$ since $x^4 \cdot x^4 = x^8$.

Note in Example 21d that if $x = 6$, then $\sqrt{x^2} = \sqrt{36} = 6 = x$. But in Example 21e, if $x = -6$, then $\sqrt{x^2} = \sqrt{36} = 6 = -(-6) = -x$. Thus, $\sqrt{x^2}$ will be x if x is positive, and will be $-x$ if x is negative. (Remember that if x is negative, then $-x$ is positive.) More simply, for the general case we can say that

$$\sqrt{x^2} = |x|.$$

Thus,

$$\sqrt{6^2} = |6| = 6 \qquad \text{and} \qquad \sqrt{(-6)^2} = |-6| = 6.$$

Unless otherwise stated, we shall assume that all literal numbers appearing under radical signs are positive and, as a result, we shall feel free to write, for example, $\sqrt{y^2} = y$.

We wish to point out that difficulties arise in considering even roots of negative numbers. These difficulties are related to the fact that such roots are not real numbers. For example, $\sqrt{-4}$ is not a real number since no real number squared will result in a negative number. Such numbers are called *imaginary numbers* and will be discussed in Chapter 11. Therefore, we shall for the present avoid any questions concerning such roots.

EXERCISE 1-10

In Problems 1–20, evaluate the given expression.

1. $\sqrt{81}$
2. $\sqrt[5]{32}$

3. $\sqrt[3]{-64}$
4. $\sqrt[3]{-27}$

5. $-\sqrt[3]{-1}$
6. $\sqrt{.25}$

7. $|-2| - \sqrt[4]{16}$
8. $\sqrt[3]{x^9}$

9. $\sqrt{\dfrac{1}{5} \cdot \dfrac{1}{5}}$
10. $-\sqrt{36} + \sqrt{81}$

11. $\sqrt[4]{x^8}$

$\dfrac{1-9}{4+4} = \dfrac{-8}{8} = -1$

12. $\dfrac{-\sqrt{144}}{|12|} = \dfrac{-12}{12} = -1$

13. $2(-\sqrt{9})(-|2|)$
14. $\sqrt[3]{|-8|}$

15. $\sqrt{(-3)^2} - \sqrt{3^2}$
16. $\dfrac{\sqrt{49} - \sqrt{36}}{\sqrt{100} - \sqrt[3]{1}} = \dfrac{1}{9}$

17. $\dfrac{(-1)^2 - (3\sqrt{9})}{|-4| + \sqrt{16}} =$
18. $\dfrac{(-2)^2 + (-3)^4}{\sqrt[5]{1} + |2 - 3|} = \dfrac{4+81}{1+1} = \dfrac{85}{2}$

19. $\sqrt{.01} + \sqrt{.0025}$
20. $2(\sqrt[3]{.008}) - (.01)^2 = 2(.2) - .0001 =$
$\qquad\qquad .4 - 0001 = .3999$

1-11

FRACTIONAL EXPONENTS AND RULES FOR RADICALS

We are now in a position to attach a meaning to a fractional exponent such as appears in the symbol $5^{1/2}$. Since $\sqrt{5}$ is a number whose square is 5, it follows that $\sqrt{5} \cdot \sqrt{5} = 5$. However, the laws of exponents, which have been stated as true for all types of exponents, imply

$$5^{1/2} \cdot 5^{1/2} = 5^{(1/2)+(1/2)} = 5^1 = 5.$$

Thus the square of $5^{1/2}$ is 5, as is the square of $\sqrt{5}$. It follows that for continued consistency it is natural to define $5^{1/2}$ to be $\sqrt{5}$. More generally, $x^{1/n}$ will denote the principal nth root of x (where n is a positive integer).

Definition. $x^{1/n} = \sqrt[n]{x}$.

EXAMPLE
22

a. $3^{1/5} = \sqrt[5]{3}$.

b. $4^{1/2} = \sqrt{4} = 2$.

c. $27^{1/3} = \sqrt[3]{27} = 3$.

d. $9^{-1/2} = \dfrac{1}{9^{1/2}} = \dfrac{1}{\sqrt{9}} = \dfrac{1}{3}$.

e. $(12^{-1/2})^4 = 12^{-4/2} = 12^{-2} = \dfrac{1}{12^2} = \dfrac{1}{144}$.

f. $(-1000)^{1/3} = \sqrt[3]{-1000} = -10$.

Consider the expression $(x^3)^{1/4}$. By previous rules this can be written in two ways:

$$(x^3)^{1/4} = x^{3/4}$$

or

$$(x^3)^{1/4} = \sqrt[4]{x^3}.$$

More generally, for $x \geq 0$ we have the following rule:

Rule 8. $x^{m/n} = \sqrt[n]{x^m} = (\sqrt[n]{x})^m$ where m and n are integers and n is positive.

EXAMPLE
23

a. $8^{2/3} = \sqrt[3]{8^2} = (\sqrt[3]{8})^2 = 2^2 = 4$.

b. $\sqrt[3]{x^6} = x^{6/3} = x^2$.

c. $\sqrt[6]{x^3} = x^{3/6} = x^{1/2} = \sqrt{x}$.

d. $(\sqrt[8]{7})^8 = (7^{1/8})^8 = 7$. *More generally,* $(\sqrt[n]{x})^n = x$.

e. $(\sqrt{3})^4 = (3^{1/2})^4 = 3^2 = 9$.

f. The reason we require $x \geq 0$ in Rule 8 is clear by considering

$$\sqrt{(-2)^2} = \sqrt{4} = 2,$$

which is true. We should *not* write

$$\sqrt{(-2)^2} = [(-2)^2]^{1/2} = (-2)^1 = -2,$$

which is **false**.

We shall now state some further rules for radicals which can essentially be derived from those of exponents. Keep in mind that the index of a radical is always a positive integer.

Rule 9. $\sqrt[n]{x}\, \sqrt[n]{y} = \sqrt[n]{xy}$.

EXAMPLE
24

a. $\sqrt[3]{9}\, \sqrt[3]{2} = \sqrt[3]{9 \cdot 2} = \sqrt[3]{18}$.

b. $\sqrt{50} = \sqrt{25 \cdot 2} = \sqrt{25}\, \sqrt{2} = 5\sqrt{2}$.

c. $\sqrt{x^8 y^5} = \sqrt{x^8}\, \sqrt{y^5} = (x^8)^{1/2}(y^5)^{1/2} = x^4 y^{5/2} = x^4 y^2 y^{1/2} = x^4 y^2 \sqrt{y}$.

d. $\sqrt{\frac{3}{2}}\sqrt{6} = \sqrt{\frac{3}{2}\cdot 6} = \sqrt{9} = 3.$

e. $\sqrt[3]{x^6 y} = \sqrt[3]{x^6}\sqrt[3]{y} = x^2\sqrt[3]{y}.$

f. $\sqrt[4]{48} = \sqrt[4]{16\cdot 3} = \sqrt[4]{16}\sqrt[4]{3} = 2\sqrt[4]{3}.$

g. It is interesting to note the trouble that arises if we neglect the warning to stay in the real number system. Since $(\sqrt[n]{x})^n = x$, we have $(\sqrt{-1})^2 = -1$. But by Rule 9, $\sqrt{-1}\sqrt{-1} = \sqrt{(-1)(-1)} = \sqrt{1} = 1.$ *Moral:* Obey the warning and stick to real numbers.

Rule 10. $\dfrac{\sqrt[n]{x}}{\sqrt[n]{y}} = \sqrt[n]{\dfrac{x}{y}}.$

EXAMPLE 25

a. $\dfrac{\sqrt[3]{90}}{\sqrt[3]{10}} = \sqrt[3]{\dfrac{90}{10}} = \sqrt[3]{9}.$

b. $\sqrt[4]{\dfrac{x^{16}}{y^8}} = \dfrac{\sqrt[4]{x^{16}}}{\sqrt[4]{y^8}} = \dfrac{x^{16/4}}{y^{8/4}} = \dfrac{x^4}{y^2}.$

Rule 11. $\sqrt[m]{\sqrt[n]{x}} = \sqrt[mn]{x}.$

EXAMPLE 26

a. $\sqrt[3]{\sqrt[4]{2}} = \sqrt[12]{2}.$

b. $\sqrt{\sqrt[4]{x}} = \sqrt[8]{x}.$

c. $\sqrt{\sqrt{16}} = \sqrt[4]{16} = 16^{1/4} = (2^4)^{1/4} = 2.$

d. $\sqrt[3]{\sqrt{x^{12}}} = \sqrt[6]{x^{12}} = (x^{12})^{1/6} = x^2.$

e. $\sqrt[3]{\sqrt[5]{x}} = (\sqrt[5]{x})^{1/3} = (x^{1/5})^{1/3} = x^{1/15} = \sqrt[15]{x}.$

When a fraction involves a radical in its denominator, it is possible to multiply the fraction by a quantity that although leaving the value of the fraction unchanged will, indeed, remove the radical from the denominator. The resulting fraction may then be in a more desirable form. This procedure is referred to as **rationalizing the denominator**.

Consider, for example,

$$\frac{1}{\sqrt{5}} = \frac{1}{5^{1/2}}.$$

If we were to multiply the denominator by $5^{1/2}$, the result would be $5^{1/2}\cdot 5^{1/2} = 5$. In order not to change the value of the fraction, we must also multiply the numerator by $5^{1/2}$. Thus, we are merely multiplying the original fraction by $5^{1/2}/5^{1/2}$, which is just 1:

$$\frac{1}{\sqrt{5}} = \frac{1}{5^{1/2}} = \frac{1}{5^{1/2}}\cdot 1 = \frac{1}{5^{1/2}}\cdot\frac{5^{1/2}}{5^{1/2}} = \frac{5^{1/2}}{5^1} = \frac{\sqrt{5}}{5}.$$

EXAMPLE
27

a. $\dfrac{2}{\sqrt[3]{x}} = \dfrac{2}{x^{1/3}} = \dfrac{2}{x^{1/3}} \cdot \dfrac{x^{2/3}}{x^{2/3}} = \dfrac{2x^{2/3}}{x} = \dfrac{2\sqrt[3]{x^2}}{x}.$

b. $\dfrac{1}{\sqrt[4]{2}} = \dfrac{1}{2^{1/4}} = \dfrac{1}{2^{1/4}} \cdot \dfrac{2^{3/4}}{2^{3/4}} = \dfrac{2^{3/4}}{2} = \dfrac{\sqrt[4]{2^3}}{2} = \dfrac{\sqrt[4]{8}}{2}.$

c. $\dfrac{y}{\sqrt[4]{x^5}} = \dfrac{y}{x^{5/4}} = \dfrac{y}{x^{5/4}} \cdot \dfrac{x^{3/4}}{x^{3/4}} = \dfrac{yx^{3/4}}{x^{8/4}} = \dfrac{y\sqrt[4]{x^3}}{x^2}.$

One illustration of the use of the process of rationalizing the denominator concerns the approximations of certain irrational expressions. For example, if you were given that $\sqrt{2}$ is approximately 1.4142 and you were to compute $1/\sqrt{2} = 1/1.4142$ by long division, the process would be rather tedious due to the denominator. However, by rationalizing the denominator, you reduce the problem to division by 2:

$$\frac{1}{\sqrt{2}} = \frac{1}{2^{1/2}} \cdot \frac{2^{1/2}}{2^{1/2}} = \frac{\sqrt{2}}{2} = \frac{1.4142}{2} = .7071.$$

It is sometimes preferred, following a calculation or algebraic manipulation, to express the answer in rationalized form.

The following examples illustrate various applications of the rules for exponents and radicals.

EXAMPLE
28

a. $\dfrac{\sqrt{20}}{\sqrt{5}} = \sqrt{\dfrac{20}{5}} = \sqrt{4} = 2.$

b. $\sqrt{2 + 5x} = (2 + 5x)^{1/2}.$

c. $\sqrt{\dfrac{2}{7}} = \sqrt{\dfrac{2}{7} \cdot \dfrac{7}{7}} = \sqrt{\dfrac{14}{49}} = \dfrac{\sqrt{14}}{\sqrt{49}} = \dfrac{\sqrt{14}}{7}.$

d. $\left(\dfrac{x^{1/5}y^{6/5}}{x^{2/5}}\right)^5 = \dfrac{xy^6}{x^2}.$

e. $(x^{5/9}y^{4/3})^{18} = x^{10}y^{24}.$

EXAMPLE
29

The frequency of vibration of a string of length L, fixed at both ends and vibrating in its fundamental mode, is

$$f = \frac{1}{2L}\sqrt{\frac{T}{\mu}},$$

where f is frequency, μ ($= $ mu) is mass per unit length, and T is the tension in the string. If the tension in the string is quadrupled, what happens to the frequency?

Let f_0 denote the frequency when the tension is quadrupled, that is, when it is $4T$. Then

$$f_0 = \frac{1}{2L}\sqrt{\frac{4T}{\mu}} = \frac{1}{2L}(2)\sqrt{\frac{T}{\mu}} = 2\left(\frac{1}{2L}\sqrt{\frac{T}{\mu}}\right)$$
$$= 2f.$$

Thus, the frequency is doubled.

EXERCISE 1-11

In Problems 1–6, write the exponential forms in equivalent forms involving radicals. Simplify if possible.

1. $25^{1/2}$
2. $7^{1/6}$
3. $61^{1/8}$
4. $125^{1/3}$
5. $x^{3/5}$
6. $y^{11/13}$

In Problems 7–66, simplify the given expression. Express all answers in terms of positive exponents and rationalize the denominator where necessary to avoid fractional exponents in the denominator. For example, $y^{-1}\sqrt{x} = \dfrac{x^{1/2}}{y}.$

7. $\sqrt[3]{x} = x^{1/3}$
8. $(x^{1/2})^3$
9. $16^{-1/2}$
10. $2(4)^{-1/2}$
11. $(x^{-1/2})^4$
12. $(y^{-1/4})^8$
13. $(x^3)^{1/5} = x^{3/5}$
14. $(x^7)^{1/49}$
15. $\sqrt[3]{x^{12}}$
16. $\sqrt{x^7}$
17. $\sqrt[5]{x^2}$
18. $\sqrt[4]{x^9}$
19. $x^{-2}\sqrt[3]{y}$
20. $y^{-3}\sqrt{x^3}$
21. $\sqrt[6]{7^6}$
22. $\sqrt[9]{x^9}$
23. $\sqrt{x^{16}y^{18}}$
24. $\sqrt[3]{x^4 z^5}$
25. $\sqrt[5]{xy^2}$
26. $\sqrt{18}$
27. $\sqrt{75}$
28. $(\sqrt[5]{5})^5$
29. $\sqrt[3]{x^3 y^6 z^9}$
30. $(\sqrt[3]{x})^6$
31. $(\sqrt{x})^3$
32. $(x^2 y^4)^{2/3}$
33. $(\sqrt[3]{x^2 y})^4$
34. $\sqrt[5]{\dfrac{x^2}{y^{10}}}$
35. $\sqrt[4]{\dfrac{x^8}{y^{16}}}$
36. $[(x^{1/2})^3]^2$
37. $\dfrac{\sqrt[3]{24}}{\sqrt[3]{3}} = \sqrt[3]{\dfrac{24}{3}} = \sqrt[3]{8} = 2$
38. $\dfrac{\sqrt{50}}{\sqrt{2}} = \sqrt{\dfrac{50}{2}} = \sqrt{25} = 5$
39. $\sqrt[3]{\sqrt[4]{x^{24}}} = \sqrt[3]{x \times \frac{24}{4}} = \sqrt[3]{x^6} = x^2$
40. $\sqrt{\sqrt{7}} = 7^{\frac{1}{4}}$
41. $\sqrt[5]{\sqrt{2}}$
42. $\sqrt[3]{\sqrt{x}}$
43. $(ab^2 c^3)^{3/4}$
44. $[(x^{-4})^{1/5}]^{-1/6} = (x^{-4})^{-1/30} = x^{4/30} = x^{2/15}$
45. $\sqrt{75x^4}$
46. $(\sqrt[5]{x^2 y})^{2/5} = (x^{\frac{2}{5}} y^{\frac{1}{5}})^{2/5} = x^{\frac{4}{25}} y^{\frac{2}{25}}$

47. $\dfrac{1}{\sqrt{3}}$

48. $\dfrac{1}{\sqrt{7}}$

49. $\dfrac{1}{\sqrt[3]{6}}$

50. $\dfrac{2}{\sqrt[4]{5}}$

51. $\dfrac{2}{\sqrt[3]{x^2}}$

52. $\dfrac{1}{x^{5/4}}$

53. $\dfrac{1}{x^{12/7}}$

54. $\dfrac{1}{2\sqrt{x}}$

55. $81^{-3/4}$

56. $\sqrt{\sqrt[3]{\sqrt{x}}}$

57. $\sqrt[3]{\dfrac{x^4 y^3}{z^3 w^6}}$

58. $\sqrt{x^3}\sqrt{xy^5}\sqrt{x^2 y} = $

59. $\sqrt{3x^2 yw}\,\sqrt{27y^3}$

60. $\sqrt{(a^{-1/2})^{-4} a^4}$

61. $\dfrac{1}{\sqrt{144x}}$

62. $(2x^{-1/2}y^{1/4})^3$

63. $(x^m)^{1/2}$

64. $x^{-4/5}$

65. $\sqrt[4]{xy^{-2}z^3}$

66. $\sqrt[5]{xy^{-3}}$

67. In atomic and nuclear physics it is shown that, at $0°K$ (the Kelvin scale), theory predicts an upper limit for the energy E of conduction electrons. This energy is given by

$$E = \frac{h^2}{2m}\left(\frac{3n}{8\pi}\right)^{2/3}.$$

Perform the indicated operations and express the energy in rationalized form with positive exponents.

68. The grating spacing, the distance between ions, of a sodium chloride crystal can be shown to be given by the expression,

$$\sqrt[3]{\frac{M}{2\rho N_0}}.$$

Express this distance in rationalized form with positive exponents.

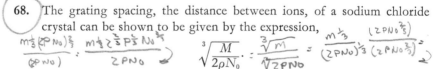

69. In the formula of Example 29, if the mass per unit length is quadrupled, what happens to the frequency?

1-12
REVIEW

REVIEW QUESTIONS

The sets $A = \{1, 2, 3, 4, a\}$ and $B = \{1, 2, 3, 3, 4\}$ $\underline{\text{(are)(are not)}}$ equal.
(1)

(1) **are not**

The previous sets A and B would be equal if the element _____(2)_____ were not in A.

(2) *a*

A number which can be expressed in the form p/q where p and q are integers and $q \neq 0$ is called a _____(3)_____ number.

(3) **rational**

The number $\dfrac{\frac{2}{3} \cdot 1000}{\frac{1}{2} + \frac{1}{3}}$ is (rational)(irrational) .

(4) **rational**

$7^0 = $ _____(5)_____ but 0^0 is _____(6)_____ .

(5) **1**   (6) **undefined**

$|-6| = $ _____(7)_____ , $|0| = $ _____(8)_____ , and $|6| = $ _____(9)_____ .

(7) **6**   (8) **0**   (9) **6**

True or false: There is no number x such that $|x| = -x$. _____(10)_____

(10) **false: for example, if x = −10, then $|-10|$ = 10 = −(−10) = −x.**

True or false: $|x| > 0$ for any real number x. _____(11)_____

(11) **false: $|x| \geq 0$.**

The product of an odd number of negative numbers is (positive)(negative) , and the product of an even number of negative numbers is (positive)(negative) .

(12) **negative**   (13) **positive**

Although $2^2 = 4$ and $(-2)^2 = 4$, by the symbol $\sqrt{4}$ we denote (2) or (−2) .

(14) **2**

In exponential form, the expression $\sqrt[6]{x}$ is written _____(15)_____ .

(15) $x^{1/6}$

$x^2(x^3) =$ _____(16)_____ and $(x^3)^2 =$ _____(17)_____ .

(16) x^5 (17) x^6

Arrange the following numbers in order of increasing absolute value: $-6, -3,$
$0, 4, 5.$ _____(18)_____

(18) **0, −3, 4, 5, −6**

The statement $a < b$ means that b lies to the _____(19)_____ of a on the
real number line.

(19) **right**

The _____(20)_____ axiom of addition asserts that $a + b = b + a$.

(20) **commutative**

N denotes the set of _____(21)_____ and **Q** the set of _____(22)_____ .

(21) **natural numbers or positive integers**
(22) **rational numbers**

$\frac{0}{7} =$ _____(23)_____ but $\frac{7}{0}$ is _____(24)_____ .

(23) **0**   (24) **undefined**

The expression x^2/x^3 may be written as x raised to what negative power?
_____(25)_____

(25) **−1**

In scientific notation, the number .026 would be written as _____(26)_____ .

(26) **2.6 × 10⁻²**

The distributive axiom asserts that $a(b + c) =$ _____(27)_____ .

(27) **ab + ac**

To rationalize the denominator of $\dfrac{1}{\sqrt[3]{2}}$ we may multiply both the numerator and denominator by _____(28)_____.

(28) $2^{2/3}$

Arrange the sets **Z**, **Q**, **N**, and **R** in an order such that each set is a subset of all the sets to the right of it. _____(29)_____

(29) **N, Z, Q, R**

By the definition of division, $\dfrac{a}{b} = c$ if $a =$ _____(30)_____, or a/b means a times _____(31)_____.

(30) $\boldsymbol{bc}$ (31) $\boldsymbol{1/b}$

The multiplicative inverse of 3 is _____(32)_____ and the additive inverse of 3 is _____(33)_____.

(32) $\frac{1}{3}$ (33) $\boldsymbol{-3}$

The reciprocal of $\frac{1}{7}$ is _____(34)_____.

(34) **7**

The result of subtracting -7 from 0 is _____(35)_____.

(35) **7**

The interval $(3, 4)$ is a(n) $\underset{(36)}{\text{(open)(closed)}}$ interval and $[3, 4]$ is a(n) $\underset{(37)}{\text{(open)(closed)}}$ interval.

(36) **open**   (37) **closed**

The set $\{x \mid x \le -4\}$ can also be written as _____(38)_____.

(38) $(-\infty, \boldsymbol{-4}]$

The interval $(3, 4]$ can be written in set-builder notation as _____(39)_____.

(39) $\{\boldsymbol{x} \mid \boldsymbol{3 < x \le 4}\}$

REVIEW PROBLEMS

Simplify the expressions in Problems 1–18, if possible.

1. $\dfrac{(-2)[-(-3)^2](2) + [-(-1)] - (\sqrt{4^2})}{(8 - \sqrt[3]{64})^{1/2}}$

2. $(x^{3/2}y^{4/5}z^{2/3})^{30}$

3. $\sqrt[3]{\sqrt{64}}$

4. $-[-(-1)]$

5. $\dfrac{1}{\sqrt[5]{4}}$

6. $\sqrt[3]{2}\,\sqrt[3]{4}$

7. $(-2x)^2(-x)^3$

8. $x^a x^b x^r x^t$

9. $(\sqrt[3]{x^4})^{3/2}$

10. $\sqrt{x^{20}y^4 z^8}$

11. $|-1| + |1| - |-2|$

12. $2 + (-|3|) + (-2 + 3)\cdot|3|$

13. $\left(\dfrac{\sqrt{x}}{\sqrt[3]{y}}\right)^{12}$

14. $\dfrac{7}{0}$

15. $\sqrt{\tfrac{1}{4}} - \sqrt[3]{.027}$

16. $\{4[\sqrt[4]{\sqrt[3]{\sqrt[5]{4}}}]^0\}^2$

17. $(x^2 y^{-3} z^{12})^4$

18. $\dfrac{x^2 x^{-5} x^{10}}{x^{12}}$

Write the intervals in Problems 19–22 in set-builder notation.

19. $(3, 5)$

20. $[2, 6)$

21. $[2, \infty)$

22. $(-6, 6]$

23. Write .000876 in scientific notation.

2

fundamental operations
with algebraic expressions

2-1
ALGEBRAIC EXPRESSIONS

If numbers, represented by symbols, are combined by the operations of addition, subtraction, multiplication, division, or extraction of roots, then the resulting expression is called an **algebraic expression**. A constant or variable by itself is also considered to be an algebraic expression.

EXAMPLE
1

The following are algebraic expressions:

a. $5ax^3 - 2bx + 3$.

b. $4 - \sqrt{y} + \dfrac{2}{3 + y^2}$.

c. $\sqrt{x^2 + 2 + y^2}$.

d. mc^2.

e. 7.

f. x.

In an algebraic expression, the parts which are connected by plus or minus signs are, along with the associated signs, called the **terms** of the expression. For example, the algebraic expression in Example 1a has three terms: the first is $+5ax^3$, the second is $-2bx$, and the third is $+3$. Likewise, Example 1b consists of three terms. However, the remaining expressions each consist of one term.

If a term can be expressed as a product of two or more expressions, then each expression is called a **factor** of the term. For example, since the term $5ax^3$ can be written as $(5a)(x)(x^2)$, we can say that $5a$, x, and x^2 are factors of $5ax^3$. Some other factors are 5, a, x^3, $5x$, and ax^3. Moreover, since $5ax^3 = (1)(5ax^3)$, certainly $5ax^3$ can be considered a factor of $5ax^3$. Indeed, any term is a factor of itself.

If a term of an algebraic expression is considered a product of two factors, each factor is called the **coefficient** of the other factor. In particular, a numerical factor is called a **numerical coefficient**. For example, given the term $6xy^2$ we would call $6x$ the coefficient of y^2, and $6y^2$ would be the coefficient of x. Also, 6 is the numerical coefficient of xy^2. Similarly, the numerical coefficient in the term x^2 is 1.

Algebraic expressions, such as $3x^2$, which have only one term are called **monomials**. Those, such as $x^2 + 2x + y$, which have more than one term are called **multinomials**. Two special cases of multinomials, namely those containing exactly two or exactly three terms, are called **binomials** or **trinomials**, respectively. Thus, the multinomial $3x^2 - 2$ is a binomial, and the multinomial $2x^2 + \sqrt{x} + 2$ is a trinomial.

The word *polynomial* refers to an algebraic expression of a special form. A **polynomial** in one variable, say x, is an algebraic expression in which every term is either a constant or the product of a constant and a positive integral power of x. The greatest exponent of x which appears in the polynomial is called the **degree** of the polynomial. Thus, $2x^3 - 6x$ is a polynomial in x of degree 3, and $8y^5 - 6y^4 - 2y + 3$ is a polynomial in y of degree 5. A *nonzero* constant is regarded as being a polynomial of degree zero. Thus, the expression 7 is a polynomial of degree zero.

EXAMPLE
2

a. The expression $8x^4 - 2x + 6$ is a multinomial, a trinomial, and a polynomial (in x) of degree 4.

b. The expression $\frac{6}{7}y$ is a monomial and a polynomial (in y) of degree 1.

c. The expression $\dfrac{2}{x^2} + 6$ is a multinomial and a binomial, but it is not a polynomial since the term $\dfrac{2}{x^2}$, which can be written $2x^{-2}$, is *not* the product of a constant and a *positive* integral power of x.

d. $x^2 + 2x + \sqrt{x} + 2$ is not a polynomial due to the term $\sqrt{x}$ $(= x^{1/2})$.

EXERCISE 2-1

In Problems 1–18, classify the given expression as (a) a monomial, (b) a multinomial, (c) a binomial, (d) a trinomial, or (e) a polynomial in x. *More than one classification may apply. For those which are polynomials in* x, *give the degree.*

1. $2x^2$

2. $2x^2 + 6$

3. $x^2 - 2x + 3$

4. $3yz$

5. $3x^4 - \sqrt{3}$

6. $2 - x^2 - x$

7. $3x$

8. $a^2 + b^2 + c^2 + a + b + c$

9. $\dfrac{1}{x^2} + y^2 + 2y$

10. $a^2 + 2a - 4$

11. $\frac{4}{5}x - x^4$

12. $x - 2$

13. $2 + \dfrac{1}{x} + x - x^2$

14. 7

15. $x^5 - x^4 - x^3 + x^2 - x + 1$

16. $x^2(x^3)$

17. $\dfrac{1}{x^2} + \dfrac{1}{y^2} + \dfrac{1}{z^2}$

18. $\dfrac{4}{3}\pi r^3$

In Problems 19–26, for the given expression state the coefficient of the x^2-*term and give the numerical coefficient of the* x-*term.*

19. $2x^3 + 3x^2 + 4x$

20. $3x^2 - x$

21. $2x^2y - 2x + 1$

22. $4x - 6x^2y^2$

23. $4ax^2 + x - 2$

24. x^2

25. $2 + \dfrac{7x}{3}$

26. $-x + \sqrt{2}\,x^2$

2-2
SYMBOLS OF GROUPING

Sometimes a number of terms in an algebraic expression are grouped together to indicate that they are to be treated as a single number or quantity. The most common symbols of grouping are **parentheses** ( ), **brackets** [ ], and **braces** { }. A consideration of various algebraic expressions will reveal the fundamental ways of handling such symbols of grouping.

Consider the expression

$$2ax + (3bc - 4a + 5c).$$

This expression can be written as

$$2ax + (1)(3bc - 4a + 5c),$$

and by application of the distributive axiom it becomes

$$2ax + 3bc - 4a + 5c.$$

Observe that we would arrive at the same result by merely removing the parentheses that appear in the original expression. Generalizing our observation, we state the *first rule of grouping symbols*:

> **If a symbol of grouping preceded by a plus sign is to be removed, the sign of every term that is within the grouping symbol must remain the same.**

EXAMPLE
3

   a. $3x + (2y + 3z) = 3x + 2y + 3z$.

   b. $(2b - 3c - 4d) + 6 = 2b - 3c - 4d + 6$.

   c. $2x^2 + [-8b + 4d] - 4 = 2x^2 - 8b + 4d - 4$.

   d. When removing the parentheses in $x^2 + 2(-16x + xy)$, we must remember, by the distributive axiom, to multiply each term within the parentheses by 2:

$$x^2 + 2(-16x + xy) = x^2 - 32x + 2xy.$$

Consider the expression

$$x - (8a + b - 2),$$

which may be written

$$x + (-1)(8a + b - 2).$$

Applying the distributive axiom we obtain

$$x + (-8a - b + 2)$$

or

$$x - 8a - b + 2.$$

Thus

$$x - (8a + b - 2) = x - 8a - b + 2.$$

The same result would be obtained from the original expression by removing the minus sign, removing the parentheses, and changing the sign of *every* term within the parentheses. Generalizing this observation, we state the *second rule of grouping symbols*:

> **If a symbol of grouping preceded by a minus sign is to be removed, then the minus sign should be removed and the sign of every term within the grouping symbol must be changed.**

EXAMPLE
4

a. $2x - (3y - z + 1) = 2x - 3y + z - 1.$

b. $(3a + b) - (-x^4 + 3x^2 - 1) = 3a + b + x^4 - 3x^2 + 1.$

c. $-[x - y] - [a + b] = -x + y - a - b.$

d. $ab - 3(x^2y - 2x) + z = ab - 3x^2y + 6x + z.$ Here we have multiplied every term within the parentheses by -3.

In some cases, one or more types of grouping symbols will be contained *within* other grouping symbols. Although the order in which the symbols are removed has no bearing on the result, mistakes are less likely if you develop the habit of removing the *innermost* grouping symbols first, then the next innermost, etc., with the outermost grouping symbols being removed last.

EXAMPLE
5

Simplify $2x - [2w - 3(y + 3)].$

Since the parentheses are the innermost grouping symbols, they are removed first; then the brackets are removed:

$$2x - [2w - 3(y + 3)] = 2x - [2w - 3y - 9]$$
$$= 2x - 2w + 3y + 9.$$

EXAMPLE
6

Simplify $-(-3xy - 2[3x - y - 3\{w + 5\} + a]) + z.$

Since the braces are the innermost grouping symbols, they are removed first:

$$-(-3xy - 2[3x - y - 3\{w + 5\} + a]) + z$$
$$= -(-3xy - 2[3x - y - 3w - 15 + a]) + z$$

Now remove the brackets:

$$= -(-3xy - 6x + 2y + 6w + 30 - 2a) + z$$

Finally, remove the parentheses:

$$= 3xy + 6x - 2y - 6w - 30 + 2a + z.$$

EXAMPLE
7

For each of the following expressions, enclose the last three terms by parentheses preceded by a (1) plus sign and (2) a minus sign so that in each case the result is equivalent to the original form.

a. $x^3 + 2x^2 - 3x + 1$:

(1) $x^3 + 2x^2 - 3x + 1 = x^3 + (2x^2 - 3x + 1)$ by reversing the procedure of the first rule of grouping symbols.

(2) $x^3 + 2x^2 - 3x + 1 = x^3 - (-2x^2 + 3x - 1)$ by reversing the procedure in the second rule of grouping symbols.

b. $2 - x - x^2 + x^3 - 6$:

(1) $2 - x - x^2 + x^3 - 6 = 2 - x + (-x^2 + x^3 - 6).$

(2) $2 - x - x^2 + x^3 - 6 = 2 - x - (x^2 - x^3 + 6).$

EXERCISE 2-2

In Problems 1–20, write the given expression in an equivalent form in which no symbols of grouping appear.

1. $7 + (x - y)$

2. $2 - (x^2 - y^2 + 4)$

3. $a + 2(b + c)$

4. $2x + 3(1 - x^2)$

5. $x - (y - z)$

6. $-(x + y) - 1$

7. $(a - b) - (c - d)$

8. $(8x - y) + 2(y^2 - x^2)$

9. $(2x + 3) - (-2y + x^2)$

10. $(-1)^2(x - y) + (-1)^3(x^2 - y^2)$

11. $3(a - b) - 2(x + 3y) + 4$

12. $-3[x^2 - (-y^2)]$

13. $(-1)^5 x - 3(2 - x^2)$

14. $2[y + 3(-x + 2) + x^2]$

15. $2ab + 2[3 - (x^2 - y^2)]$

16. $2(2 - 3\{x - [y + 2z] - w\} + z^2)$

17. $3x - 3[2 - 2\{x^2 - 2y\}] - x^3$

18. $2a^2b - [ab + 3(a - \{b + c\})]$

19. $-2(x - y) + 3(1 - 2[x^2 - \{y^2 - xy\} - y^3] - x^3)$

20. $(-2)^3(x - y) + 3[[x^2 - 2(y^2 - 1) - xy] + w]$

In Problems 21–24, write the given expression in an equivalent form in which no symbols of grouping appear and then enclose all terms that contain x to any power within one set of grouping symbols preceded by a minus sign. For example,

$$3x^2 - (x - 5) = 3x^2 - x + 5 = -(-3x^2 + x) + 5.$$

21. $x + 2(x^2 + b) - 3x^3 + 1$

22. $3(x^2y + b - x) - 4(x^2 + 2)$

23. $[-2x^2 + 3x^2y - (5x + 6y + 1)] - 3xy$

24. $3(x - y) - 4(w - x^2 + xy)$

25. In a mathematical development of the relativity of time, the following expression was encountered:

$$K\left\{(t_1 - t_2) - \frac{v}{c^2}(x_1 - x_2)\right\}.$$

Remove all symbols of grouping.

2-3
ADDITION AND SUBTRACTION OF ALGEBRAIC EXPRESSIONS

The addition and subtraction of algebraic expressions involves first performing the necessary operations to remove all symbols of grouping and then, by the commutative and distributive axioms, combining all *similar terms*. **Similar terms**, or *like terms*, are those terms which differ at most only in their numerical coefficients. For example, the terms $3x^2y$ and $-42x^2y$ are similar, as well as the pairs $-2x$ and $6x$, and 2 and 7. However, the terms $2x^2y^3$ and $2x^3y^2$ are not similar since $x^2y^3 \neq x^3y^2$.

EXAMPLE 8

Simplify $(6x^2y - 8x + 1) - (4x^2y - 6x - 5)$.

Removing parentheses we have

$$(6x^2y - 8x + 1) - (4x^2y - 6x - 5)$$
$$= 6x^2y - 8x + 1 - 4x^2y + 6x + 5$$

and then gathering similar terms together by the commutative axiom,

$$= 6x^2y - 4x^2y - 8x + 6x + 1 + 5.$$

By the distributive axiom we combine similar terms:

$$6x^2y - 4x^2y = (6 - 4)x^2y = 2x^2y,$$

and

$$-8x + 6x = (-8 + 6)x = -2x,$$

and

$$1 + 5 = 6.$$

Thus,

$$(6x^2y - 8x + 1) - (4x^2y - 6x - 5) = 2x^2y - 2x + 6.$$

EXAMPLE 9

Simplify $-(3x + y^2 - 2) + (y^2 - 2x + 1)$.

$$-(3x + y^2 - 2) + (y^2 - 2x + 1)$$
$$= -3x - y^2 + 2 + y^2 - 2x + 1 \qquad \text{(removing parentheses)}$$
$$= (-3 - 2)x + (-1 + 1)y^2 + (2 + 1) \qquad \text{(distributive axiom)}$$
$$= -5x + 3.$$

EXAMPLE 10

Simplify $2[x - y^2 - 3(2x - y^2)]$.

$$2[x - y^2 - 3(2x - y^2)]$$
$$= 2[x - y^2 - 6x + 3y^2] \qquad \text{(removing parentheses)}$$
$$= 2[-5x + 2y^2] \qquad \text{(combining similar terms)}$$
$$= -10x + 4y^2 \qquad \text{(removing brackets)}.$$

Here we chose to combine similar terms before removing the brackets.

EXAMPLE 11

Simplify $-[3w^2x^2y - (2xy^2 - 3w^2) - 3(w^2x^2y + w^2)]$.

$$-[3w^2x^2y - (2xy^2 - 3w^2) - 3(w^2x^2y + w^2)]$$
$$= -[3w^2x^2y - 2xy^2 + 3w^2 - 3(w^2x^2y + w^2)]$$
$$= -[3w^2x^2y - 2xy^2 + 3w^2 - 3w^2x^2y - 3w^2]$$
$$= -[(3 - 3)w^2x^2y + (3 - 3)w^2 - 2xy^2]$$
$$= -[-2xy^2]$$
$$= 2xy^2.$$

EXAMPLE   a. Subtract $-2x^3 + 6x^2 - 8x$ from $x^4 - 8x^3 + 6x^2 - 2x + 1$.
12
$$(x^4 - 8x^3 + 6x^2 - 2x + 1) - (-2x^3 + 6x^2 - 8x)$$
$$= x^4 - 8x^3 + 6x^2 - 2x + 1 + 2x^3 - 6x^2 + 8x$$
$$= x^4 - 6x^3 + 6x + 1.$$

   b. Subtract $-2x + y$ from 0.
$$(0) - (-2x + y) = 2x - y.$$

EXERCISE 2-3

In Problems 1–16, simplify each of the expressions by removing all symbols of grouping and combining similar terms.

1. $(3x + 2y - 5) + (8x - 4y + 2)$
2. $(6x^2 - 10xy + 7) + (2z - xy + 4)$
3. $(3x + 2y - 5) - (8x - 4y + 2)$
4. $(6x^2 - 10xy + 7) - (2z - xy + 4)$
5. $3 - (b - a) + 2$
6. $(2 + b) - (4 + a)$
7. $x^2 - (y^2 + x^2 + 1)$
8. $2(-a - 2) + 3$
9. $-(a - b - c) - (-a - b)$
10. $-(-a + b - c) + (-a + b)$
11. $(2a - 3 - 4c + x^2) - (-3 - 4c)$
12. $8 + (2 + c) - (2 + c) - 8$
13. $[3x + 2y - (5x + 6y - 1)] - 3xy$
14. $3(x + 2) - [5x - 6y - (x + y - w)]$
15. $3(3x + 2y - 5) - 2(8x + 4y + 2)$
16. $2(x - y + 2) - 3(y - x - 3) + 2(-x - y + 1)$

In each of Problems 17–22 two algebraic expressions are given. In each case (a) find their sum, (b) subtract the first expression from the second, and (c) subtract the second expression from the first.

17. $3a^2x - 3a^2y + 3a^2bz; \ 2a^2x + 3a^2y - a^2bz$
18. $3x^5 + 2x^4 - x^3 - 2x^2 + 1; \ 2x^6 - 3x^5 - x^4 + 3x^3 + 2x^2 - 1$
19. $9z^2xy - 8z^2 + 3x + 4y; \ 7z^2xy + 8z^2 - 2x - 7y$
20. $xy + yz + zw; \ xz + yz + yw$
21. $a^2x + \frac{1}{2}bc - 3f + g; \ \frac{1}{2}a^2x + 3bc + 2f$

22. $(x^2 + y^2 + 1)^0 - 1$; $-a^2x^2y - x^2 + 3x + 2$

In each of Problems 23–26 three algebraic expressions are given. In each case (a) subtract the sum of the first and second expressions from the third, (b) find the sum of the three expressions, and (c) subtract the first expression from the sum of the second and third expressions.

23. $3x^4 - 2x^2 + 7x + 6$
$x - x^4 - x^2 - 1$
$7x^2 - 3x^4 - 2x + 1$

24. $14x^3 + 12x^2y - 16xy^2$
$3x^2y - 4xy^2 + 6y^3$
$x^3 + y^3$

25. $18xy - 9x + 7y + z$
$16xy + 1$
$13xy + 9x - 7z + 1$

26. $21x^2y^2w + 8x^2yw^2 - 7x^2y^2w^2$
$13x^2y^2w - 6x^2yw^2 + 2x^2y^2w^2$
1

27. A rectangle has dimensions $2a$ and $3(b + c)$, while a smaller rectangle has dimensions a and $b + c$. Write an expression for the difference in the perimeters of the rectangles, the larger minus the smaller perimeter; perform all indicated operations and simplify.

2-4
MULTIPLICATION OF ALGEBRAIC EXPRESSIONS

Finding the product of algebraic expressions involves the application of the distributive, associative, and commutative axioms and the rules of exponents previously developed. We shall consider, one by one, the three types of products that occur most frequently.

A. PRODUCT OF TWO MONOMIALS

Consider $(2x^2yz)(xy^3z^6)$, which is the product of two monomials. By the associative law of multiplication, the order of grouping is unimportant and we can remove the parentheses:

$$(2x^2yz)(xy^3z^6) = 2x^2yzxy^3z^6.$$

By the commutative axiom of multiplication we can rearrange the factors so that those involving the same letters are adjacent:

$$2x^2yzxy^3z^6 = 2x^2xyy^3zz^6.$$

Finally, by the rules of exponents,

$$(2x^2yz)(xy^3z^6) = 2x^3y^4z^7.$$

Similarly,

$$2x^2y(-4xy^3) = (2)(-4) \cdot x^2x \cdot yy^3$$
$$= (-8) \cdot x^3 \cdot y^4$$
$$= (-8)x^3y^4$$
$$= -8x^3y^4.$$

This procedure can be extended to products of more than two factors:

$$2x^2y^2(-3xyz^2)(3xz) = (2)(-3)(3) \cdot (x^2xx) \cdot (y^2y) \cdot (z^2z)$$
$$= (-18)(x^4)(y^3)(z^3)$$
$$= -18x^4y^3z^3.$$

In the following example we shall consider problems of a more difficult nature. Intermediate steps will be omitted.

EXAMPLE
13

Determine each of the following products.

a. $(8ab^2)(4a^2b^3)^3$.

First we apply the rules of exponents to $(4a^2b^3)^3$:

$$(8ab^2)(4a^2b^3)^3 = 8ab^2(64a^6b^9)$$
$$= 512a^7b^{11}.$$

b. $(-2x^2yz)^4(-2xy)^3$.

$$(-2x^2yz)^4(-2xy)^3 = (16x^8y^4z^4)(-8x^3y^3)$$
$$= -128x^{11}y^7z^4.$$

We point out that it is *incorrect* to write $a(ac)^2 = (a^2c)^2$, because this would mean we are multiplying a by ac, which we are not. Rather, we are multiplying a by $(ac)^2$ or a^2c^2. Thus, $a(ac)^2 = a(a^2c^2) = a^3c^2$.

B. PRODUCT OF A MONOMIAL AND A MULTINOMIAL

The distributive axiom is the key tool that is employed in determining the product of a monomial and a multinomial. Consider

$$2x^2y(3x^2 - 4xy + 9w).$$

Here $2x^2y$ is a monomial and $3x^2 - 4xy + 9w$ is a multinomial. By considering $2x^2y$ as a single number and employing the distributive axiom we have

$$(2x^2y)(3x^2 - 4xy + 9w) = (2x^2y)(3x^2) - (2x^2y)(4xy) + (2x^2y)(9w).$$

The problem is thus reduced to multiplying monomials. Hence, by the rules of exponents,

$$(2x^2y)(3x^2 - 4xy + 9w) = 6x^4y - 8x^3y^2 + 18x^2yw.$$

EXAMPLE
14

a. Determine $(ab)^2(2aby - 8ab + 2)$.

$$(ab)^2(2aby - 8ab + 2) = (ab)^2(2aby) - (ab)^2(8ab) + (ab)^2(2)$$
$$\text{(distributive axiom)}$$
$$= (a^2b^2)(2aby) - (a^2b^2)(8ab) + (a^2b^2)(2)$$
$$\text{(rules of exponents)}$$
$$= 2a^3b^3y - 8a^3b^3 + 2a^2b^2 \quad \text{(rules of exponents)}.$$

b. Simplify $2(2x^3 + 3x) - 3x(x^2 - 2x)$.

$$2(2x^3 + 3x) - 3x(x^2 - 2x) = 4x^3 + 6x - 3x^3 + 6x^2$$
$$= x^3 + 6x^2 + 6x.$$

c. Simplify $2\{2x(2x^3 + 3x^2) - 3x(x^2 - 2x) + 5[4x^2 - (3 - 4x)]\}$.

$$2\{2x(2x^3 + 3x^2) - 3x(x^2 - 2x) + 5[4x^2 - (3 - 4x)]\}$$
$$= 2\{4x^4 + 6x^3 - 3x^3 + 6x^2 + 5[4x^2 - 3 + 4x]\}$$
$$= 2\{4x^4 + 3x^3 + 6x^2 + 20x^2 - 15 + 20x\}$$
$$= 2\{4x^4 + 3x^3 + 26x^2 + 20x - 15\}$$
$$= 8x^4 + 6x^3 + 52x^2 + 40x - 30.$$

C. PRODUCT OF TWO MULTINOMIALS

To find $(2x + 3)(6x + 7)$ which is the product of two multinomials, we once again appeal to the distributive axiom: $(b + c)a = ba + ca$. If we consider $6x + 7$ as a single number, then by the distributive axiom we take the first term in the left factor times the right factor, plus the second term in the left factor times the right factor:

$$(2x + 3)(6x + 7) = 2x(6x + 7) + 3(6x + 7).$$

Using the distributive axiom again,

$$2x(6x + 7) + 3(6x + 7) = 2x(6x) + 2x(7) + 3(6x) + 3(7)$$
$$= 12x^2 + 14x + 18x + 21$$
$$= 12x^2 + 32x + 21.$$

Thus,

$$(2x + 3)(6x + 7) = 12x^2 + 32x + 21.$$

EXAMPLE
15

Find the product $(2x - 3)(5x^2 + 3x - 1)$.

$$(2x - 3)(5x^2 + 3x - 1) = 2x(5x^2 + 3x - 1) - 3(5x^2 + 3x - 1)$$
$$= 10x^3 + 6x^2 - 2x - 15x^2 - 9x + 3$$
$$= 10x^3 - 9x^2 - 11x + 3.$$

EXAMPLE
16

Find the product $3x(2x + 1)(x - 2)$.

$$3x(2x + 1)(x - 2) = (6x^2 + 3x)(x - 2)$$
$$= 6x^2(x - 2) + 3x(x - 2)$$
$$= 6x^3 - 12x^2 + 3x^2 - 6x$$
$$= 6x^3 - 9x^2 - 6x.$$

EXAMPLE
17

Find the product $(2a + b + 1)(3a + 2b + c + d)$.

$$(2a + b + 1)(3a + 2b + c + d)$$
$$= 2a(3a + 2b + c + d) + b(3a + 2b + c + d) + 1(3a + 2b + c + d)$$
$$\text{(distributive axiom)}$$

$$= 6a^2 + 4ab + 2ac + 2ad + 3ab + 2b^2 + bc + bd + 3a + 2b + c + d$$
$$\text{(distributive axiom)}$$
$$= 6a^2 + 7ab + 2ac + 2ad + 2b^2 + bc + bd + 3a + 2b + c + d.$$

Before concluding this section, we leave you with an important note. Remember that $a + b(x + y) = a + bx + by$, but $(a + b)(x + y) = a(x + y) + b(x + y) = ax + ay + bx + by$. Similarly, $(a + b)x + y = ax + bx + y$, but $(a + b)(x + y) = a(x + y) + b(x + y) = ax + ay + bx + by$.

EXERCISE 2-4

In Problems 1–40, find the product of the given expressions.

1. $2x^2(4xy^2z^2)$ **2.** $3xy^2(2x^2y)$

3. $xy^2z(3x^2y^2z)$ **4.** $-4xy(x^2y^2)$

5. $\frac{1}{2}x^2yz^2(-2xz^2)$ **6.** $-xy^2(-2yz)$

7. $3a^2bc(2ab^2c)(bc)$ **8.** $xy(xz)(xw)$

9. $(-2xy)^3(18xy^2z)(x^6)$ **10.** $(2x^2y)^2(xy)^3$

11. $(3xy^2)^2(-2x^2y)^2(xy)^2$ **12.** $x(x^2y^3)^2(4xy^2z^2)^2$

13. $a(bc)^2(cd)^3$ **14.** $(-a)^3(-abc)^3$

15. $x(x^2 - 2x + 4)$ **16.** $(x - 2y - 4)(-3)$

17. $a^2b(-2ab^2 + ab - 2)$ **18.** $x^2yz(xy - xz + yz)$

19. $(-2xy^2)(xy^2 - xy + 2x + 1)$ **20.** $(2xy)^2(x - y + xy)$

21. $(a^2xy)^2(ab - ax - ay^2)$ **22.** $(-3ab)^3(a^3 - b^3 + x^3)$

23. $(x + 2)(x + 3)$ **24.** $(x + 4)(x + 5)$

25. $(y - 4)(2y + 3)$ **26.** $(y - 3)(y + 3)$

27. $(2x + 3)(5x + 2)$ **28.** $(2x + 7)(7x + 2)$

29. $(x^2 - 3)(x + 4)$ **30.** $(x^3 + 2)(x^4 + 2)$

31. $(2x - 1)(3x^3 + 7x^2 - 5)$ **32.** $(z - 4)(z^2 + 4z + 16)$

33. $2(x + 6)(x^2 + 4)$ **34.** $(x + 4)(x^2 - x + 1)$

35. $(x + y + 2)(3x + 2y - 4)$ **36.** $(2x + 3y + 1)(8x + 4y + 2)$

37. $(3ab^3 - 2rt)^2$ **38.** $(2ab - b^2c)^2$

39. $(2x + 1)^2(3x^2 - 5)$ **40.** $(x + 3y)^2(6x + 2y)$

In Problems 41–50, perform the indicated operations and simplify.

41. $x(x - 1) - 2(3 - x)$

42. $x^2y - 2x - (x^2y)(x + 1)$

43. $3x(x^2y - xy^2) + 3y(x^3 + 4x^2y)$
44. $2x^2(x^2 - 2xy) + 2(x^3y - 2x^2)$
45. $3xy[w(x - y - w)] - 3w(x^2y - xy^2)$
46. $3(x - x^2) - (x + 1)(x - 1)$
47. $x\{3(x - 1)(x - 2) + 2[x(x + 7)]\}$
48. $(x + 1)(x^2 - x + 1) - x^3 - 1$
49. $-abc[d + e - (f + g)] + bc[ad + ae - (af + ag)]$
50. $s[s\{s[s(s - 1)] - 2\} - 1]$
51. A rectangle has dimensions $2a$ and $3(b + c)$, while a smaller rectangle has dimensions $a - 1$ and $b + c$. Write an expression for the difference in the areas of the rectangles, the larger area minus the smaller; perform all indicated operations and simplify.

2-5
DIVISION OF ALGEBRAIC EXPRESSIONS

A. DIVISION OF MONOMIALS

To divide one monomial by another, the rules of exponents are employed. Consider, for example,

$$\frac{ax^2y^5}{x^3y^3}.$$

This can be written

$$\frac{ax^2y^5}{x^3y^3} = a \cdot \frac{x^2}{x^3} \cdot \frac{y^5}{y^3}$$

$$= a \cdot \frac{1}{x} \cdot y^2 = \frac{a}{x} \cdot y^2 = \frac{ay^2}{x}.$$

Similarly,

$$\frac{(2a^3b^2d)(3a^5b^2c)}{-3a^4b^2cd} = \frac{6a^8b^4cd}{-3a^4b^2cd}$$

$$= \frac{6}{-3} \cdot \frac{a^8}{a^4} \cdot \frac{b^4}{b^2} \cdot \frac{c}{c} \cdot \frac{d}{d}$$

$$= -2a^4b^2.$$

In actual practice, intermediate steps are done mentally. Thus we write

$$\frac{-35x^3y^2}{7xy^2} = -5x^2.$$

EXAMPLE
18

Simplify $\dfrac{(2ax^2y)^3(-2a^4xy)}{(axy^2)^3}$.

$$\frac{(2ax^2y)^3(-2a^4xy)}{(axy^2)^3} = \frac{(8a^3x^6y^3)(-2a^4xy)}{a^3x^3y^6}$$

$$= \frac{-16a^7x^7y^4}{a^3x^3y^6} = -\frac{16a^4x^4}{y^2}.$$

B. DIVISION OF A MULTINOMIAL BY A MONOMIAL

By the definition of division and the distributive axiom,

$$\frac{a+b+c}{d} = (a+b+c)\left(\frac{1}{d}\right) = a\cdot\frac{1}{d} + b\cdot\frac{1}{d} + c\cdot\frac{1}{d}.$$

But

$$a\cdot\frac{1}{d} + b\cdot\frac{1}{d} + c\cdot\frac{1}{d} = \frac{a}{d} + \frac{b}{d} + \frac{c}{d}$$

and hence,

$$\boxed{\frac{a+b+c}{d} = \frac{a}{d} + \frac{b}{d} + \frac{c}{d}.}\qquad\qquad(1)$$

*This important result **does not** mean that* $\dfrac{a}{b+c+d} = \dfrac{a}{b} + \dfrac{a}{c} + \dfrac{a}{d}$, *a very common error.*

From result (1) we make the general statement that any fraction in the form of a multinomial divided by a monomial can be expressed as a sum of fractions; therefore,

To divide a multinomial by a monomial, divide each term of the multinomial by the monomial.

Thus,

$$\frac{3x^6 - x^3}{x} = \frac{3x^6}{x} - \frac{x^3}{x} = 3x^5 - x^2,$$

and

$$\frac{20x^2y^2 + 5xy^2 - 2x}{2xy} = \frac{20x^2y^2}{2xy} + \frac{5xy^2}{2xy} - \frac{2x}{2xy}.$$

$$= 10xy + \frac{5y}{2} - \frac{1}{y}.$$

EXAMPLE
19

a. The reciprocal of the total resistance of three resistors, $R_1, R_2,$ and R_3, connected in parallel in an electric circuit is

$$\frac{R_2R_3 + R_1R_3 + R_1R_2}{R_1R_2R_3}.$$

Simplify this expression.

$$\frac{R_2R_3 + R_1R_3 + R_1R_2}{R_1R_2R_3} = \frac{R_2R_3}{R_1R_2R_3} + \frac{R_1R_3}{R_1R_2R_3} + \frac{R_1R_2}{R_1R_2R_3}$$

$$= \frac{1}{R_1} + \frac{1}{R_2} + \frac{1}{R_3}.$$

b. Determine $\dfrac{2x^2y + (2xy^2w)^2 - 3x^3y^3w^3}{2xy}$.

$$\frac{2x^2y + (2xy^2w)^2 - 3x^3y^3w^3}{2xy}$$

$$= \frac{2x^2y}{2xy} + \frac{(2xy^2w)^2}{2xy} - \frac{3x^3y^3w^3}{2xy}$$

$$= x + \frac{4x^2y^4w^2}{2xy} - \frac{3}{2}x^2y^2w^3$$

$$= x + 2xy^3w^2 - \frac{3}{2}x^2y^2w^3.$$

EXAMPLE 20

Determine $\dfrac{6a^4b^2c^3 + 3a^5b^2c - 12a^6bc}{-6a^4b^2c^2}$.

$$\frac{6a^4b^2c^3 + 3a^5b^2c - 12a^6bc}{-6a^4b^2c^2}$$

$$= \frac{6a^4b^2c^3}{-6a^4b^2c^2} + \frac{3a^5b^2c}{-6a^4b^2c^2} - \frac{12a^6bc}{-6a^4b^2c^2}$$

$$= -c - \frac{a}{2c} + \frac{2a^2}{bc}.$$

C. DIVISION OF A POLYNOMIAL BY A POLYNOMIAL

For the division of a polynomial by a polynomial, we shall employ the so-called *long division* process. But first, consider the following long division:

$$\begin{array}{r} 14 \\ 16\overline{)239} \\ \underline{16} \\ 79 \\ \underline{64} \\ 15 \end{array}$$

Here, 239 is the *dividend,* 16 is the *divisor,* the *quotient* is 14, and the *remainder* is 15. A method of checking a division is to verify that

$$\text{(Quotient)(Divisor)} + \text{Remainder} = \text{Dividend}.$$

In our case

$$(14)(16) + 15 = 239$$

checks. The answer may also be expressed as $14 + \frac{15}{16}$, and you should realize that by the definition of division this means that

$$(16)(14 + \tfrac{15}{16}) = 239.$$

That is, the answer is that number which when multiplied by the divisor will give the dividend.

We shall now consider dividing a polynomial by a polynomial.

EXAMPLE
21

Divide $2x^3 - 5x^2 + 3x + 4$ by $x - 2$.

Here the polynomial $2x^3 - 5x^2 + 3x + 4$ is the dividend and the polynomial $x - 2$ is the divisor.

$$
\begin{array}{r}
2x^2 - x + 1 \quad \longleftarrow \text{quotient} \\
x - 2 \overline{)\, 2x^3 - 5x^2 + 3x + 4} \\
\underline{2x^3 - 4x^2} \\
-x^2 + 3x \\
\underline{-x^2 + 2x} \\
x + 4 \\
\underline{x - 2} \\
6 \quad \longleftarrow \text{remainder}
\end{array}
$$

Note that the powers of x in the dividend and divisor are written in decreasing order. Here we have divided x, the first term of the divisor, into $2x^3$, the first term of the dividend, which resulted in $2x^2$, the first term of the quotient. Then we multiplied this quotient term by the divisor $x - 2$, obtaining $2x^3 - 4x^2$, which is written below the similar terms of the dividend and subtracted from them. We obtained $-x^2$ and then "brought down" $3x$, the next term of the dividend. This process was continued until we arrived at 6, the remainder. More generally, *we end the division process when the remainder is a polynomial whose degree is less than the degree of the divisor.* The division can be checked by verifying that

$$(\text{Quotient})(\text{Divisor}) + (\text{Remainder}) = \text{Dividend}.$$

Doing this we have

$$
\begin{aligned}
&(2x^2 - x + 1)(x - 2) + (6) \\
&= 2x^2(x - 2) - x(x - 2) + 1(x - 2) + 6 \\
&= 2x^3 - 4x^2 - x^2 + 2x + x - 2 + 6 \\
&= 2x^3 - 5x^2 + 3x + 4,
\end{aligned}
$$

which is equal to the dividend. The answer may be given in the form

$$2x^2 - x + 1 + \frac{6}{x - 2}.$$

EXAMPLE
22

Divide $x^3 + 8$ by $x + 2$.

When the dividend is arranged in order of decreasing exponents, there are times when certain terms do not appear. In the present case, for example, in the dividend $x^3 + 8$ there is no term containing x^2 and no term containing x. We may, however, consider such terms as being part of the dividend with numerical coefficients of zero. That is, we write the dividend as $x^3 + 0x^2 + 0x + 8$. Since the process of division of polynomials involves the subtraction of similar terms, you will find it less confusing and possibly avoid error if, when the dividend is arranged in order of decreasing exponents, a "space" is left for each of the "missing" terms:

$$
\begin{array}{r}
x^2 - 2x + 4 \\
x + 2 \overline{)\ x^3 + 0x^2 + 0x + 8.} \\
\underline{x^3 + 2x^2} \\
-2x^2 + 0x \\
\underline{-2x^2 - 4x} \\
4x + 8 \\
\underline{4x + 8} \\
0
\end{array}
$$

Check:

$$(x^2 - 2x + 4)(x + 2) + 0$$
$$= x^3 + 2x^2 - 2x^2 - 4x + 4x + 8$$
$$= x^3 + 8.$$

EXAMPLE 23

Divide $-2x^3 + 3x^4 - 3x^2 + 7$ by $3x^2 + x - 1$.

Remember to write the dividend in order of decreasing exponents.

$$
\begin{array}{r}
x^2 - x - \frac{1}{3} \\
3x^2 + x - 1 \overline{)\ 3x^4 - 2x^3 - 3x^2 + 0x + 7.} \\
\underline{3x^4 + x^3 - x^2} \\
-3x^3 - 2x^2 + 0x \\
\underline{-3x^3 - x^2 + x} \\
-x^2 - x + 7 \\
\underline{-x^2 - \frac{1}{3}x + \frac{1}{3}} \\
-\frac{2}{3}x + \frac{20}{3}
\end{array}
$$

Note that here the remainder has an x-term. The division should not be continued because the remainder is of lower degree than the divisor. You should perform the necessary operations to check the work of this example. The answer can be written as

$$x^2 - x - \frac{1}{3} + \frac{-\frac{2}{3}x + \frac{20}{3}}{3x^2 + x - 1}.$$

EXERCISE 2-5

In Problems 1–28, perform the indicated operations.

1. $\dfrac{4x + 2y}{2}$

2. $\dfrac{6x^2y^2z^3}{2xy^3z^2}$

3. $\dfrac{2x^4yw^2}{4x^3yw}$

4. $\dfrac{25ab^3c^2}{-5ab^2}$

5. $\dfrac{-9x^2y^3w}{3x^2w}$

6. $\dfrac{-x(x^2y)}{-y(y^2x)}$

7. $\dfrac{5xy^2z^4}{\frac{1}{2}x^2yz^2} = \dfrac{10yz^2}{x}$

8. $\dfrac{(3x^2y^2z)^2}{x^4y^8z}$

9. $\dfrac{x^3y^2}{(x^2y)^2}$

10. $\dfrac{(a^2bc^2)(ab^2c^2)}{a^2b^4c}$

11. $\dfrac{(-2xy)^2(x^2y)}{(xyz)^3}$

12. $\dfrac{(abc^2)^2(ab)}{a^2b^2}$

13. $\dfrac{(2a^2b^2)^4(2abc)}{8ab^2c}$

14. $\dfrac{(5u^2v^2w)(-3u^4v^2w^5)^2}{2v^3u^2w^2}$

15. $\dfrac{(-2x)^2(-xy^2)^3}{(-2x^2y)^2}$

16. $\dfrac{(-x)(-y)^2(xy)^3(y^4x)}{-xy^2}$

17. $\dfrac{xy+x^2}{xy}$

18. $\dfrac{2x^2y^2-4xz^2+4x}{2xy^2z}$

19. $\dfrac{3x^2y-xy^2+1}{xy^2}$

20. $\dfrac{x(xy)+y(-x^2)+x^3y^3}{-x^2y}$

21. $\dfrac{2x(x^3y^2)^2+3x^2y^3-x}{-xy^4}$

22. $\dfrac{p^2r(1+pr+pr^2)}{pr}$

23. $\dfrac{(4xy^2z)^2+8x^5y^4z^3-4x^6y^3z^2}{4x^2y^2z^2}$

24. $\dfrac{(3p^2qr)^2(-p)^{15}}{3pqr}$

25. $\dfrac{6x^2y-2y^2+7x-4}{-3x}$

26. $\dfrac{2x^2-6y^2+5z^2+3w^2}{-30xyz^2}$

27. $\dfrac{(25s^4r^3z^{12})^{-2}(-25srz)^{12}(-25srz^2)^2}{(25srz)^{14}}$

28. $\dfrac{[(-2xy)^3(-3xy^2)^2]^2}{2x^2y^2} =$

In Problems 29–36, divide the first polynomial by the second.

29. $x^2+5x+6;\ x+2$

30. $x^2;\ x-8$

31. $x^4+2x^2+1;\ x-1$

32. $3x^3-2x^2+x-3;\ x+2$

33. $x^3+x^2+x+3;\ x^2-x+1$ **34.** $x^3-1;\ x-1$

35. $5x^4-18x^3+7x^2-3x+4;\ x^2-x+6$

36. $6x^3+5x^2-2;\ 2x+4$

37. Divide x^4-y^4 by $x+y$.

38. Divide x^5+y^5 by $x+y$.

39. The efficiency $\mathcal{E}$ of a reversible heat engine operating between a high-temperature reservoir at an absolute temperature T_1 and a low-temperature reservoir at temperature T_2 is given by

$$\varepsilon = \frac{T_1 - T_2}{T_1}.$$

Find another form for ε. The absolute temperature scale has a range from zero degrees through all positive values. From purely mathematical considerations, is there any restriction imposed on the value of T_1?

40. When three capacitors $C_1, C_2,$ and C_3 are connected in series, the total capacitance of the combination, C_t, is given by

$$C_t = \frac{C_1 C_2 C_3}{C_2 C_3 + C_1 C_3 + C_1 C_2}$$

Find the reciprocal of C_t and perform the indicated division.

2-6
REVIEW

REVIEW QUESTIONS

An algebraic expression composed of more than one term can *always* be called a _____(1)_____ ; however, expressions containing exactly two and three terms are usually referred to as _____(2)_____ and _____(3)_____ , respectively.

(1) multinomial   (2) binomials   (3) trinomials

An algebraic expression in which every term is either a constant or the product of a constant and a positive integral power of x is called a _____(4)_____ .

(4) polynomial (in x)

Of the multinomials

a. $-5x^2 + 2x^3 - \sqrt{2}$,

b. $4x^{1/2} + 2x^3 - 5$,

c. $\dfrac{-3}{x^{-2}} + x + 7$,

d. $\sqrt{19}x^5 - \sqrt{2}x + 7$, and

e. $19x^5 - 2\sqrt{x} + 9$,

those which are polynomials are _____(5)_____ , _____(6)_____ , and _____(7)_____ . Specifically, _____(8)_____ is not a polynomial because of the term _____(9)_____ , and _____(10)_____ is not a polynomial because of the term _____(11)_____ .

(5) **a**   (6) **c**   (7) **d**   (8) **b**   (9) **$4x^{1/2}$**   (10) **e**

(11) **$-2\sqrt{x}$**

For the polynomial $3x^2 - 4x + 7$, the degree is equal to _____(12)_____,
the numerical coefficient of the second term is _____(13)_____, and the
constant term is _____(14)_____.

(12) **2**   (13) **-4**   (14) **7**

If a polynomial in y of degree k were to be cubed, would the resulting expression be a polynomial in y, and, if so, what would be its degree?

(15) **yes**   (16) **$3k$**

In the simplification of algebraic expressions, removal of a grouping symbol
preceded by a minus sign requires _____(17)_____.

(17) **the sign of every term within the symbol to be changed**

In cases containing combinations of grouping symbols, it is preferable to remove
the _____(18)_____ symbol first, although it is not necessary to do so.

(18) **innermost**

Similar terms can differ only in their _____(19)_____.

(19) **numerical coefficients**

In the division of one polynomial by another, the process is carried on until
the _____(20)_____ is of _____(21)_____ degree than the _____(22)_____.

(20) **remainder**   (21) **lower**   (22) **divisor**

The expression $-\{1 - [(-1)^2(-1)] - 1\}$ is equal to _____(23)_____.

(23) **-1**

The product of $x + 7$ and $a + b$ is equal to the sum of x times _____(24)_____
and 7 times _____(25)_____.

(24) **$a + b$**   (25) **$a + b$**

A_{LL} **REVIEW PROBLEMS**

1. Simplify $2a - \{3b + (4c - [3b + 2a])\}$.

2. Simplify $3a - \{-[-(-\{4a + 2b\} + 1) - 2] - 1\}$.

In Problems 3–13, perform the indicated operations and simplify.

3. $(9x^4 + 2x^3 - x^2 + 12x + 1) + (x^5 - 3x^4 - 7x^3) - (-2x^4 + 12x^3 - 7)$

4. $-2x(8x^2 + x - y) + [(-3x^3 + yx) - 6x^3 + 1]$

5. $[7a^2b^3c^2 - 7a(ab^3c^2) - 7] - [8 + 4(abc)^2b]$

6. $(5u^2v^4w)(-3uv^2w^3)(-2u^5v^4w^2)$

7. $[(7x^{-4}bw^3)(x^3b^2y)]^2$

8. $3x^2y(xy - \frac{1}{3}x^2y^{-1} + 6y)$

9. $(3p^2qr)^2(p^2 + r^3 - prq) = 9p^6q^2r^2 + 9p^4q^2r^5 - 9p^5q^3r^3$

10. $(x + 1)(x^2 + x - 3)$

11. $(x - y + 3)(x + y - 3)$

12. $\dfrac{[(-xy)^3(-2xy^2)^2]^3}{4x^2y} = -16x^{13}y^{20}$

13. $\dfrac{(x^2y)^3 - 3x^2(x^2y^4) + 2y^{13}}{-\frac{1}{3}xyz} = -3\dfrac{x^5y^2}{z} + \dfrac{9xy^3}{z} - \dfrac{6y^{12}}{xy}$

14. Divide $x^3 - 6x^2 + x - 4$ by $x + 2$.

15. Divide $x^5 - 3x^4 + x + 1$ by $x - 1$.

3

special products
and factoring

3-1

SPECIAL PRODUCTS

In Sec. 2-4 various axioms and the laws of exponents were applied to determine products of multinomials. There is, however, a group of products that occur so frequently that we find it worthwhile to memorize their patterns, thereby avoiding repeated and time-consuming evaluation by the basic methods. These special products are presented here with verbal interpretations and examples. Just "knowing" these relationships, however, is insufficient. You must be so familiar with these products that you can easily recognize them in every form. Each of these products can be obtained by the distributive axiom and you are encouraged to verify them for yourself. We point out that the letters, such as a, x, y, etc., which appear in the statements of the special products are understood to denote any expression or quantity representing a number.

A.
$$a(b + c) = ab + ac.$$

This is simply the distributive axiom.

EXAMPLE
1

a. $3(x + y) = 3x + 3y.$

b. $-2x(3x - 4) = -6x^2 + 8x.$

c. $a^2b(2ab^2c + 3a^2xy) = 2a^3b^3c + 3a^4bxy.$

B. $$(a + b)(a + b) = (a + b)^2 = a^2 + 2ab + b^2$$

To square a binomial, square the first term, add to this two times the product of the two terms, and add the square of the second term.

It is important to note that, in general,

$$(a + b)^2 \neq a^2 + b^2.$$

EXAMPLE
2

a. $(x + 3)^2 = (x)^2 + 2(x)(3) + (3)^2 = x^2 + 6x + 9.$

b. $(x - 4)^2 = [x + (-4)]^2$
$$= (x)^2 + 2(x)(-4) + (-4)^2$$
$$= x^2 - 8x + 16.$$

c. $(-x + a)^2 = (-x)^2 + 2(-x)(a) + (a)^2$
$$= x^2 - 2ax + a^2.$$

d. $(2x + 3y^2)^2$. We use Rule B where $2x$ takes the role of a, and $3y^2$ the role of b:

$$(2x + 3y^2)^2 = (2x)^2 + 2(2x)(3y^2) + (3y^2)^2$$
$$= 4x^2 + 12xy^2 + 9y^4.$$

e. $-2x(2x - 4y)^2 = (-2x)[(2x)^2 + 2(2x)(-4y) + (-4y)^2]$
$$= (-2x)[4x^2 - 16xy + 16y^2]$$
$$= -8x^3 + 32x^2y - 32xy^2.$$

Note, as in Examples 2b and c, that the signs of the terms in the binomial do not affect the squaring process. In all cases we must be careful to abide by the rules for signed numbers. Actually, Example 2b could be generalized by writing

$$(a - b)^2 = a^2 - 2ab + b^2,$$

which is a frequently occurring product.

C. $$(a + b)(a - b) = a^2 - b^2$$

The product of the sum and difference of two numbers is equal to the square of the first number minus the square of the second number.

EXAMPLE
3

a. $(x + 4)(x - 4) = (x)^2 - (4)^2 = x^2 - 16.$

b. $(-7 + x)(-7 - x) = (-7)^2 - (x)^2 = 49 - x^2.$

c. $(4x - 3y)(4x + 3y)$. We use Rule C where $4x$ takes the role of a and $3y$ the role of b:

$$(4x - 3y)(4x + 3y) = (4x)^2 - (3y)^2 = 16x^2 - 9y^2.$$

d. $(3x - 4)(-4 - 3x)$ does not appear to fit the form in Rule C. However, we can write $3x - 4$ as $-4 + 3x$. Thus,

$$(3x - 4)(-4 - 3x) = (-4 + 3x)(-4 - 3x)$$
$$= (-4)^2 - (3x)^2 = 16 - 9x^2.$$

D. $$(ax + b)(cx + d) = acx^2 + (ad + bc)x + bd$$

As a special case of Rule D, when $a = 1$ and $c = 1$ we have

E. $$(x + b)(x + d) = x^2 + (b + d)x + bd$$

As you can see in Rules D and E, the product of two binomials may be a trinomial. These two forms, which occur very frequently, conveniently lend themselves to a schematic representation. Let's consider the more general case $(ax + b)(cx + d)$ as given in Rule D. As shown in Fig. 3-1, the product is a trinomial where the first term is the product of

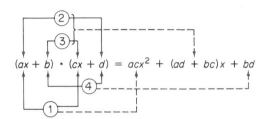

FIGURE 3-1

the first terms of the binomials, ①; the middle term is the sum of the product of the *outer* terms, ②, and the product of the *inner* terms, ③; and the third term is the product of the second terms of the binomials, ④. The computations indicated by Rules D and E are usually done mentally, keeping in mind the schematic representation of Fig. 3-1. For example, for the product $(x + 2)(3x - 7)$ we have the schematic representation in Fig. 3-2, or equivalently,

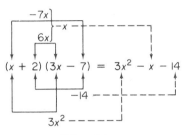

FIGURE 3-2

$$(x + 2)(3x - 7) = 3x^2 + (-7x + 6x) - 14 = 3x^2 - x - 14.$$

EXAMPLE

4

In a, b, and c below, the product of two binomials is determined by (1) direct application of Rule D or E and (2) application of the schematic technique. In d, e, and f, only the application of the schematic technique is shown.

a. (1) $(x + 7)(x + 4) = x^2 + (7 + 4)x + 7(4)$
$= x^2 + 11x + 28.$
   (2) $(x + 7)(x + 4) = x^2 + (4x + 7x) + 28$
$= x^2 + 11x + 28.$

b. (1) $(x + 2)(x - 5) = (x + 2)[(x + (-5)]$
$= x^2 + (2 - 5)x + 2(-5)$
$= x^2 - 3x - 10.$
   (2) $(x + 2)(x - 5) = x^2 + (-5x + 2x) + (-10)$
$= x^2 - 3x - 10.$

c. (1) $(2x - 3)(2x - 4) = [2x + (-3)][2x + (-4)]$
$= (2)(2)x^2 + (-8 - 6)x + (-3)(-4)$
$= 4x^2 - 14x + 12.$
   (2) $(2x - 3)(2x - 4) = (2x)^2 + (-8x - 6x) + (12)$
$= 4x^2 - 14x + 12.$

d. $(x - 2a)(x + 3a) = x^2 + (3ax - 2ax) - 6a^2 = x^2 + ax - 6a^2.$

e. $(3x + 2)(2x - 1) = 6x^2 + (-3x + 4x) - 2 = 6x^2 + x - 2.$

f. $(-1 + x)(1 - x) = -1 + (x + x) - x^2 = -1 + 2x - x^2.$

F. $$(a + b)^3 = a^3 + 3a^2b + 3ab^2 + b^3$$

To cube a binomial, cube the first term, add three times the product of the square of the first term and the second term, add three times the product of the first term and the square of the second term, and add the cube of the second term.

Remember that the signs of the terms of the binomial are of no concern as long as the proper techniques are applied. We could state, for example, that $(a - b)^3 = a^3 - 3a^2b + 3ab^2 - b^3$; however, in Example 5a it will be seen that this same result is obtainable by the use of Rule F and the laws of signed numbers.

EXAMPLE

5

a. $(a - b)^3 = [a + (-b)]^3 = a^3 + 3a^2(-b) + 3a(-b)^2 + (-b)^3$
$= a^3 - 3a^2b + 3ab^2 - b^3.$

b. $(2x + 3)^3 = (2x)^3 + 3(2x)^2(3) + 3(2x)(3)^2 + (3)^3$
$= 8x^3 + 36x^2 + 54x + 27.$

c. $(x - 1)^3 = (x)^3 + 3(x)^2(-1) + 3(x)(-1)^2 + (-1)^3$
$= x^3 - 3x^2 + 3x - 1.$

d. $(a + 2b)^3 = a^3 + 3(a)^2(2b) + 3(a)(2b)^2 + (2b)^3$
$$= a^3 + 6a^2b + 12ab^2 + 8b^3.$$

e. $(a + b + c)^3$ can be evaluated by writing $(a + b + c)$ as $[(a + b) + c]$.
Then

$(a + b + c)^3 = [(a + b) + c]^3$
$$= (a + b)^3 + 3(a + b)^2(c) + 3(a + b)(c)^2 + c^3$$
$$= a^3 + 3a^2b + 3ab^2 + b^3 + 3(a^2 + 2ab + b^2)c +$$
$$3(a + b)c^2 + c^3$$
$$= a^3 + 3a^2b + 3ab^2 + b^3 + 3a^2c + 6abc + 3b^2c +$$
$$3ac^2 + 3bc^2 + c^3.$$

EXERCISE 3-1

Dᴏ   Aʟʟ   Oᴅ

In Problems 1–64, find the indicated products by direct use of the special product relationships. It should not be necessary to refer to the relationships.

1. $3(x - y)$ 2. $-4x(a + b)$

3. $(x + 5)^2$ 4. $(x - 2)^2$

5. $(x - 4)^2$ 6. $(x + 12)^2$

7. $(x + \frac{1}{2})^2$ 8. $(x - \frac{7}{3})^2$

9. $(x + 2y)^2$ 10. $(t - 3s)^2$

11. $3x(x - 2b)^2$ 12. $4xy(x - y)^2$

13. $(2x + 3)^2$ 14. $(3 - 4x)^2$

15. $(3x - 4y)^2$ 16. $(3x + \frac{1}{3}y)^2$

17. $(-8x - 2y)^2$ 18. $(-2x + 5y)^2$

19. $(3abc - 4ef)^2$ 20. $\left(\dfrac{x}{y} + 3\right)^2$

21. $(-6x + 5y)^2$ 22. $(-x - 3y)^2$

23. $(x + 8)(x + 3)$ 24. $(x - 6)(x - 1)$

25. $(x - 2)(x + 1)$ 26. $(x + 5)(x + 4)$

27. $(x + 7)(x - 5)$ 28. $(x + 14)(x - 2)$

29. $(x - 2)(x - 3)$ 30. $(x + 4y)(x - 2y)$

31. $(x - 3)(x + 3)$ 32. $(y - 8)(y + 8)$

33. $(3x + 4)(3x - 4)$ 34. $(x - 2y)(x + 2y)$

35. $(\frac{1}{2}x + 2)(\frac{1}{2}x - 2)$ 36. $(3x - 4y)(3x + 4y)$

37. $(2x + 3)(5x + 2)$ 38. $(y^2 - 4)(y^2 + 4)$

39. $(4x - 3)(3x - 4)$ 40. $(3x - 2)(-x + 7)$

41. $(x^2 - 3)(x^2 + 3)$ **42.** $(3x^2 - 4)(3 - 4x^2)$

43. $(\frac{2}{3}a + 1)(\frac{1}{3}a - 2)$ **44.** $\left(\frac{ab}{c} + d\right)\left(\frac{ab}{c} - d\right)$

45. $(2t - 1)(t + 3)$ **46.** $(6x + b)(-b + 6x)$

47. $(xyz + a)(xyz - a)$ **48.** $(x - 2)^3$

49. $(-x + 7)(7 + x)$ **50.** $18ab(x - 3)(2x - 3)$

51. $3(x - 5)(2x + 8)$ **52.** $(4a^2b^3 - 3m^2n)^2$

53. $(x + 5)^3$ **54.** $(x + y - 3)^3$

55. $(3y - b)(b + 2y)$ **56.** $[(x - y) + 2][(x - y) - 2]$

57. $2x(x - 3)^3$ **58.** $(a + b + c)(a + b - c)$

59. $(3xyz + a)(4xyz - 2a)$ **60.** $[(x + y) - (a + b)]^2$

61. $(x - y - 2)^2$ **62.** $(x + 1)(x - 1)(x - 3)$

63. $(x + \sqrt{3})(x - \sqrt{3})$ **64.** $(\sqrt{6} - \sqrt{7})(\sqrt{6} + \sqrt{7})$

65. In the study of X-ray spectroscopy it is shown that the frequency of the K_α line of any element is given by

$$\tfrac{3}{4}Rc(z - 1)^2,$$

where c is the speed of light, z is the atomic number of the element, and R is a constant called the Rydberg constant. Perform the indicated multiplication.

66. When a flexible chain of length l and density w pounds per foot is released with c feet overhanging a smooth table, energy considerations result in the following expression for the square of the velocity of the chain as it leaves the table:

$$\frac{2g}{wl}\left[wc(l - c) + \frac{1}{2}w(l - c)^2\right].$$

Perform all indicated operations and simplify.

67. For a body moving in a straight line with constant acceleration, it is shown in mechanics that

$$s = v_{ave}\cdot t, \qquad\qquad (1)$$

$$v_{ave} = \frac{v_f + v_0}{2}, \qquad\qquad (2)$$

and

$$t = \frac{v_f - v_0}{a}, \qquad\qquad (3)$$

where s = displacement of the body, v_{ave} = average velocity, t = time, v_0 = initial velocity, v_f = final velocity, and a = acceleration. Show by substituting (2) and (3) into (1) that

$$s = \frac{v_f^2 - v_0^2}{2a}.$$

X

The process whereby an algebraic expression is resolved into a product of its factors is called **factoring**. That is, in factoring we are given an expression and we must find the factors whose product is that expression. In essence, factoring is the reverse of the process of evaluating the special products in Sec. 3-1. Hence, the relationships presented there will form the basis of our discussion on factoring as well. For your convenience here is a list of basic relationships which you should memorize.

-LERN

(1)	$ab + ac = a(b + c)$	(common factor)
(2)	$a^2 + 2ab + b^2 = (a + b)^2$	(perfect-square
(3)	$a^2 - 2ab + b^2 = (a - b)^2$	trinomial)
(4)	$a^2 - b^2 = (a + b)(a - b)$	(difference of two squares)
(5)	$x^2 + (a + b)x + ab = (x + a)(x + b)$	
(6)	$a^3 + b^3 = (a + b)(a^2 - ab + b^2)$	(sum of two cubes)
(7)	$a^3 - b^3 = (a - b)(a^2 + ab + b^2)$	(difference of two cubes)

When factoring a polynomial, we shall adopt the practice of choosing only those factors which are themselves polynomials. For example,

$$x^2 - 1 = (x + 1)(x - 1)$$

is appropriate, but we shall not factor $x - 1$ as $(\sqrt{x} + 1)(\sqrt{x} - 1)$ since these factors are not polynomials. Also, unless otherwise specified, we shall consider only factors having rational coefficients. Thus we shall not write $x^2 - 5 = (x + \sqrt{5})(x - \sqrt{5})$, since $\sqrt{5}$ is not rational.

We usually factor a polynomial so that no factor can be written as a product of other factors other than itself and 1 or -1. When this has been accomplished we shall consider the polynomial to be *completely factored*. Thus $x^2 - 1 = (x + 1)(x - 1)$ would be completely factored since $x + 1$ and $x - 1$ cannot be broken down any further.

We can consider an analogous situation in arithmetic. Recall that a positive integer, greater than 1, is called a *prime* if and only if its only positive integral divisors are itself and 1. Thus,

$$2, \quad 3, \quad 5, \quad 7, \quad 11, \quad 13, \quad 17, \quad \text{and} \quad 19$$

are primes. On the other hand, positive integers greater than 1 which are not primes are said to be *composite*. Thus,

$$4, \quad 6, \quad 8, \quad 9, \quad 10, \quad \text{and} \quad 12$$

are composite numbers. The *fundamental theorem of arithmetic* states that every composite number can be written uniquely as a product of primes. Thus, for example,

$$42 = 2\cdot3\cdot7, \qquad 72 = 2^3\cdot3^2, \qquad \text{and} \qquad 825 = 3\cdot5^2\cdot11.$$

Similarly, in $x^2 - 1 = (x + 1)(x - 1)$ both $x + 1$ and $x - 1$ are considered *prime polynomials*. In essence, factoring polynomials is simply resolving them into their prime factors. We point out, however, that not all expressions are factorable in the real number system. The polynomials $x^2 + 1$ and $x^2 + x + 1$ are of this type and we consider them to be prime polynomials.

The following examples should be studied thoroughly. In fact, you should read them with pencil and paper in hand so that you may reproduce the algebraic manipulations for yourself.

EXAMPLE
6

Common factor in each term. This type of factoring follows Rule (1), the distributive axiom, and is the simplest type.

a. Completely factor $3x^2y^2 - 9x^3y$.

Since $3x^2y^2 = (3x^2y)y$ and $9x^3y = (3x^2y)(3x)$, each term of the original expression contains the common factor $3x^2y$. Thus, by the distributive axiom,

$$3x^2y^2 - 9x^3y = 3x^2y(y - 3x).$$

Note that although $3x^2y^2 - 9x^3y = 3(x^2y^2 - 3x^3y)$, we do not say that the expression is completely factored since $x^2y^2 - 3x^3y$ can yet be factored. Also note that in this and every other example, the special products along with the various axioms and the rules of exponents afford a method of checking the factored result.

b. Completely factor $6xy^2 + xy$.

$$6xy^2 + xy = xy(6y) + xy(1) = xy(6y + 1).$$

Don't forget the 1.

c. Completely factor $8a^5x^2y^3 - 6a^2b^3yz - 2a^4b^4xy^2z^2$.

$$8a^5x^2y^3 - 6a^2b^3yz - 2a^4b^4xy^2z^2 = 2a^2y(4a^3x^2y^2 - 3b^3z - a^2b^4xyz^2).$$

d. Completely factor $6ac - 6bc + (a - b)$.

The first two terms have $6c$ as their common factor. Factoring we obtain

$$(6ac - 6bc) + (a - b) = 6c(a - b) + (a - b)(1).$$

But in each of these terms $a - b$ is a common factor. If we treat $a - b$ as we would a monomial factor, then we can write the right member as

$$(a - b)(6c + 1).$$

EXAMPLE

7

Completely factor $x^2 + 3x - 18$.

In factoring a trinomial such as this which is not a perfect square, we usually appeal to Rule (5). If, indeed, this expression factors into the form

$$x^2 + 3x - 18 = (x + a)(x + b),$$

which is the product of two binomials, then all we must do is determine a and b. Since

$$(x + a)(x + b) = x^2 + (a + b)x + ab,$$

then

$$x^2 + (3)x + (-18) = x^2 + (a + b)x + ab.$$

By equating corresponding coefficients, we want

$$a + b = 3 \quad \text{and} \quad ab = -18.$$

Consequently the sum of a and b must be 3 and their product must be -18. The factors of -18 are

-18	and	1
18	and	-1
-2	and	9
2	and	-9
-6	and	3
6	and	$-3.$

If $a = 6$ and $b = -3$, then both conditions are satisfied, and hence,

$$x^2 + 3x - 18 = (x + 6)(x - 3).$$

There is no clear procedure in handling such problems in factoring, but proficiency in factoring is, in part, based on the ability to recognize special products; this ability can come only with sufficient practice.

EXAMPLE

8

a. $x^2 - 7x + 12 = (x - 3)(x - 4)$.

b. $x^2 - 9x + 18 = (x - 6)(x - 3)$.

c. $x^2 - 4x - 32 = (x - 8)(x + 4)$. Here we must choose -8 and $+4$ and not -4 and $+8$, for $(x - 4)(x + 8) = x^2 + 4x - 32$, which is not equal to the given expression.

d. $x^2 + 12x + 20 = (x + 10)(x + 2)$.

e. $5x^2 + 25x + 30 = 5(x^2 + 5x + 6) = 5(x + 3)(x + 2)$. As we have done here, it is emphasized that *whenever a factor is common to all the terms of an expression it should immediately be factored out before proceding further.*

EXAMPLE

9

Factor $6x^2 - 5x - 6$.

We first note that this trinomial is not the square of a binomial and also that Rule (5) does not apply. In attempting to factor this, we shall appeal to the schematic approach in Fig. 3-1 which essentially involves trial and error.

That is, we try different pairs of binomial factors until we find a pair such that: (a) the product of the first terms is the first term of our trinomial; (b) the product of the second terms is the third term of our trinomial; and (c), the middle term of our trinomial is the product of the outer terms of the binomials plus the product of the inner terms. After trying various combinations such as

$$(3x - 1)(2x + 6),$$
$$(x + 3)(6x - 2), \quad \text{and}$$
$$(3x - 2)(2x + 3),$$

we find that

$$6x^2 - 5x - 6 = (3x + 2)(2x - 3).$$

EXAMPLE
10

Listed below are algebraic expressions that are completely factored. The numbers in parentheses refer to the rules used, and, in addition to studying the factoring processes involved, you should mentally verify that the product of the factors is equivalent to the given expression.

a. $x^2 + 8x + 16 = (x + 4)^2$ (2).

b. $9x^2 + 9x + 2 = (3x + 1)(3x + 2)$ (trial and error).

c. $3x^2 - 12 = 3(x^2 - 4)$ (1)
    $\qquad\qquad = 3(x + 2)(x - 2)$ (4).

d. $6y^2 + 3y - 18 = 3(2y^2 + y - 6)$ (1)
    $\qquad\qquad\quad = 3(2y - 3)(y + 2)$ (trial and error).

e. $4x^2 + 12x + 9 = (2x + 3)^2$ (2).

f. $4x^2 - 16b^2 = 4(x^2 - 4b^2)$ (1)
    $\qquad\qquad = 4(x + 2b)(x - 2b)$ (4).

g. $6x^2 - 24 = 6(x^2 - 4)$ (1)
    $\qquad\qquad = 6(x + 2)(x - 2)$ (4).

h. $x^4 - y^4 = (x^2)^2 - (y^2)^2 = (x^2 + y^2)(x^2 - y^2)$ (4)
    $\qquad = (x^2 + y^2)(x + y)(x - y)$ (4).

Note that $x^2 - y^2$ is factorable but that $x^2 + y^2$ is prime.

i. $2x^2 - 20x + 50 = 2(x^2 - 10x + 25)$ (1)
    $\qquad\qquad\quad = 2(x - 5)^2$ (3).

j. $8 - x^3 = (2)^3 - (x)^3 = (2 - x)(4 + 2x + x^2)$ (7).

k. $x^6 - y^6 = (x^3)^2 - (y^3)^2 = (x^3 + y^3)(x^3 - y^3)$ (4)
    $\qquad = (x + y)(x^2 - xy + y^2)(x - y)(x^2 + xy + y^2)$ (6) and (7).

l. $ax^2 - ay^2 + bx^2 - by^2 = (ax^2 - ay^2) + (bx^2 - by^2)$
    $\qquad\qquad\qquad = a(x^2 - y^2) + b(x^2 - y^2)$ (1)
    $\qquad\qquad\qquad = (x^2 - y^2)(a + b)$ (1)
    $\qquad\qquad\qquad = (x + y)(x - y)(a + b)$ (4).

Here factoring was achieved by making use of appropriate grouping symbols.

EXERCISE 3-2

Completely factor the expressions in Problems 1–60. If prime, so indicate.

1. $10xy + 5xz$

2. $3x^2y - 9x^3y^3$

3. $3xy + 6x^2y^2 + 9x^3y^3$

4. $2a^4y - 8y^4a$

5. $21z^2y + 7zyw^2 + 14z^3y^3w^3$

6. $-2x^2y - 4x^3y^2w + 6x^5y^5w$

7. $x^2 - 16$

8. $x^2 - 25$

9. $x^2 + 6x + 8$

10. $y^2 - 3y + 2$

11. $x^2 + 2x - 15$

12. $x^2 + 4x - 21$

13. $x^2 - 12x + 35$

14. $x^2 + 18x + 32$

15. $x^2 + 18x + 81$

16. $x^2 - 10x + 25$

17. $x^2 - 6x + 9$

18. $x^2 + 22x + 121$

19. $4x^2 - 16$

20. $x^2 + 6$

21. $x^2 + 1$

22. $2x^2 - 3x + 1$

23. $2x^2 + 7x + 3$

24. $3x^2 + 5x + 2$

25. $3x^2 - 7x - 6$

26. $4x^2 + 3x - 1$

27. $4x^2 + 12x + 9$

28. $9x^2 - 6x + 1$

29. $9x^2 - 24x + 16$

30. $25x^2 + 20x + 4 =$

31. $x^4 + 2$

32. $x^2 + x$

33. $9 - y^2$

34. $16w^2 - 9z^2$

35. $-x^2 + 5x + 14$

36. $625 - y^2$

37. $32x^2y^2m^2n^4 - 64x^3y^3m^2n^3 - 96x^2y^2mn^2$

38. $81 + x^2$

39. $49 - 4x^2$

40. $9(y - x) + 18(y - x) + (y - x)$

41. $4x^2 - 8x - 60$

42. $-81 - x^2$

43. $4x^2 + 4x + 1$

44. $-2x^2 + 11x - 12$

45. $-x^2 - x + 6$

46. $14x^2wy^3 - 7x^5w^5y^5 + 49x^4$

47. $12x^3 + 28x^2 + 8x$

48. $x^3 - 1$

49. $x^3 + 8$

50. $6x^2 + 36x + 48$

51. $4a^8b^2c^{16} - 7^0$

52. $12x^2 - 4x - 21$

53. $4x^2y^2 - w^2$

54. $x^3 + 6x^2 + 12x + 7x^2$

55. $y^3 - 3y^2 + 3y - y^2$

56. $32x^2 + 28x + 6$

57. $32x^2 - 4x - 6$

58. $8x^3 - 24x^2y + 18xy^2$

59. $2 + 5x^2$

60. $32x^2y - 2y^3$

Problems 61–90 require a bit more thought than those preceding. In some cases appropriate grouping is helpful.

61. $(x + y)^3 - b^3$

62. $x^3 + x^2 + 3x + 3$

63. $9a^{4/3}x^{5/3}y^{7/2} - 3a^{4/3}x^{2/3}y^{5/2} - 6a^{7/3}x^{2/3}y^{5/2}$

64. $16(x + y)^2 - 81$

65. $18a^2b^2 - 6abc - 4c^2$

66. $(3x - y)^2 - (2v + 1)^2$

67. $x^2 - y^2 + x - y$

68. $4x^2 + 4x + 1 - a^2$

69. $x^6 - y^6$

70. $x^6 + y^6$

71. $a^2 - b^2 - c^2 + 2bc$

72. $16x^2 - z^2 - 8x + 1$

73. $x^{2a+2b} - y^{2a-2b}$ = $x^{2(A+B)} \cdot y^{2(A+B)} = (x^{A+B} + y^{A-B})(y^{(A+B)} - y^{(A-B)})$

74. $16x^{7/3}y^{8/3}z^{7/2} - 32x^{4/3}y^{2/3}z^{5/2} - 16x^{19/3}y^{2/3}z^{5/2}$

75. $x^2 - y^2 + x + y$

76. $x^8 - y^8$

77. $9z^2 - 4x^2 + 4x - 1$

78. $x^3 - x^2 + x - 1 = (x^2 + 1)(x - 1)$

79. $x^3 + x^2 + x + 1$

80. $x^2 - 6x + 9 - 4z^2 - 4z - 1$

81. $2x^2(a + b) - 2(a + b)y^2$

82. $x^6 + x^3$

83. $4x^2 + 12x + 9 - a^2$

84. $2x^2 - 6xy - 2y + x + 4y^2$

85. $16x^4 - 68x^2 + 16$

86. $2x^2 - 2x + x^2y - xy - 2y - 4$

87. $4x^4 - 4x^3y + x^2 + 4x^3y - 4x^2y^2 - y^2$

88. $(m + n)^2(m - n) - (m + n)(m - n)^2$

89. $(y^2 - x^2)(n - m) + (m + n)(m - n)(x + y)$

90. $x^2 + nx + px + nx + mx + (n + m)(n + p)$

91. According to the Bohr theory of the hydrogen atom, the frequency of the energy radiated when an electron goes from orbit n_1 to orbit n_2 is given by

$$\frac{2\pi^2 me^4}{h^3 n_2^2} - \frac{2\pi^2 me^4}{h^3 n_1^2},$$

where m is the mass of the electron, e is the charge of the electron, and h is Planck's constant. Express this in completely factored form.

92. The total area of a right circular cylinder of radius r and altitude h is given by

$$2\pi r^2 + 2\pi rh.$$

Express the surface area in completely factored form.

93. The moment of inertia of a right circular cone of base radius r and altitude h about a certain axis is

$$\tfrac{1}{20}(\pi \sigma r^4 h) + \tfrac{1}{5}(\pi \sigma r^2 h^3),$$

where σ (sigma) is its mass density. Factor this expression.

94. A rectangle has dimensions $2a$ by $3(b + c)$, while a smaller rectangle has dimensions $a - 1$ by $b + c$. Write an expression for the difference in the areas of the rectangles, the larger minus the smaller; perform all indicated operations and express the answer in completely factored form.

95. Under certain conditions, the force that the rear wheels of an automobile of weight W must exert on the ground to cause an acceleration a is given by

$$\frac{\mu W d}{c + d} + \frac{\mu W h a}{gc + gd},$$

where μ (mu) is the coefficient of friction, g is the acceleration due to gravity, and c, d, and h are constants which specify the location of the center of gravity of the automobile. Express this force in factored form.

3-3
REVIEW

REVIEW QUESTIONS

The product of the factors of a factored algebraic expression must be equal to _____(1)_____.

(1) **the expression itself**

$x^2 + 6x + 9$ _____(is) (is not)_____ the square of a binomial.
                          (2)

(2) **is**

To factor $2x^3 + 10x^2 + 12x$, you should first factor out what expression?
_____(3)_____

(3) **2x**

$(53)(47) = (50 + 3)(50 - 3) = 2500 -$ _____(4)_____.

(4) **9**

$(22)^2 = (20 + 2)^2 = 400 + \underline{\qquad (5) \qquad} + 4.$

(5) **80**

Since $x^2 + 7$ cannot be factored further, it is said to be $\underline{\qquad (6) \qquad}$.

(6) **prime**

If $x^4 - y^4$ is written as $(x^2 + y^2)(x^2 - y^2)$, would you consider it completely factored? $\underline{\qquad (7) \qquad}$

(7) **no**

True or false: $(2 + 0)^2 = 2^2 + 0^2.$ $\underline{\qquad (8) \qquad}$

(8) **true**

REVIEW PROBLEMS

Find the special products in Problems 1–12.

1. $(x + 7)(x - 7)$
2. $(6 - 2x)(6 + 2x)$
3. $(2x + 3)^2$
4. $(x + 2y)^2$
5. $(4x - 7)^2$
6. $(2y - 3x)^2$
7. $(x + 3)(x - 1)$
8. $(x - 2)(x - 7)$
9. $(2x + 4)(3x - 2)$
10. $(5x - 6)(2x - 1)$
11. $(2x + 1)^3$
12. $(x - 2y)^3$

Completely factor the expressions in Problems 13–24.

13. $10xy + 5xz$
14. $3x^2y - 9x^3y^3$
15. $8a^3bc - 12ab^3cd + 4b^4c^2d^2$
16. $x^2 - 25$
17. $2x^2 + 7x - 15$
18. $x^2 + 6x + 9$
19. $8x^3 - 27$
20. $2x^2 + 12x + 16$
21. $x^3y^2 - 10x^2y + 25x$
22. $(x^3 - 4x) + (8 - 2x^2)$
23. $4x^2 - 9z^2$
24. $x^3y - xy + z^2x^2 - z^2$

25. When a flexible chain of length l and density w pounds per foot is released with c feet overhanging a smooth table, energy considerations result in

the following expression for the square of the velocity of the chain as it leaves the table:

$$\frac{2g}{wl}\left[wc(l - c) + \frac{1}{2}w(l - c)^2 \right].$$

Show that this can be expressed in completely factored form as

$$\frac{g}{l}(l + c)(l - c).$$

fractions

4-1

SIMPLIFICATION OF FRACTIONS

The basic property employed in the simplification of fractions is known as the *fundamental principle of fractions:*

Multiplying or dividing both the numerator and denominator of a fraction by any quantity except zero results in a fraction which is equivalent to the original fraction.

Symbolically, the fundamental principle asserts that for $c \neq 0$,

$$\frac{a}{b} = \frac{ac}{bc} \qquad \text{and} \qquad \frac{a}{b} = \frac{\dfrac{a}{c}}{\dfrac{b}{c}}.$$

Thus,

$$\frac{2}{3} = \frac{2 \cdot 4}{3 \cdot 4} = \frac{8}{12}$$

and

$$\frac{4}{10} = \frac{\dfrac{4}{2}}{\dfrac{10}{2}} = \frac{2}{5}.$$

The process of reducing a fraction to its simplest form involves the application of the fundamental principle. A fraction is said to be in *simplest form* when its numerator and denominator have no factors in common other than 1.

For example, suppose we simplify

$$\frac{(x+2)(x+3)}{(x+2)(x+1)}.$$

Noting that the numerator and denominator have the common factor $x + 2$, we multiply *both* the numerator and denominator by the multiplicative inverse of $x + 2$, namely $\dfrac{1}{x+2}$:

$$\frac{(x+2)(x+3)}{(x+2)(x+1)} = \frac{\dfrac{1}{x+2}(x+2)(x+3)}{\dfrac{1}{x+2}(x+2)(x+1)}$$

$$= \frac{1(x+3)}{1(x+1)} = \frac{x+3}{x+1}.$$

Actually, multiplying by $\dfrac{1}{x+2}$ has the same effect as dividing by $x + 2$. Here, as elsewhere in the text, the fractions are assumed to have nonzero denominators. The process we used above is commonly referred to as *cancellation,* but it is, in effect, strictly an application of the fundamental principle. Usually, this step is done mentally and we simply write

$$\frac{\overset{1}{\cancel{(x+2)}}(x+3)}{\underset{1}{\cancel{(x+2)}}(x+1)} = \frac{x+3}{x+1}.$$

Frequently, before the fundamental principle can be applied the numerator and denominator of the given fraction must be factored completely. We illustrate this in the following examples.

EXAMPLE
1

Simplify $\dfrac{x^2 + 7x + 10}{x^2 + 4x - 5}$.

First we completely factor the numerator and denominator:

$$\frac{x^2 + 7x + 10}{x^2 + 4x - 5} = \frac{(x + 2)(x + 5)}{(x + 5)(x - 1)}.$$

Next, we divide both the numerator and denominator by their common factor, $x + 5$:

$$\frac{(x + 2)\overset{1}{\cancel{(x + 5)}}}{\underset{1}{\cancel{(x + 5)}}(x - 1)} = \frac{x + 2}{x - 1}.$$

Note that we cannot cancel the x's in this final expression since they are *terms*, and **we can only cancel *factors*.**

EXAMPLE
2

a. Simplify $\dfrac{3ab - 3b^2}{a^2 - b^2}$.

$$\frac{3ab - 3b^2}{a^2 - b^2} = \frac{3b\overset{1}{\cancel{(a - b)}}}{(a + b)\underset{1}{\cancel{(a - b)}}}$$

$$= \frac{3b}{a + b}.$$

b. Simplify $\dfrac{a + b}{3a^2b}$.

This fraction cannot be reduced any further and, hence, is in simplest form. We emphasize again that cancellation can only be performed when there are *factors* common to the numerator and denominator. In this case, although a and b are factors of the denominator, they are *terms* in the numerator and therefore cannot be cancelled.

Consider the fraction $\dfrac{-b}{a}$. By the fundamental principle we can multiply both the numerator and denominator by -1 without changing the value of the fraction. Thus,

$$\frac{-b}{a} = \frac{(-b)(-1)}{a(-1)} = \frac{b}{-a}.$$

Indeed, in Chapter 1 it was stated that

$$\frac{-b}{a} = \frac{b}{-a} = -\frac{b}{a},$$

which can also be written as

$$+\frac{-b}{+a} = +\frac{+b}{-a} = -\frac{+b}{+a}. \tag{1}$$

Although in the latter forms the insertion of plus signs is certainly unnecessary, they make it clear that with each fraction we can associate three signs: the sign in front of the fraction, the sign of the numerator, and the sign of the denominator. From an observation of (1) we conclude that any two of these signs can be reversed without affecting the value of the fraction itself. Thus,

$$\frac{-2}{7} = \frac{2}{-7} = -\frac{2}{7} = -\frac{-2}{-7}.$$

Similarly, note how we can obtain three equivalent forms of the following fraction.

$$\frac{x-2}{x-3} = \frac{(-1)(x-2)}{(-1)(x-3)} = \frac{-x+2}{-x+3} = \frac{2-x}{3-x}$$

$$\frac{x-2}{x-3} = -\frac{x-2}{(-1)(x-3)} = -\frac{x-2}{-x+3} = -\frac{x-2}{3-x}$$

and

$$\frac{x-2}{x-3} = -\frac{(-1)(x-2)}{x-3} = -\frac{-x+2}{x-3} = -\frac{2-x}{x-3}.$$

Hence,

$$\frac{x-2}{x-3} = \frac{2-x}{3-x} = -\frac{x-2}{3-x} = -\frac{2-x}{x-3}.$$

In fact, if both the numerator and denominator of a fraction are in factored form, we may reverse the signs of the terms in any *even* number of *factors* without changing the value of the fraction itself. Thus,

$$\frac{(a-b)(c-d)}{(e-f)(g-h)} = \frac{(a-b)[d-c]}{[f-e](g-h)} = -\frac{[b-a](d-c)}{(f-e)(g-h)}$$

and so

$$\frac{(a-b)(c-d)}{(e-f)(g-h)} = -\frac{[b-a][d-c]}{[f-e](g-h)}.$$

EXAMPLE
3

a. Simplify $\dfrac{a^2-b^2}{b-a}$.

$$\frac{a^2-b^2}{b-a} = \frac{(a+b)(a-b)}{b-a}$$

$$= -\frac{(a+b)(a-b)}{[a-b]} = -(a+b).$$

b. Simplify $\dfrac{(3-a)(a-2)(a-4)}{(a-3)(a+2)(4-a)}$.

$$\frac{(3-a)(a-2)(a-4)}{(a-3)(a+2)(4-a)} = \frac{[(-1)(3-a)](a-2)[(-1)(a-4)]}{(a-3)(a+2)(4-a)}$$

$$= \frac{[a-3](a-2)[4-a]}{(a-3)(a+2)(4-a)} = \frac{a-2}{a+2}.$$

Note in this example that we strategically inserted *two* factors of -1 and, hence, did not change the value of the fraction. Equivalently, we reversed the signs of the terms in an even number (2) of factors.

c. Simplify $\dfrac{2ax^2 - 4ax + 2a}{a - ax}$.

$$\frac{2ax^2 - 4ax + 2a}{a - ax} = \frac{2a(x^2 - 2x + 1)}{a(1 - x)}$$

$$= \frac{2(x-1)^2}{(1-x)}$$

$$= -\frac{2(x-1)^2}{[x-1]} = -2(x-1)$$

$$= 2(1-x).$$

d. Simplify $\dfrac{3x^2 + 9x - 12}{8 - 4x - 4x^2}$.

$$\frac{3x^2 + 9x - 12}{8 - 4x - 4x^2} = \frac{3(x^2 + 3x - 4)}{4(2 - x - x^2)}$$

$$= \frac{3(x-1)(x+4)}{4(1-x)(2+x)}$$

$$= -\frac{3(x-1)(x+4)}{4[x-1](2+x)}$$

$$= -\frac{3(x+4)}{4(2+x)}.$$

EXERCISE 4-1

Determine whether the statements in Problems 1–10 are identically true or false.

1. $-\dfrac{x}{x-y} = \dfrac{x}{y-x}$

2. $\dfrac{(x+y)(y-z)}{z-x} = \dfrac{(y+x)(z-y)}{x-z}$

3. $\dfrac{(a-b)(b-c)}{c-a} = \dfrac{(c-b)(b-a)}{c-a}$

4. $\dfrac{x-y}{z} = \dfrac{y-x}{z}$

5. $-\dfrac{a-b}{c} = \dfrac{-(b-a)}{c}$

6. $-\dfrac{x-y}{w-z} = \dfrac{x-y}{z-w}$

7. $\dfrac{x}{y} = -\dfrac{-x}{y} = \dfrac{-x}{-y}$

8. $\dfrac{a-b}{-x} = -\dfrac{b-a}{x}$

9. $\dfrac{(x-a)(x-b)}{(x-c)(x-d)} = -\dfrac{(x-a)(b-x)}{(x-c)(d-x)}$

10. $\dfrac{(x-a)(x-b)(x-c)}{(x-d)(x-e)} = \dfrac{(x-a)(b-x)(c-x)}{(e-x)(d-x)}$

Simplify the fractions in Problems 11–34.

11. $\dfrac{8x^2}{16x}$ ⟂ $\dfrac{x}{2}$

12. $\dfrac{3x^2y^3z}{xyz}$

13. $\dfrac{(x+2)(x-1)}{(x-1)(3x+5)}$

14. $\dfrac{(x-3)(x+4)(x+6)}{3x^2(x-3)}$

15. $\dfrac{3x+12}{x+4}$

16. $\dfrac{x^2-y^2}{x+y}$

17. $\dfrac{2x+2y}{6ax+6ay}$

18. $\dfrac{3x^2-12}{3x-6}$

19. $\dfrac{x^2-81}{x^2+9x}$

20. $\dfrac{x^2y-4xy}{x^3y^2-2x^2y^2}$

21. $\dfrac{(x+3)(x^2-144)}{(x^2-9)(x+12)}$

22. $\dfrac{6a^2b^2+6ab^3y+6a^2b^2c^2}{3abc}$

23. $\dfrac{x^2+5x+6}{x^2-2x-8}$ ⟂ $\dfrac{x+3}{x+4}$

24. $\dfrac{x^2+x-12}{x^2-6x+9}$

25. $\dfrac{c-d}{d-c}$

26. $\dfrac{(x-2)(4-x)}{(x-4)(2-x)}$

27. $\dfrac{(x+5)(2-x)(x+7)(6-x)}{(x-6)(7-x)(x-2)(x+5)}$

28. $\dfrac{(x+a)(-x-b)(c-x)(x+d)}{(-x-d)(x-c)(x+b)(x-a)}$

29. $\dfrac{x^2+x-12}{-x-4}$

30. $\dfrac{16x^4-4b^4}{4x^2+2b^2}$

31. $\dfrac{x^3+8}{x^2+3x+2}$

32. $\dfrac{x^2-y^2}{x^3-y^3}$

33. $\dfrac{(x-2y)^2}{(2y-x)^3}$

34. $\dfrac{(y-x)^2y}{(x^2-xy)(x+y)}$

Simplifying the fractions in Problems 35–44 is somewhat more challenging than the preceding problems.

35. $\dfrac{a^4-b^4}{a+b}$

36. $\dfrac{3(x+y)-6x-6y}{-3(x-y)}$

37. $\dfrac{(2x^2-x-3)(4x^2-1)}{(2x^2+3x+1)(6x^2-13x+6)}$

38. $\dfrac{8a^3b^{1/2}c^{1/2}+8a^2b^{3/2}c^{1/2}}{a^2-b^2}$

39. $\dfrac{(x^2-xy)(x^2-y^2)(xy-2y^2)}{(x-y)^2(x^2-3xy+2y^2)(x^2+xy)}$

40. $\dfrac{x^2 - y^2 + x - y}{x + y + 1}$

41. $\dfrac{9w^2 - 4x^2 + 4x - 1}{3w + 2x - 1}$

42. $\dfrac{(x + y)(y - x)(x + 1)}{x^3 - xy^2 + x^2 - y^2}$

43. $\dfrac{-(a - b) + 3(a - b) - 2(a^2 - b^2)}{1 - a - b}$

44. $\dfrac{2x^3 + 2x - 7x^2 - 7}{2x^3 - 2x + 7x^2 - 7}$

45. For the arrangement of particles shown in Fig. 4-1, it is shown in mechanics

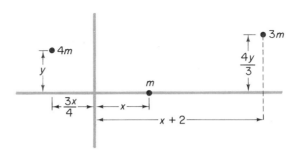

FIGURE 4-1

that the coordinates $\bar{x}$ and $\bar{y}$ of the center of mass of the particles are given by

$$\bar{x} = \frac{m(x) + 3m(x + 2) + 4m\left(-\dfrac{3x}{4}\right)}{m + 3m + 4m}$$

and

$$\bar{y} = \frac{m(0) + 3m\left(\dfrac{4y}{3}\right) + 4m(y)}{m + 3m + 4m}.$$

Simplify the expressions for $\bar{x}$ and $\bar{y}$.

4-2
ADDITION AND SUBTRACTION OF FRACTIONS

In the addition and subtraction of fractions, the general procedure is to first express the given sum or difference as a single fraction and then reduce this fraction to its simplest form.

Consider the result of adding two fractions having a common denominator, say the fractions $\dfrac{a}{c}$ and $\dfrac{b}{c}$:

$$\frac{a}{c} + \frac{b}{c} = a \cdot \frac{1}{c} + b \cdot \frac{1}{c} \qquad \text{(definition of division)}$$

$$= (a + b)\frac{1}{c} \qquad \text{(distributive axiom)}$$

$$\boxed{\frac{a}{c} + \frac{b}{c} = \frac{a + b}{c}} \qquad \text{(definition of division)}.$$

Hence, the result is a fraction whose denominator is the common denominator and whose numerator is the sum of the numerators of the original fractions. This important result *does not* mean that $\dfrac{a}{b+c} = \dfrac{a}{b} + \dfrac{a}{c}$, a very common error. In a manner similar to that above it can also be shown that

$$\boxed{\frac{a}{c} - \frac{b}{c} = \frac{a - b}{c}}.$$

We emphasize that *only fractions with the same denominator can be combined.*

EXAMPLE
4

In combining each of the following fractions, note that the numerators are added when and only when the denominators are exactly alike.

a. $\dfrac{x^2 - 5}{x - 2} + \dfrac{3x + 2}{x - 2} = \dfrac{(x^2 - 5) + (3x + 2)}{x - 2}$

$$= \frac{x^2 + 3x - 3}{x - 2}.$$

b. $\dfrac{x^2 - 5x + 4}{x^2 + 2x - 3} - \dfrac{x^2 + 2x}{x^2 + 5x + 6} = \dfrac{(x - 1)(x - 4)}{(x - 1)(x + 3)} - \dfrac{x(x + 2)}{(x + 2)(x + 3)}$

$$= \frac{x - 4}{x + 3} - \frac{x}{x + 3} = \frac{(x - 4) - x}{x + 3}$$

$$= \frac{-4}{x + 3} = -\frac{4}{x + 3}.$$

c. $\dfrac{x^2 + x - 5}{x - 7} - \dfrac{x^2 - 2}{x - 7} + \dfrac{-4x + 8}{x^2 - 9x + 14}$

$$= \frac{x^2 + x - 5}{x - 7} - \frac{x^2 - 2}{x - 7} + \frac{(-4)(x - 2)}{(x - 2)(x - 7)}$$

$$= \frac{(x^2 + x - 5) - (x^2 - 2) + (-4)}{x - 7}$$

$$= \frac{x - 7}{x - 7} = 1.$$

To add or subtract two or more fractions with different denominators, we first transform the fractions, using the fundamental principle of fractions, into equivalent fractions which have the same denominator. We then combine the fractions in the manner described previously.

For example, to add

$$\frac{2}{x^3(x-3)} + \frac{3}{x(x-3)^2}$$

we can convert the first fraction into the equivalent fraction

$$\frac{2(x-3)}{x^3(x-3)^2}$$

by multiplying its numerator and denominator by $x-3$. The second fraction can be converted into

$$\frac{3x^2}{x^3(x-3)^2}$$

by multiplying its numerator and denominator by x^2. Since these fractions have the same denominator, they can be combined:

$$\frac{2}{x^3(x-3)} + \frac{3}{x(x-3)^2} = \frac{2(x-3)}{x^3(x-3)^2} + \frac{3x^2}{x^3(x-3)^2}$$
$$= \frac{2(x-3)+3x^2}{x^3(x-3)^2}$$
$$= \frac{3x^2+2x-6}{x^3(x-3)^2}.$$

Although we could have converted the original fractions into equivalent fractions with *any* common denominator (such as the product of the denominators), we chose to convert them into fractions with the denominator $x^3(x-3)^2$, which is called the **least common denominator** (L.C.D.) of the fractions $2/[x^3(x-3)]$ and $3/[x(x-3)^2]$. It can be shown that if any other common denominator were chosen, it must have the L.C.D. as a factor. In this sense the L.C.D. is the *least* such common denominator. For this reason, our work may be simplified by using the L.C.D.

In general, *the L.C.D. of two (or more) fractions is the product of each of the distinct factors appearing in the denominators, each raised to the highest power to which that factor occurs in any one denominator.*

EXAMPLE
5

Find the L.C.D. of the fractions

$$\frac{3x}{(x-1)(x+2)}, \qquad \frac{6x}{(x+1)^2}, \qquad \text{and} \qquad \frac{3}{(x-1)^3(x+2)^2}.$$

Three different factors appear in the denominators: $x - 1$, $x + 2$, and $x + 1$. The L.C.D. is the product of these factors, each raised to the highest power to which it occurs in the denominator of any fraction. The factor $x - 1$ occurs at most three times (in the third fraction), the factor $x + 2$ occurs at most two times (in the third fraction), and the factor $x + 1$ occurs at most two times (in the second fraction). Thus, the L.C.D. in this case is

$$(x - 1)^3(x + 2)^2(x + 1)^2.$$

Now, to add (or subtract) fractions with different denominators,

For each fraction multiply both its the numerator and denominator by a quantity which will make its denominator equal to the L.C.D. of the fractions. Then combine the fractions and simplify if possible.

Symbolically, this means that

$$\frac{a}{x} + \frac{b}{y} = \frac{a(y)}{x(y)} + \frac{b(x)}{y(x)} = \frac{ay + bx}{xy}.$$

EXAMPLE 6

Find $\dfrac{7}{x + 1} - \dfrac{4}{x - 1}$.

The L.C.D. is $(x + 1)(x - 1)$:

$$\frac{7}{x + 1} - \frac{4}{x - 1} = \frac{7(x - 1)}{(x + 1)(x - 1)} - \frac{4(x + 1)}{(x - 1)(x + 1)}$$

$$= \frac{7(x - 1) - 4(x + 1)}{(x + 1)(x - 1)}$$

$$= \frac{3x - 11}{(x + 1)(x - 1)}.$$

EXAMPLE 7

a. Find $\dfrac{2}{x^2 - 5x + 6} + \dfrac{1}{x - 3} - 2$.

The denominators of the three terms are $(x - 3)(x - 2)$, $x - 3$, and 1, respectively. Hence the L.C.D. is $(x - 3)(x - 2)$.

$$\frac{2}{x^2 - 5x + 6} + \frac{1}{x - 3} - 2$$

$$= \frac{2}{(x - 3)(x - 2)} + \frac{(x - 2)}{(x - 3)(x - 2)} - \frac{2(x - 3)(x - 2)}{(x - 3)(x - 2)}$$

$$\overbrace{x^2 - 5x + 6}$$

$$= \frac{2 + (x - 2) - 2(x - 3)(x - 2)}{(x - 3)(x - 2)} = \frac{2 + x - 2 - 2x^2 + 10x - 12}{(x - 3)(x - 2)}$$

$$= \frac{-2x^2 + 11x - 12}{(x - 3)(x - 2)}.$$

b. Find $\dfrac{6}{x-2} + \dfrac{7}{2-x}$.

You might be tempted to consider the L.C.D. to be $(x-2)(2-x)$. However, the sum will be much simpler to obtain if we rewrite the second term so that the denominators are alike:

$$\frac{6}{x-2} + \frac{7}{2-x} = \frac{6}{x-2} - \frac{7}{-(2-x)}$$

$$= \frac{6}{x-2} - \frac{7}{x-2} = \frac{6-7}{x-2} = \frac{-1}{x-2}$$

$$= -\frac{1}{x-2}.$$

c. Find $\dfrac{x}{2} - \dfrac{y}{6(x-y)} + \dfrac{2}{x^2-y^2}$.

Since the denominator x^2-y^2 can be written $(x+y)(x-y)$, it should be clear that the L.C.D. is $6(x+y)(x-y)$.

$$\frac{x}{2} - \frac{y}{6(x-y)} + \frac{2}{(x+y)(x-y)}$$

$$= \frac{x(3)\overbrace{(x+y)(x-y)}^{x^2-y^2} - y(x+y) + 2(6)}{6(x+y)(x-y)}$$

$$= \frac{3x^3 - 3xy^2 - xy - y^2 + 12}{6(x+y)(x-y)}.$$

EXERCISE 4-2

In Problems 1–24 perform the indicated operations and simplify as much as possible.

1. $\dfrac{x^2}{x-2} + \dfrac{x-6}{x-2}$

2. $\dfrac{1}{x} + \dfrac{2}{3x}$

3. $\dfrac{4}{2x-1} + \dfrac{x}{x+3}$

4. $\dfrac{x+1}{x-1} - \dfrac{x-1}{x+1}$

5. $\dfrac{x}{2} + \dfrac{x+1}{x} - \dfrac{1}{x}$

6. $\dfrac{5c}{x^2} - \dfrac{2a}{xy}$

7. $\dfrac{x}{x-y} + \dfrac{y}{x+y} =$

8. $\dfrac{2}{x+1} - \dfrac{3}{x-1}$

9. $\dfrac{1}{x-y} + \dfrac{1}{y-x}$

10. $\dfrac{3x}{2(x-2)} + \dfrac{3}{2-x}$

11. $\dfrac{1}{x^2-1} - \dfrac{1}{x-1} + \dfrac{1}{x+1}$

12. $\dfrac{x}{x+1} - \dfrac{2x}{x^2+3x+2}$

13. $\dfrac{1}{x-1} - \dfrac{x^2+2}{x^3-1}$

14. $\dfrac{2}{x^2-9} + \dfrac{x-1}{x^2+6x+9}$

15. $\dfrac{x+1}{x^2+7x+10} - \dfrac{2x}{x^2+6x+5}$ **16.** $x^2+2 - \dfrac{x^4}{x^2-2}$

17. $2x+3 + \dfrac{2}{x-1}$ **18.** $\dfrac{x}{a^2} - \dfrac{y}{ab}$

19. $\dfrac{y}{3y^2-5y-2} - \dfrac{2}{3y^2-7y+2}$ **20.** $\dfrac{4}{x-1} - 3 + \dfrac{-3x^2}{5-4x-x^2}$

21. $\dfrac{y}{x^2+2xy+y^2} + \dfrac{3x}{x^2-y^2} - \dfrac{2}{x+y}$

22. $\dfrac{2}{x^2-5x+6} + \dfrac{1}{x^2-3x+2} + \dfrac{4}{x^2-4x+3}$

23. $1 - \dfrac{2y^2}{x^2-y^2} + \dfrac{2xy}{x^2+y^2}$ **24.** $\dfrac{2x+1}{x+3} - \dfrac{3x+4}{x-2} - \dfrac{1}{x+4}$

Problems 25–32 are more challenging than the previous ones. Again, perform the indicated operations and simplify as much as possible.

25. $\left(2x - \dfrac{x+3}{2} - 2a\right) + \dfrac{a-4}{3} - a$

26. $\dfrac{3m+n}{m-n} + \dfrac{m+3n}{n-m}$

27. $\dfrac{ay}{a-y} + \dfrac{a^2}{y-a} + a$

28. $\dfrac{2y+3}{2-y} - \dfrac{2-3y}{y+2} + \dfrac{16y-y^2}{(y+2)(y-2)}$

29. $\dfrac{2}{x-3a} + \dfrac{2a}{x^2-4ax+4a^2} - \dfrac{x-a}{6a^2-5ax+x^2}$

30. $\dfrac{1}{8-6x+x^2} - \dfrac{x+2}{(x-3)(4-x)} + \dfrac{x+2}{x^2-5x+6}$

31. $\dfrac{3}{a-x} + \dfrac{3}{x+a} - \dfrac{1}{x+3a} + \dfrac{1}{x-3a}$

32. $\dfrac{x+y}{(z-x)(z-y)} + \dfrac{y+z}{(x-y)(x-z)} - \dfrac{z+x}{(y-z)(y-x)}$

33. For the arrangement in Fig. 4-2, it is shown in physics that the potential at point O due to the point charges q_1, q_2, and q_3 is given by

$$\dfrac{kq_1}{x} + \dfrac{kq_2}{x+2} + \dfrac{kq_3}{x+3}.$$

Express the potential as a single fraction.

34. For the arrangement in Fig. 4-2, if the distance of charge q_2 from point O were doubled, theory shows that the potential at point O would be

$$\dfrac{kq_1}{x} + \dfrac{kq_2}{2(x+2)} + \dfrac{kq_3}{x+3}.$$

Express the potential as a single fraction.

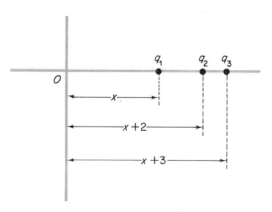

FIGURE 4-2

35. In studies of electric potential, the expression

$$\frac{1}{4\pi\epsilon_0}\left(\frac{q_1}{r_1} - \frac{q_2}{r_2}\right)$$

was encountered. Simplify this expression.

4-3
MULTIPLICATION AND DIVISION OF FRACTIONS

To multiply fractions, we need only to multiply the numerators and multiply the denominators:

$$\boxed{\frac{a}{b} \cdot \frac{c}{d} \cdot \frac{e}{f} = \frac{ace}{bdf}.}$$

EXAMPLE
8

a. $\dfrac{2}{x} \cdot \dfrac{3}{y} = \dfrac{2 \cdot 3}{x \cdot y} = \dfrac{6}{xy}.$

b. $\dfrac{2x^2y}{z^2} \cdot \dfrac{3z^2}{7} = \dfrac{6x^2yz^2}{7z^2} = \dfrac{6x^2y}{7}.$

c. $\dfrac{2x^2y}{3x} \cdot \dfrac{4xy^2}{xy} = \dfrac{2xy}{3} \cdot \dfrac{4y}{1} = \dfrac{8xy^2}{3}.$

d. $\dfrac{x}{x+2} \cdot \dfrac{x+3}{x-5} = \dfrac{x(x+3)}{(x+2)(x-5)}.$

e. $\dfrac{2x^2+3x+1}{x+1} \cdot \dfrac{x-1}{4x^2-1} = \dfrac{[(2x+1)(x+1)](x-1)}{(x+1)[(2x+1)(2x-1)]}$

$\qquad\qquad\qquad\qquad = \dfrac{x-1}{2x-1}.$

f. $\dfrac{x^2 - 4x + 4}{x^2 + 2x - 3} \cdot \dfrac{6x^2 - 6}{x^2 + 2x - 8} = \dfrac{(x-2)^2[6(x^2-1)]}{[(x+3)(x-1)][(x+4)(x-2)]}$

$$= \dfrac{(x-2)^2[6(x+1)(x-1)]}{(x+3)(x-1)(x+4)(x-2)}$$

$$= \dfrac{6(x-2)(x+1)}{(x+3)(x+4)}.$$

The division of fractions is accomplished by applying the fundamental principle of fractions:

$$\dfrac{\dfrac{a}{b}}{\dfrac{c}{d}} = \dfrac{\dfrac{a}{b} \cdot \dfrac{d}{c}}{\dfrac{c}{d} \cdot \dfrac{d}{c}} = \dfrac{\dfrac{a}{b} \cdot \dfrac{d}{c}}{\dfrac{cd}{cd}} = \dfrac{\dfrac{a}{b} \cdot \dfrac{d}{c}}{1} = \dfrac{a}{b} \cdot \dfrac{d}{c} = \dfrac{ad}{bc}.$$

Thus,

$$\boxed{\dfrac{\dfrac{a}{b}}{\dfrac{c}{d}} = \dfrac{a}{b} \div \dfrac{c}{d} = \dfrac{a}{b} \cdot \dfrac{d}{c} = \dfrac{ad}{bc}.}$$

Hence, in effect, to divide a/b by c/d we can multiply a/b by the reciprocal of the divisor, d/c. This procedure is commonly referred to as *inverting* c/d and multiplying. In general, any factoring that is possible should be performed before any final multiplications, since this may simplify your work.

Some consequences of the above discussion are as follows:

$$\boxed{\dfrac{a}{\dfrac{b}{c}} = a \div \dfrac{b}{c} = a \cdot \dfrac{c}{b} = \dfrac{ac}{b}}$$

and

$$\boxed{\dfrac{\dfrac{a}{b}}{c} = \dfrac{a}{b} \div c = \dfrac{a}{b} \cdot \dfrac{1}{c} = \dfrac{a}{bc}}$$

EXAMPLE
9

a. $\dfrac{x}{x+2} \div \dfrac{x+3}{x-5} = \dfrac{x}{x+2} \cdot \dfrac{x-5}{x+3}$

$$= \dfrac{x(x-5)}{(x+2)(x+3)}.$$

b. $\dfrac{x^2 - 4}{x} \div (x + 2)^2 = \dfrac{(x + 2)(x - 2)}{x} \cdot \dfrac{1}{(x + 2)^2}$

$$= \dfrac{x - 2}{x(x + 2)}.$$

c. $\dfrac{x^2 + 4x + 4}{x - 1} \div \dfrac{(x + 2)^2}{6x - 6} = \dfrac{(x + 2)^2}{(x - 1)} \cdot \dfrac{6(x - 1)}{(x + 2)^2} = 6.$

d. $\dfrac{\dfrac{x}{7}}{3y} = \dfrac{x}{7} \div 3y = \dfrac{x}{7} \cdot \dfrac{1}{3y} = \dfrac{x}{21y}.$

e. $\dfrac{4}{\dfrac{y}{x}} = 4 \div \dfrac{y}{x} = \dfrac{4}{1} \cdot \dfrac{x}{y} = \dfrac{4x}{y}.$

f. $\dfrac{\dfrac{x - 5}{x - 3}}{2x} = \dfrac{x - 5}{x - 3} \div 2x = \dfrac{x - 5}{x - 3} \cdot \dfrac{1}{2x} = \dfrac{x - 5}{2x(x - 3)}.$

EXERCISE 4-3

ALL ODD

In Problems 1–32, perform the indicated operations and simplify.

1. $\dfrac{7}{y} \cdot \dfrac{1}{x}$

2. $\dfrac{2}{x} \cdot \dfrac{y}{5}$

3. $\dfrac{2x^2}{3} \cdot \dfrac{6}{x^5}$

4. $\dfrac{1}{x} \cdot \dfrac{3xy^2z^3}{x^2y} \cdot \dfrac{x^2}{3yz^3}$

5. $\dfrac{x^2y^2z}{abc} \cdot \dfrac{b^2c}{y^3z} \cdot \dfrac{a^2b}{x^3y}$

6. $\dfrac{3x^2y}{a^4} \cdot \dfrac{2ab^2x}{y} \cdot \dfrac{a^2x^3}{6} \cdot \dfrac{a^2}{b^2x^6}$

7. $\dfrac{y^2}{y - 3} \cdot \dfrac{-1}{y + 2}$

8. $\dfrac{x - 3}{x + 4} \cdot \dfrac{x - 5}{x - 3}$

9. $\dfrac{x^2 - 5x + 6}{x - 1} \cdot \dfrac{x - 1}{x - 3} \cdot \dfrac{x + 3}{(x - 2)^2}$

10. $\dfrac{(x - 2)^2}{x + 4} \cdot \dfrac{(x + 4)^2}{x - 2} \cdot \dfrac{(x - 1)^2}{(x + 4)^2}$

11. $\dfrac{2x - 3}{x - 2} \cdot \dfrac{2 - x}{2x + 3}$

12. $\dfrac{x^2 - 3x + 2}{x^2 - 7x + 12} \cdot \dfrac{(x - 3)(x + 2)}{x^2 + x - 2}$

13. $\dfrac{x^2 - y^2}{x + y} \cdot \dfrac{x^2 + 2xy + y^2}{y - x}$

14. $\dfrac{b(6a - 6b)}{a^2 + ab} \cdot \dfrac{a^3 - ab^2}{2b^2}$

15. $\dfrac{x + 3}{x - 4} \div \dfrac{x + 3}{x + 2}$

16. $\dfrac{x + 5}{x - 3} \div \dfrac{x - 5}{x - 3}$

17. $\dfrac{x^2 - 9}{x^2} \div \dfrac{x + 3}{x}$

18. $\dfrac{x}{x^2 - 1} \div \dfrac{2}{(x + 1)^2}$

19. $\dfrac{2x - 2}{x^2 - 2x - 8} \div \dfrac{x^2 - 1}{x^2 + 5x + 4}$

20. $\dfrac{x^2 - 2xy + y^2}{x + y} \div \dfrac{x^2 - xy}{y}$

21. $\dfrac{(x-3)(x+2)}{x^2+x-6} \div \dfrac{x^2+4x+4}{x^2+2x-3}$

22. $\dfrac{x^3+8}{3x^2-18x+24} \div \dfrac{x^2-x-6}{x^2-4x+4}$

23. $\dfrac{\dfrac{x}{6}}{y}$

24. $\dfrac{6}{\dfrac{x}{y}}$

25. $\dfrac{\dfrac{2}{9x}}{\dfrac{3y}{4}}$

26. $\dfrac{\dfrac{1}{x}}{\dfrac{2z}{y}}$

27. $\dfrac{\dfrac{2x-2y}{3z}}{\dfrac{x-y}{6z^3}}$

28. $\dfrac{\dfrac{x^2-y^2}{xy}}{\dfrac{x+y}{xy}}$

29. $\dfrac{\dfrac{x-1}{x(x^2-y^2)}}{\dfrac{1-x}{x+y}}$

30. $\dfrac{\dfrac{3xz-15z}{2x^2y}}{\dfrac{5-x}{2(x+1)}}$

31. $2x \cdot \dfrac{\dfrac{1}{x}}{x^2}$

32. $\dfrac{\dfrac{x}{2y}}{\dfrac{x^2}{y^2}} \div \dfrac{\dfrac{2}{x}}{\dfrac{y}{2}}$

33. For a stretched steel wire, stress $= F/A$ and strain $= e/L$, where F is force, A is cross-sectional area, e is elongation, and L is the original length. Young's modulus for the material is stress/strain. Find an expression for this modulus and simplify.

4-4
COMPLEX FRACTIONS

A **complex fraction** is one in which the numerator and/or denominator are themselves fractions or combinations thereof. In order to reduce such a fraction to its simplest form, you should first perform any indicated additions and subtractions in both the numerator and denominator. This will result in one fraction divided by another. Follow carefully the steps in Example 10.

EXAMPLE
10

a. $\dfrac{x+\dfrac{1}{2x}}{2+\dfrac{x}{x+1}} = \dfrac{\dfrac{x(2x)+1}{2x}}{\dfrac{2(x+1)+x}{x+1}} = \dfrac{\dfrac{2x^2+1}{2x}}{\dfrac{3x+2}{x+1}}$

$= \dfrac{2x^2+1}{2x} \cdot \dfrac{x+1}{3x+2} = \dfrac{(2x^2+1)(x+1)}{2x(3x+2)}.$

b. $\dfrac{x+7+\dfrac{x-1}{x-2}}{2-\dfrac{x-2}{x}} = \dfrac{\dfrac{(x+7)(x-2)}{x-2}+\dfrac{x-1}{x-2}}{2-\left(\dfrac{3}{x-2}\cdot\dfrac{1}{x}\right)}$

$$= \dfrac{\dfrac{(x+7)(x-2)+(x-1)}{x-2}}{2-\dfrac{3}{x(x-2)}}$$

$$= \dfrac{\dfrac{x^2+6x-15}{x-2}}{\dfrac{2x(x-2)-3}{x(x-2)}}$$

$$= \dfrac{x^2+6x-15}{x-2}\cdot\dfrac{x(x-2)}{2x^2-4x-3}$$

$$= \dfrac{x(x^2+6x-15)}{2x^2-4x-3}.$$

c. $\left[\left(\dfrac{2}{x}+\dfrac{2}{y}\right)(xy)\right]\div(2wx+2wy)=\left[\left(\dfrac{2y+2x}{xy}\right)(xy)\right]\cdot\dfrac{1}{2wx+2wy}$

$$= \dfrac{2(y+x)(xy)}{(xy)}\cdot\dfrac{1}{(2)(w)(x+y)}$$

$$= \dfrac{1}{w}.$$

EXERCISE 4-4

Perform the indicated operations and simplify.

1. $\dfrac{1+\dfrac{1}{x}}{2-\dfrac{1}{y}}$

2. $\dfrac{1+\dfrac{1}{x}}{2+\dfrac{1}{xy}}$

3. $\dfrac{a-\dfrac{b^2}{a}}{b-\dfrac{a^2}{b}}$

4. $\dfrac{\dfrac{x}{y}+\dfrac{y}{x}}{\dfrac{x}{y}-\dfrac{y}{x}}$

5. $\dfrac{3-\dfrac{1}{2x}}{x+\dfrac{x}{x+2}}$

6. $\dfrac{2y+\dfrac{y}{y-2}}{2y-\dfrac{y}{y-2}}$

7. $\dfrac{\dfrac{1}{x^4}+\dfrac{1}{x^2}+1}{\dfrac{1}{x^2}+\dfrac{1}{x}+1}$

8. $\dfrac{x-\dfrac{x^2+1}{x+1}}{3-x}$

9. $\dfrac{\dfrac{x}{y} - \dfrac{y}{x}}{\dfrac{x+y}{xy}}$

10. $\dfrac{\dfrac{a+b}{a-b} - \dfrac{a-b}{a+b}}{1 - \dfrac{a^2+b^2}{(a+b)^2}}$

11. $\dfrac{5 - \dfrac{y^2 - 19x^2}{y^2 - 4x^2}}{3 - \dfrac{y - 5x}{y - 2x}}$

12. $\dfrac{1}{2^{-1} + 1^{-1}} - \dfrac{3^{-1}}{2^{-1}}$

13. $\dfrac{\dfrac{x}{y} + x}{y - \dfrac{1}{y}} - 1$

14. $\left(x + \dfrac{y^2}{x-y}\right) \div \left(x - \dfrac{(x-y)y^2}{x^2 + y^2}\right)$

15. $\dfrac{a}{1 + \dfrac{c}{d + \dfrac{1}{2}}}$

16. $\left(\dfrac{\dfrac{x + xy}{y}}{y - \dfrac{1}{y}} - 1\right) \div \left(\dfrac{x^2 - 2x}{1 - x} - x - 1\right)$

4-5
RATIO—CONVERSION OF UNITS

Throughout the engineering sciences are found many physical quantities that are expressed as quotients. For example, pressure (p) is equal to force (F) divided by area (A); that is, $p = F/A$. In quite general terms, by definition the **ratio of a to b** is the quotient a/b, also denoted $a:b$. Hence, we speak of pressure as the ratio of force to area. Similarly, average power ($\bar{P}$) is the ratio of work to time, $\bar{P} = W/t$, and acceleration (a) is the ratio of force to mass, $a = F/m$.

Physical quantities like those mentioned above, however, usually require a *number* and a *unit* if they are to be completely specified. There is no meaningful interpretation, for instance, for the statement that a given area is 3. Does this mean 3 square feet (3 ft²), 3 square meters (3 m²), or 3 square centimeters (3 cm²)? The appropriate units must be specified.

If units such as feet, square feet, meters, centimeters, etc., are considered as algebraic quantities, ratios provide a means of conversion from one unit to another, as the following examples show.

EXAMPLE
11

Convert 8 yards (yd) to inches (in.).

Since 3 ft = 1 yd, the ratio $\dfrac{3 \text{ ft}}{1 \text{ yd}}$ is equivalent to 1. Thus,

$$8 \text{ yd} = 8 \text{ yd} \cdot (1) = 8 \text{ yd} \cdot \frac{3 \text{ ft}}{1 \text{ yd}} = 24 \text{ ft.}$$

Note that the cancellation of yards was a result of treating the units as algebraic quantities, just as

$$8x \cdot \frac{3y}{1x} = 24y.$$

Now, since 12 in. = 1 ft,

$$8 \text{ yd} = 24 \text{ ft} \cdot (1) = 24 \text{ ft} \cdot \frac{12 \text{ in.}}{1 \text{ ft}} = 288 \text{ in.}$$

It is usual to perform such unit conversions in a single step; thus,

$$8 \text{ yd} = 8 \text{ yd} \cdot \frac{3 \text{ ft}}{1 \text{ yd}} \cdot \frac{12 \text{ in.}}{1 \text{ ft}} = 288 \text{ in.}$$

EXAMPLE
12

Convert 8 yd² to ft².

$$8 \text{ yd}^2 = 8 \text{ yd}^2 \cdot \left(\frac{3 \text{ ft}}{1 \text{ yd}}\right)^2 = 8 \text{ yd}^2 \cdot \left(\frac{3^2 \text{ ft}^2}{1^2 \text{ yd}^2}\right) = 8 \text{ yd}^2 \cdot \left(\frac{9 \text{ ft}^2}{1 \text{ yd}^2}\right) = 72 \text{ ft}^2.$$

Note that here it was necessary to square the ratio 3 ft/1 yd to permit cancellation of the unit yd². Operations such as this are often pitfalls for students, and we caution you to take particular care in performing them. It should be clear that when converting units of volume—cubic feet, cubic meters, cubic inches—it will be necessary to cube the basic ratios used as conversion factors.

In the examples which follow, and in the exercises as well, we shall make use of units from the metric system—the predominant system in science and engineering, and, by the way, the system used routinely in most of the world. The metric system is especially convenient because it is essentially a *power of 10* system. Some of the prefixes associated with the powers of 10 are

$$\text{micro} \longrightarrow 10^{-6} = \frac{1}{1,000,000}$$

$$\text{milli} \longrightarrow 10^{-3} = \frac{1}{1000}$$

$$\text{centi} \longrightarrow 10^{-2} = \frac{1}{100}$$

$$\text{kilo} \longrightarrow 10^3 = 1000$$

$$\text{mega} \longrightarrow 10^6 = 1,000,000.$$

Thus, using the standard abbreviations of m for meter (the basic unit of length), mm for millimeter, cm for centimeter, and km for kilometer, we can write

$$1 \text{ m} = 100 \text{ cm} = 1000 \text{ mm} = \frac{1}{1000} \text{ km},$$

or, equivalently,

$$1 \text{ km} = 1000 \text{ m} = 10^5 \text{ cm} = 10^6 \text{ mm}.$$

We point out that

$$1 \text{ in.} = 2.54 \text{ cm}.$$

Similarly, with the gram, a basic unit of mass, which is abbreviated gm, we have

$$1 \text{ gm} = 1000 \text{ milligrams (mg)} = \frac{1}{1000} \text{ kilograms (kg)}$$

and

$$1 \text{ kg} = 1000 \text{ gm} = 10^6 \text{ mg}.$$

EXAMPLE 13

Convert 1 mile (mi) to mm.

$$1 \text{ mi} = 1 \text{ mi}\left(\frac{5280 \text{ ft}}{1 \text{ mi}}\right)\left(\frac{12 \text{ in.}}{1 \text{ ft}}\right)\left(\frac{2.54 \text{ cm}}{1 \text{ in.}}\right)\left(\frac{10 \text{ mm}}{1 \text{ cm}}\right) = 1,609,344 \text{ mm.}$$

EXAMPLE 14

Convert 30 mi/hr (miles per hour) to ft/sec (feet per second).

$$30 \frac{\text{mi}}{\text{hr}} = 30 \frac{\text{mi}}{\text{hr}}\left(\frac{5280 \text{ ft}}{1 \text{ mi}}\right)\left(\frac{1 \text{ hr}}{60 \text{ min}}\right)\left(\frac{1 \text{ min}}{60 \text{ sec}}\right) = 44 \text{ ft/sec.}$$

EXAMPLE 15

The volume of an oil storage tank is 10^3 m^3. Express the volume in cubic millimeters.

$$10^3 \text{ m}^3 = 10^3 \text{ m}^3\left(\frac{1000 \text{ mm}}{1 \text{ m}}\right)^3 = 10^3 \text{ m}^3\left(\frac{10^9 \text{ mm}^3}{1 \text{ m}^3}\right) = 10^{12} \text{ mm}^3.$$

EXAMPLE 16

The very large distances with which astronomers deal have necessitated the introduction of the *light year* as a unit of distance. One light year is the distance traveled by light in a vacuum in 1 year. Given that the speed of light is 3×10^8 m/sec in vacuum, determine the number of meters in a light year to three significant figures.

$$\text{Distance} = [\text{Rate}][\text{Time}]$$
$$= \left[\frac{3 \times 10^8 \text{ m}}{\text{sec}}\right][1 \text{ yr}]\left(\frac{365 \text{ days}}{1 \text{ yr}}\right)\left(\frac{24 \text{ hr}}{1 \text{ day}}\right)\left(\frac{60 \text{ min}}{1 \text{ hr}}\right)\left(\frac{60 \text{ sec}}{1 \text{ min}}\right)$$
$$= 9.46 \times 10^{15} \text{ m.}$$

EXAMPLE

17

The farad (f) is a unit of capacitance, the property of a device called a capacitor which is frequently encountered in electrical circuits. When capacitors are connected in parallel, the total capacitance of the combination is the sum of the individual capacitances. If a .05 microfarad (μ f) capacitor is connected in parallel with a 1.0 microfarad capacitor, how many farads of capacitance does the combination have?

Let C_t be the total capacitance. Then

$$C_t = C_1 + C_2$$
$$= (.05 \times 10^{-6}) + (1.0 \times 10^{-6})$$
$$= 1.05 \times 10^{-6} \text{ farad.}$$

EXERCISE 4-5

1. Convert 63 in. to meters.

2. Convert 3 yd to centimeters.

3. Convert 2540 mm to yards.

4. Convert 10^8 mm^2 to square meters.

5. Convert 20 ft^2 to square centimeters.

6. Convert 60 mi/hr to feet per second.

7. Convert 60 mi/hr to meters per second.

8. The speedometers in European automobiles are usually calibrated to read in kilometers per hour. If such a speedometer reads 60 km/hr, what would the speedometer read in an American-made automobile traveling at the same speed?

9. The density of iron is 450 lb/ft^3. Express the density in pounds per cubic meter.

10. In Problem 9, express the density of iron in grams per cubic centimeter given that 1 lb = 454 gm.

11. Normal atmospheric pressure, taken to be 14.7 lb/in.^{2}, is also called 1 atmosphere. If a gas is compressed in a tank until the pressure is 18.9 $\times 10^3$ lb/ft^2, what is the pressure in atmospheres?

12. Einstein's famous mass-energy equivalence formula is $E = mc^2$, where m is a mass and c is the speed of light. Given that $c = 3 \times 10^8$ m/sec, and that 1 gm-cm^2/sec^2 = 1 erg, a unit of energy, find the energy in ergs associated with a mass of 1 kg.

13. Given that light travels at a speed of 1.86×10^5 mi/sec in vacuum, determine the number of miles in a light year.

14. If 1 Angstrom (Å) equals 10^{-7} mm and the green line of the mercury

spectrum has a wavelength of .00005461 cm, find this wavelength in Angstrom units.

4-6
REVIEW

REVIEW QUESTIONS

The basic property employed in the simplification of fractions asserts that if $c \neq 0$, then $\dfrac{a}{b} = \dfrac{ac}{bc}$. What name is given to this property? _____(1)_____

(1) fundamental principle of fractions

Insert $+$ or $-$ in the parentheses to make the equalities true in the following.

$$\frac{a-b}{c-d} = (\ \ )\frac{-(a-b)}{c-d} = (\ \ )\frac{b-a}{d-c} = (\ \ )\frac{b-a}{(-)(c-d)}. \qquad \text{(2)}$$

(2)   −, +, +

True or false:

$$\frac{a+bx}{a+cy} = \frac{bx}{cy} \qquad\qquad \underline{\hspace{2cm}} \quad \text{(3)}$$

$$\frac{1}{x} + \frac{1}{y} = \frac{xy}{x+y} \qquad\qquad \underline{\hspace{2cm}} \quad \text{(4)}$$

$$\frac{abx}{acy} = \frac{bx}{cy} \qquad\qquad \underline{\hspace{2cm}} \quad \text{(5)}$$

$$\frac{1}{x} - \frac{x-y}{x^2} = \frac{x-x-y}{x^2} \qquad \underline{\hspace{2cm}} \quad \text{(6)}$$

$$\frac{1}{a} + \frac{1}{b} = \frac{2}{ab} \qquad\qquad \underline{\hspace{2cm}} \quad \text{(7)}$$

$$\frac{\dfrac{1}{x}}{y} = \frac{1}{xy} \qquad\qquad \underline{\hspace{2cm}} \quad \text{(8)}$$

$$\frac{1}{x} - \frac{1}{y} = \frac{x-y}{xy} \qquad\qquad \underline{\hspace{2cm}} \quad \text{(9)}$$

$$\frac{x+y}{y+x} + \frac{x-y}{y-x} = 0 \qquad\qquad \underline{\hspace{2cm}} \quad \text{(10)}$$

$$\frac{1}{x} + \frac{1}{y} = \frac{1}{x+y} \qquad\qquad \underline{\hspace{2cm}} \quad \text{(11)}$$

$$\frac{1}{x} \cdot \frac{1}{y} = \frac{1}{xy} \qquad\qquad \underline{\hspace{2cm}} \quad \text{(12)}$$

(3) **false**　　(4) **false**　　(5) **true**　　(6) **false**　　(7) **false**
(8) **true**　　(9) **false**　　(10) **true**　　(11) **false**　　(12) **true**

The L.C.D. of the fractions

$$\frac{6}{x^2 - 1}, \quad \frac{8}{x + 1}, \quad \text{and} \quad \frac{7}{2(x + 1)}$$

is _____(13)_____ .

(13) **2(x² − 1)**

FOR EVERY EVEN DONE CORECT 3 PTS

ALL ODD

REVIEW PROBLEMS

For Problems 1–14, perform the indicated operations and simplify.

1. $\dfrac{6}{y + 3} - 2 - \dfrac{2y}{3 - y} - \dfrac{12y}{y^2 - 9}$

2. $\dfrac{3x}{2(x - 2)} + \dfrac{3}{2 - x}$

3. $\dfrac{-2abc + x^2 + b^2}{2abc - b^2 - x^2}$

4. $\dfrac{x^2 - 9}{x(x + 1)} \cdot \dfrac{x^2 - 1}{x - 3}$

5. $\dfrac{x + 1}{y} \div \dfrac{x^2 - 1}{4y^2}$

6. $x + 2 - \dfrac{x^2 + x - 6}{x - 3}$

7. $\dfrac{xy}{x^4 - y^4}\left(\dfrac{x}{y} + \dfrac{y}{x}\right)\left(\dfrac{x^2 - y^2}{x}\right)$

8. $\left(3x - 5 - \dfrac{2}{x}\right)\left(3x + \left\{-\dfrac{2}{x}\right\}\right)$

9. $\dfrac{\dfrac{x^2 + 9x + 20}{x^2 - 4}}{\dfrac{x^2 + 10x + 25}{x + 2}}$

10. $\dfrac{\dfrac{x^2 - 12x + 27}{x^2 - 81}}{x^2 + 81}$

11. $\dfrac{x^2 - 3x + 2}{x^2 + 5x - 6} \cdot \dfrac{1}{x - 2}$

12. $\dfrac{2}{x - 1} + \dfrac{3}{x^2 - 1} - \dfrac{2}{x + 1}$

13. $\dfrac{x^2 + 2x - 15}{x^2 + x - 2} \div \dfrac{x^2 + 4x - 5}{x^2 - x - 6}$

14. $\dfrac{\dfrac{1}{x} + \dfrac{1}{y}}{\dfrac{1}{x} - \dfrac{1}{y}} \cdot \dfrac{x - y}{x + y}$

15. Convert 2 ft² to square millimeters to three significant figures.

16. Convert a speed of 1 mi/min to meters per second to three significant figures.

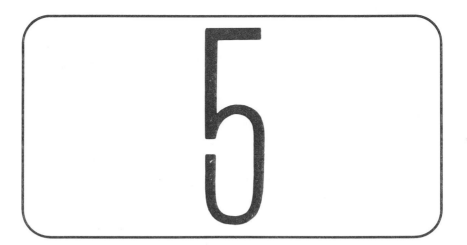

linear equations

5-1
TYPES OF EQUATIONS

A beginning student in the technological disciplines soon faces elementary equations to solve. The purpose of this chapter is to develop techniques to accomplish this task.

Definition. An **equation** is a statement that two expressions are equal.

The two expressions that comprise an equation are called its **members**, or **sides**. They are separated by the equality sign, $=$.

EXAMPLE
1

The following are equations:

a. $x + 2x = 3x$.

b. $x + 3 = 4$.

c. $x^2 + 2 = x(2x - 5)$.

d. $\dfrac{x - a}{b - x} = \dfrac{x - b}{a - x}$.

e. $y^6 + 9y^3 + 8 = 0$.

f. $\dfrac{1}{y-4} = 0.$

g. $w + z = 7.$

h. $2 + 1 = 5.$

In Example 1, each of equations a–g contains at least one variable. A **variable** is a symbol, such as x, which can represent any one of a set of numbers, this set being called the **domain** of the variable. Usually, variables are denoted by letters from the latter part of the alphabet such as x, y, z, w, s, t, etc. Hence, equations a and e are in the variables x and y, respectively; equation g is in the variables w and z.

Unless otherwise specified, we consider the domain of a variable in an equation to consist of all real numbers for which every expression containing the variable is defined. Thus, the domain of x in $x + 3 = 4$ is the set of real numbers, **R**, because the sum of any real number and 3 is always defined; however, in the equation $\dfrac{1}{y-4} = 0$ we assume $y \neq 4$ since the left member is not defined when y is 4. We emphasize that the domain of a variable is not simply the values of the variable which make the equation true.

Constants, such as 7 and π, are symbols which represent only one number in a given discussion. Equations in which some of the constants are represented by letters such as a, b, c, d, etc., are called **literal equations**. For example, in the literal equation $x + a = b$ we consider a and b to be constants. Formulas, such as $C = 2\pi r$ which gives the circumference C of a circle of radius r, express a relationship between certain quantities and are also considered literal equations.

Conditions imposed by an equation can be expressed verbally. For example, the equation $x + 3 = 4$ states that when a number x is added to 3, the result is 4. On the other hand, the equation $7 - (4x)^2 = 3 + 2x$ expresses that when a number x is multiplied by 4, the result squared, and this subtracted from 7, the final result is equal to the sum of 3 and twice the number x.

Consider the equation of Example 1a:

$$x + 2x = 3x.$$

Regardless of what values are assigned to x, the left member $x + 2x$ is always equal to the right member $3x$, and hence, the equation is always a true statement. We refer to this equation as an *identity* in the sense of the following definition:

Definition. An **identity** is an equation which is true for all permissible values of the variables employed.

EXAMPLE
2

The following equations are identities:

a. $2 + 2x^2 = x^2 + 2 + x^2$ (both members are identically equal to $2x^2 + 2$).

b. $7y = y(6 + 1)$ (both members are identically equal to $7y$).

c. $(w + z)(w - z) = w^2 - z^2$ (both members are identically equal to $w^2 - z^2$).

Observe that equation b of Example 1, namely

$$x + 3 = 4,$$

is true if and only if x is assigned the value 1; that is, the equation is true *on the condition* that x is 1. Such an equation is called a *conditional equation*.

Definition. A **conditional equation** is an equation which is true for some, but not all, of the permissible values of the variables employed.

An equation may not be true regardless of the value of the variable involved. For example, in the equation $x + 2 = x$, there is no number x which when added to 2 is equal to the original number x. Such an equation is called an **impossible equation.** Example 1f is another impossible equation since $1/(y - 4)$ can never be zero. Likewise, Example 1h is an impossible equation.

Our objective in dealing with an equation is to find its *solution set*.

Definition. The **solution set** of an equation is the set of all values of the variable employed for which the equation is a true statement.

To *solve an equation* means to find its solution set. The elements in the solution set are called the **roots** of the equation and these roots are said to *satisfy the equation*. Frequently, in an equation a letter representing an unknown quantity whose value is to be found is simply referred to as an *unknown*. Let us now illustrate the terminology we have introduced.

EXAMPLE
3

a. In the equation $x + 3 = 4$, the variable x is the unknown. The only value of x which satisfies the equation is 1. Hence 1 is a root of the equation and the solution set is $\{1\}$.

b. $W_1 + W_2 = 7$ lb is an equation which indicates that the sum of the weights of two bodies, the unknowns W_1 and W_2, is equal to 7 lb. A particular solution to this equation is $W_1 = 1$ lb and $W_2 = 6$ lb. However, there are infinitely many solutions. Can you think of another?

EXAMPLE
4

a. For the equation $x^2 = 4$, the only numbers which when squared are equal to 4 are 2 and -2. Hence, the solution set is $\{2, -2\}$. More simply, we write $+2$ and -2 collectively as ±2 and thus the solution set can be written as $\{\pm2\}$.

b. For the equation $x + 2a = 2a$, clearly the only number which when added to $2a$ gives $2a$ is zero. Thus, the solution set is $\{0\}$.

c. The solution set of $x - 4 = 0$ is $\{4\}$.

d. The solution set of $x + 2b = 0$ is $\{-2b\}$.

In the equation $\dfrac{1}{y - 4} = 0$, the variable y is the unknown. Since there is no number y for which the equation is true, the solution set is a set having no elements in it, $\{\ \}$, called the *empty set* (or *null set*), which we denote by $\varnothing$. Do not confuse the set $\varnothing$ with $\{0\}$, which is not empty; $\{0\}$ has the number 0 in it.

Definition. The **empty set**, written $\varnothing$, is the set having no elements in it, $\{\ \}$.

EXERCISE 5-1

1. In each of the following determine which of the given numbers, if any, when substituted for x satisfies the given equation.

   a. $x^2 - 2x = 0;\ 0, 2$

   b. $20 - 9x = -x^2;\ 5, 4$

   c. $x + 2(x - 3) = 4;\ \frac{10}{3}, 1$

   d. $2x + x^2 - 8 = 0;\ 2, -4$

   e. $x(7 + x) - 2(x + 1) - 3x = -2;\ -3$

   f. $x^4 + 2x^3 - x^2 - 4x - 2 = 0;\ \sqrt{2}, -\sqrt{2}, 3$

   g. $(x + 1)^2(x + 2)(x - 3) = 0;\ -1, -2, 3$

2. Which of the following equations are identities?

   a. $x - 5 = 2(x - 10) - x + 15$

   b. $x - 3(x + 2) + 4x - 7 = -13 + 2x$

   c. $3x(7) - 5 + x = 21x$

d. $2\left(\dfrac{x}{2} + 3\right) - x = 6$

e. $x^2 + 6x + 9 = -(3 - x)(x - 3)$

3. Symbolize the following verbal statements and in a–f also determine whether the result is an identity. Do not attempt to solve your result.

   a. When 10 is subtracted from a number, the result is zero.

   b. When five times a number is decreased by 3, the result is 14.

   c. The product of x and the sum of a and b is equal to the sum of b times x and a times x.

   d. If 3 is decreased by 15 times a number, the result is 3 plus 4 times the number.

   e. Twice the sum of x and 3 is equal to five times the result of subtracting 9 from x.

   f. The sum of a number and 2 is multiplied by the value of the number decreased by 2. The result is equal to -4 plus the square of the number.

   g. The reciprocal of t_1 plus the reciprocal of t_2 is equal to the reciprocal of T.

   h. The sum of three consecutive integers is 36.

   i. The distance s traveled by a particle along a path is equal to the product of its velocity v and the time t it travels.

   j. The simple interest earned in 1 year from an investment of $8000—part of it, x, at 5 percent and the rest at 6 percent—is $420.

   k. The number z exceeds $y - 4z$ by $5(z - \frac{1}{5}y)$.

   l. Fahrenheit temperature F is equal to $\frac{9}{5}$ Celsius temperature C plus 32.

   m. If a mass were converted into energy, the amount E of energy liberated is equal to the product of the mass m and the square of the speed of light c.

   n. An interpretation of Newton's second law leads to the result that W, the weight of an object, is equal to the mass m of the object times the acceleration g due to gravity.

   o. The product of the pressure p of a gas times the volume v of the gas is equal to a constant k.

4. A spring streches .3 in. for each pound of weight that is suspended from it. If the initial length of the spring is 11 in., write an equation that relates the length L (in inches) of the streched spring to the load W suspended from it. From this equation, find the length of the spring when 20 lb are suspended from it.

EQUIVALENT EQUATIONS

In solving an equation, it is desirable that any operation on the equation result in another equation whose solution set contains *all* the roots of the given equation and *only* those roots. Equations having the same solution set are said to be **equivalent**. For example, it is easily seen by inspection that the equations

$$2x = 4 \quad \text{and} \quad 3x = 6$$

have the same solution set, $\{2\}$; thus, these equations are equivalent.

An equation will be transformed into an equivalent equation by any of the following three operations:

1. **Adding (subtracting) the same polynomial to (from) both members of an equation.**

2. **Multiplying (dividing) both members of an equation by the same non-zero constant.**

3. **Replacing either member of an equation by an equal expression.**

We repeat, applying operations 1–3 on an equation will *guarantee* that the result is an equivalent equation. These operations are demonstrated in the following example.

EXAMPLE
5

The following equations are equivalent:

$$6x - 15 = 3x - 6$$

$$6x - 15 + (-3x) = 3x - 6 + (-3x) \qquad \text{(adding } -3x \text{ to both members)}$$

$$3x - 15 = -6 \qquad \text{(simplifying [operation 3])}$$

$$3x - 15 + 15 = -6 + 15 \qquad \text{(adding 15 to both members)}$$

$$3x = 9 \qquad \text{(simplifying)}$$

$$\frac{3x}{3} = \frac{9}{3} \qquad \text{(dividing both members by 3)}$$

$$x = 3. \qquad \text{(simplifying)}$$

Since $\{3\}$ is the solution set of the last equation, and since each equation is equivalent to the equation that precedes it, $\{3\}$ is also the solution set of $6x - 15 = 3x - 6$.

As should be evident from Example 5, solving an equation essentially involves applying operations on the equation until the solution set is evident. Sometimes, however, in solving an equation it becomes necessary to apply operations other than 1–3. Although preserving the equality, these operations may not necessarily result in equivalent equations.

For example, by inspection we see that the equation

$$x - 1 = 0$$

has the solution set {1}. Multiplying both members by x yields

$$x^2 - x = 0,$$

which is satisfied if $x = 0$ or $x = 1$ (you should verify this by substitution). But $x = 0$ *does not* satisfy the *original* equation. Thus, the equations are not equivalent. This situation arose because we multiplied both members of the given equation by x, which is really zero for one of the roots of the resulting equation. The "root" $x = 0$ which was introduced in the transformed equation is sometimes referred to as being an *extraneous root*.

Continuing, you may easily verify that the equation $(x - 4)(x - 3) = 0$ is satisfied when x is 4 or when x is 3. But upon dividing both members by $x - 4$ we obtain $x - 3 = 0$, which does not have 4 as a root. Again we do not have equivalence since, in this case, a root has been "lost." Note that when x is 4, which is a root of the given equation, division by $x - 4$ is actually division by zero.

Finally, if we square each member of the equation $x = 2$, we get $x^2 = 4$, which is true if $x = 2$ or $x = -2$. But $x = -2$ does not satisfy the original equation.

Summarizing, there are operations that when applied to an equation will not change the equality, but will, under certain conditions, lead to nonequivalent equations. They include

4. **Multiplying both members of an equation by an expression involving the variable.**

5. **Dividing both members of an equation by an expression involving the variable.**

6. **Raising both members of an equation to equal powers.**

It is evident that when operations 4–6 are performed, we must be cautious about drawing conclusions concerning the solution set of the given equation. Operations 4 and 6 *can* produce an equation with more roots. Thus, you should determine whether each "solution" obtained by these operations satisfies the *original* equation. Operation 5 *can* produce an equation with fewer roots. In this case, any "lost" roots may never be determined.

EXERCISE 5-2

*For each of the following, determine what operations were applied to the first equation to obtain the second equation. State whether or not the operations **guarantee** that the equations will be equivalent. Do not solve.*

1. $x - 5 = 4x + 10$; $x = 4x + 15$

2. $8x - 4 = 16$; $x - \frac{1}{2} = 2$

3. $x = 4$; $x^2 = 16$

4. $\frac{1}{2}x^2 + 3 = x - 9$; $x^2 + 6 = 2x - 18$

5. $x^2 - 2x = 0$; $x - 2 = 0$

6. $\dfrac{2}{x - 2} + x = x^2$; $2 + x(x - 2) = x^2(x - 2)$

7. $\dfrac{x^2 - 1}{x - 1} = 3$; $x^2 - 1 = 3(x - 1)$

8. $x(x + 5)(x + 9) = x(x + 1)$; $(x + 5)(x + 9) = x + 1$

9. $\dfrac{x(x + 1)}{x - 5} = x + 9$; $x + 1 = \dfrac{(x + 9)(x - 5)}{x}$

10. $2x^2 - 9 = x$; $x^2 - \dfrac{x}{2} = \dfrac{9}{2}$

5-3
LINEAR EQUATIONS

The principles developed in the preceding sections will now be used to solve a *linear equation*.

Definition.   A **linear equation** in the variable x is an equation that can be written in the form

$$ax + b = 0, \tag{1}$$

where a and b are constants and $a \neq 0$.

Equation (1) is also called a *first-degree equation* or an *equation of degree 1* since the highest power of the variable that occurs is the first.

In solving a linear equation, we perform operations so that the equation's solution set is obvious, as the following examples illustrate.

EXAMPLE
6

Solve the following equations.

a. $3x + 5 = 0$.

This is Eq. (1) with $a = 3$ and $b = 5$; hence, it is a linear equation.

$$3x + 5 = 0$$
$$3x + 5 + (-5) = 0 + (-5) \quad \text{(adding } -5 \text{ to both members)}$$
$$3x = -5 \quad \text{(simplifying)}$$
$$\frac{3x}{3} = \frac{-5}{3} \quad \text{(dividing both members by 3)}$$
$$x = -\frac{5}{3}.$$

Since $\{-\frac{5}{3}\}$ is the solution set of the last equation and since each equation is equivalent to the one preceding it, we conclude that $\{-\frac{5}{3}\}$ is the solution set of $3x + 5 = 0$. We can check our answer by substitution. If $x = -\frac{5}{3}$, the left member of the original equation is $3(-\frac{5}{3}) + 5 = -5 + 5 = 0$, which agrees with the right member.

b. $5x - 2 = 0$.

This is Eq. (1) with $a = 5$ and $b = -2$.

$$5x - 2 = 0$$
$$5x = 2 \quad \text{(adding 2 to both members)}$$
$$x = \frac{2}{5} \quad \text{(dividing both members by 5)}.$$

The solution set is $\{\frac{2}{5}\}$. *Check:* $5(\frac{2}{5}) - 2 = 2 - 2 = 0$.

c. $-10x - 9 = 0$.

This is Eq. (1) with $a = -10$ and $b = -9$.

$$-10x - 9 = 0$$
$$-10x = 9 \quad \text{(adding 9 to both members)}$$
$$x = \frac{9}{-10} = -\frac{9}{10} \quad \text{(dividing both members by } -10\text{)}.$$

The solution set is $\{-\frac{9}{10}\}$, which you may check by substitution.

d. $6x - 4 = 3x$.

$$6x - 4 = 3x$$
$$3x - 4 = 0 \quad \text{(subtracting } 3x \text{ from both members)}$$
$$3x = 4 \quad \text{(adding 4 to both members)}$$
$$x = \frac{4}{3} \quad \text{(dividing both members by 3)}.$$

Let us check our result. When $x = \frac{4}{3}$, the left member is $6(\frac{4}{3}) - 4 = 8 - 4 = 4$ and the right member is $3(\frac{4}{3}) = 4$. The solution set is $\{\frac{4}{3}\}$. Also, the original equation can be put in the form $3x + (-4) = 0$ and thus it is a linear equation.

EXAMPLE
7

Solve the following equations:

a. $2(x + 4) = 7x + 2$.

$2(x + 4) = 7x + 2$

$2x + 8 = 7x + 2$ (distributive axiom)

$2x = 7x - 6$ (subtracting 8 from both members)

$-5x = -6$ (subtracting $7x$ from both members)

$x = \dfrac{-6}{-5}$ (dividing both members by -5)

$x = \dfrac{6}{5}$ $\left(\text{solution set } \left\{\dfrac{6}{5}\right\}\right)$.

Check: $2\left(\dfrac{6}{5} + 4\right) \overset{?}{=} 7\left(\dfrac{6}{5}\right) + 2$

$2\left(\dfrac{26}{5}\right) \overset{?}{=} \dfrac{42}{5} + 2$

$\dfrac{52}{5} = \dfrac{52}{5}$.

b. $\dfrac{8y}{3} = \dfrac{7y + 5}{6} + 8$.

We first clear the equation of fractions by multiplying *both* members by the constant 6, which is the least common demoninator (L.C.D.) of the fractions involved.

$6\left(\dfrac{8y}{3}\right) = 6\left(\dfrac{7y + 5}{6} + 8\right)$

$2(8y) = (7y + 5) + 6(8)$

$16y = 7y + 5 + 48$

$9y = 53$

$y = \dfrac{53}{9}$ $\left(\text{solution set } \left\{\dfrac{53}{9}\right\}\right)$.

c. $x(7x + 5) - x^2 = 4x(x + 4) + 2x^2 + 1$.

$x(7x + 5) - x^2 = 4x(x + 4) + 2x^2 + 1$

$7x^2 + 5x - x^2 = 4x^2 + 16x + 2x^2 + 1$

$6x^2 + 5x = 6x^2 + 16x + 1$

$-1 = 11x$

$-\dfrac{1}{11} = x$ $\left(\text{solution set } \left\{-\dfrac{1}{11}\right\}\right)$.

Each equation in Example 7 has one and only one root. **It is typical that every linear equation in one variable has one and only one root.** Now let us look at the solution of some literal equations.

EXAMPLE
8

If $(a + c)x + x^2 = (x + a)^2$, express x in terms of a and c.

$$(a + c)x + x^2 = (x + a)^2$$

$$ax + cx + x^2 = x^2 + 2ax + a^2$$

$$cx - ax = a^2$$

$$x(c - a) = a^2.$$

If $c - a \neq 0$, then

$$x = \frac{a^2}{c - a}.$$

Note that a and c are literal constants.

If we want to express a particular letter in a formula in terms of the remaining letters, this particular letter is considered to be the unknown.

EXAMPLE
9

a. The equation $p_1 v_1 = p_2 v_2$ is a statement of Boyle's law of gases, where p_1 and p_2 are variables denoting pressures and the variables v_1 and v_2 denote volumes. Express v_1 in terms of $p_1, p_2,$ and v_2.

Here, v_1 is considered to be the unknown.

$$p_1 v_1 = p_2 v_2$$

$$v_1 = \frac{p_2 v_2}{p_1} \qquad \text{(dividing both members by } p_1\text{)}.$$

b. The relationship between Fahrenheit and Celsius temperature readings can be expressed by the equation

$$\frac{F - 32}{180} = \frac{C}{100},$$

where F and C represent the corresponding temperatures. Solve for F.

$$\frac{F - 32}{180} = \frac{C}{100}.$$

Multiplying both members by 180,

$$180\left(\frac{F - 32}{180}\right) = 180\left(\frac{C}{100}\right)$$

$$F - 32 = \frac{9}{5}C$$

$$F = \frac{9}{5}C + 32.$$

EXERCISE 5-3

In Problems 1–30, find the solution set of the given equation.

1. $8x = 36$

2. $.2x = 7$

3. $8y = 0$

4. $2x - 3 = 4$

5. $3 - 5x = 9$

6. $2(x + 3) = 5$

7. $6y + 5y - 3 = 41$

8. $7x + 4 = 2(x - 1)$

9. $2(x - 1) - 3(x - 4) = 4x$

10. $x = 2 - 2[2x - 3(1 - x)]$

11. $\dfrac{x}{5} = 2x - 6$

12. $\dfrac{5y}{7} - \dfrac{6}{7} = 2 - 4y$

13. $5 + \dfrac{4x}{9} = \dfrac{x}{2}$

14. $\dfrac{x}{3} - 4 = \dfrac{x}{5}$

15. $y = \dfrac{3}{2}y - 4$

16. $\dfrac{x}{2} + \dfrac{x}{3} = 7$

17. $3x + \dfrac{x}{5} - 5 = \dfrac{1}{5} + 5x$

18. $y - \dfrac{y}{2} + \dfrac{y}{3} - \dfrac{y}{4} = \dfrac{y}{5}$

19. $w + \dfrac{w}{2} - \dfrac{w}{3} + \dfrac{w}{4} = 5$

20. $\dfrac{z}{3} + \dfrac{3}{4}z = \dfrac{9}{2}(z - 1)$

21. $\dfrac{x + 2}{3} - \dfrac{2 - x}{6} = x - 2$

22. $\dfrac{x}{5} + \dfrac{2(x - 4)}{10} = 7$

23. $\dfrac{3}{4}(z - 3) = \dfrac{9}{5}(3 - z)$

24. $\dfrac{2y - 7}{3} + \dfrac{8y - 9}{14} = \dfrac{3y - 5}{21}$

25. $2[x + 3(x - 4) - 2] + 4(x + 1) = 5$

26. $2y + \{-y - 6[2 - 3(y - 4)]\} = -8$

27. $w + 3(w + 2) - 4(w - 1) = w + 2(w - 5)$

28. $\dfrac{3}{2}(4x - 3) = 2[x - (4x - 3)]$

29. $(3x - 1)^2 - (5x - 3)^2 = -(4x - 2)^2$

30. $\dfrac{2(1 - 2x)}{3} - 2x = -\dfrac{2}{5} + \dfrac{4(2 - 3x)}{3}$

The relationships in Problems 31–58 occur in physics, chemistry, and various branches of engineering technology. Express the indicated symbol(s) in terms of the remaining symbols.

31. $P_1 V_1 = P_2 V_2$; V_2

32. $PV = nRT$; R

33. $v = v_0 - at$; a

34. $E = mc^2$; m

35. $K = \frac{1}{2}mv^2$; m

36. $mgh + \frac{1}{2}mv^2 = c$; m

37. $V = \pi r^2 h$; h

38. $S = v_0 t + \frac{1}{2}at^2$; v_0

39. $P = i^2 R$; R

40. $I = \dfrac{n\mathcal{E}}{r + nR}$; $\mathcal{E}$

41. $2R_1 i + 3R_2 i - 4R_4 i = -(E_1 - E_2)$; i

42. $T^2 = 4\pi^2\left(\dfrac{L}{g}\right)$; L

43. $F = k\dfrac{QQ'}{r^2}$; Q'

44. $P = \dfrac{E^2}{R + r} - \dfrac{E^2 r}{(R + r)^2}$; E^2

45. $V = V_0\left(\dfrac{P_1}{P_2}\right)\left(\dfrac{T_2}{T_1}\right)$; T_2

46. $F = \dfrac{1}{2\pi}\left(\dfrac{e}{m}\right)b$; b

47. $I = \dfrac{E}{R}(1 - e^{-Rt/L})$; E

48. $F = \tfrac{9}{5}C + 32$; C

49. $Q = mc(t_2 - t_1) + mL$; t_2, t_1, c

50. $Q = kA\left(\dfrac{t_2 - t_1}{d}\right)T$; t_1, t_2

51. $mgh = \tfrac{1}{2}mv^2 + \tfrac{1}{2}I\omega^2$; m, I

52. $y = mx + b$; x, m

53. $V = 2\pi r^2 + 2\pi rh$; h

54. $y = \dfrac{(B + D)\lambda}{2B(N - 1)\alpha}$; B, D, λ

55. $U = \left(\dfrac{f - f_0}{f_0}\right)\dfrac{\epsilon}{v_0}$; f

56. $n - 1 = C + \dfrac{C'}{\lambda^2}$; C, C'

57. $\sigma = \dfrac{n_0 - n_e}{\lambda}L$; n_0, n_e

58. $P = \dfrac{m(\rho_s - \rho_p)}{1 + (m - 1)(\rho_s + \rho_p) - (2m - 1)(\rho_s\rho_p)}$; m

59. In a certain calorimetry (that is, measurement of heat) experiment, 2 kg of water at 79°C is mixed with 4 kg of water at 40°C. The final temperature of the mixture is given by t, where

$$\text{Heat loss} = \text{Heat gain}$$
$$2000(1)(79 - t) = 4000(1)(t - 40).$$

Find t.

60. If a ball of mass m is tied to a string and whirled around in a vertical circle, at its lowest point Newton's second law indicates that

$$T - mg = m\dfrac{v^2}{r},$$

where T is the tension in the string, g is acceleration due to gravity, v is the speed of the ball, and r is the radius of the circle. Solve the equation for m.

5-4
FRACTIONAL EQUATIONS

Some equations which are not linear may lead to linear equations. Of this type are *fractional equations*, that is, equations involving fractions in which the variable occurs in one or more of the denominators involved. A method of solving fractional equations is explained in the following examples. *In each case we must exclude from our considerations those values of x for which any denominators are zero.*

EXAMPLE

10

Solve the following equations.

a. $\dfrac{6}{x-3} = \dfrac{5}{x-4}$.

Here we assume that $x \neq 3$ and $x \neq 4$ so that both members are defined. Multiplying both members by the L.C.D., which is $(x-3)(x-4)$, we have

$$(x-3)(x-4)\left(\frac{6}{x-3}\right) = (x-3)(x-4)\left(\frac{5}{x-4}\right),$$

$$6(x-4) = 5(x-3)$$

$$6x - 24 = 5x - 15$$

$$x = 9.$$

Since we multiplied each member of the original equation by an expression involving the variable x, we should check that when x is 9, the *original* equation is satisfied. If 9 is substituted for x in that equation, the left member is

$$\frac{6}{9-3} = \frac{6}{6} = 1$$

and the right member is

$$\frac{5}{9-4} = \frac{5}{5} = 1.$$

Thus, the solution set is $\{9\}$.

b. $\dfrac{4}{x-1} = \dfrac{7}{x-2} - \dfrac{3}{x+1}$.

Assuming that $x \neq 1, 2$, and -1, we multiply both members by the L.C.D., which is $(x-1)(x-2)(x+1)$:

$$4(x-2)(x+1) = 7(x-1)(x+1) - 3(x-1)(x-2),$$

$$4(x^2 - x - 2) = 7(x^2 - 1) - 3(x^2 - 3x + 2)$$

$$4x^2 - 4x - 8 = 7x^2 - 7 - 3x^2 + 9x - 6$$

$$-13x = -5$$

$$x = \frac{5}{13}.$$

You may verify by substitution that $\{\frac{5}{13}\}$ is, indeed, the solution set.

EXAMPLE
11

Solve the following equations.

a. $\dfrac{9}{x-3} = \dfrac{3x}{x-3}$.

Assuming that $x \neq 3$, we multiply both members by the L.C.D., which is $x - 3$:

$$9 = 3x$$
$$3 = x.$$

But the original equation is not defined when x is 3 (division by zero) and so there are no roots. The solution set is a set having no elements in it. Recall that we denote this set by $\varnothing$, the empty set.

b. $\dfrac{3x+4}{x+2} - \dfrac{3x-5}{x-4} = \dfrac{12}{x^2-2x-8}$.

Noting that $x^2 - 2x - 8 = (x+2)(x-4)$, we conclude that the L.C.D. of the fractions appearing in the equation is $(x+2)(x-4)$. Multiplying both members by the L.C.D. we have

$$(3x+4)(x-4) - (3x-5)(x+2) = 12, \qquad x \neq -2, 4$$
$$(3x^2 - 8x - 16) - (3x^2 + x - 10) = 12$$
$$3x^2 - 8x - 16 - 3x^2 - x + 10 = 12$$
$$-9x - 6 = 12$$
$$-9x = 18$$
$$x = -2.$$

But the original equation is not defined for $x = -2$ (division by zero) and so the solution set is $\varnothing$.

EXAMPLE
12

a. Solve $\dfrac{1}{a} + \dfrac{1}{x} = \dfrac{1}{b}$ for x.

Multiplying both members by abx, which is the L.C.D.,

$$bx + ab = ax$$
$$bx - ax = -ab$$
$$x(b - a) = -ab$$
$$x = \frac{-ab}{b-a} = \frac{ab}{a-b} \qquad \text{(assuming } b - a \neq 0\text{)}.$$

b. If n cells, each having an internal resistance r and electromotive force $\mathcal{E}$, are connected in series to a load resistance R, the current i in the circuit is

$$i = \frac{n\mathcal{E}}{R + nr}.$$

Solve for r.

$$i = \frac{n\mathcal{E}}{R + nr}$$

$$i(R + nr) = n\mathcal{E}$$

$$iR + nir = n\mathcal{E}$$

$$nir = n\mathcal{E} - iR$$

$$r = \frac{n\mathcal{E} - iR}{ni}.$$

c. An important equation for lenses is

$$\frac{1}{f} = \frac{1}{p} + \frac{1}{q},$$

where f is focal length, p is object distance, and q is image distance. Suppose that for a converging lens the focal length is 12.0 cm and the object distance is 24.0 cm. Find the image distance.

$$\frac{1}{f} = \frac{1}{p} + \frac{1}{q}$$

$$\frac{1}{12.0} = \frac{1}{24.0} + \frac{1}{q}$$

$$\frac{1}{12.0} - \frac{1}{24.0} = \frac{1}{q}$$

$$\frac{2}{24.0} - \frac{1}{24.0} = \frac{1}{q}$$

$$\frac{1}{24.0} = \frac{1}{q}$$

$$24.0 \text{ cm} = q.$$

EXERCISE 5-4

Find the solution sets in Problems 1–18.

1. $\dfrac{3}{x} = 12$

2. $\dfrac{4}{x - 1} = 2$

3. $\dfrac{x}{3x - 4} = 3$

4. $\dfrac{4x}{7 - x} = 1$

5. $\dfrac{1}{x - 1} = \dfrac{2}{x - 2}$

6. $\dfrac{2x - 3}{4x - 5} = 6$

7. $\dfrac{1}{x} + \dfrac{1}{5} = \dfrac{4}{5}$

8. $\dfrac{4}{x - 3} = \dfrac{3}{x - 4}$

9. $\dfrac{3x - 2}{2x + 3} = \dfrac{3x - 1}{2x + 1}$

10. $\dfrac{9}{x - 3} = \dfrac{3x}{x - 3}$

11. $\dfrac{y-6}{y} - \dfrac{6}{y} = \dfrac{y+6}{y-6}$

12. $\dfrac{y-3}{y+3} = \dfrac{y-3}{y+2}$

13. $\dfrac{-4}{x-1} = \dfrac{7}{2-x} + \dfrac{3}{x+1}$

14. $\dfrac{1}{x-3} - \dfrac{3}{x-2} = \dfrac{4}{1-2x}$

15. $\dfrac{x+2}{x-1} + \dfrac{x+1}{2-x} = 0$

16. $\dfrac{x}{x+3} - \dfrac{x}{x-3} = \dfrac{3x-4}{x^2-9}$

17. $\dfrac{2x}{x-1} - \dfrac{3}{x+2} = \dfrac{4x}{(x+2)(x-1)} + 2$

18. $\dfrac{2}{x} + \dfrac{3}{x+1} = \dfrac{x}{x+1} - \dfrac{x+1}{x}$

In Problems 19–32, express the indicated letter(s) in terms of the remaining letters.

19. $2 + \dfrac{b}{abx} \neq \dfrac{4}{x} - \dfrac{a}{b}; x$

20. $\dfrac{x-a}{b-x} = \dfrac{x-b}{a-x}; x$

21. $\dfrac{p_1 v_1}{t_1} = \dfrac{p_2 v_2}{t_2}; t_1$

22. $h = kat\left(\dfrac{T}{L}\right); L$

23. $V = V_0\left(\dfrac{P_1}{P_2}\right)\left(\dfrac{T_2}{T_1}\right); P_2$

24. $F = \dfrac{1}{2\pi}\left(\dfrac{e}{m}\right)b; m$

25. $\dfrac{1}{p} + \dfrac{1}{q} = \dfrac{1}{f}; q, f$

26. $\dfrac{1}{C_t} = \dfrac{1}{C_1} + \dfrac{1}{C_2} + \dfrac{1}{C_3}; C_t$

27. $R_t = \dfrac{R_1 R_2}{R_1 + R_2}; R_1$

28. $\dfrac{x}{a} + \dfrac{y}{b} = 1; a, b$

29. $\dfrac{1}{2} mv^2 - \dfrac{p^2}{2m} = 0; m^2$

30. $S = \dfrac{\frac{W_s}{V}}{\frac{W_w}{V}}; W_w, W_s$

31. $\dfrac{1}{f} = (n-1)\left(\dfrac{1}{R_1} - \dfrac{1}{R_2}\right); R_1, R_2$

32. $V = \dfrac{1}{4\pi\epsilon_0}\left(\dfrac{q_1}{r_1} + \dfrac{q_2}{r_2}\right); r_1, r_2$

33. The combined resistance R of two resistors R_1 and R_2 connected in parallel is given by

$$R = \dfrac{1}{\dfrac{1}{R_1} + \dfrac{1}{R_2}}.$$

If the combined resistance is 10 ohms and one resistor has a resistance of 60 ohms, find the resistance of the other resistor.

34. The total capacitance C_T of a circuit network containing two capacitors C_1 and C_2 in series is given by

$$\dfrac{1}{C_T} = \dfrac{1}{C_1} + \dfrac{1}{C_2}.$$

If in a particular circuit $C_1 = 2$ (microfarads) and $C_T = \frac{2}{3}$ (microfarads), find C_2.

35. The reciprocal of the focal length f of a spherical lens is

$$\frac{1}{f} = (n - 1)\left(\frac{1}{R_1} - \frac{1}{R_2}\right)$$

(called the lens-makers equation), where n is the index of refraction of the lens material and R_1 and R_2 are radii of curvature of the surfaces. For a certain double-convex lens, $n = 1.5$, $R_1 = 10$ cm, and $R_2 = -20$ cm. Find f.

36. If a space capsule of mass m were launched from the surface of the earth with speed v, it would rise, neglecting the effects of air resistance, to a height h such that

$$\frac{1}{2}mv^2 = \frac{mghR}{R + h},$$

where R is the radius of the earth and g is the acceleration due to gravity at the earth's surface. Solve for h.

5-5
PROPORTION

In physics and chemistry, a type of equation that occurs frequently is that of a *proportion*.

Definition. A **proportion** is an equation asserting the equality of two ratios. That is, it is an equation of the form

$$\frac{a}{b} = \frac{c}{d} \qquad \text{or} \qquad a : b = c : d.$$

The numbers a and d in the above definition are called the **extremes** of the proportion, while b and c are called the **means.** Either of the proportions above can be read "a is to b as c is to d." To say that the three numbers a, b, and c are in the ratio $2 : 3 : 5$ means that $\dfrac{a}{b} = \dfrac{2}{3}$, $\dfrac{b}{c} = \dfrac{3}{5}$, and $\dfrac{a}{c} = \dfrac{2}{5}$. Thus 8, 12, and 20 are in the ratio $2 : 3 : 5$.

EXAMPLE
13

Solve each of the following proportions for x.

a. $\dfrac{7}{5} = \dfrac{2}{x}$.

Multiplying both members by $5x$, the L.C.D.,

$$7x = 10 \tag{1}$$

$$x = \frac{10}{7}.$$

The solution set is $\{\frac{10}{7}\}$, as can be verified. Note that step (1) indicates that the product of the means is equal to the product of the extremes.

b. $(3-x):4 = (x+2):5$.

$$\frac{3-x}{4} = \frac{x+2}{5}$$

$$5(3-x) = 4(x+2)$$

$$15 - 5x = 4x + 8$$

$$7 = 9x$$

$$\frac{7}{9} = x \qquad \left(\text{solution set } \left\{\frac{7}{9}\right\}\right).$$

EXAMPLE
14

a. A triangle whose shortest side has a length of 14 units (see Fig. 5-1) is similar to a triangle having sides of lengths $10, 15,$ and 20 units. Find the lengths of the other two sides of the triangle.

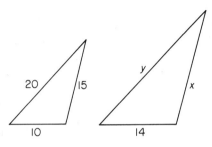

FIGURE 5-1

From geometry, recall that if two triangles are similar, the lengths of their corresponding sides are proportional. Let x be the side corresponding to the side 15. Then

$$x:15 = 14:10$$

$$\frac{x}{15} = \frac{14}{10}$$

$$x = 15\left(\frac{14}{10}\right)$$

$$= 21.$$

Let y be the side corresponding to the side 20. Then

$$\frac{y}{20} = \frac{14}{10}$$

$$y = 20\left(\frac{14}{10}\right)$$

$$= 28.$$

Hence, the other two sides of the triangle have lengths of 21 and 28 units.

b. In a scale drawing, the scale is $\frac{1}{4}$ in. = 2 ft. Find the scale length of an object if its true length is 56 ft.

Let x denote the scale length of the object in inches. Then

$$\frac{x}{\frac{1}{4}} = \frac{56}{2}$$

$$x = \frac{1}{4}\left(\frac{56}{2}\right)$$

$$x = 7 \text{ in.}$$

c. Find the ratio of a to b given the proportion

$$\frac{a+b}{a-b} = \frac{5}{2}.$$

Clearing of fractions,

$$2(a+b) = 5(a-b)$$
$$2a + 2b = 5a - 5b$$
$$7b = 3a$$
$$7 = \frac{3a}{b}$$
$$\frac{7}{3} = \frac{a}{b}.$$

Thus, the ratio of a to b is $\frac{7}{3}$. This does *not* mean that a must be 7 and b must be 3. Why?

EXERCISE 5-5

In Problems 1–10, solve the given proportion.

1. $\dfrac{x+1}{5} = \dfrac{3}{7}$

2. $\dfrac{x-3}{6} = \dfrac{x}{5}$

3. $\dfrac{4}{8-x} = \dfrac{3}{4.}$

4. $\dfrac{2y-3}{4} = \dfrac{6y+7}{3}$

5. $x : \frac{3}{2} = 2\frac{1}{3} : 3\frac{1}{6}$

6. $3\frac{1}{2} : 2\frac{1}{2} = 6 : x$

7. $\dfrac{2(x+3)^2}{4} = \dfrac{x^2+1}{2}$

8. $\dfrac{7 + 2(x+1)}{3} = \dfrac{8x}{5}$

9. $\dfrac{\frac{1}{\frac{3}{2}}}{2} = \dfrac{\frac{1}{4}}{x}$

10. $\dfrac{\frac{1}{2}}{x} = \dfrac{8}{\frac{2}{3}}$

In Problems 11–14, find the ratio of a to b.

11. $\dfrac{a-b}{a+b} = \dfrac{4}{13}$

12. $\dfrac{a+2b}{b-a} = \dfrac{7}{2}$

13. $\dfrac{2(a+2b)}{13b-a} = \dfrac{1}{2}$

14. $\dfrac{2(2b-a)}{3(a-b)} = \dfrac{1}{3}$

15. A triangle whose longest side has a length of 14 units is similar to a triangle having sides of lengths 3, 5, and 6 units. Find the lengths of the other two sides of the triangle.

16. The length and width of a sheet metal plate must be in the ratio of 7 : 5, respectively. If the length of plate is to be 56 cm, what must its width be?

17. A Wheatstone bridge is a device that can be used to measure an unknown resistance. In a balanced condition it is governed by the proportion

$$\frac{R_1}{R_2} = \frac{R_3}{R_4},$$

where the R's are the resistances in ohms in each arm of the bridge. If $R_1 = 3$ ohms, $R_2 = 5$ ohms, $R_3 = 7$ ohms, and $R_4 = 1 + x$ ohms, find x.

18. If a block is on a frictionless inclined plane as shown in the diagram, the following proportion is applicable:

$$\frac{F}{W} = \frac{h}{l},$$

where F is the force to just start the block moving up the inclined plane,

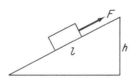

W is the weight of the block, and h and l are the height and length, respectively, of the inclined plane. Find l if $W = 50$ lb, $F = 5$ lb, and $h = 2$ ft.

19. Graham's law of diffusion of two gases is given by the proportion

$$\frac{\text{Rate of diffusion of gas } A}{\text{Rate of diffusion of gas } B} = \frac{\sqrt{\text{Density of } B}}{\sqrt{\text{Density of } A}}.$$

Suppose gas A is nine times as dense as B (that is, take the density of B to be x and that of A to be $9x$). In a certain diffusion apparatus, and under constant temperature and pressure, gas A diffuses 8 cm/sec. Under the same conditions, what is the rate of diffusion of B?

20. For the proportion in Problem 19, suppose gas A is 25 times as dense as gas B. How do their rates of diffusion compare?

21. The relationship between the temperature F on the Fahrenheit scale and the corresponding temperature C on the Celsius scale is given by the proportion

$$\frac{C}{100} = \frac{F - 32}{180}.$$

Convert 77°F into the equivalent Celsius temperature. Do the same for $-40°F$.

22. Under conditions of constant pressure, Charles's law of gases is given by the proportion

$$\frac{V_1}{V_2} = \frac{T_1}{T_2},$$

where V_1 and V_2 are the volumes of the gas at temperatures T_1 and T_2, respectively. Suppose that at a particular pressure gas A has a volume of 200 cc at a temperature of 273°K (the Kelvin scale). What volume will it occupy at 373°K? Give your answer to the nearest cubic centimeter.

23. Under conditions of constant temperature, Boyle's law for gases is given by the proportion

$$\frac{P_1}{V_2} = \frac{P_2}{V_1},$$

where P_1 and P_2 are the pressures of the gas at volumes V_1 and V_2, respectively. If 10 liters of hydrogen gas at a pressure of 1 atmosphere is allowed to expand at constant temperature to a new volume of 18 liters, find the new pressure of the gas.

24. A step–up transformer having 100 turns on the input winding is connected to a 10-volt generator. The desired output voltage is to be 40 volts. To find the number of turns, N_2, that the output winding must have, it is necessary to solve the proportion

$$\frac{V_1}{V_2} = \frac{N_1}{N_2},$$

where $V_1 = 10$, $V_2 = 40$, and $N_1 = 100$. Find N_2.

25. At a temperature of 0°C, the frequency f_s emitted by a source and the frequency f_0 heard by a stationary observer are related by the proportion

$$\frac{f_0}{f_s} = \frac{331}{331 \pm v_s}.$$

In this special case of the Doppler effect, v_s is the speed of the source in meters per second. The plus sign is used if the source recedes from the observer, and the minus sign if it approaches the observer. What frequency would you hear in a laboratory if a tuning fork with a frequency of 256 cycles/sec were moved *toward* you at a speed of 4 m/sec? Give your answer to the nearest cycle per second.

26. In Problem 25, what frequency would you hear if the same tuning fork were moved *away* from you at 4 m/sec?

27. Referring to Problem 25, if the *source* is stationary and the observer is moving, then, under the same conditions, the Doppler effect gives

$$\frac{f_0}{f_s} = \frac{331 \pm v_0}{331}.$$

Here the plus sign is used if the observer moves toward the source. If you were approaching a stationary fire siren emitting a frequency of 512 cycles/sec at a speed of 20 m/sec, what frequency would you hear?

28. In Problem 27, if you were moving *away* from the given source at the given speed, what frequency would you hear?

29. In analyzing a negative feedback amplifier the following equation was obtained:

$$V_0 = \alpha(V_s - \beta V_0),$$

where V_0 and V_s are output and input voltages, respectively. Find the ratio of V_0 to V_s.

5-6
VERBAL PROBLEMS

The solution of a verbal problem requires the ability to express, in mathematical symbols, the relationships stated in the problem. No prescribed procedure exists for the solution of such problems and proficiency can come only with practice. The following examples illustrate basic techniques and concepts. You should study them carefully before proceeding to the exercises.

EXAMPLE
15

a. How many gallons of antifreeze that is 70-percent alcohol must be added to 10 gallons of a 35-percent solution to yield a 50-percent solution?

Let x denote the required amount of the 70-percent solution in gallons. The total amount of solution we start with is 10 gallons; afterwards there are $10 + x$ gallons. The total *amount of alcohol* we start with is $(35\%)(10) = .35(10) = 3.5$ gallons; afterwards there are $3.5 + (70\%)(x) = 3.5 + .7x$ gallons. Since at the end the total amount of alcohol must equal 50 percent of the total amount of solution, we have

$$\text{Total alcohol} = 50\% \text{ of total solution}$$
$$3.5 + .7x = .5(10 + x)$$
$$3.5 + .7x = 5 + .5x$$
$$.2x = 1.5$$
$$x = \frac{15}{2} \text{ gallons.}$$

b. A chemical manufacturer mixes a 30-percent acid solution (30 percent by volume is acid) with an 18-percent acid solution. How much of each solution should be used to obtain 500 gallons of a 25-percent acid solution?

Let x and $500 - x$ denote the number of gallons of the 30-percent solution and the 18-percent solution used, respectively. The total number of gallons of acid in the 500 gallons of the 25-percent acid solution is $.25(500) = 125$. This acid comes from two sources; $.30x$ gallons come from the 30-percent solution and $.18(500 - x)$ gallons come from the 18-percent solution. Hence,

$$.30x + .18(500 - x) = 125$$
$$.30x + 90 - .18x = 125$$
$$.12x = 35$$
$$x = \frac{35}{.12} = \frac{3500}{12} = \frac{875}{3}$$
$$= 291\frac{2}{3}.$$

Thus $500 - x = 208\frac{1}{3}$. The manufacturer should mix $291\frac{2}{3}$ gallons of the 30-percent solution and $208\frac{1}{3}$ gallons of the 18-percent solution.

EXAMPLE
16

If a man can row 7 mi/hr in still water and the rate of a stream is 2 mi/hr, how far upstream can he row if he is to be back at his starting point in 2 hr?

We use the basic relation that distance = (rate)(time). Upstream, the rate of the boat is $7 - 2 = 5$ mi/hr; downstream it is $7 + 2 = 9$ mi/hr. Let t denote the time, in hours, the man can row upstream; then $2 - t$ is the time he rows downstream. Thus,

$$\text{Distance upstream} = (\text{Rate})(\text{Time})$$
$$= 5t,$$
$$\text{Distance downstream} = 9(2 - t).$$

Since distance upstream = distance downstream,

$$5t = 9(2 - t)$$
$$5t = 18 - 9t$$
$$14t = 18$$
$$t = \frac{18}{14} = \frac{9}{7} \text{ hr.}$$

Thus the distance upstream is $5(\frac{9}{7}) = \frac{45}{7} = 6\frac{3}{7}$ mi.

EXAMPLE
17

Entering into a certain storage tank are three pipes: A, B, and C. Pipe A can fill the tank in 2 hr, pipe B in 3 hr, and pipe C in 4 hr. How long will it take to fill the tank if all three pipes are used together?

In 1 hr, pipe A fills $\frac{1}{2}$ of the tank, pipe B fills $\frac{1}{3}$ of the tank, and pipe C fills $\frac{1}{4}$ of the tank. Thus, if all three pipes are used, in 1 hr a total of $\frac{1}{2} + \frac{1}{3} + \frac{1}{4}$ of the tank is filled. Let x denote the total number of hours it will take if the three pipes are used together. Then in 1 hr, $1/x$ of the tank is filled. Thus,

$$\frac{1}{2} + \frac{1}{3} + \frac{1}{4} = \frac{1}{x}$$

$$\frac{13}{12} = \frac{1}{x}.$$

Hence $x = \frac{12}{13}$. The time required is $\frac{12}{13}$ hr.

EXAMPLE
18

In order to produce 84 gm of a certain compound, chemicals A and B must be combined in the ratio of 2 : 5, respectively (by weight). Find the amount of each chemical that must be used.

Let $2x$ and $5x$ denote the amounts of chemical A and chemical B, respectively, that must be combined. Then the ratio of these amounts is $\frac{2x}{5x} = \frac{2}{5}$ as required. Thus,

$$2x + 5x = 84$$

$$7x = 84$$

$$x = 12.$$

Thus $2x$, or 24 gm, of A are required, and $5x$, or 60 gm, of B are required.

EXAMPLE
19

If two downward forces F_1 and F_2 are made to act on a beam (Fig. 5-2), the beam will balance on a knife edge placed directly below the center of gravity of the beam when $F_1 d_1 = F_2 d_2$. Here d_1 and d_2 are the distances of

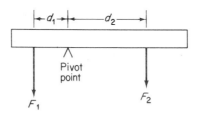

FIGURE 5-2

F_1 and F_2, respectively, from the pivot point created by the knife edge. The distances d_1 and d_2 are called the *moment* or *lever arms* and the product of a force and associated moment arm is called a *torque*.

a. If $F_1 = 20$ lb, $F_2 = 13$ lb, and $d_1 = .5$ ft, find d_2.

$$F_1 d_1 = F_2 d_2$$

$$(20)(.5) = (13)d_2$$

$$\frac{20(.5)}{13} = d_2$$

$$\frac{10}{13} = d_2.$$

Thus, the force of 13 lb should be applied $\frac{10}{13}$ ft to the right of the pivot point.

b. If the beam in Fig. 5-3 is in balance on a knife edge placed at the point shown, how large is the force F?

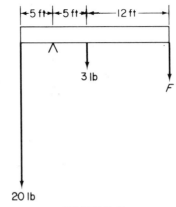

FIGURE 5-3

For the beam to be in equilibrium, the sum of the clockwise torques, that is, the torques associated with forces tending to cause clockwise rotation about the pivot point, must equal the sum of the counterclockwise torques.

Sum of clockwise torques = Sum of counterclockwise torques

$$3(5) + F(12 + 5) = 20(5)$$

$$15 + 17F = 100$$

$$17F = 85$$

$$F = 5 \text{ lb.}$$

EXERCISE 5-6

1. How many gallons of a 60-percent acid solution must be added to 12 gallons of a 35-percent acid solution so that the resulting solution is 50-percent acid?

2. A 6-gallon car radiator is two-thirds full of water. How much of a 95-percent alcohol solution must be added to make a 10-percent alcohol solution in the radiator?

3. A chemical manufacturer mixes a 20-percent acid solution with a 30-percent acid solution so as to obtain 700 gallons of a 24-percent acid solution. How many gallons of each does he use?

4. To produce iron for miscellaneous castings, how many lb of ferrosilicon (86-percent silicon) should be added to 2000 lb of the base iron (2.17-percent silicon) to give the iron a 2.25-percent silicon content to make it easier to machine?*

5. A lunar rover traveled from point A to point B at the rate of 3 mi/hr and returned to point A along the same path at the rate of 9 mi/hr. The total traveling time was 2 hr. Determine the distance from A to B.

6. A pilot, flying against a headwind, traveled from A to B at 250 mi/hr. From B to A he had the wind to his advantage and returned at 300 mi/hr. Moreover, the trip from B to A took 1 hr less than from A to B. Find the distance between A and B.

7. Water is flowing into a tank by means of pipes A and B. Pipes A and B can fill the tank, individually, in 2 hr and 5 hr, respectively. However, water is also flowing out of the tank into another tank by a pipe C by which the original tank can be completely emptied in 4 hr. How long would it take to fill the original tank if it were initially empty and pipes A, B, and C were all opened?

8. If Dick can do a job in 4 days, Len in 3 days, and Ernie in 6 days, how long will it take them to do the job if they work together?

9. In order to produce a certain compound, chemicals A and B must be combined, by weight, in the ratio of 3: 11, respectively. If 175 gm of the compound are needed, how many grams of A and B, respectively, must be used?

10. How many grams each of chemicals A, B, and C must be combined to obtain 93 gm of a compound which requires that A, B, and C be in the ratio 1: 2: 3, respectively, by weight?

11. If the beam shown in Fig. 5-4 is balanced on a knife edge placed below the center of gravity, how large is the force F?

12. If the beam shown in Fig. 5-5 is balanced on a knife edge placed at the point shown, how large is the force F?

13. If the beam shown in Fig. 5-6 is to be balanced about its pivot point, determine d.

Mathematics at Work, Number 1 (Detroit, Michigan: General Motors Corp., 1963).

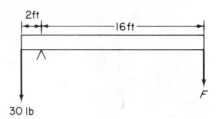

FIGURE 5-4

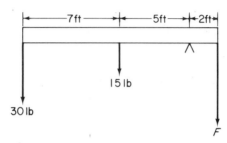

FIGURE 5-5

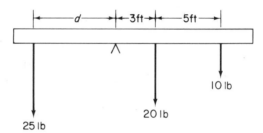

FIGURE 5-6

14. A satellite orbits the earth in a circular path of radius 4100 mi at constant speed. If the period of its motion, the time to complete one orbit, is 98 min, what is the speed of the satellite to the nearest mile per hour? (Approximate π by 3.14.)

15. A marksman hears the bullet strike the target 3 sec after the report of his rifle. If the bullet travels 1925 ft/sec and sound travels 1100 ft/sec, find the distance to the target and the time the bullet was in the air.

16. If you travel 120 mi from A to B at an average speed of 60 mi/hr, and then you return to A at an average speed of 40 mi/hr, what is your average speed for the entire trip? *Note:* The answer is not 50 mi/hr.

17. On a certain day the temperature rose 3 degrees hourly from 3 A.M. until noon. If the average temperature for this period was $43\frac{1}{2}$ degrees, what was the temperature at 8 A.M.?

18. Assuming that the velocity of sound in air increases 60 cm/sec for each Celsius degree of increase in temperature, find the velocity of sound in meters per second at 0°C from the following data: When the temperature is -3°C, a sound produced at A is heard at B after an interval of $5\frac{1}{5}$ sec; When the temperature is 19°C, the interval is 5 sec.

19. The water in a certain reservoir is 6 ft deep, but the level is sinking at the rate of 4 in./day. The water in another reservoir is 2 ft 9 in. deep and is rising $5\frac{3}{4}$ in./day. When will the depths of the two reservoirs be the same, and what will this depth be?

20. A certain type of concrete must be made up, by volume, of 1 part cement, 3 parts sand, and 5 parts stone. If a total of 104 ft³ of material was used in preparing such concrete, how many cubic feet of each were used?

21. The top section of a transmitting antenna tower was blown over by the wind. This top section, still attached to the bottom section, touched the ground at a point 20 ft from the base of the tower. If the top section was 5 ft longer than the bottom section, how high was the original tower?

22. Two airplanes leave two airports that are 300 mi apart at the same instant and fly toward each other. One plane flys 50 mi/hr faster than the other and the planes pass each other in 30 min. Find the speed of each plane.

23. A beam has weights of 50 lb and 100 lb attached to its ends. A knife edge placed directly below the center of gravity of the beam is located 10 in. from the center of the beam in the direction of the 100-lb weight. If the beam balances on the knife edge, how long is the beam?

5-7
REVIEW

REVIEW QUESTIONS

An equation that can be written in the form $ax + b = 0$, where $a \neq 0$, is called a _____ (1) _____ equation.

  (1) **linear or first-degree**

In the equation $x + 3 = 3(x - 5)$, we call $x + 3$ the _____ (2) _____ member and $3(x - 5)$ the _____ (3) _____ member.

  (2) **left**   (3) **right**

The equation $x + 4 = 5$ _____(is)(is not)_____ equivalent to $x + 6 = 7$. The
                              (4)
equation $x + 5 = 7$ _____(is)(is not)_____ equivalent to $2x + 10 = 9$.
                            (5)

(4) **is**   (5) **is not**

The equation $7 - 4x = 9$ is of the _____(6)_____ degree and its solution set
is _____(7)_____.

(6) **first**   (7) $\{-\frac{1}{2}\}$

With regard to being an identity or a conditional equation, we would classify
$2(x + 3) + 1 = 7 + 2x$ as a(n) _____(8)_____.

(8) **identity**

If $\frac{p_1}{p_2} = \frac{t_1}{t_2}$, then $t_2 =$ _____(9)_____.

(9) $\dfrac{t_1 p_2}{p_1}$

The number of roots of a linear equation is _____(10)_____.

(10) **1**

If the solution set of the equation $ax + b = 0$ is $\{-a\}$, then b must be equal to
_____(11)_____.

(11) a^2

If the solution set of the equation $ax + b = 0$ is $\left\{\frac{1}{a}\right\}$, then b must be equal to
_____(12)_____.

(12) **−1**

The statement "2 is to x as 3 is to 4" can be written as _____(13)_____.

(13) $\dfrac{2}{x} = \dfrac{3}{4}$ or $2 : x = 3 : 4$

The equation $a : b = c : d$ is referred to as a(n) _____(14)_____.

(14) **proportion**

REVIEW PROBLEMS

In Problems 1–12, find the solution sets.

1. $4 - 3x = 2 + 5x$

2. $3[2 - 4(1 + y)] = 5 - 3(3 - y)$

3. $\dfrac{5}{7}x - \dfrac{2}{3}x = \dfrac{3}{21}x$

4. $3(x + 4)^2 + 6x = 3x^2 + 7$

5. $2 - x = 3 + x$

6. $x = 2x$

7. $\dfrac{2x}{x - 3} - \dfrac{x + 1}{x + 2} = 1$

8. $\dfrac{x + 3x + 4}{7 - x} = 14$

9. $x = 2x - (7 + x)$

10. $3x - 8 = 4(x - 2)$

11. $2\left(4 - \dfrac{3}{5}x\right) = 5$

12. $\dfrac{x - 4}{4 - x} = 0$

13. If 2 is to $x + 5$ as 4 is to $x - 5$, find x.

14. Into a graduated storage tank two-thirds full, 10 gallons of fluid were poured and it was found to be five-sixths full. How many gallons did the storage tank hold?

15. How many gallons of a 60-percent acid solution must be added to 40 gallons of a 6-percent acid solution to obtain a 15-percent solution?

16. A chromium content of .15-percent is desired in iron to be used in a cylinder head. How many lb of V-5 ferroalloy (containing 40-percent chromium) should be added to 3000 lb of a base iron that has .10-percent chromium?

6

functions, graphs, and straight lines

6-1
FUNCTIONS

In 1694 the word *function* was first introduced into the mathematical vocabulary, and today the concept of a function is one of the most fundamental to all of mathematics. Although there are various equivalent ways in which a function can be defined, we shall take the approach in which a function is viewed as a special type of *correspondence* or *mapping* from one set to another—for in the final analysis this, essentially, is the main role of a function.

The table in Fig. 6-1 gives the data obtained in a study of the elongation of a vertical coil spring. F is the load, in pounds, which is suspended on the spring and S is the elongation, in inches, that it produces. Now, suppose we consider two sets: the set $A = \{1, 4, 6, 9\}$ whose elements are the loads (in pounds) in the experiment, and the set $B = \{.4, 1.6, 2.4, 3.6\}$ whose elements are the elongations (in inches) produced by the loads. Then to each load in A we can associate the corresponding elongation of the spring which is an element of B (Fig. 6-2).

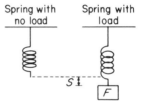

FIGURE 6-1

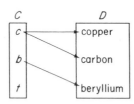

FIGURE 6-2

Note that the arrows designate a rule whereby there is associated with each element of A one and only one element of B. We say that a *function* from A into B has been defined in the sense of the following definition.

Definition.   A **function** is a rule which associates with each element of a given set, called the **domain**, one and only one element of another set.

In the example above, the set A is the domain of the function.

Not every rule of correspondence defines a function. In Fig. 6-3 a rule, indicated by the arrows, is established which associates with elements in

```
  C              D
┌───┐         ┌──────────┐
│ c ├───────► copper     │
│   │  ╲   ╱               │
│ b ├───╳─────► carbon     │
│   │  ╱   ╲               │
│ t │         beryllium    │
└───┘         └──────────┘
```

FiGURE 6-3

set $C = \{c, b, t\}$ elements in set $D = \{$copper, carbon, beryllium$\}$. However, this rule is not a function from C into D for two reasons. First, with the element $c \in C$ there are associated *two* elements, copper and carbon, in D; second, no element in D is associated with $t \in C$.

Customarily, functions are denoted by letters such as f, g, h, F, G, etc. Suppose we give the function in Fig. 6-2 a name, say f. Then to indicate that the number paired with 1 by f is .4 we shall simply write $f(1) = .4$, read "f of 1 equals .4." Similarly, $f(4) = 1.6$, $f(6) = 2.4$, and $f(9) = 3.6$. Generalizing we have:

Definition.   If x is an element in the domain of a function f, then the unique element which f associates with x is denoted $f(x)$, read "f of x."

Occasionally, $f(x)$ is called the **functional value of f at x**. As another example, suppose f is the function that associates with each real number the square of that number. Then with the number 2 the function f associates the number 2^2 or 4; thus $f(2) = 4$. Similarly, $f(3) = 3^2 = 9$, $f(-3) = (-3)^2 = 9$, $f(\frac{3}{2}) = \frac{9}{4}$, and $f(0) = 0$. In fact, for any real number x we have $f(x) = x^2$. Actually, f can be defined by the equation $f(x) = x^2$, where x is any real number. For brevity, we can speak of the function $f(x) = x^2$, although it is understood we mean the function defined by that equation.

EXAMPLE
1

For the function $f(x) = -x^2 + 2x + 3$ some functional values are

$$f(0) = -(0)^2 + 2(0) + 3 = 3$$
$$f(2) = -(2)^2 + 2(2) + 3 = 3$$
$$f(-3) = -(-3)^2 + 2(-3) + 3 = -12$$
$$f(z) = -z^2 + 2z + 3$$
$$\begin{aligned}f(x + h) &= -(x + h)^2 + 2(x + h) + 3 \\ &= -(x^2 + 2xh + h^2) + 2x + 2h + 3 \\ &= -x^2 - 2xh - h^2 + 2x + 2h + 3.\end{aligned}$$

If x is a real number, then the equation $y = x + 2$ also defines a function, since it expresses the rule whereby with each real number x there is associated one and only one real number $x + 2$, which is y. If we denote this function by g, then $y = g(x) = x + 2$ (Fig. 6-4). Thus, $g(0) = 0 + 2$

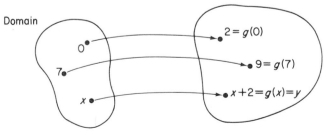

FIGURE 6-4

$= 2$, $g(7) = 7 + 2 = 9$, $g(t) = t + 2$, and $g(x + h) = (x + h) + 2$. For simplicity, we can speak of the function $y = x + 2$, although, as before, we mean the function defined by the equation, and we say that $y = x + 2$ is of the *form* $y = g(x)$, read "y equals g of x." The variable x which represents elements in the domain of the function is referred to as an **independent variable,** and the variable y is called a **dependent variable** in the sense that its value depends on the value of the independent variable. We also speak of the dependent variable as being a "function of" the independent variable. Thus, for $y = x + 2$ we say y is a function of x.

An equation which defines the function for the elongated spring in Fig. 6-1 is $S = .4F$. Here, the load F is the independent variable, the elongation S is the dependent variable, and we say that S is a function of F. It follows that we can write $S = f(F) = .4F$.

Up to this point we have seen that a function may be defined by an equation. If such an equation is in terms of x and y, it is customary—unless otherwise stated—to assume x denotes an independent variable and y a dependent variable. Thus, for the function $y = x^2 + 2x + 3$, x is the independent variable, y is the dependent variable, and y is a function of x. Similarly, if $s = 2t$ and t is the independent variable, then s is a function of t. Denoting this function by f, we can write $s = f(t) = 2t$. Moreover, since $t = s/2$ we can view t as a function of the independent variable s.

If the domain of a function (or domain of the independent variable) is not explicitly stated, it is understood to consist of all real numbers for which the defining equation has meaning. For example, for the function defined by $f(x) = \dfrac{1}{x - 6}$, the domain is understood to be all real numbers except 6, since the equation has no meaning if $x = 6$ (division by zero). Similarly, the equation $y = \sqrt{x}$ defines a function since to each x in the domain there corresponds a unique y, namely $\sqrt{x}$. For $\sqrt{x}$ to be a real number, x cannot be negative. Thus, the domain is $\{x \mid x \geq 0\}$. In the formula for the area of a circle, $A = \pi r^2$, A can be considered as a function, say f, of the length r of the radius; that is, $A = f(r)$. Since the length of the radius is a nonnegative number, the domain is the set of all nonnegative numbers.

At this time, two points should be noted. First, not every equation in x and y defines a function of x. For example, consider the equation $y^2 = x$. If x is 4, then y can be 2 or -2 and this violates our definition of a function. In short, the variables x and y are related by the equation $y^2 = x$, but this relation is not a function of x. However, it is interesting to note that since each value of y determines a unique value of x, the equation $y^2 = x$ *does* define x as a function of y. The second point we wish to stress is that the letters used to define a function are arbitrary. That is, the equation $w = f(z) = z^2$ defines the same function as $y = f(x) = x^2$. In both cases, with each real number we associate its square; that is, $f(1) = 1$, $f(2) = 4$, etc.

EXAMPLE
2

a. For the function $g(x) = x^3 + x^2$, some functional values are

$$g(-4) = (-4)^3 + (-4)^2 = -48$$
$$g(1) = 1^3 + 1^2 = 2$$
$$g\left(\frac{1}{2}\right) = \left(\frac{1}{2}\right)^3 + \left(\frac{1}{2}\right)^2 = \frac{1}{8} + \frac{1}{4} = \frac{3}{8}.$$

The domain of g is the set of all real numbers, **R**.

b. The function $f(x) = 2x + 3$ is called a **linear function**, one of the form $f(x) = ax + b$ where a and b are real numbers and $a \neq 0$. The domain of f is **R**. Some functional values are

$$f(-5) = 2(-5) + 3 = -7$$
$$f(3) = 2(3) + 3 = 9$$
$$f(t + 7) = 2(t + 7) + 3 = 2t + 17$$
$$f[f(x)] = 2(2x + 3) + 3 = 4x + 9.$$

c. The equation $y = 2$ defines a **constant function** g, one of the form $g(x) = c$ where c is a fixed real number. The domain of g is **R**.

$$g(.10) = 2$$
$$g(-420) = 2$$
$$g(0) = 2.$$

Although the number 2 is paired with every number in the domain, our definition of a function is not violated.

d. The equation $y = h(x) = -3x^2 + x - 5$ defines a **quadratic function**, one of the form $h(x) = ax^2 + bx + c$ where a, b, and c are real numbers and $a \neq 0$. The domain of h is **R** and y is a function of x.

$$h(2) = -3(2)^2 + 2 - 5 = -15$$
$$h\left(\frac{1}{t}\right) = -3\left(\frac{1}{t}\right)^2 + \left(\frac{1}{t}\right) - 5 = -\frac{3}{t^2} + \frac{1}{t} - 5$$
$$h(0) = -3(0)^2 + (0) - 5 = -5$$
$$h(x + t) - h(x) = [-3(x + t)^2 + (x + t) - 5] - [-3x^2 + x - 5]$$
$$= -6xt - 3t^2 + t.$$

e. The equation

$$z = \frac{2t}{(t-1)(t+1)} = \frac{2t}{t^2 - 1}$$

defines z as a function of t, where z and t are the dependent and independent variables, respectively. The right member is defined if and only if the denominator is not zero. Thus, for $z = f(t)$, the domain of f is $\{t \mid t \neq 1, -1\}$.

$$f\left(\frac{1}{2}\right) = \frac{2(\frac{1}{2})}{(\frac{1}{2})^2 - 1} = \frac{1}{\frac{1}{4} - 1} = \frac{1}{-\frac{3}{4}} = -\frac{4}{3}$$

$$f(-t) = \frac{2(-t)}{(-t)^2 - 1} = -\frac{2t}{t^2 - 1}.$$

f. The function F defined by

$$F(s) = \begin{cases} 1, & \text{if } s > 0 \\ 0, & \text{if } s = 0 \\ -1, & \text{if } s < 0 \end{cases}$$

is interpreted as follows: If s is positive, $F(s) = 1$; if s is zero, $F(s) = 0$; if s is negative, $F(s) = -1$. Thus,

$$F(0) = 0$$
$$F(23.4) = 1 \qquad \text{(since } 23.4 > 0)$$
$$F(-6\pi) = -1 \qquad \text{(since } -6\pi < 0).$$

g. The function defined by $y = |x|$ is called the **absolute value function.** Its domain is **R**.

$$f(0) = |0| = 0$$
$$f(6) = |6| = 6$$
$$f(-2) = |-2| = 2.$$

h. The formula $C = \frac{5}{9}(F - 32)$ relates Fahrenheit temperature F and Celsius temperature C. Here C can be considered a function of F. If $C = g(F)$, then

$$g(32) = \frac{5}{9}(32 - 32) = 0$$

$$g(212) = \frac{5}{9}(212 - 32) = \frac{5}{9}(180) = 100.$$

Thus, 0°C corresponds to 32°F and 100°C corresponds to 212°F.

EXAMPLE
3

In a laboratory experiment a potential difference of 10 volts is applied across an initially uncharged capacitor in series with a resistor as shown in Fig. 6-5. The readings obtained for the potential difference V, in volts, across the capacitor at various times t, in seconds, are given in the indicated table.

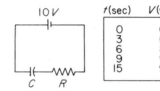

t(sec)	V(volts)
0	0
3	6.3
6	8.7
9	9.5
15	9.9

FIGURE 6-5

This table gives rise to a function, namely the one defined by the correspondence between the time t and the potential difference V at that time. If t is the independent variable, then V is a function of t, $V = f(t)$, and

$$f(0) = 0$$
$$f(3) = 6.3$$
$$f(6) = 8.7$$
$$f(9) = 9.5$$
$$f(15) = 9.9.$$

EXERCISE 6-1

In Problems 1–16 a function is defined. State the independent variable and the domain of the function. Determine the indicated functional values.

1. $f(x) = x + 7$; $f(4), f(-\frac{1}{2}), f(12), f(0)$

2. $g(x) = -3x$; $g(0), g(7), g(-3)$

3. $h(x) = -\frac{2}{3}$; $h(0), h(-120), h[(12)^2], h(w)$

4. $f(x) = 2x + 1$; $f(0), f(2), f(w), f\left(\dfrac{1}{w}\right)$

5. $f(x) = x^2 + 1$; $f(0), f(1), f(-1), f(2), f(x + h)$

6. $f(t) = t^2 + 3t - 2$; $f(0), f(2), f(-3)$

7. $f(s) = s^2 + 2s$; $f(0), f(1), f(2), f(-5), f(6)$

8. $f(z) = 3z + z^3 + 1$; $f(0), f(2z), f(z^2)$

9. $F(t) = 2(4 - t)$; $F(0), F\left(\dfrac{4}{3}\right), F\left(\dfrac{t}{2}\right), F\left(\dfrac{1}{t}\right)$

10. $y = G(t) = (t + 4)^2$; $G(0), G(2), G(2h)$

11. $y = H(t) = |t - 3|$; $H(0), H(3), H(-3)$

12. $f(z) = \dfrac{z}{z - 1}$; $f(2), f(-2), f(-1), f(0)$

13. $y = f(x) = \dfrac{1}{\sqrt{x}}$; $f(1), f(\sqrt{16})$

14. $y = g(x) = \dfrac{1}{x-2} + \dfrac{1}{x+3}; g(-2), g(3), g(0)$

15. $h(r) = \begin{cases} 3r - 1, & \text{if } r \geq 1 \\ r^2 - 2, & \text{if } r < 1 \end{cases}; h(0), h(-2), h(3), h(1)$

16. $y = f(x) = x^3 + x; f(x + h)$

17. If $z = 4x^2$, can z be considered a function of x? Can x be considered a function of z?

18. If $2x = 3y - 2$, can x be considered a function of y? Can y be considered a function of x?

19. The distance s, in feet, that an object will fall from rest in a vacuum in t sec is given by $s = f(t) = 16t^2$. Find $f(0)$, $f(1)$, and $f(2)$. From a practical standpoint, what would you define the domain of f to be?

20. Express the circumference C of a circle as a function f of its radius r.

21. A metal plate which is to be used in the construction of an electromechanical device is shown in Fig. 6-6. Express the area A of the metal surface as a function f of a. Give your answer in simplest form.

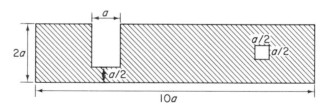

FIGURE 6-6

22. Express the length l of a side of a square as a function f of its perimeter P.

23. An open-top box is to be made from a rectangular piece of sheet metal, 12 by 16 in., by cutting out equal squares from each corner and folding up the sides. Express the volume V of the box as a function f of the length x of a side of the squares cut out.

24. The period T (in seconds) of a simple pendulum is given by

$$T = f(l) = 2\pi \sqrt{\dfrac{l}{g}},$$

where l is the length of the pendulum in feet and g is the acceleration due to gravity taken to be 32 ft/sec². Find $f(2)$.

6-2
GRAPHS IN RECTANGULAR COORDINATES

The set $\{5, 3\}$ consists of a pair of real numbers, 5 and 3. Since the order in which we list the elements of a set does not affect the set itself, we can say that $\{5, 3\} = \{3, 5\}$. However, we shall often find it necessary to consider the order in which a pair of real numbers is to be regarded. This notion leads to the following definition.

Definition. An **ordered pair** of real numbers, denoted (a, b), is a pair of real numbers a and b in which a is considered to be the *first element* of the pair and b the *second element*.

Some examples of ordered pairs are $(3, 10)$, $(2, -3)$, $(0, 0)$, and $(-\sqrt{3}, 7)$. Although the symbol $(3, 10)$ can indicate an open interval as well as an ordered pair, its correct interpretation will be clear from the context in which it appears.

To avoid any ambiguity, we shall be precise as to when two ordered pairs are equal.

Definition. Two ordered pairs (a, b) and (c, d) are equal if and only if $a = c$ and $b = d$—that is, when they have the same first element and the same second element.

Thus, $(5, 3) = (\frac{10}{2}, \frac{12}{4})$, but $(5, 3) \neq (3, 5)$. Ordered pairs of real numbers are an integral part of any discussion of a rectangular coordinate system.

A **rectangular** (or **Cartesian**) **coordinate system** allows us to specify and locate points in a plane and also provides a geometrical way to represent a function. Such a system is obtained as follows. In a plane two real number lines, called **coordinate axes**, are constructed perpendicular to each other so that their origins coincide. Their point of intersection is called the **origin** of the coordinate system (Fig. 6-7). Usually, the lines are oriented horizontally and vertically so that the positive numbers on

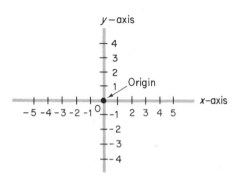

FIGURE 6-7

the horizontal line (or x-axis) are to the right of the origin and the positive numbers on the vertical line (or y-axis) lie above the origin.

The plane on which the coordinate axes are imposed is called a **rectangular coordinate plane** or, more simply, the **xy-plane.** The unit distance on the axes need not necessarily be the same and convenience should dictate an appropriate choice.

Suppose P is a point in the xy-plane. From P we construct perpendiculars to the x- and y-axes intersecting them at points P_x and P_y, respectively (Fig. 6-8). Let x and y be the directed distances P_yP and P_xP,

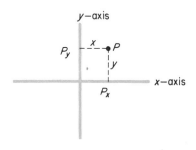

FIGURE 6-8

respectively. We say "directed" in the sense that only when P is to the right of the y-axis is P_yP considered positive; only when P lies above the x-axis is P_xP considered positive. With point P we associate the ordered pair (x, y), where x is called the **abscissa** or **x-coordinate** of P, and y is the **ordinate** or **y-coordinate** of P. The abscissa is *always* written before the ordinate. Together x and y are called the **coordinates** of P.

Thus, with each point in the coordinate plane we can associate a unique ordered pair (x, y) of real numbers. On the other hand, it should be obvious that with each ordered pair (x, y) of real numbers we can associate a unique point in the plane. Since there is a one-to-one correspondence between the points in the plane and all ordered pairs of real numbers, we shall refer to a point P with abscissa x and ordinate y simply as the point (x, y), or as $P(x, y)$. Moreover, we shall use the words *point* (in a plane) and *ordered pair* interchangeably. In the following example, the locations of various points in the xy-plane are indicated.

EXAMPLE

4

Refer to Fig. 6-9.

a. The point $(4, 0)$ is located on the x-axis four units to the right of the y-axis.

b. The point $(1, 2)$ is located one unit to the right of the y-axis and two units above the x-axis.

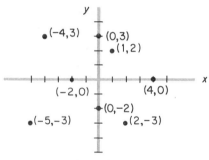

FIGURE 6-9

c. The point $(0, 3)$ is located on the y-axis three units above the x-axis.

d. The point $(-4, 3)$ is located four units to the left of the y-axis and three units above the x-axis.

e. The point $(-2, 0)$ is located on the x-axis two units to the left of the y-axis.

f. The point $(-5, -3)$ is located five units to the left of the y-axis and three units below the x-axis.

g. The point $(0, -2)$ is located on the y-axis two units below the x-axis.

h. The point $(2, -3)$ is located two units to the right of the y-axis and three units below the x-axis.

Observe that the coordinate axes divide the plane into four regions, called **quadrants** (Fig. 6-10). Quadrant I consists of all points (x_1, y_1) such that $x_1 > 0$ and $y_1 > 0$, and the other quadrants can be characterized

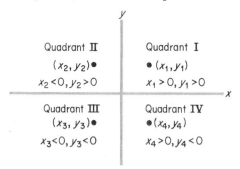

FIGURE 6-10

as indicated. The points on the axes do not lie in any quadrant. The x-coordinate of every point on the y-axis is 0, and similarly the y-coordinate of every point on the x-axis is 0.

By means of a rectangular coordinate system we can represent geometrically the solution set of an equation in two variables. For example, in the equation

$$y = x^2 + 2x - 3,$$

if $x = 1$, then $y = 1^2 + 2(1) - 3 = 0$. Thus, one solution is $x = 1$, $y = 0$. We shall represent this solution geometrically by the point $(1, 0)$ in the coordinate plane (Fig. 6-11). Note that the abscissa of this point is the x-value of the solution, while the ordinate is the y-value. If $x = -2$, then $y = (-2)^2 + 2(-2) - 3 = -3$. This solution is represented by the point $(-2, -3)$ in Fig. 6-11.

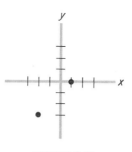

FIGURE 6-11

Certainly there are infinitely many solutions to the equation $y = x^2 + 2x - 3$. In Fig. 6-12, a table listing some of the solutions appears

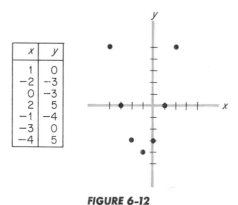

x	y
1	0
-2	-3
0	-3
2	5
-1	-4
-3	0
-4	5

FIGURE 6-12

together with the corresponding points in the plane.

Definition.   The **graph of an equation** in two variables is the geometrical representation of the set of all points in the plane whose coordinates satisfy the equation.

Essentially, the graph is a representation of the solution set of an equation. Thus, the points in Fig. 6-12 lie on the graph of $y = x^2 + 2x - 3$ since their coordinates satisfy that equation. Since the equation $y = x^2 + 2x - 3$ has infinitely many solutions, it would seem an impossible task to determine precisely its graph. However, our interest is only in the general nature of the graph. For this reason, we locate only enough points to determine the general behavior of the graph, and then join these points by a *smooth curve* wherever conditions permit. This process is called *sketching the graph of the equation*. Care should be taken to observe those values for which the equation is not defined, for there will be no corresponding points in the plane.

Certainly, consideration of only the points $(1, 0)$ and $(-2, -3)$ in Fig. 6-11 gives us little idea of the graph of $y = x^2 + 2x - 3$, but those in Fig. 6-12 are suitable since they make the general behavior apparent. A sketch of a portion of the graph appears in Fig. 6-13. It should be evi-

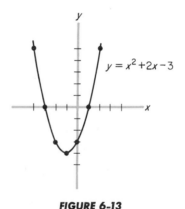

FIGURE 6-13

dent that the graph extends indefinitely. We can think of this graph as all points in the plane associated with the ordered pairs of the set

$$\{(x, y) \mid y = x^2 + 2x - 3\}.$$

At times we may refer to the graph of an equation simply as a *curve*; thus we speak of the curve $y = x^2 + 2x - 3$.

Since the equation $y = x^2 + 2x - 3$ determines one and only one value of y for each value of x, it defines a function of x, say f, and we can say

$$y = f(x) = x^2 + 2x - 3.$$

Figure 6-13 is then the geometrical representation of the set

$$\{(x, f(x)) \mid f(x) = x^2 + 2x - 3\}$$

or the graph of the function (equation)

$$f(x) = x^2 + 2x - 3.$$

Extending this concept we have the following definition.

Definition.   The **graph of a function** f is the set of all points (x, y) in the plane such that x is in the domain of f, and x and y are related by the equation $y = f(x)$.

There is a relatively simple geometrical technique to determine whether the graph of an equation is indeed that of a function. If a *vertical* line L can be constructed which intersects the graph of an equation in at least two distinct points, then the equation *does not* define a function of x. Such a condition implies the existence of two distinct points with the same first coordinate, and this would pair a value of x with more than one value of y. However, if no such line can be drawn, then that graph *does* represent a function of x. Thus the graphs in Fig. 6-14 are graphs of functions of x, but those of Fig. 6-15 are graphs which *do not* represent functions of x.

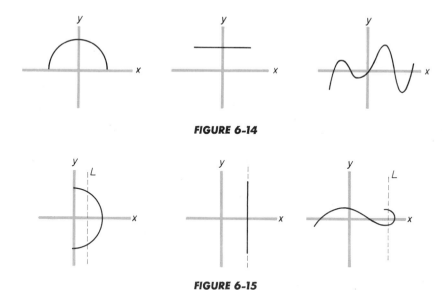

FIGURE 6-14

FIGURE 6-15

EXAMPLE
5

a. We have indicated that for the spring in Fig. 6-1 the equation relating the elongation S to the load F is $S = f(F) = .4F$. The graph of this equation is given in Fig. 6-16(a). Note that since the independent variable F represents the load on the spring, which must be positive or zero, we restrict the graph to nonnegative values of F.

F	1	4	6	9
S	.4	1.6	2.4	3.6

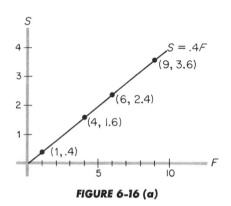

FIGURE 6-16 (a)

b. In Example 3 a table was given in which the potential difference V across a capacitor was given as a function of time t. This table, also given here, gives rise to ordered pairs, namely each time-volt pair: $(0, 0)$,

t (sec)	V (volts)
0	0
3	6.3
6	8.7
9	9.5
15	9.9

$(3, 6.3)$, $(6, 8.7)$, $(9, 9.5)$, and $(15, 9.9)$. In Fig. 6-16(b) we have sketched each pair, choosing the horizontal axis for the possible times, and connected the pairs by a smooth curve. If we observe the graph, a relationship appears evident: As time increases, the voltage increases, while at the same time getting closer and closer to 10. Further data and theoretical considerations would show that this is indeed the case. Note that as t gets larger, the curve "settles down" near the broken line. Such a line is called a *horizontal* **asymptote** of the curve.

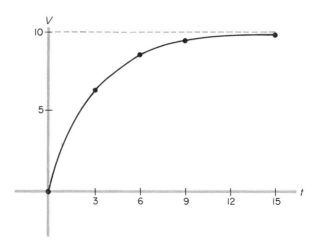

FIGURE 6-16 (b)

In Example 5 note that *the horizontal axis corresponds to the independent variable.* This is usually the case.

EXAMPLE
6

a. Sketch the graph of $y = 2x^2 - 3$. See Fig. 6-17(a).

x	0	1	2	-1	-2
y	-3	-1	5	-1	5

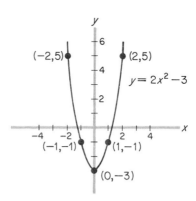

FIGURE 6-17 (a)

b. Sketch the graph of $f(x) = |x|$ (absolute value function).
Here we set $y = f(x)$. See Fig. 6-17(b).

x	0	1	−1	2	−2	3	−3
y	0	1	1	2	2	3	3

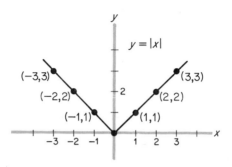

FIGURE 6-17 (b)

c. Sketch the graph of $y = x^3 + 1$. See Fig. 6-17(c).

x	−2	−1	$-\frac{1}{4}$	0	$\frac{1}{4}$	1	2
y	−7	0	$\frac{63}{64}$	1	$\frac{65}{64}$	2	9

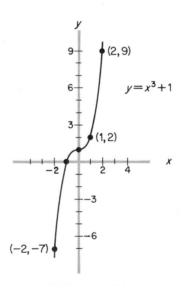

FIGURE 6-17 (c)

Although the points $(-1/4, 63/64)$ and $(1/4, 65/64)$ in the table in Fig. 6-17(c) are difficult to show in the graph, their location serves to reassure us as to the behavior of the curve.

d. Sketch the graph of $z = f(r) = \dfrac{100}{r}$. See Fig. 6-17(d).

r	5	−5	10	−10	20	−20	25	−25	50	−50
z	20	−20	10	−10	5	−5	4	−4	2	−2

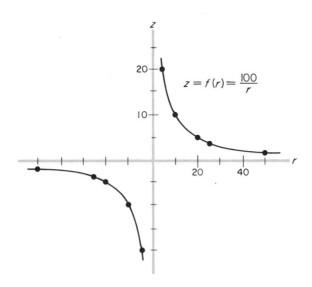

FIGURE 6-17 (d)

This function is not defined when $r = 0$. Observe the convenient choice of unit distance on each axis. The independent variable is r and the dependent variable is z.

e. The electric potential V, in volts, at a distance r, in meters, from a small object having an electric charge of $\frac{1}{9} \times 10^{-7}$ coulombs is given by

$$V = V(r) = \frac{100}{r}.$$

Sketch the graph of this equation.

You should observe that this equation is identical in form to that of part d above. Here, however, the variable r has a physical significance— it represents a measurable distance. For this reason the values of r must be restricted to positive values. The graph of this electric potential function would be identical to the portion of the graph of part d which lies in the first quadrant.

EXAMPLE
7

a. The current i in a 1-ohm resistor as a function of the power P developed in the resistor is given by

$$i = f(P) = \sqrt{P}.$$

Sketch the graph of this equation.

Here the domain is $\{P \mid P \geq 0\}$ since the square root of a negative number is *not* a real number.

P	0	$\frac{1}{4}$	1	4	9
i	0	$\frac{1}{2}$	1	2	3

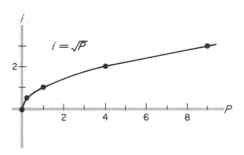

b. Sketch the graph of $x = 2$.

Since there is no restriction on y, the graph of this equation consists of all points whose abscissa is 2. This equation does *not* define a function of x.

x	2	2	2
y	1	2	-3

c. Sketch the graph of $x = 2y^2$.

Here it is convenient to choose arbitrary values of y and find the corresponding values of x. The equation does *not* define a function of x.

x	0	2	2	8	8	18	18
y	0	1	-1	2	-2	3	-3

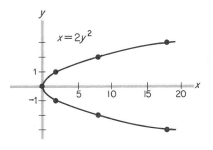

EXERCISE 6-2

In Problems 1 and 2, locate and label each of the given points and state, if possible, the quadrant in which each point lies.

1. $(3, 4)$, $(7, -5)$, $(-\frac{1}{2}, 1)$, $(0, 0)$

2. $(2, 0)$, $(3, 3)$, $(-2, -3)$, $(0, -4)$

3. The following results were obtained in an experiment to measure the current i (in amperes) as a function of time t (in seconds) in an electrical circuit containing a resistor and an inductor:

t (in sec)	0	2	4	6	8	10	12
i (in amp)	0	6.3	8.7	9.5	9.8	9.9	10.0

Represent these data graphically, taking t as the independent variable.

4. In a certain experiment, a mass m was suspended from a vertical spring and allowed to come to rest. It was then pulled down 10 cm below its equilibrium position and released. The position s (in centimeters) of the mass as a function of time t (in seconds) was found to be as follows:

t (in sec)	0	0.5	1	1.5	2	2.5	3	3.5	4
s (in cm)	-10	-8	0	8	10	8	0	-8	-10

Represent these data graphically, taking t to be the independent variable.

In Problems 5–36, sketch graph of the given equation or function. In each case, determine whether the graph is that of a function of x or t.

5. $y = x$ **6.** $y = 3x$

7. $y = -2x + 1$ **8.** $y = 3x - 5$

9. $y = f(t) = t^2$ **10.** $y = 3x^2 - 1$

11. $y = -x^2 + 2$ 12. $y = \frac{1}{2}x^2 + 4$

13. $y = |x| + 1$ 14. $y = x^2 + x + 1$

15. $f(x) = 2x^2 + 2x + 1$ 16. $y = x + |x|$

17. $y = x^3$ 18. $y = x^3 - 2x^2 + x$

19. $y = x^3 - 3x$ 20. $x = -4$

21. $y = \dfrac{1}{x}$ 22. $f(x) = x(2 - x)$

23. $y = \sqrt{4 - x^2}$ 24. $y = -8$

25. $y = 1 + x$, where $1 \le x \le 5$ 26. $y = \dfrac{x + 1}{x}$

27. $y = g(x) = 2$ 28. $y = |x - 3|$

29. $x = 0$ 30. $x + y = 1$

31. $x = -3y^2$ 32. $s = f(t) = \sqrt{t - 5}$

33. $2x + y - 2 = 0$ 34. $y = \dfrac{x}{x}$

35. $F(x) = x^2, \quad 0 < x \le 2$
$\qquad = 2x, \quad 2 < x \le 3$
$\qquad = 6, \quad x > 3$

36. $y = x + 1$, where $x \in \{1, 2\}$

37. If a horizontal force F acts on a 20-lb block resting on a horizontal surface, and the coefficients of static and kinetic friction are 0.3 and 0.1, respectively, then the frictional force f (in pounds) acting on the block can be written as a function g of F defined by

$$g(F) = F, \quad 0 \le F \le 6$$
$$\qquad = 2, \quad F > 6.$$

Sketch the graph of g for values of F from 0 to 10 lb.

38. While monitoring a radioactive sample with a detection instrument, a technician observed the disintegration rate R (in counts per minute) as a function of time t (in hours) to be as follows:

t (hr)	R (counts/min)	t (hr)	R (counts/min)
0.5	9535	6.0	1800
1.0	8190	7.0	1330
1.5	7040	8.0	980
2.0	6050	9.0	720
3.0	4465	10.0	535
4.0	3300	11.0	395
5.0	2430	12.0	290

Represent the given data graphically.

39. For a tuned circuit in a radio, the output voltage V (in volts) as a function of frequency f (in kilohertz) was observed to be as follows:

f (in Khz)	1300	1325	1375	1400	1425	1475	1500
V (in volts)	0.01	0.03	0.25	0.95	0.35	0.10	0.05

Represent the given data graphically.

6-3
THE SLOPE OF A STRAIGHT LINE

Suppose L is the nonvertical straight line which passes through the points (1, 3) and (4, 8). As indicated in Fig. 6-18, we have constructed line

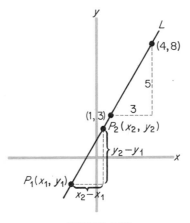

FIGURE 6-18

segments resulting in a right triangle having sides of lengths 5 and 3 units. If $P_1(x_1, y_1)$ and $P_2(x_2, y_2)$ are two other distinct points on L, we can construct another right triangle whose sides will have lengths $y_2 - y_1$ and $x_2 - x_1$. Since the triangles are similar, the ratios of the lengths of their corresponding sides are equal and we can write

$$\frac{5}{3} = \frac{y_2 - y_1}{x_2 - x_1}.$$

Clearly, for line L the number $(y_2 - y_1)/(x_2 - x_1)$ is independent of the choice of P_1 and P_2 used to evaluate it and is always equal to $\frac{5}{3}$. This number can be associated with L in the following sense.

Definition. If (x_1, y_1) and (x_2, y_2) are two distinct points on a nonvertical straight line L, the **slope** m of L is the real number

$$m = \frac{y_2 - y_1}{x_2 - x_1}. \tag{6-1}$$

In the formula for the slope, the order of the subscripts in the denominator *must* be the same as their order in the numerator. It is *incorrect* to write $m = (y_2 - y_1)/(x_1 - x_2)$, but correct to write $m = (y_1 - y_2)/(x_1 - x_2)$, since the latter is obtained by multiplying the fraction in the definition by $(-1)/(-1)$.

For a vertical line—that is, one parallel to the y-axis—the number $x_2 - x_1$ would be zero since any two distinct points on the line would have the same abscissa (Fig. 6-19). Hence, the ratio $(y_2 - y_1)/(x_2 - x_1)$

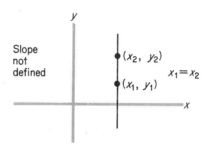

FIGURE 6-19

would have no meaning (division by zero) and, for this reason, **the slope of a vertical line is not defined**.

EXAMPLE
8
   a. Determine the slope of the line passing through (3, 4) and (5, 1) [Fig. 6-20(a)].

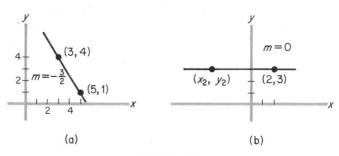

(a)   (b)

FIGURE 6-20

If we let $(x_1, y_1) = (3, 4)$ and $(x_2, y_2) = (5, 1)$, then

$$m = \frac{y_2 - y_1}{x_2 - x_1} = \frac{1 - 4}{5 - 3} = \frac{-3}{2} = -\frac{3}{2}.$$

If we let $(x_1, y_1) = (5, 1)$ and $(x_2, y_2) = (3, 4)$, then

$$m = \frac{y_2 - y_1}{x_2 - x_1} = \frac{4 - 1}{3 - 5} = \frac{3}{-2} = -\frac{3}{2},$$

which agrees with the original result.

b. Determine the slope of the horizontal line passing through $(2, 3)$ [Fig. 6-20(b)].

Let $(x_1, y_1) = (2, 3)$. If (x_2, y_2) is any other point on the line, then y_2 must be 3 and

$$m = \frac{y_2 - y_1}{x_2 - x_1} = \frac{3 - 3}{x_2 - 2} = 0.$$

In fact, **the slope of every horizontal line is zero.**

Suppose we analyze the notion of slope more closely. Refer to Fig. 6-21(a). As the independent variable x takes increasing values from x_1

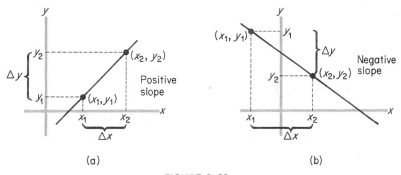

(a)   (b)

FIGURE 6-21

to x_2, we see that the dependent variable y undergoes a corresponding change that characterizes the orientation of the line. More specifically, the change in x, namely $x_2 - x_1$, produced a *positive* change in y of $y_2 - y_1$. For example, if $(x_1, y_1) = (4, 2)$ and $(x_2, y_2) = (9, 8)$, then the change in x of five units produced a change in y of six units. The slope is simply the ratio of the change in y to the change in x. That is, if we denote the change in x by the symbol Δx (read "delta x") and the change in y by Δy, then

$$m = \frac{\Delta y}{\Delta x} = \frac{y_2 - y_1}{x_2 - x_1} = \frac{6}{5}.$$

In Fig. 6-21(b), the positive change in x from x_1 to x_2 caused a *decrease* in y from y_1 to y_2—that is, a negative change. Hence, $\Delta y/\Delta x$ is negative. In short, **a line must rise from left to right if its slope is positive; it must fall from left to right if its slope is negative.** If the slope is zero, there is no change in y for a change in x and the line is horizontal.

Moreover, the slope is a measure of the steepness of a line. If two lines have slopes of 4 and 2, respectively, then in the first case we go up four units for a one-unit change in x to the right, while in the second case we would go up two units for the same change in x [Fig. 6-22(a)]. In this sense the first line is steeper than the second. Similarly, if two lines had slopes of -3 and $-\frac{2}{3}$, respectively, then in the first case we would go down three units while moving one unit to the right, while in the second

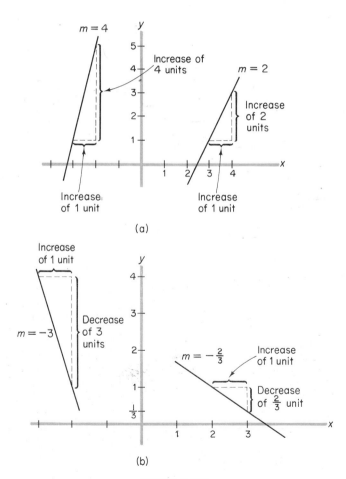

(a)

(b)

FIGURE 6-22

case we would go down two-thirds unit [Fig. 6-22(b)]. Thus, there is a much sharper decline in the first line than in the second.

EXERCISE 6-3

In Problems 1–6, find the slope of the straight line which passes through the given points.

1. (5, 2), (7, 5)   **2.** (−3, 4), (0, 1)

3. (−2, 3), (3, −1)   **4.** (2, −4), (3, −4)

5. (−2, 4), (−2, 8)   **6.** (0, 6), (−3, 0)

7. For a metallic conductor, the graph of the voltage V (dependent variable), in volts, as a function of current i, in amperes, is a straight line. The slope of this line gives the resistance R of the conductor. If $i = 4$ amps, then $V = 2$ volts; if $i = 12$ amps, then $V = 6$ volts. Determine the resistance (in ohms).

8. For an object moving in a straight line without accelerating, the graph of the displacement s (dependent variable), in meters, as a function of time t, in seconds, is a straight line. The slope of this line gives the velocity of the object in meters per second. Find the velocity if $s = 13$ m when $t = 2$ sec, and $s = 65$ m when $t = 10$ sec.

9. In an experiment, a steel rod is found to have a length of 1.0000 m at a temperature of 0°C, and a length of 1.0012 m at a temperature of 100°C. The graph of the length L as a function of temperature T is a straight line. The slope of this line is the coefficient of linear expansion of steel. From the given data, determine this coefficient.

10. A capacitor, a device commonly found in radios, consists of two parallel metal plates. When these plates are given equal and opposite electric charges, the graph of the potential difference V, in volts, as a function of the separation s of the plates, in meters, is a straight line. The slope of this line gives the strength of the electric field (in volts per meter) between the plates. For a particular capacitor, it is found that $V = .09$ volts when $s = .001$ m, and $V = .45$ volts when $s = .005$ m. Find the electric field strength between the plates of this capacitor.

6-4
STRAIGHT LINES

Our purpose in this section will be to determine equations of straight lines which lie in a rectangular coordinate plane—that is, to determine equations whose graphs are those lines.

Suppose a line L with slope m passes through the point (x_1, y_1). If

(x, y) is any other point on L, then

$$\frac{y - y_1}{x - x_1} = m$$

and so

$$y - y_1 = m(x - x_1). \tag{6-2}$$

That is, the coordinates of every point on L satisfy Eq. (6-2). On the other hand, it can be shown that every point whose coordinates satisfy Eq. (6-2) must lie on L. Thus we say that Eq. (6-2) is an equation for L. More specifically, $y - y_1 = m(x - x_1)$ is the *point-slope form* of an **equation of the line passing through (x_1, y_1) and having slope m.**

EXAMPLE
9

Determine and sketch an equation of the line that has slope 2 and passes through $(2, -3)$.

Here $m = 2$ and $(x_1, y_1) = (2, -3)$. Using Eq. (6-2) for a point-slope form,

$$y - (-3) = 2(x - 2)$$
$$y + 3 = 2x - 4,$$

which equivalently can be written

$$y = 2x - 7.$$

Note that in sketching the line, only two points need be plotted since two points are sufficient to determine a straight line. See Fig. 6-23.

x	2	0
y	-3	-7

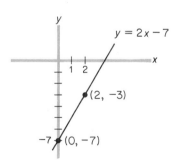

FIGURE 6-23

An equation of the line passing through two given points can be found by first determining the slope of the line and then using a point-slope form with either of the points as (x_1, y_1).

EXAMPLE 10

Determine an equation of the line passing through $(4, 1)$ and $(-5, 4)$.

We first find the slope:

$$m = \frac{4 - 1}{-5 - 4} = \frac{3}{-9} = -\frac{1}{3}.$$

Choosing $(4, 1)$ as (x_1, y_1) in a point-slope form,

$$y - 1 = -\frac{1}{3}(x - 4),$$

which equivalently can be written

$$y - 1 = -\frac{1}{3}x + \frac{4}{3}$$

$$y = -\frac{1}{3}x + \frac{7}{3}.$$

Choosing $(-5, 4)$ as (x_1, y_1) produces the same result:

$$y - 4 = -\frac{1}{3}[x - (-5)] = -\frac{1}{3}[x + 5]$$

$$y - 4 = -\frac{1}{3}x - \frac{5}{3}$$

$$y = -\frac{1}{3}x + \frac{7}{3}.$$

Any point at which a graph intersects the y-axis is called a **y-intercept**. At such a point, $x = 0$. If the slope m and y-intercept $(0, b)$ of a line L are given, an equation for L is (using a point-slope form)

$$y - y_1 = m(x - x_1)$$
$$y - b = m(x - 0)$$

or

$$y = mx + b. \tag{6-3}$$

Thus, **$y = mx + b$ is an equation of the line with slope m and y-intercept $(0, b)$ and is called the *slope-intercept form*.** (Refer to Fig. 6-24.) We remark that the equation $y = mx + b$ defines a so-called *linear function*.

EXAMPLE 11

a. An equation of the line with slope -2 and y-intercept $(0, -3)$ is (see Fig. 6-25)

$$y = -2x + (-3)$$
$$y = -2x - 3.$$

x	0	1
y	-3	-5

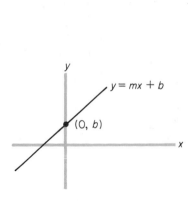

FIGURE 6-24   **FIGURE 6-25**

b. The line whose equation is $y = \frac{8}{9}x + 7$ has slope $\frac{8}{9}$ and y-intercept $(0, 7)$ since $m = \frac{8}{9}$ and $b = 7$.

c. When an object moves in a straight line with constant acceleration, its velocity, v, as a function of time, t, is a straight line. Moreover, the slope of this line is the acceleration of the object. Find the slope-intercept form of an equation for the velocity of such an object whose acceleration is $\frac{1}{4}$ ft/sec², if $v = 3$ ft/sec when $t = 2$ sec.

Here we want an equation of the line passing through the point $(2, 3)$ and having slope $\frac{1}{4}$. A point-slope form is

$$v - 3 = \frac{1}{4}(t - 2).$$

Solving for v gives the slope-intercept form:

$$v = \frac{1}{4}t + \frac{5}{2}.$$

Note that the y-intercept, $(0, \frac{5}{2})$, gives the initial velocity of the object, the velocity at $t = 0$.

Any point at which a graph intersects the x-axis is called an **x-intercept**. At such a point, $y = 0$. If a line L has x-intercept $(a, 0)$ and y-intercept $(0, b)$(see Fig. 6-26), its slope is

$$m = \frac{b - 0}{0 - a} = -\frac{b}{a}$$

and its equation, using the slope-intercept form, is

$$y = -\frac{b}{a}x + b.$$

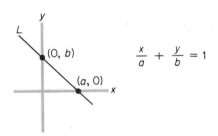

FIGURE 6-26

Multiplying both members by a,

$$ay = -bx + ab$$

$$bx + ay = ab.$$

Dividing both members by ab,

$$\frac{x}{a} + \frac{y}{b} = 1, \tag{6-4}$$

which is called the ***intercept form*** of an equation for L. Note that if L passes through $(0, 0)$, then $a = b = 0$ and Eq. (6-4) is not defined.

EXAMPLE
12

a. Determine an equation of the line passing through $(3, 0)$ and $(0, 5)$.

Since $(3, 0)$ and $(0, 5)$ are the x- and y-intercepts, respectively, an equation, using the intercept form, is

$$\frac{x}{3} + \frac{y}{5} = 1.$$

b. Find the intercept form of the line $y = -3x - 15$.

$$y = -3x - 15$$

$$3x + y = -15$$

$$\frac{x}{-5} + \frac{y}{-15} = 1.$$

We observe that the intercepts are $(-5, 0)$ and $(0, -15)$.

If a *vertical* line L passes through the point (a, b), then any other point (x, y) lies on L if and only if $x = a$. Hence, an equation for L is $x = a$ [see Fig. 6-27(a)]. Similarly, an equation for the horizontal line passing through (a, b) is $y = b$ [see Fig. 6-27(b)].

EXAMPLE
13

An equation of the vertical line passing through $(5, 3)$ is $x = 5$ (Fig. 6-28). An equation of the horizontal line passing through $(5, 3)$ is $y = 3$ (Fig. 6-29).

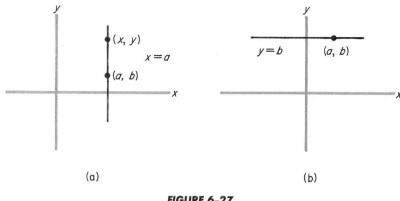

FIGURE 6-27

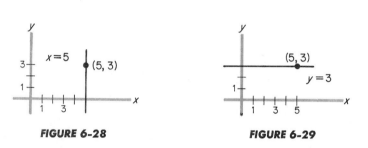

FIGURE 6-28   FIGURE 6-29

Based on our discussions, it can easily be shown that every straight line is the graph of an equation of the form $Ax + By + C = 0$, where A and B are not both zero. We call this form a **general linear equation** (or an *equation of the first degree*) **in the variables x and y**, and x and y are said to be **linearly related**. For example, a general linear equation for the line $y = 3x - 4$ is $(-3)x + (1)y + (4) = 0$. Conversely, the graph of every general linear equation is a straight line. For example, since $3x + 4y + 5 = 0$ is equivalent to $y = (-\frac{3}{4})x + (-\frac{5}{4})$, its graph is a straight line with slope $-\frac{3}{4}$ and y-intercept $(0, -\frac{5}{4})$.

EXAMPLE
14

Sketch the graph of $3x - 4y + 12 = 0$.

Since this is a general linear equation, its graph is a straight line; hence, it suffices to determine two points on the graph, for example, its intercepts. If $x = 0$, then $y = 3$, and if $y = 0$, then $x = -4$. We therefore draw the line passing through $(0, 3)$ and $(-4, 0)$ (see Fig. 6-30).

EXAMPLE
15

Assuming that Fahrenheit temperature F and the corresponding Celsius temperature C are linearly related, find the linear equation relating F and C using the facts that $32°F = 0°C$ and $212°F = 100°C$. Use F as the independent variable.

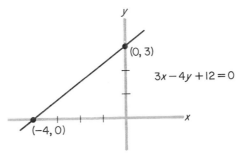

FIGURE 6-30

Since the points $(32, 0)$ and $(212, 100)$ lie on the graph of the equation, its slope is

$$\frac{100 - 0}{212 - 32} = \frac{100}{180} = \frac{5}{9}.$$

Using a point-slope form with the point $(32, 0)$,

$$C - 0 = \frac{5}{9}(F - 32)$$

$$C = \frac{5}{9}(F - 32).$$

In the following table, for your convenience we summarize the various forms of equations of straight lines.

Forms of Equations of Straight Lines

Point-slope form	$y - y_1 = m(x - x_1)$
Slope-intercept form	$y = mx + b$
Intercept form	$\dfrac{x}{a} + \dfrac{y}{b} = 1$
General linear form	$Ax + By + C = 0$
Vertical line	$x = a$
Horizontal line	$y = b$

The linear equation $ax + b = 0$ may be solved graphically by carefully sketching the line $y = ax + b$ and observing for what value of x the line meets the x-axis, for at such a point $y = 0$. For example, the line $y = -2x - 3$ in Example 11a intersects the x-axis only when x is $-\frac{3}{2}$. Thus the solution set of $-2x - 3 = 0$ is $\{-\frac{3}{2}\}$.

Before concluding, one point is worthy of additional emphasis. The equation $f(x) = mx + b$ defines a linear function of x. If $y = f(x)$, then $y = mx + b$ and thus *the graph of a linear function is a straight line.* We say the slope of the linear function is m and the y-intercept is $(0, b)$. Thus, the slope of the linear function $f(x) = 3x + 5$ is 3 and its y-intercept is $(0, 5)$.

Finally, it should be intuitively obvious that if L_1 and L_2 are straight lines, then L_1 is parallel to L_2 if and only if L_1 and L_2 are vertical lines or if they have the same slope. Thus, the line $y = 3x + 5$ is parallel to the line $y = 3x - 7$ since they both have a slope of 3. Moreover, it can be shown that if L_1 and L_2 have slopes of m_1 and m_2, respectively, and $m_1 = -\dfrac{1}{m_2}$, then L_1 and L_2 are perpendicular to each other. Thus, the line $y = -\dfrac{1}{3}x - 9$ is perpendicular to $y = 3x + 5$ since the slope of the first line ($m_1 = -1/3$) is the negative reciprocal of the slope of the second line ($m_2 = 3$).

EXERCISE 6-4

In Problems 1–14, determine a general linear equation of the straight line that has the indicated properties and sketch each.

1. Passes through $(2, -5)$ and $(3, 4)$

2. Passes through $(0, 0)$ and $(2, 3)$

3. Passes through $(-2, 5)$ and $(3, 5)$

4. Passes through $(2, 3)$ and $(2, 4)$

5. Passes through $(2, 4)$ and has y-intercept $(0, 3)$

6. Passes through $(-3, 2)$ and has x-intercept $(-3, 0)$

7. Passes through $(1, 2)$ and has slope 6

8. Passes through $(5, -2)$ and has slope $-\frac{1}{5}$

9. Passes through the origin and has slope -5

10. Passes through $(0, 2)$ and has slope $\frac{3}{5}$

11. Passes through $(2, 1)$ and is parallel to the line $y = 2x + 3$

12. Passes through $(2, 4)$ and is parallel to the line $y = 2$

13. Passes through $(7, 5)$ and is parallel to the y-axis

14. Passes through $(-5, 4)$ and is perpendicular to the line $2y = 6x + 1$

15. If a straight line passes through $(1, 2)$ and $(-3, 8)$, find the coordinates of the point on it that has an abscissa of 5.

16. If a straight line has slope -3 and passes through $(4, -1)$, find the coordinates of the point on it that has an ordinate of -2.

17. Suppose s and t are linearly related such that $s = 40$ when $t = 12$, and $s = 25$ when $t = 18$. Find an equation that satisfies these conditions. Choose t as the independent variable.

18. Determine whether the point $(0, -7)$ lies on the graph of the straight line passing through $(1, -3)$ and $(4, 9)$

In Problems 19–28, determine, if possible, the slope and x-and y-intercepts of the straight line determined by the equation, and sketch the graph.

19. $y = 2x - 1$ **20.** $(x - 1) + (y - 2) = 0$

21. $3x - 8y = 8$ **22.** $2y - 3 = 1$

23. $x = -5$ **24.** $x = y$

25. $y = 1$ **26.** $x - 1 = 5y + 3$

27. $x + 2y - 3 = 0$ **28.** $y - 7 = 3(x - 4)$

In Problems 29–32, determine a general linear form, the slope-intercept form, and the intercept form of the given equation and sketch its graph.

29. $x = -2y + 4$ **30.** $4x + 9y - 5 = 0$

31. $\dfrac{3}{4}x = \dfrac{7}{3}y + \dfrac{1}{4}$ **32.** $\dfrac{x}{2} - \dfrac{y}{3} = -4$

33. Geometrically solve the linear equation $8 - 4x = 0$ by sketching the graph of $y = 8 - 4x$.

34. The slope of the line joining $(2, 5)$ and $(3, k)$ is 4. Find k.

35. Sketch the graphs of the following linear functions.
    a. $f(x) = x + 2$
    b. $f(x) = -3x$
    c. $f(x) = 2x + 3$

36. If a ball is thrown straight up in the air with an initial velocity of 160 ft/sec, the velocity at the end of t sec is given by $v(t) = 160 - 32t$. Sketch this linear function for $0 \le t \le 5$.

37. The graph of the pressure P of a fixed volume of gas, in centimeters of mercury, as a function of temperature T, in degrees Celsius, is a straight line. In an experiment using dry air, it was found that $P = 90$ when $T = 40$ and that $P = 100$ when $T = 80$. Determine an equation for P as a function of T.

38. When a graph of the terminal potential difference V, in volts, of a Daniell cell is plotted as a function of the current i, in amperes, delivered to an external resistor, a straight line is obtained. The slope of this line is the negative of the internal resistance of the cell. For a particular cell having an internal resistance of 0.06 ohm, it was found that $V = 0.6$ volt when $i = 0.12$ amp. Find an equation for V as a function of i.

39. The velocity-time graph shown below was constructed by observing the first 10 sec of motion of a racing car.

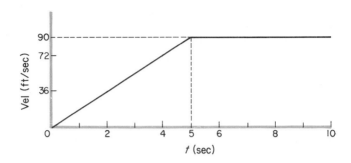

a. Given that the slope of a velocity-time graph is the acceleration, determine the acceleration of the racing car for the intervals $0 < t < 5$ and $5 < t < 10$.

b. Find equations for the velocity v of the car as functions of time t for these same intervals.

6-5

A FINAL COMMENT ON GRAPHS*

In engineering and science it is not uncommon to encounter graphical displays which show more information than simply how one quantity varies with another. As one example, a *psychrometric chart,* Fig. 6-31, can be used to determine the *relative humidity, absolute humidity,* and *dew point* from wet-bulb and dry-bulb temperature readings.

A *sling psychrometer* is a device containing two thermometers, one kept moist by wrapping its bulb with wet gauze and the other kept dry. When the device is whirled through the air, the reading of the wet-bulb thermometer decreases. From the reading of the dry-bulb thermometer and the decrease in the reading (depression) of the wet-bulb thermometer, the absolute humidity, in grains of water vapor per cubic foot, and the relative humidity, in percent, can be determined with the aid of the chart. Let us see how such a chart can be used.

Across the top of the psychrometric chart we find the dry-bulb thermometer readings. Arranged vertically at the left side of the chart is a scale called "depression of wet bulb $T - T'$ (Fahrenheit)." These depression values are possible differences between dry-bulb and wet-bulb readings.

*This discussion and the accompanying diagram are adapted from A. Joseph, K. Pomerantz, J. Prince, and D. Sacher, *Physics for Engineering Technology,* 1966, by permission of John Wiley & Sons, Inc., New York.

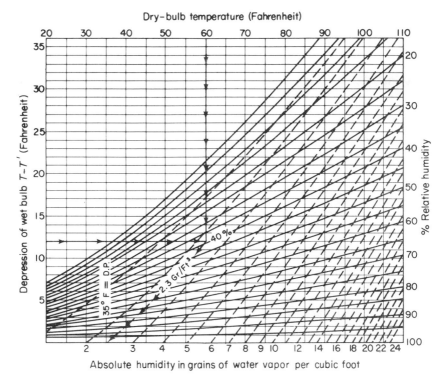

FIGURE 6-31 (Reprinted by permission of John Wiley
& Sons, Inc.)

Suppose that after we use a psychrometer the dry-bulb reading is
60°F and the wet-bulb reading is 48°F. The depression would then
be 60° − 48° or 12°. The heavy lines with the arrowheads indicate how
to find the point on the graph corresponding to the 60° reading and the
12° depression. From the point where these heavy lines intersect, we pro-
cede to the right along the solid line that slants upward. Where this line
ends on the right side of the chart, we read the relative humidity and find
that it is 40 percent. The *relative humidity* is the ratio of mass of water
vapor per unit volume present in the air to the mass of water vapor the
air could contain per unit volume if saturated at that temperature. To find
the absolute humidity, we go back to the intersection point of the lines
marked with arrows and follow the *broken line* which slants downward
and to the left. Where this line ends at the bottom of the chart, we read
the absolute humidity, which in our example is 2.3 grains/ft³. The *absolute
humidity* is the mass of water vapor in the air per unit volume. The dew
point lies directly above the absolute humidity on the chart. The *dew point*
is simply the dry-bulb temperature at which air during cooling becomes

saturated and, as a result of condensation, water droplets are formed. The dew point in our example is 35°F.

6-6
REVIEW

REVIEW QUESTIONS

The function defined by $f(x) = x + \dfrac{1}{x}$ is not defined when x is _____(1)_____.

(1) **0**

The sign of the abscissa of a point in the third quadrant is _____(2)_____.

(2) **negative**

The point three units to the right of the y-axis and two units below the x-axis is represented by the ordered pair _____(3)_____.

(3) **(3, −2)**

A linear equation in two variables is one which can be written in the general form _____(4)_____.

(4) ***Ax* + *By* + *C* = 0, where *A* and *B* are not both zero**

The line whose graph is given below has a (positive) (negative) slope.
$$\overline{}$$
(5)

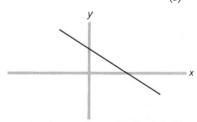

(5) **negative**

The intercept form of an equation of the line passing through (0, 2) and (3, 0) is _____(6)_____.

(6) $\dfrac{x}{3} + \dfrac{y}{2} = 1$

If f is the function defined by $f(x) = -x^2 - 1$, then $f(-1) = $ _____(7)_____ .
If $g(x) = 2$, then $g(3) = $ _____(8)_____ .

(7) **−2**   (8) **2**

The graph of the equation $x = 7$ is a line parallel to the _____(9)_____
-axis.

(9) **y**

If the points $(4, 5)$ and $(2, 1)$ lie on the graph of a straight line, then the line
has a slope equal to _____(10)_____ .

(10) **2**

The x- and y-intercepts of the line $y = x$ are _____(11)_____ and
_____(12)_____ , respectively.

(11) **(0, 0)**   (12) **(0, 0)**

A point-slope form of an equation of the line passing through $(2, -3)$ and
having slope 5 is _____(13)_____ .

(13) **y + 3 = 5(x − 2)**

The point $(2, -6)$ lies in quadrant _____(14)_____ , while the point $(-2, 6)$
lies in quadrant _____(15)_____ .

(14) **IV**   (15) **II**

The abscissa of the point $(2, 3)$ is _____(16)_____ and its ordinate is
_____(17)_____ .

(16) **2**   (17) **3**

The slope of the straight line $y = 2x - 1$ is _____(18)_____ .

(18) **2**

An equation of the vertical line passing through $(5, -3)$ is _____(19)_____ .

(19)   **x = 5**

A variable representing elements in the domain of a function is called a(n) (dependent) (independent) variable.
$\overline{(20)}$

(20) **independent**

Which of the following graphs represent(s) functions of x? $\underline{(21)}$

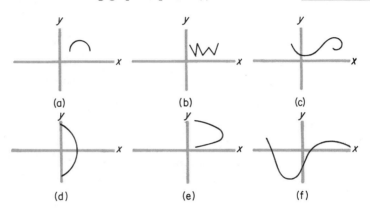

(21)   **(a), (b), and (f)**

A straight line whose slope is not defined has an equation of the form $\underline{(x=c)(y=c)}$.
(22)

(22) **x = c**

The slope of the line whose equation is $2y = 3x + 2$ is $\underline{(23)}$.

(23) $\dfrac{3}{2}$

True or false: The value obtained for the slope of a straight line will vary, depending on which two points on the line are used for its computation.
$\underline{(24)}$.

(24) **false**

A line whose slope is zero is (parallel) (perpendicular) to the y-axis.
$\overline{(25)}$

(25) **perpendicular**

If the line $y = kx + 4$ is parallel to the line $y = 7x + 16$, then $k =$ _____(26)_____ .

(26) **7**

REVIEW PROBLEMS

1. If $f(x) = 3x^2 - 4x + 7$, find $f(0)$, $f(-3)$, $f(5)$, and $f(x^2)$.

2. If $g(x) = (x - 3)/(x + 4)$, find $g(3)$, $g(-1)$, $g(2)$, and $g(x + h)$.

In Problems 3–12, sketch the graph of the given equation.

3. $y = -3x + 4$ 4. $y = 3x$

5. $2x = -y + 3$ 6. $3x - 2y = 4$

7. $\dfrac{x}{2} + \dfrac{y}{3} = 2$ 8. $y = x^2 - 2$

9. $y = -x^2 + 2x - 1$ 10. $2xy = 1$

11. $x = (y^2 - 1)^{1/2}$ 12. $y = \dfrac{3}{2x - 1}$

In Problems 13–15, transform each of the given equations to the point-slope form and the intercept form.

13. $3x - 2y = 4$ 14. $x = -3y + 4$

15. $-2x - 2y = 3$

16. Find an equation of the line passing through $(-2, 3)$ and $(4, 5)$.

17. Find an equation of the line through $(1, 2)$ that is perpendicular to the line $-3y + 5x = 7$.

18. Find an equation of the line which is parallel to the line $y = 3 + 2x$ and passes through the point $(3, 5)$.

19. Find an equation of the line passing through $(-6, 2)$ and having a slope of 3.

20. State the domain of the following functions:

     a. $y = \dfrac{3}{x - 2}$ b. $y = x^2 + 4$

21. Is the graph of $y = \dfrac{x}{x}$ the same as the graph of $y = 1$? Give a reason for your answer.

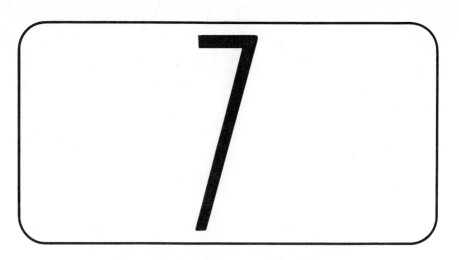

trigonometry

7-1

INTRODUCTION

Trigonometry, which is concerned in part with the relations between the angles and sides of triangles and in part with functions based on these relations, is an indispensable mathematical tool in practically all branches of science and engineering. For example, many of the problems from physics and mechanics involving forces, velocities, and displacements are simplified by the application of trigonometry. You may be familiar with the indirectness of trigonometric measurement. For instance, with trigonometry we can measure the height of a cliff without ever climbing it, and we can measure the width of a river without ever crossing it. Even in business and economics, in fact in any situation where the available data seem to suggest periodic phenomena, trigonometry can be a useful aid.

Our study of trigonometry begins with descriptions of angles and angular measurement and extends those concepts to include certain ratios, called trigonometric functions, which are dependent on an angle. We

conclude this chapter with the application of the trigonometric functions
to the solution of right triangles, followed by a discussion of *vectors*.

7-2
ANGLES AND ANGULAR MEASUREMENT

To begin we consider the *half-line* OA shown in Fig. 7-1(a). The term
half-line refers to that part of a straight line which extends indefinitely
to one side of a point O on the line. If, in a plane, this half-line is

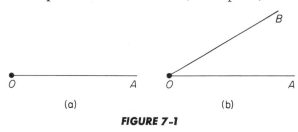

(a)   (b)

FIGURE 7-1

rotated about its *endpoint* O to a new position OB [Fig. 7-1(b)], an **angle**
is said to be generated. Essentially, the term *angle* refers to the amount
of this rotation. When the rotation is counterclockwise, the generated
angle is said to be a **positive angle**; when the rotation is clockwise, the
angle is said to be a **negative angle**. Figure 7-2(a) shows a positive angle
θ (the Greek letter theta), and Fig. 7-2(b) shows a negative angle θ. In

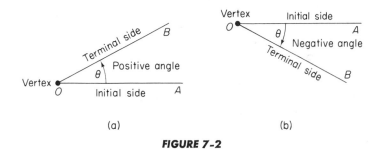

(a)   (b)

FIGURE 7-2

both cases the direction of rotation of the half-line from its original position to its new position is indicated by a curved arrow. Moreover, we
call OA the **initial side** of angle θ, OB the **terminal side** of θ, and
point O the **vertex** of θ.

When an angle has its vertex at the origin of a rectangular coordinate
system and its initial side coincides with the positive x-axis, we say the
angle is in **standard position**. Examples of angles in standard position

are given in Fig. 7-3. If the terminal side of an angle in standard position lies in the first quadrant, the angle is said to be a *first-quadrant angle*, and similarly for the other quadrants. Thus, in Fig. 7-3(b), θ is a third-quadrant angle, and in Fig. 7-3(c), θ is a second-quadrant angle.

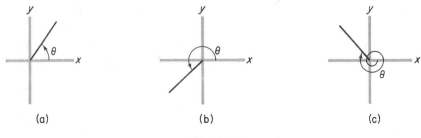

(a)   (b)   (c)

FIGURE 7-3

A comment on "reading" angles seems appropriate before we continue. When no confusion results, an angle may be referred to by the letter associated with its vertex. In Fig. 7-4(a) the angle can be called angle A. Similarly, in Fig. 7-4(b) the three angles of the triangle are angle A, angle B, and angle C. However, we can, at any time, designate a par-

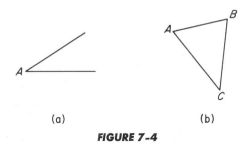

(a)   (b)

FIGURE 7-4

ticular angle by three letters, where the letter associated with the vertex is the middle letter. For example, in Fig. 7-4(b) angle A could be called angle BAC or angle CAB. You will find that regardless of how complex a figure might be, the three-letter designation will completely eliminate confusion in identifying a particular angle. In Fig. 7-5 we can identify three angles: angle ABC, angle ABD, and angle CBD. Angle ABC has been labeled α (the Greek letter alpha), angle ABD has been labeled β (the Greek letter beta), and angle CBD has been labeled γ (the Greek letter gamma). These angles may be referred to by either designation.

The term angle was said to refer to an amount of rotation. We are now ready to consider how we measure that rotation. There are two generally

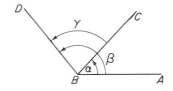

FIGURE 7-5

accepted units for measuring angles—*degrees* and *radians*. If a circle is divided into 360 equal arcs, the central angle subtended by each such arc is said to have a measure of *one degree*. Equivalently, we have the following definition:

Definition. One **degree**, denoted 1°, is the measure of the angle generated by $\frac{1}{360}$ of a complete revolution of a half-line.

In Fig. 7-6, various angles in standard position are indicated along with their corresponding measure in degrees.

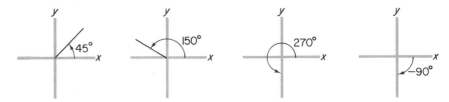

FIGURE 7-6

Certain angles are given special names. A positive angle less than 90° is called an **acute angle,** an angle of 90° is called a **right angle,** and an angle greater than 90° but less than 180° is an **obtuse angle.**

Since it is often necessary to deal with more precise measurements, a degree is subdivided into 60 equal parts called **minutes,** and each minute is further subdivided into 60 equal parts called **seconds.** The symbol ′ is used to denote minutes and the symbol ″ to denote seconds. Thus, 60′ = 1° and 60″ = 1′.

EXAMPLE 1 If angle ABC in Fig. 7-7 is 110° 47′ 32″ and angle DBC is 28° 58′ 51″, find the measure of angle ABD.

Clearly,

$$\measuredangle ABD = \measuredangle ABC - \measuredangle DBC \qquad (\measuredangle = \text{angle})$$
$$\measuredangle ABC = 110°\ 47′\ 32″$$
$$\measuredangle DBC = 28°\ 58′\ 51″$$

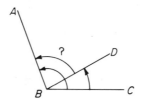

FIGURE 7-7

Since we cannot subtract 51″ from 32″, or 58′ from 47′, it is necessary to make a conversion from degrees to minutes and from minutes to seconds:

$$110° \ 47′ \ 32″ = 109° \ 107′ \ 32″ = 109° \ 106′ \ 92″.$$

Thus,

$$\sphericalangle ABD = 109° \ 106′ \ 92″ - 28° \ 58′ \ 51″ = 81° \ 48′ \ 41″.$$

In certain areas of science and technology it is somewhat inconvenient to use the degree as the unit of angular measurement. The other commonly used unit for measuring angles is the *radian*.

Definition. One **radian** is the measure of a central angle of a circle whose subtended arc is equal in length to the radius of the circle (see Fig. 7-8).

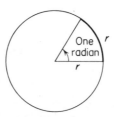

FIGURE 7-8

Since for a circle of radius r, a central angle of 1 radian subtends an arc of length r, a central angle of 2π radians subtends an arc of length $2\pi \cdot r$. But $2\pi r$ is the length of arc subtended by a central angle of 360° (that is, $2\pi r$ is the circumference). We conclude that

$$2\pi \text{ radians} = 360°$$

or

$$\boldsymbol{\pi \text{ radians} = 180°.} \tag{1}$$

Thus,

$$1 \text{ radian} = \frac{180°}{\pi}$$

$$\approx 57° \ 17′ \ 45″,$$

where the symbol $\approx$ means "is approximately equal to." Also, from Eq. (1) it follows that

$$1° = \frac{\pi}{180} \text{ radian}$$

$$\approx .0175 \text{ radian.}$$

When an angle is measured in radians, the word radian is frequently omitted. Thus, to speak of an angle of π it is understood to mean π radians; similarly, an angle of 1 means 1 radian. You must, however, always include the degree symbol if that unit of measure is used.

It is by relationship (1) that we convert degree measure to radian measure, and vice versa, as the following example shows.

EXAMPLE 2

a. Convert 30° to radian measure.

$$30° = 30°(1) = 30° \left(\frac{\pi \text{ radians}}{180°} \right)$$

$$= \frac{\pi}{6} \text{ radian} \approx .5236.$$

Note that $\pi/6$ is an exact answer, while .5236 radian is an approximation.

b. Convert $-240°$ to radian measure.

$$-240° = -240° \left(\frac{\pi}{180°} \right) = -\frac{4\pi}{3}.$$

c. Convert $\pi/4$ radians to degree measure.

$$\frac{\pi}{4} = \frac{\pi}{4} \left(\frac{180°}{\pi} \right) = 45°.$$

d. Convert $-\frac{5}{6}\pi$ radians to degree measure.

$$-\frac{5}{6}\pi = -\frac{5}{6}\pi \left(\frac{180°}{\pi} \right) = -150°.$$

e. Convert an angle of 10 to degree measure.

$$10 = 10 \left(\frac{180°}{\pi} \right) = \left(\frac{1800}{\pi} \right)°.$$

When the terminal sides of two or more angles in standard position coincide, the angles are said to be **coterminal angles**. For example, some pairs of coterminal angles are 0° and 360°, 10° and 370°, and 270° and $-90°$. Clearly, for any angle θ there are infinitely many positive angles, as well as infinitely many negative angles, that are coterminal with θ; these angles can be obtained by adding (or subtracting) integral multiples of 360° to (from) θ.

EXAMPLE
3

Find all angles θ, where $-1100° < \theta < 1100°$, which are coterminal with 10°.

One complete counterclockwise revolution beyond 10° yields an angle of 370°, an additional revolution yields 730° (Fig. 7-9), and a third revolution gives 1090°.

One complete clockwise revolution from a position of 10° gives an angle of $-350°$, a second revolution yields $-710°$ (Fig. 7-10), and a third revolution gives $-1070°$.

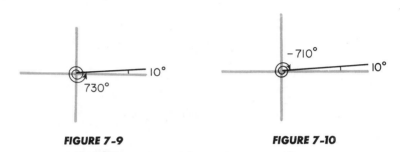

FIGURE 7-9   **FIGURE 7-10**

An additional revolution in either direction would give rise to an angle outside the specified interval. Hence, the required angles are $-1070°$, $-710°$, $-350°$, 10°, 370°, 730°, and 1090°.

EXERCISE 7-2

In Problems 1–20, convert the given angle to radians or degrees as required and draw the angle in standard position.

1. 60° to radians

2. $\frac{5\pi}{6}$ to degrees

3. $\frac{\pi}{2}$ to degrees

4. 135° to radians

5. $\frac{3\pi}{4}$ to degrees

6. 210° to radians

7. 330° to radians

8. $\frac{3\pi}{2}$ to degrees

9. 225° to radians

10. $\frac{2}{3}\pi$ to degrees

11. $\frac{7\pi}{6}$ to degrees

12. 15° to radians

13. $\frac{-\pi}{8}$ to degrees

14. $\frac{14\pi}{16}$ to degrees

15. $-60°$ to radians

16. 22° to radians

17. 45 to degrees

18. $\dfrac{-3\pi}{4}$ to degrees

19. 6π to degrees

20. 186° to radians

In Problems 21–24, determine θ.

21.

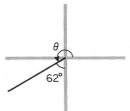

22.

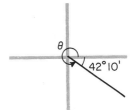

23.

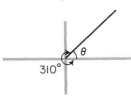

24.

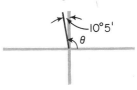

25. Through how many radians does the minute hand of a clock rotate in 40 min? In 40 hr?

26. Convert 7.26 radians/sec to revolutions per minute.

In Problems 27–34, find two positive angles and two negative angles that are coterminal with the given angle θ.

27. $\theta = 35°$

28. $\theta = 480°$

29. $\theta = 221°\ 51'\ 5''$

30. $\theta = -15°\ 27'\ 1''$

31. $\theta = -70°\ 20'\ 20''$

32. $\theta = \pi/4$

33. $\theta = \pi/3$

34. $\theta = 2000°$

In Problems 35 and 36, determine θ.

35.

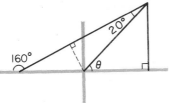

36.

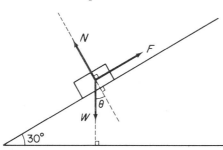

37. When a spectrometer is used to view the spectral lines of a light source, a doublet (two spectral lines representing light of nearly equal wavelengths) is observed at angular positions 121° 59′ 23″ and 122° 02′ 23″. What is the exact angular separation of the two lines, in radians?

7-3
TRIGONOMETRIC FUNCTIONS OF ACUTE ANGLES

The entire realm of trigonometry is based on the definition of six trigonometric functions. We shall first consider these functions for an acute angle in standard position; later in this chapter we shall give a complete treatment for any angle whatsoever.

Consider the angle θ in standard position shown in Fig. 7-11. On the terminal side of θ we arbitrarily select a point P, different from the origin,

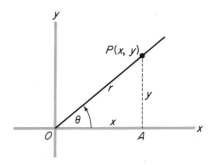

FIGURE 7-11

whose coordinates are (x, y). From P we construct a perpendicular to the x-axis, intersecting it at A. Note that right triangle OAP is formed, this triangle being called the **triangle of reference** for θ. The length of line segment $\overline{OA}$ is x; the length of $\overline{AP}$ is y. If we denote the length of $\overline{OP}$ by r, then by the Pythagorean theorem,

$$r^2 = x^2 + y^2$$

and

$$r = \sqrt{x^2 + y^2}.$$

The number r is often called the **radius vector** of P and is always taken to be positive.

Thus with point P there are associated three numbers—its abscissa x, its ordinate y, and the radius vector r that P determines—and from these numbers six ratios can be formed. It is these ratios that define the trigonometric functions of θ.

Definition. Let $P(x, y)$ be an arbitrary point on the terminal side of an angle θ in standard position, where $P(x, y) \neq (0, 0)$. If $r = \sqrt{x^2 + y^2}$, then the trigonometric functions of θ are

$$\text{sine } \theta = \sin \theta = \frac{y}{r}$$

$$\text{cosine } \theta = \cos \theta = \frac{x}{r}$$

$$\text{tangent } \theta = \tan \theta = \frac{y}{x}$$

$$\text{cotangent } \theta = \cot \theta = \frac{x}{y}$$

$$\text{secant } \theta = \sec \theta = \frac{r}{x}$$

$$\text{cosecant } \theta = \csc \theta = \frac{r}{y}.$$

(7-1)

You should completely familiarize yourself with these definitions of the trigonometric functions and with the indicated standard abbreviations of their names. The above definitions depend only on the angle θ and not on the choice of P for the following reason. If $P_1(x_1, y_1)$ is any other point on the terminal side of θ, then in Fig. 7-12 the triangles OAP and OBP_1

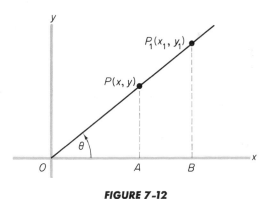

FIGURE 7-12

are similar triangles. Since, from geometry, the lengths of corresponding sides of similar triangles are proportional, it follows that for any ratio formed by the sides of triangle OAP, the corresponding ratio in triangle OBP_1 will be equal to it. Thus the ratios are independent of the choice of

P. As a result, the above definition associates, for example, the unique number sin θ with a given angle θ and in this sense defines a function, that is, the function $f(\theta) = \sin \theta$, and similarly for the other trigonometric functions.

EXAMPLE
4

If the terminal side of an acute angle θ in standard position passes through the point $(6, 8)$, find the values of the six trigonometric functions of θ (see Fig. 7-13).

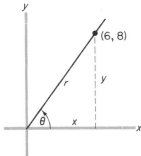

FIGURE 7-13

We construct a perpendicular from the point $(6, 8)$ to the x-axis (Fig. 7-13). Since $x = 6$ and $y = 8$, by the Pythagorean theorem $r^2 = 6^2 + 8^2$ or

$$r = \sqrt{6^2 + 8^2} = \sqrt{100} = 10.$$

Applying the definitions of the trigonometric functions of θ,

$$\sin \theta = \frac{y}{r} = \frac{8}{10} = \frac{4}{5}$$

$$\cos \theta = \frac{x}{r} = \frac{6}{10} = \frac{3}{5}$$

$$\tan \theta = \frac{y}{x} = \frac{8}{6} = \frac{4}{3}$$

$$\cot \theta = \frac{x}{y} = \frac{6}{8} = \frac{3}{4}$$

$$\sec \theta = \frac{r}{x} = \frac{10}{6} = \frac{5}{3}$$

$$\csc \theta = \frac{r}{y} = \frac{10}{8} = \frac{5}{4}.$$

From the definitions of the trigonometric functions in Eq. (7-1), it follows that

$$\cot \theta = \frac{x}{y} = \frac{1}{\dfrac{y}{x}} = \frac{1}{\tan \theta}.$$

In fact, we have the following identities:

$$\cot \theta = \frac{1}{\tan \theta}$$

$$\sec \theta = \frac{1}{\cos \theta} \tag{7-2}$$

$$\csc \theta = \frac{1}{\sin \theta}.$$

To describe these results we say the sine and cosecant, the cosine and secant, and the tangent and cotangent are pairs of **reciprocal functions**. This means, for example, that if $\sin \theta = .5$, then $\csc \theta = 1/.5 = 2$; if $\cos \theta = 2/3$, then $\sec \theta = 3/2$. Similarly, if $\cot \theta = 3$, then $\tan \theta = 1/3$, etc. We encourage you to verify Eq. (7-2) for the results of Example 4.

Suppose we wish to find the trigonometric functions of θ given the right triangle in Fig. 7-14(a). By performing certain movements we can

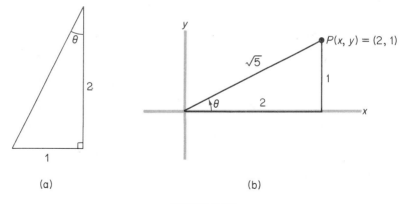

(a)   (b)

FIGURE 7-14

place θ in standard position as shown in Fig. 7-14(b). By the Pythagorean theorem, the length r of the hypotenuse is given by

$$r^2 = 2^2 + 1^2 = 5$$
$$r = \sqrt{5}.$$

We can now find the trigonometric functions of θ from their definitions:

$$\sin \theta = \frac{y}{r} = \frac{1}{\sqrt{5}} = \frac{\sqrt{5}}{5}$$

$$\cos \theta = \frac{x}{r} = \frac{2}{\sqrt{5}} = \frac{2\sqrt{5}}{5}$$

$$\tan \theta = \frac{y}{x} = \frac{1}{2}$$

$$\cot \theta = \frac{x}{y} = \frac{2}{1} = 2$$

$$\sec \theta = \frac{r}{x} = \frac{\sqrt{5}}{2}$$

$$\csc \theta = \frac{r}{y} = \frac{\sqrt{5}}{1} = \sqrt{5}.$$

Based on the above results, you should note that, in effect, the values of the trigonometric functions of an acute angle θ in a right triangle can be determined from the right triangle *directly* if the definitions of those functions are interpreted as follows:

$$\sin \theta = \frac{\textbf{Opposite side}}{\textbf{Hypotenuse}} \qquad \cot \theta = \frac{\textbf{Adjacent side}}{\textbf{Opposite side}}$$

$$\cos \theta = \frac{\textbf{Adjacent side}}{\textbf{Hypotenuse}} \qquad \sec \theta = \frac{\textbf{Hypotenuse}}{\textbf{Adjacent side}} \qquad (7\text{-}3)$$

$$\tan \theta = \frac{\textbf{Opposite side}}{\textbf{Adjacent side}} \qquad \csc \theta = \frac{\textbf{Hypotenuse}}{\textbf{Opposite side}}$$

The usual reciprocal relationships of Eq. (7-2) are still valid, and it is of great importance that the correctness of Eq. (7-3) as a special interpretation of the basic definitions of the trigonometric functions given in Eq. (7-1) be verified by you.

EXAMPLE
5

Find the six trigonometric functions of the acute angle θ given that $\cos \theta = \frac{5}{7}$.

An appropriate triangle is given in Fig. 7-15. The third side is y, where

$$5^2 + y^2 = 7^2$$
$$y^2 = 49 - 25 = 24$$
$$y = \sqrt{24} = \sqrt{4}\sqrt{6} = 2\sqrt{6}.$$

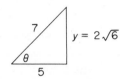

FIGURE 7-15

Thus the side opposite θ has length $2\sqrt{6}$, the adjacent side has length 5, and the hypotenuse is 7. Using Eq. (7-3) we obtain

$$\sin \theta = \frac{\text{Opposite side}}{\text{Hypotenuse}} = \frac{2\sqrt{6}}{7}$$

$$\cos \theta = \frac{\text{Adjacent side}}{\text{Hypotenuse}} = \frac{5}{7}$$

$$\tan \theta = \frac{\text{Opposite side}}{\text{Adjacent side}} = \frac{2\sqrt{6}}{5}$$

$$\cot \theta = \frac{\text{Adjacent side}}{\text{Opposite side}} = \frac{5}{2\sqrt{6}} = \frac{5\sqrt{6}}{12}$$

$$\sec \theta = \frac{\text{Hypotenuse}}{\text{Adjacent side}} = \frac{7}{5}$$

$$\csc \theta = \frac{\text{Hypotenuse}}{\text{Opposite side}} = \frac{7}{2\sqrt{6}} = \frac{7\sqrt{6}}{12}.$$

Alternatively, you can find $\sin \theta$, $\cos \theta$, $\tan \theta$, and use the reciprocal relationships of Eq. (7-2) for the remaining functions.

EXERCISE 7-3

In Problems 1–16, determine values for all trigonometric functions of angle θ if θ is in standard position and the terminal side of θ passes through the given point.

1. $(8, 6)$

2. $(3, 3)$

3. $(5, 1)$

4. $(3, 4)$

5. $(1, \sqrt{3})$

6. $(\sqrt{3}, 1)$

7. $(4\sqrt{3}, 4)$

8. $(2, 3)$

9. $(3, \sqrt{3})$

10. $(2, 1)$

11. $(2, 2)$

12. $(5, 12)$

13. $(\sqrt{2}, 1)$

14. $(\sqrt{15}, 1)$

15. (b, a)

16. $(\sqrt{c^2 - a^2}, a)$

In Problems 17–22, a trigonometric function of an acute angle θ is given. Find the remaining trigonometric functions.

17. $\sin \theta = 2/5$

18. $\cos \theta = 2/3$

19. $\tan \theta = 4$

20. $\sec \theta = 5/4$

21. $\csc \theta = 1.25$

22. $\cot \theta = .75$

7-4

TRIGONOMETRIC FUNCTIONS OF 0°, 30°, 45°, 60°, AND 90°

For certain so-called *special angles,* among which are 0°, 30°, 45°, 60°, and 90°, we can make use of basic geometrical concepts to evaluate their trigonometric functions. Figure 7-16 shows an angle of 0° in standard

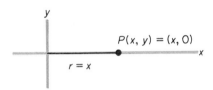

FIGURE 7-16

position. In this case, the initial and terminal sides coincide. If $P(x, y)$ is any point on the terminal side, it must lie on the x-axis and thus $y = 0$. Hence $r^2 = x^2 + 0^2 = x^2$ and $r = x$. Therefore,

$$\sin 0° = \frac{y}{r} = \frac{0}{r} = 0$$

$$\cos 0° = \frac{x}{r} = 1, \quad \text{since } r = x$$

$$\tan 0° = \frac{y}{x} = \frac{0}{x} = 0$$

$$\cot 0° = \frac{x}{y} = \frac{x}{0}, \quad \text{undefined}$$

$$\sec 0° = \frac{r}{x} = 1, \quad \text{since } r = x$$

$$\csc 0° = \frac{r}{y} = \frac{r}{0}, \quad \text{undefined.}$$

In a similar fashion as above, an analysis of Fig. 7-17 gives the trigonometric functions of 90°.

$$\sin 90° = \frac{y}{r} = 1, \quad \text{since } r = y$$

$$\cos 90° = \frac{x}{r} = \frac{0}{r} = 0$$

$$\tan 90° = \frac{y}{x} = \frac{y}{0}, \quad \text{undefined}$$

$$\cot 90° = \frac{x}{y} = \frac{0}{y} = 0$$

$$\sec 90° = \frac{r}{x} = \frac{r}{0}, \qquad \text{undefined}$$

$$\csc 90° = \frac{r}{y} = 1, \qquad \text{since } r = y.$$

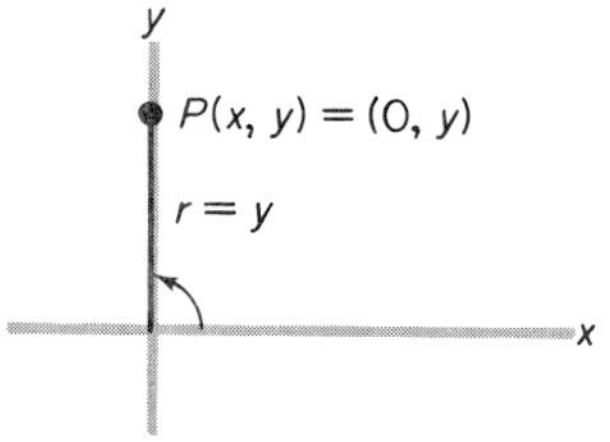

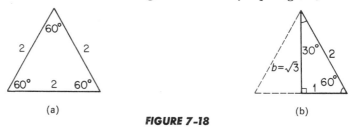

FIGURE 7-17

Figure 7-18(a) shows an equilateral triangle, each side of which has a length of two units. The triangle is necessarily equiangular, each angle

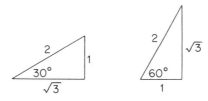

(a)   (b)

FIGURE 7-18

being 60°. The bisector of any of these angles will also be the perpendicular bisector of the side opposite the angle, and Fig. 7-18(b) shows the result of constructing such a bisector. The length b of the bisector is found by the Pythagorean theorem:

$$2^2 = 1^2 + b^2 \qquad \text{and thus} \qquad b = \sqrt{4-1} = \sqrt{3}.$$

If we now consider the triangles in Fig. 7-19 [obtained from Fig.

FIGURE 7-19

7-18(b)], we can find the trigonometric functions of 30° and 60° immediately from Eq. (7-3).

$$\sin 30° = \frac{1}{2} \qquad\qquad \sin 60° = \frac{\sqrt{3}}{2}$$

$$\cos 30° = \frac{\sqrt{3}}{2} \qquad\qquad \cos 60° = \frac{1}{2}$$

$$\tan 30° = \frac{1}{\sqrt{3}} = \frac{\sqrt{3}}{3} \qquad \tan 60° = \frac{\sqrt{3}}{1} = \sqrt{3}$$

$$\cot 30° = \frac{\sqrt{3}}{1} = \sqrt{3} \qquad \cot 60° = \frac{1}{\sqrt{3}} = \frac{\sqrt{3}}{3}$$

$$\sec 30° = \frac{2}{\sqrt{3}} = \frac{2\sqrt{3}}{3} \qquad \sec 60° = \frac{2}{1} = 2$$

$$\csc 30° = \frac{2}{1} = 2 \qquad\qquad \csc 60° = \frac{2}{\sqrt{3}} = \frac{2\sqrt{3}}{3}.$$

Figure 7-20 is a square, each side of which has a length of one unit. Recall that the diagonal of a square forms angles of 45° with the sides of the square. By the Pythagorean theorem, the length d of the diagonal is given by

$$d^2 = 1^2 + 1^2$$
$$d = \sqrt{2}.$$

From Fig. 7-20(b) we obtain

$$\sin 45° = \frac{1}{\sqrt{2}} = \frac{\sqrt{2}}{2}$$

$$\cos 45° = \frac{1}{\sqrt{2}} = \frac{\sqrt{2}}{2}$$

$$\tan 45° = \frac{1}{1} = 1$$

$$\cot 45° = \frac{1}{1} = 1$$

$$\sec 45° = \frac{\sqrt{2}}{1} = \sqrt{2}$$

$$\csc 45° = \frac{\sqrt{2}}{1} = \sqrt{2}.$$

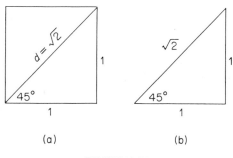

(a)   (b)

FIGURE 7-20

The results of the preceding discussion of special angles are summarized in Table 7-1. These angles occur so frequently that you should become totally familiar with these results. Rather than have you memorize them, however, we suggest that you memorize the triangles, etc. from which the results were obtained and use them in conjunction with the definitions of the trigonometric functions.

Table 7-1	0°	30°	45°	60°	90°
$\sin \theta$	0	$\dfrac{1}{2}$	$\dfrac{\sqrt{2}}{2}$	$\dfrac{\sqrt{3}}{2}$	1
$\cos \theta$	1	$\dfrac{\sqrt{3}}{2}$	$\dfrac{\sqrt{2}}{2}$	$\dfrac{1}{2}$	0
$\tan \theta$	0	$\dfrac{\sqrt{3}}{3}$	1	$\sqrt{3}$	Undefined
$\cot \theta$	Undefined	$\sqrt{3}$	1	$\dfrac{\sqrt{3}}{3}$	0
$\sec \theta$	1	$\dfrac{2\sqrt{3}}{3}$	$\sqrt{2}$	2	Undefined
$\csc \theta$	Undefined	2	$\sqrt{2}$	$\dfrac{2\sqrt{3}}{3}$	1

One very useful relationship that exists between the acute angles of a right triangle can be observed from Fig. 7-21. You may recall from geo-

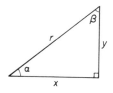

FIGURE 7-21

metry that the sum of the angles of any triangle must be 180°. Hence, in the figure the sum of angles α and β must be 90°; that is, angle α and angle β are **complementary angles**. Using Eq. (7-3) you may verify that

$$\sin \alpha = \frac{y}{r} = \cos \beta$$

$$\cos \alpha = \frac{x}{r} = \sin \beta$$

$$\tan \alpha = \frac{y}{x} = \cot \beta$$

$$\cot \alpha = \frac{x}{y} = \tan \beta$$

$$\sec \alpha = \frac{r}{x} = \csc \beta$$

$$\csc \alpha = \frac{r}{y} = \sec \beta.$$

(7-4)

The pairs of functions sine and cosine, tangent and cotangent, and secant and cosecant are called **cofunctions**. Since we considered angle α to be *any* acute angle, the results of Eq. (7-4) are completely general. Hence, we may conclude:

A trigonometric function of any acute angle is equal to the corresponding cofunction of its complementary angle.

For example, $\sin 10° = \cos 80°$ and $\sec 30° = \csc 60°$. Do not confuse the cofunction relationships of Eq. (7-4) with the reciprocal relationships of Eq. (7-2). For example, $\tan 30° = \cot 60°$ and $\tan 30° = \dfrac{1}{\cot 30°}$.

EXERCISE 7-4

In Problems 1–4, determine the values, if any, for all trigonometric functions of angle θ if θ is in standard position and the terminal side of θ passes through the given point.

1. $(2, 0)$
2. $(0, \frac{1}{2})$
3. $(0, 5)$
4. $(\frac{1}{4}, 0)$

In Problems 5–18, state the value, if any, of the indicated quantity.

5. $\sin (\pi/4)$
6. $\cos (\pi/3)$
7. $\tan (\pi/6)$
8. $\tan (\pi/4)$
9. $\cos (\pi/2)$
10. $\csc (\pi/2)$

11. $\sin(\pi/3)$ **12.** $\tan(\pi/2)$

13. $\sec(\pi/4)$ **14.** $\cos(\pi/6)$

15. $\left(\cos\dfrac{\pi}{4}\right)^2$ **16.** $\left(\sin\dfrac{\pi}{4}\right)\left(\cos\dfrac{\pi}{4}\right)$

17. $\left(\tan\dfrac{\pi}{3}\right)(\sin 0)$ **18.** $(\cos 0°)\left(\sin\dfrac{\pi}{6}\right)$

In Problems 19–24, use your knowledge of cofunctions and reciprocal functions to classify the given statements as true or false.

19. $\sin 35° = \csc 55°$ **20.** $\csc 50° = \dfrac{1}{\sin 50°}$

21. $\tan 2° = \dfrac{1}{\cot 88°}$ **22.** $\cos 4° = \sec 86°$

23. $\cos 20° = \sin 70°$ **24.** $\tan 45° = \dfrac{1}{\cot 45°}$

25. The index of refraction n of a prism with apex angle A is given by

$$n = \frac{\sin\left(\dfrac{A+\delta}{2}\right)}{\sin\left(\dfrac{A}{2}\right)},$$

where δ (delta) is called the angle of minimum deviation. If a prism has an angle of minimum deviation of $30°$ and an apex angle of $60°$, determine its index of refraction.

7-5
THE TRIGONOMETRIC TABLES—INTERPOLATION

The geometric approach to finding the trigonometric functions of an angle is quite limited. For instance, there is no convenient geometric figure we could use that would allow us to determine the trigonometric functions of such angles as $32°$, $11° 20'$, or $89°$. Fortunately, however, extensive tables of trigonometric functions of various angles have been prepared. A typical table is given in Appendix C. Most entries are approximations to the true values. In this section, the use of the tables will be illustrated.

By the use of the table in Appendix C, it is possible to find trigonometric functions of certain angles from $0°$ to $90°$ directly. The angles from $0°$ to $45°$ are listed, in increments of $10'$, in the **left-hand columns**, and the names of the functions for these angles are listed along the **top rows**. There are also columns which give the radian measure of an indicated angle. The angles from $45°$ to $90°$ are listed in the **right-hand columns**, and the names of the functions for these angles are listed along the **bottom rows**.

Note that from 0° to 45° the angles are listed in increasing order, but from 45° to 90° they are in decreasing order. For example, the entry under 89° 00' refers to 88° 50', not 89° 50'. To help avoid errors, *always read the entries for 0°–45° in the direction of top to bottom and the entries for 45°–90° in the direction from bottom to top.*

In the following example we show how to find a trigonometric function of an angle given in the tables.

EXAMPLE
6

a.　Find sin 37°.

　　Locate 37° in the **left-hand** column and move to the right until vertically under the *sin* function heading at the **top** of the page. We find that sin 37° = .6018.

b.　Find csc 61° 20'.

　　Locate 61° 20' in the **right-hand** column and move to the left until vertically above the *csc* function heading at the **bottom** of the page. We find that csc 61° 20' = 1.140. Note that the integer part of the number, along with the decimal point, was carried over from a prior entry.

c.　Find tan 15° 20'.

　　Find 15° 20' in the **left-hand** column and move to the right until vertically under the *tan* function heading at the **top** of the page. We find that tan 15° 20' = .2742. Here again, part of the number was carried over from a prior entry.

We now show how to find an angle when we are given the value of one of its trigonometric functions.

EXAMPLE
7

a.　Find the acute angle θ given that cot $\theta = 1.419$.

　　Here we must reverse the procedure of Example 6. Examining the cotangent columns of the trigonometric tables, we find that 1.419 is the cotangent of the angle $\theta = 35°\ 10'$.

b.　When a beam of electrons moving in a transparent medium travels at a speed greater than the speed of light in that medium, visible radiation, called Cerenkov radiation, is emitted at an angle θ to the direction of motion of the electrons, where

$$\cos \theta = \frac{c}{nv}.$$

Here c is the speed of light in vacuum, n is the index of refraction of the medium, and v is the speed of the electrons. For a particular crystal, $n = 1.785$ and $v = .5710c$; find the angle θ.

$$\cos \theta = \frac{c}{nv} = \frac{c}{(1.785)(.5710c)} = .9811.$$

From the tables we find

$$\theta = 11° \, 10'.$$

Many instances occur when the trigonometric functions of an unlisted angle such as 27° 13′ are needed. It is possible by using a method of *linear interpolation* to determine approximately the functions of such an angle. Linear interpolation is based on the assumption that the values of a trigonometric function vary uniformly with a small change in the angle; that is, it assumes that over a small interval the graph of a trigonometric function is a straight line. This assumption is, in general, *completely false*; however, when the change in the angle is quite small, a matter of a few minutes or so, then the interpolation technique yields a very good approximation.

EXAMPLE 8

Find sin 27° 13′.

The angle 27° 13′ lies between 27° 10′ and 27° 20′, both of which can be located in the tables. We can determine the sine of each angle directly. This information is tabulated below and the general technique explained.

$$10' \left\{ 3' \left\{ \begin{matrix} \sin 27° \, 10' = .4566 \\ \sin 27° \, 13' = \quad ? \end{matrix} \right\} x \atop \sin 27° \, 20' = .4592 \right\} .0026$$

As an angle increases 10′ from 27° 10′ to 27° 20′, the sine of the angle increases by .0026 from .4566 to .4592. The angle 27° 13′ is $\frac{3}{10}$ "of the way" from 27° 10′ to 27° 20′, and, *assuming linear or uniform behavior,* sin 27° 13′ is $\frac{3}{10}$ "of the way" from .4566 to .4592. Now, solving the proportion

$$\frac{x}{.0026} = \frac{3'}{10'}$$

gives

$$x = \tfrac{3}{10}(.0026) = .00078,$$

and, rounding off to four decimal places, we have $x = .0008$. Thus, as the angle 27° 10′ increased by 3′, the sine of the angle changed by .0008. Since the sine of an angle **increases** as the angle goes from 27° 10′ to 27° 20′, the adjustment number .0008 must be **added** to sin 27° 10′. Thus,

$$\sin 27° \, 13' = \sin 27° \, 10' + x$$
$$= .4566 + .0008 = .4574.$$

We remind you that the use of most mathematical tables employing decimal notation involves approximations, and although we accept the statement sin 27° 13′ = .4574, you should realize that for mathematical exactness the statement should be written sin 27° 13′ ≈ .4574.

EXAMPLE

9

Find cos 78° 46'.

The angle 78° 46' lies between 78° 40' and 78° 50'. As in the previous example, the necessary data are tabulated below. **We always write the smallest angle first.**

$$10' \left\{ 6' \left\{ \begin{matrix} \cos 78° 40' = .1965 \\ \cos 78° 46' = \quad ? \end{matrix} \right\} x \right\} .0028 \\ \cos 78° 50' = .1937 \right\}$$

Assuming cos 78° 46' is $\frac{6}{10}$ "of the way" from cos 78° 40' to cos 78° 50', we first find x, where

$$\frac{x}{.0028} = \frac{6'}{10'}$$

$$x = \frac{6}{10}(.0028) = .00168 \approx .0017.$$

Since $\cos \theta$ **decreases** as θ goes from 78° 40' to 78° 50',

$$\cos 78° 46' = \cos 78° 40' - x$$
$$= .1965 - .0017$$
$$= .1948.$$

EXAMPLE

10

a. Find the acute angle θ if $\tan \theta = 1.168$.

Examining the tangent columns of the trigonometric tables, we find that 1.168 lies between the entries 1.164 and 1.171, which correspond to angles of 49° 20' and 49° 30', respectively. We first tabulate the necessary data, *writing the smallest angle first*.

$$10' \left\{ x \left\{ \begin{matrix} \tan 49° 20' = 1.164 \\ \tan \theta \quad\;\; = 1.168 \end{matrix} \right\} .004 \right\} .007 \\ \tan 49° 30' = 1.171 \right\}$$

Since $\tan \theta$ is .004/.007 "of the way" from tan 49° 20' to tan 49° 30', we reason that θ is .004/.007 "of the way" from 49° 20' to 49° 30', which differ by 10'. Equivalently,

$$\frac{x}{10'} = \frac{.004}{.007}$$

$$x = \frac{4}{7}(10') \approx 5.7' \approx 6'.$$

Therefore, $\theta = 49° 20' + x = 49° 20' + 6' = 49° 26'$.

b. Find the acute angle θ if $\cos \theta = .8360$.

Examining the cosine columns of the trigonometric tables, we find that .8360 lies between the entries .8371 and .8355, which correspond to angles of 33° 10' and 33° 20', respectively. We first tabulate the data, *writing the smallest angle first*.

$$10' \left\{ x \left\{ \begin{matrix} \cos 33° \ 10' = .8371 \\ \cos \ \theta \quad = .8360 \end{matrix} \right\} .0011 \right\} .0016$$
$$\begin{matrix} \cos 33° \ 20' = .8355 \end{matrix}$$

Using the same reasoning given in part a, we find

$$\frac{x}{10'} = \frac{.0011}{.0016}$$

$$x = \frac{11}{16}(10') \approx 6.9' \approx 7'.$$

Therefore, $\theta = 33° \ 10' + 7' = 33° \ 17'$.

EXERCISE 7-5

Find the values of the trigonometric functions in Problems 1–24. Use the method of linear interpolation where needed.

1. $\sin 32°$ **2.** $\cos 49°$

3. $\cot 11°$ **4.** $\sec 58°$

5. $\tan (28° \ 30')$ **6.** $\cot (15° \ 40')$

7. $\cos \dfrac{4\pi}{9}$ **8.** $\sin \dfrac{\pi}{12}$

9. $\sin (53° \ 12')$ **10.** $\tan (12° \ 5')$

11. $\sec (84° \ 48')$ **12.** $\cos (79° \ 16')$

13. $\cot (33° \ 24')$ **14.** $\csc (43° \ 36')$

15. $\tan (76° \ 22')$ **16.** $\sin (88° \ 8')$

17. $\cos (28° \ 46')$ **18.** $\csc 15'$

19. $\cot \dfrac{\pi}{4}$ **20.** $\tan \dfrac{\pi}{18}$

21. $\cos (53.6°)$ **22.** $\sin (61° \ 38')$

23. $\cot (64° \ 32')$ **24.** $\sin \dfrac{\pi}{5}$

In Problems 25–38, determine the acute angle θ to the nearest minute subject to the given condition.

25. $\cos \theta = .7470$ **26.** $\sin \theta = .3201$

27. $\sin \theta = .9628$ **28.** $\tan \theta = 2.628$

29. $\cos \theta = .7071$ **30.** $\sin \theta = .2120$

31. $\sec \theta = 1.660$ **32.** $\cot \theta = .7683$

33. $\tan \theta = .7601$ **34.** $\cos \theta = .4741$

35. $\cot \theta = 1.180$

36. $\sin \theta = .8350$

37. $\tan \theta = 4.022$

38. $\csc \theta = 2.136$

39. For total internal reflection at a glass-air interface, the critical angle of incidence, θ_c, is given by the expression

$$1.5 \sin \theta_c = 1.$$

Determine the value of θ_c to the nearest minute.

40. Without the use of trigonometric tables, find the value of the tangent of the critical angle of Problem 39.

41. When a beam of electrons moves at a speed of $0.6c$ in a crystal whose index of refraction is 1.75, at what angle to the electron beam is the Cerenkov radiation emitted? Refer to Example 7b.

42. The angle of deviation θ for which maxima occur in the pattern produced by a plane diffraction grating can be found from the equation

$$\sin \theta = \frac{m\lambda}{d},$$

where m is the order of the observed spectrum, λ is the wavelength of light, and d is the grating spacing. If a particular grating has a spacing of $d = 1.69 \times 10^{-4}$ cm, what is the angular deviation of violet light of wavelength $\lambda = 4 \times 10^{-5}$ cm in the first-order $(m = 1)$ visible spectrum?

7-6
TRIGONOMETRIC FUNCTIONS OF ANY ANGLE

Our study of the trigonometric functions has been, to this point, somewhat restricted. Thus far, we have essentially considered the values of the trigonometric functions only of acute angles. We shall now consider the case for any angle in general.

To find the trigonometric functions of any angle θ we follow the same procedure as before:

a. Place θ in standard position.

b. Construct a perpendicular from any point $P(x, y)$ (other than the origin) on the terminal side of θ to the x-axis, thus forming a reference triangle.

c. If $r = \sqrt{x^2 + y^2}$, the values of x, y, and r are used to define the trigonometric functions of θ:

$$\sin \theta = \frac{y}{r}$$

$$\cos \theta = \frac{x}{r}$$

$$\tan \theta = \frac{y}{x}$$

$$\cot \theta = \frac{x}{y}$$

$$\sec \theta = \frac{r}{x}$$

$$\csc \theta = \frac{r}{y}.$$

EXAMPLE
11

Find the trigonometric functions of θ given that θ is in standard position and the point $(-3, 2)$ lies on its terminal side (Fig. 7-22).

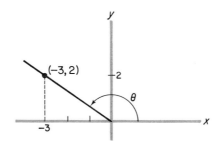

FIGURE 7-22

Letting $x = -3$ and $y = 2$, we first find r:

$$r = \sqrt{(-3)^2 + 2^2} = \sqrt{13}.$$

Thus,

$$\sin \theta = \frac{y}{r} = \frac{2}{\sqrt{13}} = \frac{2\sqrt{13}}{13}$$

$$\cos \theta = \frac{x}{r} = \frac{-3}{\sqrt{13}} = -\frac{3\sqrt{13}}{13}$$

$$\tan \theta = \frac{y}{x} = \frac{2}{-3} = -\frac{2}{3}$$

$$\cot \theta = \frac{x}{y} = \frac{-3}{2} = -\frac{3}{2}$$

$$\sec \theta = \frac{r}{x} = \frac{\sqrt{13}}{-3} = -\frac{\sqrt{13}}{3}$$

$$\csc \theta = \frac{r}{y} = \frac{\sqrt{13}}{2}.$$

Observe in Example 11 that some of the trigonometric functions took positive values and others negative values. Suppose we now consider the various possibilities for the signs of the trigonometric functions of any angle θ. Let $P(x, y)$ be a point on the terminal side of θ and let $r = \sqrt{x^2 + y^2}$. Since in the first quadrant x, y, and r are positive, all trigonometric functions of a first-quadrant angle are positive. For a second-quadrant angle, x is negative, while y and r are positive; hence, only the sine and cosecant are positive. By similar reasoning you should verify the signs of the functions in the various quadrants as given in Table 7-2.

Table 7-2	Quadrant	$\sin \theta$	$\cos \theta$	$\tan \theta$	$\cot \theta$	$\sec \theta$	$\csc \theta$
	I	+	+	+	+	+	+
	II	+	−	−	−	−	+
	III	−	−	+	+	−	−
	IV	−	+	−	−	+	−

The results summarized in Table 7-2 can be most easily remembered by observing those functions that are *positive*. In the first quadrant we note that all the functions are positive. In the second quadrant the only functions that are positive are the sine and the cosecant (reciprocal of the sine). In the third quadrant the tangent and its reciprocal function, the cotangent, are positive, while in the fourth quadrant the cosine and its reciprocal, the cosecant, are positive. If you thoroughly familiarize yourself with these results before continuing, some of the difficulties usually encountered in determining the trigonometric functions of an arbitrary angle θ will be eliminated. The following table may prove helpful to you.

Quadrant	Functions That Are Positive
I	All functions
II	Sine, cosecant
III	Tangent, cotangent
IV	Cosine, secant

Figure 7-23(a), (b), and (c) illustrates various configurations for second-, third-, and fourth-quadrant angles, respectively. In each case we have selected an arbitrary point $P(x, y)$ on the terminal side of θ and have constructed a perpendicular from $P(x, y)$ to the x-axis, thus forming a reference triangle. If you now examine Fig. 7-23, you will note that in each diagram an *acute angle* α is formed between the terminal side of angle θ and the x-axis. We shall call α the **reference angle** for θ. It should be clear that regardless of θ, the reference angle α can always be determined

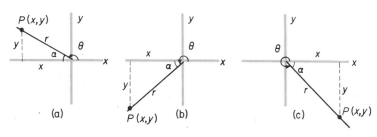

FIGURE 7-23

and it will always be an *acute angle* whose trigonometric functions can be found by using the trigonometric tables. Why do we want the trigonometric functions of the reference angle α? If you will study Fig. 7-23 in conjunction with the definitions of the trigonometric functions, you should realize that **the absolute values of the trigonometric functions of angle θ are equal to the corresponding values of the trigonometric functions of the acute reference angle α,** the latter values being positive. Thus, the only difference that might exist between a trigonometric function of θ and the corresponding function of α is in their signs, whether positive or negative. However, this presents no problem since we have already determined the signs of all the functions in all four quadrants.

We can now summarize the sequence of operations used to evaluate the trigonometric functions of any angle θ:

a. *Place the angle θ in standard position.*

b. *Construct a perpendicular from the terminal side of θ to the x-axis, thus forming a reference triangle.*

c. *Determine the size of the acute reference angle α between the x-axis and the terminal side of the angle θ.*

d. *Determine the values of the trigonometric functions of the acute reference angle α.*

e. *Affix to the value of each function the proper sign, depending on which quadrant contains the terminal side of angle θ.*

EXAMPLE
12

a. Find sin 135° [Fig. 7-24(a)].

Since 135° is a second-quadrant angle, sin 135° is positive. The reference angle α is $180° - 135° = 45°$. Thus,

$$\sin 135° = +\sin 45° = \frac{\sqrt{2}}{2}.$$

b. Find cos 225° [Fig. 7-24(b)].

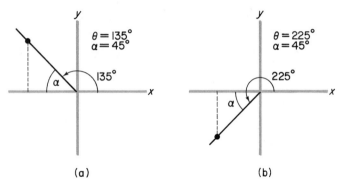

FIGURE 7-24

Since 225° is a third-quadrant angle, cos 225° is negative. The reference angle α is 225° − 180° = 45°. Thus,

$$\cos 225° = -\cos 45° = -\frac{\sqrt{2}}{2}.$$

c. Find cot (−18°) (Fig. 7-25).

Since −18° is a fourth-quadrant angle, cot (−18°) is negative. The reference angle α is clearly 18°. From trigonometric tables we find that cot 18° = 3.078. Therefore,

$$\cot (-18°) = -\cot 18° = -3.078.$$

d. Find tan (−510°) (Fig. 7-26).

Since −510° is a third-quadrant angle, tan (−510°) is positive. The reference angle α is 30°. Thus,

$$\tan (-510°) = +\tan 30° = \frac{\sqrt{3}}{3}.$$

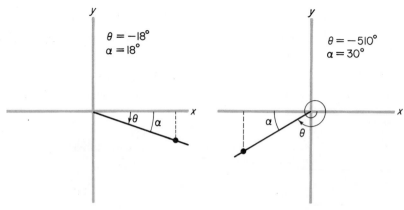

FIGURE 7-25

FIGURE 7-26

A comment seems appropriate at this time. You should be aware that from the definitions of the trigonometric functions, ***trigonometric functions of coterminal angles are equal*** since such angles lie in the same quadrant and have the same reference angle associated with them. For example, $\sin(-190°) = \sin(170°) = \sin(530°)$ since $-190°$, $170°$, and $530°$ are coterminal angles.

EXAMPLE
13

a. Find all angles θ, where $0° \leq \theta < 360°$, such that $\tan \theta = -.3640$.

From the trigonometric tables it can be determined that $\tan 20° = .3640$, and, therefore, the reference angle α of θ is $20°$. Now the tangent function is negative in the second and fourth quadrants and, clearly, there must be an angle θ in each quadrant to satisfy the given conditions. (See Fig. 7-27.) Thus, $\theta = 160°$ and $\theta = 340°$ each meet the condition that $\tan \theta = -.3640$.

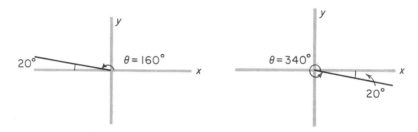

FIGURE 7-27

b. Find all angles θ, where $0° \leq \theta < 360°$, such that $\csc \theta = 1.464$ and $\cot \theta$ is negative.

We note that $\csc \theta$ is positive, and this can occur only if θ is in the first or second quadrants. However, the condition that $\cot \theta$ be negative restricts the solution to an angle in the second quadrant. From the trigonometric tables and interpolation we find that $\csc 43° 05' = 1.464$. Therefore, the reference angle α of θ is $43° 05'$ and, as seen in Fig. 7-28, the required angle is $136° 55'$; that is, we have $\csc 136° 55' = 1.466$ and $\cot 136° 55'$ is negative.

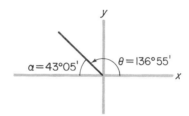

FIGURE 7-28

We have, to this point, dealt primarily with angles that lie *in* a quadrant, although we have determined the trigonometric functions of 0° and 90°. These two angles, as well as 180° and 270°, are called *quadrantal angles.* More precisely, any angle which in standard position has its terminal side on an axis is called a **quadrantal angle.** It is left as an exercise for you to determine the trigonometric functions of 180° and 270° by applying the methods used to evaluate the functions of 0° and 90°. The tabulated results of finding the values of the functions of these quadrantal angles appear in Table 7-3.

Table 7-3

Angle	sin	cos	tan	cot	sec	csc
0°	0	1	0	Undefined	1	Undefined
90°	1	0	Undefined	0	Undefined	1
180°	0	−1	0	Undefined	−1	Undefined
270°	−1	0	Undefined	0	Undefined	−1

EXERCISE 7-6

Evaluate the trigonometric functions in Problems 1–30. Do it without tables if possible.

1. sin 150°
2. sin 120°
3. cos 240°
4. cos 225°
5. tan (−225°)
6. tan 300°
7. cot 315°
8. cot (−210°)
9. sec 210°
10. sec 330°
11. csc 135°
12. csc 150°
13. tan 337°
14. sin 190° 30′
15. csc 534°
16. cot 295°
17. sin 100° 20′
18. cos 349°
19. cot (−160°)
20. sec (−255°)
21. sec (−84°)
22. sec 500°
23. cos (−553°)
24. cos 140° 40′
25. tan 570°
26. csc 270°
27. tan 161° 35′
28. csc 187°
29. sin (−740°)
30. cos (−98°)

In Problems 31–15, determine all values of the angle θ, where $0° \leq \theta < 360°$, such that the stated conditions will be met.

31. $\tan \theta = -7.953$ and $\sin \theta$ is negative

32. $\csc \theta = 1.325$

33. $\sin \theta = -.3907$ and $\cot \theta$ is positive

34. $\csc \theta = -1.390$

35. $-\sin \theta = .500$ and $\cos \theta$ is positive

36. $\cos \theta = -.8616$ and $\tan \theta$ is negative

37. $\cos \theta = -\sqrt{2}/2$

38. $-\tan \theta = .4245$ and $\cot \theta$ is negative

39. $-\sec \theta = -1.122$

40. $\sec \theta = -1.122$

41. $\sin \theta = .6428$ and $\tan \theta$ is negative

42. $\cos \theta = 1.8192$

43. $\sin \theta = -\sqrt{3}/2$

44. $\sin \theta = .9063$

45. $-\cot \theta = 1.111$ and $\csc \theta$ is positive

46. $\tan \theta = 1.000$

47. $\cos \theta = -.9877$ and $\csc \theta$ is negative

48. $\sec \theta = 1.466$

49. $-\cot \theta = -.7813$ and $\tan \theta$ is positive

50. $-\csc \theta = -1.390$

51. $\cos \theta = -.6713$ and $\sec \theta$ is positive

52. When light traveling from air to water is incident at the boundary at an angle of 30°, the angle of refraction θ_R, which is always an acute angle, can be found from Snell's law, which gives

$$(1)(\sin 30°) = (1.33)(\sin \theta_R).$$

Find the angle of refraction to the nearest minute.

53. When a ball is thrown with an initial velocity of v_0 at an angle θ with the horizontal as shown below, the upward component of the velocity is given by the expression $v_0 \sin \theta$. If the initial velocity of the ball is 20 ft/sec, what values of θ will result in an upward component of 10 ft/sec?

SOLUTION OF THE RIGHT TRIANGLE

Six parts are associated with a right triangle: three sides and three angles. *To solve a right triangle* means to use the given information to determine all unknown parts of the triangle. This can be achieved if the information consists of the lengths of two sides or the length of one side and the size of any angle other than the right angle. Common notation is to call the angles A, B, and C and the sides opposite these angles a, b, and c, respectively, where C is the right angle (Fig. 7-29).

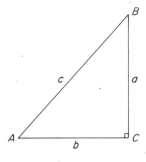

FIGURE 7-29

Certain geometrical facts are useful in the solution of a right triangle. You should recall that the sum of the measures of the angles of a triangle is 180°. Since a right triangle contains a 90° angle, the sum of the measures of the two acute angles of the triangle must be 90°; that is, the acute angles of a right triangle are *complementary angles*. Finally, when two sides are known, the third side can be found by applying the Pythagorean theorem.

There is no set procedure for solving a right triangle—indeed there are many methods of solution. Generally, the technique is to employ appropriate trigonometric relations in a step-by-step analysis that will allow the unknown parts of the triangle to be found. The special interpretation of the definitions of the trigonometric functions as given by Eq. (7-3) is often useful. The examples which follow will illustrate the basic technique.

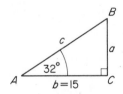

FIGURE 7-30

EXAMPLE
14

Solve the right triangle ABC given that $A = 32°$ and $b = 15$.

From Fig. 7-30 we conclude that since $A + B = 90°$, $B = 90° - 32° = 58°$ and it remains only to find a and c. We shall first find a:

$$\tan A = \frac{a}{b}$$

$$\tan 32° = \frac{a}{15}$$

$$a = 15 \, (\tan 32°).$$

From tables, $\tan 32° = .6249$. Therefore,

$$a = 15(.6249)$$
$$= 9.37$$

to two decimal places. To find c, we shall use $\sec A$, rather than $\cos A$, so that we avoid a division calculation:

$$\sec A = \frac{c}{b}$$

$$\sec 32° = \frac{c}{15}$$

$$c = 15 \, (\sec 32°)$$
$$= 15 \, (1.179) \qquad \text{(from tables)}$$
$$c = 17.7.$$

Hence $A = 32°$, $B = 58°$, $C = 90°$, $a = 9.37$, $b = 15$, and $c = 17.7$.

EXAMPLE
15

Solve the right triangle ABC given that $c = 13$ and $b = 5$ (Fig. 7-31).

$$\cos A = \frac{b}{c} = \frac{5}{13} = .3846.$$

Thus, $A = 67° \, 23'$ (from tables and interpolation). Hence

$$B = 90° - 67° \, 23' = 22° \, 37'.$$

Also, by the Pythagorean theorem,

$$a = \sqrt{13^2 - 5^2} = \sqrt{144} = 12.$$

Hence $A = 67° \, 23'$, $B = 22° \, 37'$, $C = 90°$, $a = 12$, $b = 5$, and $c = 13$.

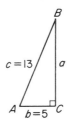

FIGURE 7-31

EXAMPLE
16

Solve the right triangle ABC given that $b = 15$ and $a = 24$ (Fig. 7-32).

$$\tan A = \frac{a}{b} = \frac{24}{15} = 1.600.$$

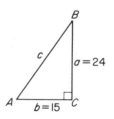

FIGURE 7-32

Thus, $A = 58°$ and

$$B = 90° - 58° = 32°.$$

Although c can be found by using the Pythogorean theorem, it is perhaps more easily done as follows:

$$\csc A = \frac{c}{a}$$

$$\csc 58° = \frac{c}{24}$$

$$c = 24 \ (\csc 58°)$$
$$= 24(1.179)$$
$$= 28.3.$$

Hence $A = 58°$, $B = 32°$, $C = 90°$, $a = 24$, $b = 15$, and $c = 28.3$.

EXAMPLE
17

A particular electrical circuit containing a pure capacitor and a pure resistor has a total impedance (Z) of 5 ohms. Regarding their magnitudes, the total impedance, the capacitive reactance (X_c), and the resistance (R) can be represented as the sides of the right triangle shown in Fig. 7-33. Find X_c and R.

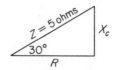

FIGURE 7-33

Since the triangle involves a special angle, we shall not appeal to tables.

$$\sin 30° = \frac{X_c}{5}$$

$$X_c = 5(\sin 30°) = 5(\tfrac{1}{2}) = 2.5 \ (\text{ohms}).$$

Also,

$$\cos 30° = \frac{R}{5}$$

$$R = 5(\cos 30°) = (5)\frac{\sqrt{3}}{2} = \frac{5\sqrt{3}}{2} \text{ (ohms).}$$

EXERCISE 7-7

In Problems 1–22, ABC is a right triangle with right angle C. Solve for the remaining parts of the triangle in each case. In Problems 1–6 it should not be necessary to refer to trigonometric tables.

1. $b = 4, c = 8$
2. $a = 2, b = 2$
3. $B = 60°, a = 3$
4. $A = 30°, b = 3$
5. $A = 45°, c = 6$
6. $B = 60°, c = 1$
7. $a = 8, b = 6$
8. $c = 15, a = 4$
9. $c = 24, A = 23°$
10. $a = 12, A = 12°$
11. $b = 24.6, B = 27° \ 25'$
12. $a = 3.5, b = 2.8$
13. $c = 10, b = 7$
14. $c = 10, B = 42° \ 20'$
15. $c = 12.4, A = 33° \ 33'$
16. $a = 82.5, B = 65° \ 44'$
17. $a = 14, b = 7$
18. $c = 24, a = 5.6$
19. $c = 80.4, B = 86° \ 29'$
20. $a = 23.1, B = 12° \ 37'$
21. $a = 23.4, b = 18.1$
22. $c = 60.3, b = 24.7$

Problems 23–27 make use of the following definition. If an observer at point P looks up at a point Q, the angle that the line PQ makes with the horizontal line containing P is

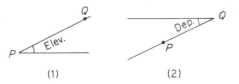

(1)   (2)

*called the **angle of elevation of Q from P**, (1). The angle that PQ makes with the horizontal line containing Q is the **angle of depression of P from Q**, (2).*

23. Find the angle of elevation of the top of a 12-ft ladder that rests on the ground and reaches a point on the wall 8 ft above the ground.

24. A man 6 ft tall casts a shadow 12.5 ft long. Find the angle of elevation of the light source.

25. From the top of a 300-ft tower, the angle of depression of a man on the ground is 39°. How far from the base of the tower is the man?

26. The length of a kite string is 625 ft and the angle of elevation of the kite is 34°. How high is the kite?

27. The angle of elevation of the top of an antenna from a point 12 ft from the base is 32°. What is the angle of elevation for a point 20 ft from the base?

28. In order to find the distance from point A to point B, the angle ABC was laid off 90°. Angle BCA was found to be 47° and $\overline{BC}$ was found to be 97 ft. Find the distance from A to B to the nearest foot.

7-8
VECTORS

The study of topics in science and engineering invariably deals with considerations of measurable physical quantities. You are undoubtedly familiar with some. If you were presented with a rectangular box and a measuring stick, for example, and were asked to determine the length and area of a side and the volume of the box, you would probably be able to supply the answers in short order. You might find the length of the side to be 5 ft, the area to be 7.5 ft², and the volume of the box to be 10 ft³. However, you might overlook the fact that each of these quantities has a common property. Each is completely specified by a *number* and a *dimension*. Such physical quantities—others are mass, speed, distance, and temperature—are called **scalar quantities**.

In science and engineering, however, there are other physical quantities that cannot be completely specified that simply. In addition to a number and a dimension which indicate magnitude, these physical quantities—force, velocity, acceleration, torque, and displacement are examples—have associated with them a characteristic *direction*. For example, we can speak of a velocity of 10 ft/sec *south*. Such quantities are called **vector quantities**.

To represent a vector quantity, we draw a directed line segment, usually as an arrow, the length of which corresponds, with some suitable scale, to the magnitude of the quantity, and whose direction, indicated by the arrow, gives the direction of the quantity. Thus, a velocity of 10 ft/sec south can be represented by a 1-in. arrow pointing south. We shall adopt the common technological terminology and call the directed line segment representing a vector quantity a **vector**.

EXAMPLE
18

The directed line segment, or arrow, shown in Fig. 7-34 is a vector. Its **initial point**, or **tail**, is at A and its **terminal point**, or **head**, is at B. The head of the arrow denotes direction and we speak of this vector as the vector from A to B, or the vector **AB**.

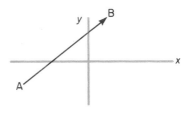

FIGURE 7-34

We shall adopt the custom of denoting a vector by boldface type and the magnitude of the vector by ordinary type. Hence, if $A = 3$ ft/sec northward, then its magnitude A is 3 ft/sec.

Definition. Two vectors are **equivalent** if and only if they have the same magnitude and the same direction.

The vectors A, B, and C in Fig. 7-35 are equivalent. Each has the same magnitude and direction; they differ, however, in their initial and ter-

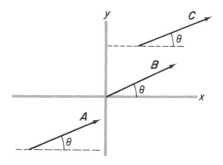

FIGURE 7-35

minal points. The notion of equivalent vectors is analogous to the notion of equivalent rational numbers. That is, just as we can replace the number $\frac{2}{3}$ by $\frac{4}{6}$ without affecting a calculation, so can we represent a vector by an equivalent vector. More specifically, at times we shall find it convenient to represent a vector by an equivalent vector whose initial point is at the origin of a rectangular coordinate system; this type of vector is said to be in **standard position**. Vector B in Fig. 7-35 is illustrative of such a vector.

Definition. If the direction of A is such that it forms an angle θ with a given reference line, then the direction of $-A$ is such that it forms an angle of $\theta + 180°$ with the same reference line. The magnitude of $-A$ is the same as that of A.

Thus, the negative of a vector is merely a vector of equal magnitude but pointing in the opposite direction.

Definition. Given a vector A, the vector nA, where n is a numerical coefficient, is a vector having magnitude $|n|$ times that of the magnitude of A and having direction

1. The same as A if $n > 0$.

2. Opposite that of A if $n < 0$.

3. Anywhere if $n = 0$. In this case the vector is called a **zero vector.**

EXAMPLE
19

Given the vector V with magnitude of 1 in the direction due northeast, Fig. 7-36 indicates $V, 2V, -V,$ and $-3V$.

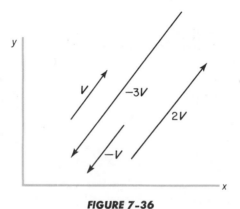

FIGURE 7-36

In Fig. 7-37(a), consider the two displacements from A to B and from B to C represented by the vectors AB and BC, respectively. The combined effect, or net displacement, is the same as the displacement from A to the head of BC, represented by the vector AC in Fig. 7-37(b). AC is called

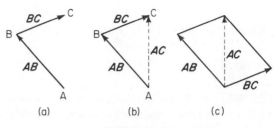

FIGURE 7-37

the **vector sum** or **resultant** of *AB* and *BC*. Symbolically, we write

$$AB + BC = AC. \tag{7-5}$$

It should be clear that Eq. (7-5) does **not**, in any sense, imply ordinary addition. We emphasize that the vector addition relation stated in Eq. (7-5) implies all the geometrical considerations necessary to account for the directions of the quantities involved.

For the case of two vectors the resultant is expressible in another manner. From Fig. 7-37(c) it can be seen that the resultant is the vector from the origin to the opposite side of the parallelogram of which two adjacent sides are the given vectors. This is a common interpretation for the addition of two vectors.

To geometrically add vectors *A, B, C, D, . . . , N,*

1. Draw *A* to scale on a diagram.

2. Draw *B* with its tail at the head of *A*.

3. Draw *C* with its tail at the head of *B*.

4. Continue in this manner until *N* has been drawn.

5. The resultant is the vector represented by the arrow drawn from the tail of *A* to the head of *N*.

We point out that any vector can be replaced by two or more vectors whose vector sum is the given vector.

Two of the fundamental properties of vector sums are given by the commutative and associative laws:

Commutative: $A + B = B + A$

Associative: $A + (B + C) = (A + B) + C.$

These laws imply that regardless of the order in which vectors are added, and regardless of how they are grouped together, the vector sum remains unchanged. Given the vectors *A, B,* and *C* as shown in Fig. 7-38(a), the commutative law is illustrated in Fig. 7-38(b) and the associative law in Fig. 7-38(c).

The geometrical method of vector addition, although helpful in establishing the framework of understanding, is useful in only the simplest cases. Indeed, in complicated problems in two dimensions and virtually any problem involving vectors in three dimensions, the techniques described thus far are rarely employed. Instead, we use an analytical method involving the resolution of vectors into so-called *components* lying along the coordinate axes. Henceforth, we shall assume that all vectors

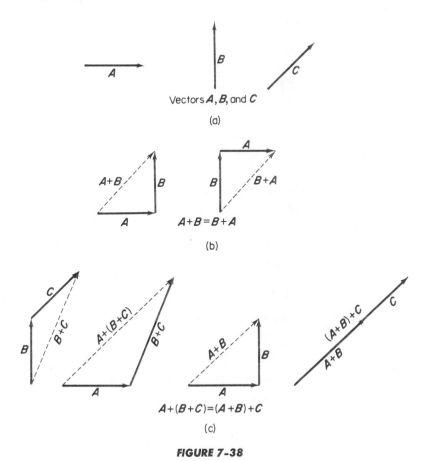

Vectors A, B, and C

(a)

$A+B=B+A$

(b)

$A+(B+C)=(A+B)+C$

(c)

FIGURE 7-38

under consideration have their tails at the origin; that is, they are in standard position.

Definition. The **vector components** V_x and V_Y of a vector V are the vectors whose heads are at the points obtained by constructing a perpendicular from the head of V to the x-axis and y-axis, respectively. [See Fig. 7-39(a).]

The vector V shown in Fig. 7-39(a) can be replaced by its vector components V_x and V_Y since the sum of V_x and V_Y is V, as shown in Fig. 7-39(b). Clearly, V_x will lie along either the positive or negative x-axis and V_Y along either the positive or negative y-axis, depending on the original orientation of V. Figure 7-40(a), (b), and (c) illustrates vectors in the second, third, and fourth quadrants, respectively, along with their vector components. In each case, $V = V_x + V_Y$.

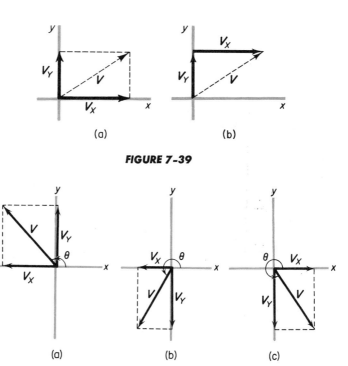

FIGURE 7-39

FIGURE 7-40

In general, if θ is the angle measured counterclockwise from the positive x-axis to a vector V with initial point at the origin, and $x = V_x$ and $y = V_y$ are the abscissa and ordinate, respectively, of the points of intersection of perpendiculars drawn from the head of V to the coordinate axes (Fig. 7-41), then

$$\cos \theta = \frac{V_x}{V} \qquad \text{or} \qquad \boxed{V_x = V \cos \theta}$$

and

$$\sin \theta = \frac{V_y}{V} \qquad \text{or} \qquad \boxed{V_y = V \sin \theta.}$$

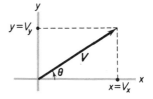

FIGURE 7-41

Observe that V_x and V_y can be looked upon as *directed* lengths associated with the vectors V_X and V_Y. Sometimes V_x and V_y are respectively called the horizontal and vertical **scalar components** of V as opposed to V_X and V_Y, which are *vector* components of V. When we speak of the components of V we may refer to either.

Note that, depending on the quadrant in which θ lies, V_x and V_y can be positive or negative. In Fig. 7-40(a), V_x would be negative and V_y positive; in Fig. 7-40(b), V_x and V_y would both be negative; and in Fig. 7-40(c), V_x would be positive and V_y negative.

EXAMPLE
20

Find the scalar components of V given that $V = 20$ and $\theta = 140°$, where θ is the angle that V makes with the positive x-axis.

An appropriate diagram is shown in Fig. 7-42.

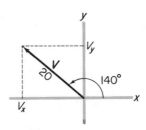

FIGURE 7-42

$$V_x = V \cos 140° = 20(-.7660) = -15.32$$
$$V_y = V \sin 140° = 20(.6428) = 12.86.$$

EXAMPLE
21

The weight, W, of an object on an inclined plane can be resolved into two vector components: one, W_x, parallel to the plane, and one, W_y, perpendicular to the plane (Fig. 7-43). It can be shown that the magnitudes of these components are given by

$$W_x = W \sin \theta$$
$$W_y = W \cos \theta.$$

(These expressions do not follow the general form for scalar components since the angle θ is not measured in the same manner.) Suppose a 150-lb weight rests on an incline which rises 4 ft vertically to every 3 ft horizontally. Find the magnitudes of the components of the weight parallel to the plane and perpendicular to it.

See Fig. 7-44.

$$W_x = W \sin \theta = 150(\tfrac{4}{5}) = 120 \text{ lb}$$
$$W_y = W \cos \theta = 150(\tfrac{3}{5}) = 90 \text{ lb}.$$

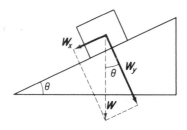

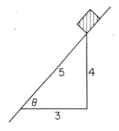

FIGURE 7-43   **FIGURE 7-44**

EXERCISE 7-8

In Problems 1–10, find the horizontal and vertical scalar components of the given vector whose magnitude is V and which makes an angle θ with the positive x-axis.

1. $V = 150, \theta = 135°$ **2.** $V = 10, \theta = 15° \, 32'$

3. $V = 10.1, \theta = 273°$ **4.** $V = .100, \theta = 226°$

5. $V = 32, \theta = 184°$ **6.** $V = 50, \theta = 316°$

7. $V = 200, \theta = 821°$ **8.** $V = 3000, \theta = 137°$

9. $V = 150, \theta = 306°$ **10.** $V = 720, \theta = 427°$

7-9
ADDITION OF VECTORS—ANALYTICAL METHOD

When a number of vectors are added together, the vector sum (or resultant) can be designated R, the scalar components of R being R_x and R_y. The basic rules governing the addition of a number of vectors by the analytical method are:

1. R_x is the algebraic sum of the horizontal scalar components of the vectors in the sum.

2. R_y is the algebraic sum of the vertical scalar components of the vectors in the sum.

Hence, to find the resultant of two or more vectors,

a. *For each vector find its horizontal and vertical scalar components.*

b. *Determine the algebraic sum R_x of the horizontal scalar components.*

c. *Determine the algebraic sum R_y of the vertical scalar components.*

d. *From R_x and R_y use the Pythagorean theorem or basic trigonometric relations to find the magnitude and direction of the resultant vector, R.*

EXAMPLE
22

Find the resultant of **A**, **B**, and **C** if

$$A = 30 \text{ lb}, \quad \theta = 30°$$
$$B = 10 \text{ lb}, \quad \theta = 270°$$
$$C = 10 \text{ lb}, \quad \theta = 135°,$$

where θ is the angle the given vector makes with the positive x-axis. The three vectors are shown in Fig. 7-45.

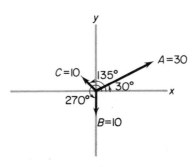

FIGURE 7-45

The scalar components of **A** are

$$A_x = A \cos 30° = (30)(.8660) = 25.98$$
$$A_y = A \sin 30° = (30)(.5000) = 15.$$

The scalar components of **B** are

$$B_x = B \cos 270° = (10)(0) = 0$$
$$B_y = B \sin 270° = (10)(-1) = -10.$$

For vector **C**, since $\theta = 135°$, the reference angle is 45°:

$$C_x = C \cos 135° = C(-\cos 45°) = (10)(-.7071) = -7.071$$
$$C_y = C \sin 135° = C \sin 45° = (10)(.7071) = 7.071.$$

The scalar components R_x and R_y of the resultant **R** are given by

$$R_x = A_x + B_x + C_x = 25.98 + 0 - 7.071 \approx 18.91$$
$$R_y = A_y + B_y + C_y = 15 - 10 + 7.071 \approx 12.07.$$

Since these scalar components are both positive, they correspond to points that lie along the positive x- and positive y-axes as shown in Fig. 7–46. By the Pythagorean theorem,

$$R^2 = R_x^2 + R_y^2$$
$$R = \sqrt{R_x^2 + R_y^2}$$
$$R = \sqrt{(18.91)^2 + (12.07)^2} \approx 22.4 \text{ lb}.$$

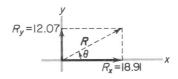

FIGURE 7-46

To determine the direction of R,

$$\tan \theta = \frac{R_y}{R_x} = \frac{12.07}{18.91} \approx .6383$$

$$\theta \approx 32° 33'.$$

We point out that the use of the Pythagorean theorem could have been avoided completely, since

$$R_x = R \cos \theta$$

$$R = \frac{R_x}{\cos \theta} = \frac{18.91}{.8429} \approx 22.4 \text{ lb}$$

or

$$R_y = R \sin \theta$$

$$R = \frac{R_y}{\sin \theta} = \frac{12.07}{.5380} \approx 22.4 \text{ lb}.$$

EXAMPLE
23

A block of weight W lb rests on a plane inclined at an angle of 30° with the horizontal (Fig. 7-47). If the frictional force is 3 lb and the block remains at rest, find the weight of the block. Assume all forces act at the center of the block.

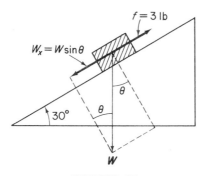

FIGURE 7-47

The weight W of the block is the force due to gravity and is directed vertically downward. The frictional force f, which always opposes motion, is directed up the plane. Since the block does not move, the net force up the plane must equal the net force down the plane. The frictional force f is the only force upward, and the component W_x of the weight is the only force downward. Using geometry we can show that $\theta = 30°$ and hence

$$W_x = f$$

$$W \sin 30° = f$$

$$W = \frac{f}{\sin 30°} = \frac{3}{\frac{1}{2}} = 6 \text{ lb.}$$

EXAMPLE
24

An object is dropped from an airplane flying horizontally at a speed of 160 ft/sec (Fig. 7-48). The vertical component of its velocity as a function of time is given by $V_y = -32t$. What is the velocity of the object after 5 sec and in what direction is the object moving?

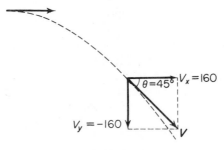

FIGURE 7-48

Assuming the effects of air resistance can be neglected, the horizontal component of velocity of such a free-falling object will be the same throughout its flight. Thus, at all times the horizontal component V_x is 160 ft/sec. At $t = 5$ the vertical component of velocity is $V_y = -32(5) = -160$, which corresponds to a velocity of 160 ft/sec downward. It is left as an exercise for you to show that at $t = 5$,

$$V = 160\sqrt{2} \approx 226.3 \text{ ft/sec}$$

at an angle of 45° below the horizontal.

EXERCISE 7-9

In Problems 1–10, add the given vectors analytically and determine the magnitude R and direction θ of the resultant.

1. $A = 220, \theta = 225°$
    $B = 100, \theta = 16°$

2. $A = 50, \theta = 30°$
    $B = 120, \theta = 30°$
    $C = 100, \theta = 90°$

3. $A = 200, \theta = 210°$
    $B = 300, \theta = 45°$
    $C = 400, \theta = 120°$

4. $A = 500, \theta = 0°$
    $B = 300, \theta = 180°$
    $C = 200, \theta = 315°$

5. $A = 300, \theta = 80°$
   $B = 10, \theta = 270°$
   $C = 100, \theta = 30°$

6. $A = 200, \theta = 200°$
   $B = 50, \theta = 50°$
   $C = 400, \theta = 400°$

7. $A = 310, \theta = 330°$
   $B = 260, \theta = 150°$
   $C = 550, \theta = 120°$

8. $A = 10, \theta = 0°$
   $B = 5, \theta = 226°$
   $C = 10, \theta = 180°$

9. $A = 80, \theta = 30°$
   $B = 50, \theta = 50°$
   $C = 60, \theta = 60°$
   $D = 30, \theta = 80°$

10. $A = 10, \theta = 10°$
    $B = 20, \theta = 20°$
    $C = 30, \theta = 30°$
    $D = 40, \theta = 40°$
    $E = 50, \theta = 50°$

11. Two vectors V_1 and V_2 are added. Suppose $V_1 = 10$, $V_2 = 14.142$, and the magnitude of the resultant is equal to 10. If the resultant R is perpendicular to V_1, determine the angle between V_1 and V_2. *Hint:* Construct V_1 along the x-axis and the resultant along the y-axis.

12. A weightless boom is shown in Fig. 7-49. A 100-lb load is suspended from its end. The forces acting at point O are T, the tension in the hori-

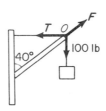

FIGURE 7-49

zontal string, the 100-lb force due to the load, and the force F due to the boom. Find T and F.

13. A block whose weight is 15 lb rests on a 45° inclined plane. If the block does not move, determine the frictional force.

14. An object is dropped from an airplane which is traveling at a speed of 100 ft/sec at a direction of 30° above the horizontal. The vertical component of its velocity is given by $V_y = 50 - 32t$. In the absence of air resistance, what is the speed of the object after 2 sec and in what direction is the object moving?

7-10
REVIEW

REVIEW QUESTIONS

Which is larger, sin (2·45°) or 2 sin 45°? _____(1)_____

(1) **2 sin 45°**

Which of the following statements is (are) true? _____(2)_____

a. $\sin 21° = 1/\sec 69°$ b. $\sin (-20°) = -\sin 20°$
c. $\cos 10° = -\sin 80°$ d. $\tan 60° = \cot 30°$
e. $\tan 45° > \cot 45°$ f. $\sec (20° + 11°) = \csc (70° - 11°)$

   (2) **a, b, d, and f**

In 3 min the second hand of a clock generates an angle of _____(3)_____ radians.

   (3) **6π**

The only trigonometric functions that have positive values for θ in quadrant III are the _____(4)_____ functions.

   (4) **tangent and cotangent**

The absolute value of $\sin 280°$ is equal to the cosine of what acute angle? _____(5)_____

   (5) **10°**

If $0° \leq \theta < 360°$, then $\sin \theta = 0$ for $\theta =$ _____(6)_____ and $\theta =$ _____(7)_____.

   (6) **0°**   (7) **180°**

250°, 610°, and $-110°$ are angles which are _____(8)_____ with one another.

   (8) **coterminal**

In the diagram below, $a =$ _____(9)_____.

   (9) **2**

In the triangle below, $\theta =$ _____(10)_____

(10) **45°**

A vector quantity is a quantity that has both magnitude and _____ (11) _____.

(11) **direction**

In the diagram below, what angle does the resultant of the given vectors make with the positive x-axis? _____ (12) _____

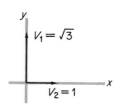

(12) **60°**

True or false: If sin A = sin B, then $A = B$. _____ (13) _____ If A and B are coterminal angles, then sin A = sin B. _____ (14) _____

(13) **false**   (14) **true**

If sin θ is negative and tan θ is positive, then θ lies in the _____ (15) _____ quadrant.

(15) **third**

Which of the following statements is (are) possible? _____ (16) _____
a.   tan θ = 1000   b.   sin θ = 1.010
c.   cos θ = −.9000   d.   cos θ = sin θ
e. $(\sin \theta)^2 = \frac{1}{4}$ f.   sin $\theta \div$ sin 0° = .7

(16) **a, c, d, and e**

720° equals _____ (17) _____ radians.

(17) **4π**

To convert x radians to degrees, all one must do is multiply x by _____ (18) _____ .

(18) **180°/π**

REVIEW PROBLEMS

In Problems 1–14, find (a) a positive angle θ, where $0° < \theta \le 360°$, *that is coterminal with the given angle; (b) a negative angle θ,* $-360° \le \theta < 0°$, *that is coterminal with the given angle; and (c) the values of the trigonometric functions of the given angle.*

1. 18° 31′

2. −234° 18′

3. 142° 12′

4. $\pi/6$

5. −16° 10′

6. 22° 22′

7. $\pi/3$

8. 66° 66′

9. 117° 45′

10. 302° 50′

11. −97° 20′

12. −452° 15′

13. 621° 30′

14. $2\pi/3$

In Problems 15–22, ABC is a right triangle with right angle C. Solve the triangle in each case.

15. $a = 4, b = 10$

16. $a = 5, c = 21$

17. $a = 6, B = 15°$

18. $b = 9, A = 24°$

19. $b = 7, A = 46°$

20. $c = 12, B = 35° 10′$

21. $c = 20, A = 32° 40′$

22. $a = 45, A = 60°$

23. A man 6 ft tall stands 12 ft from the base of a street light. He can observe the light when his line of sight makes an angle of 42° with the horizontal. How long a shadow is cast?

*Given the vectors **A**, **B**, and **C** in Problems 24 and 25, find the horizontal and vertical components of the resultant.*

24. $A = 100, \theta = 330°$
     $B = 120, \theta = 45°$
     $C = 50, \theta = 60°$

25. $A = 20, \theta = 225°$
     $B = 30, \theta = 315°$
     $C = 10, \theta = 60°$

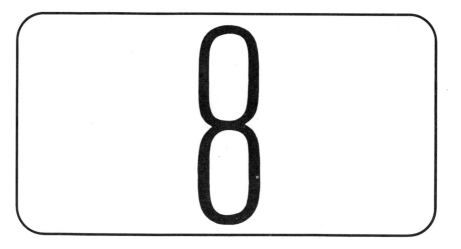

systems of
linear equations

8-1

SYSTEMS OF LINEAR EQUATIONS IN TWO
VARIABLES

Suppose an airplane flying at its top performance travels 900 mi in 3 hr with the aid of a tail wind, but takes 3 hr and 36 min for the return trip in which the pilot must fly against the same wind. Question: What is the top speed of the airplane in still air and what is the speed of the wind?

Suppose we let x denote the speed of the airplane in still air and y denote the speed of the wind. Then, with the wind the speed of the airplane is $x + y$, and against the wind the speed is $x - y$. Since we know that (rate)(time) = distance, the conditions imposed by the problem are

$$(x + y)(3) = 900$$

and

$$(x - y)(3\tfrac{3}{5}) = 900.$$

Equivalently, we have the equations

$$3x + 3y = 900$$

$$\frac{18}{5}x - \frac{18}{5}y = 900.$$

Our problem now is one of finding values of x and y for which *both* equations above are true. Let us first consider such a situation on a more general level. We shall return to our particular problem in the next section.

The set of linear equations

$$\begin{cases} a_1x + b_1y = c_1 & \text{(1)} \\ a_2x + b_2y = c_2 & \text{(2)} \end{cases}$$

is called a **system of two linear equations** in the variables (or unknowns) x and y. Here, the brace indicates that each equation is to be considered in conjunction with the other one. The solution set of the system consists of all ordered pairs (x, y) whose coordinates satisfy *both* equations *simultaneously*.

Geometrically, Eqs. (1) and (2) represent straight lines, say L_1 and L_2. Since the coordinates of any point on a line satisfy the equation of that line, the coordinates of any point of intersection of L_1 and L_2 will satisfy both equations of the system. When L_1 and L_2 are sketched on the same coordinate plane, three possibilities exist as to their relative orientations:

a. The lines may be parallel and distinct (Fig. 8-1).

b. The lines may intersect at one and only one point (x, y) (Fig. 8-2).

c. The lines may coincide (Fig. 8-3).

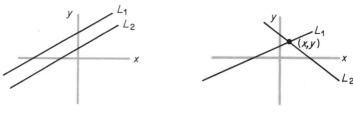

FIGURE 8-1

FIGURE 8-2

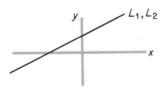

FIGURE 8-3

If the lines are parallel and distinct, as in Fig. 8-1, no point of inter-section exists. As a result, the equations represented by these lines have no common solution. That is, there is no ordered pair that will simulta-neously satisfy both equations and, hence, the solution set for the system is $\varnothing$, the empty set.

In Fig. 8-2, the point of intersection (x, y) is the only point which simultaneously satisfies the equations for L_1 and L_2. Hence, $\{(x, y)\}$ is the solution set of the system of equations which are represented by the lines L_1 and L_2.

In Fig. 8-3, *any* point on one of the lines must also lie on the other line and thus its coordinates satisfy the equations for both L_1 and L_2. Hence, there are an infinite number of ordered pairs in the solution set of such a system of equations. In this case, the equations of L_1 and L_2 must be equivalent.

Our main concern in this chapter is to determine the solution set of a system of equations. That is, we seek the values of the variables employed that will satisfy each and every equation in the system. If the solution set is not empty, the system is said to be **consistent**. Otherwise, the sys-tem is **inconsistent**. For example, the lines L_1 and L_2 in Fig. 8-1 represent an inconsistent system of equations.

EXAMPLE
1

Determine whether the system of linear equations

$$\begin{cases} 2y + 4x = 8 \\ y + 2x = -3 \end{cases}$$

is consistent or inconsistent.

We first replace the system by an equivalent system (that is, one having the same solution set) in which both equations are written in slope-intercept form:

$$\begin{cases} y = -2x + 4 \\ y = -2x - 3. \end{cases}$$

Since the lines defined by these equations have the same slope ($m = -2$) but different y-intercepts, namely $(0, 4)$ and $(0, -3)$, respectively, the system is represented by distinct parallel lines. The system is inconsistent. Figure 8-4 on the next page indicates a geometrical representation of the system.

EXAMPLE
2

The system

$$\begin{cases} y = 2x \\ y + x = 3 \end{cases}$$

is equivalent to

$$\begin{cases} y = 2x \\ y = -x + 3. \end{cases}$$

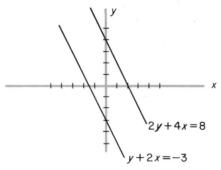

FIGURE 8-4

Since the slopes are unequal, namely $m = 2$ and $m = -1$, we assert that the lines are not parallel and that one point of intersection exists, as indicated in Fig. 8-5. By inspection, the solution set of this consistent system consists of the point $(1, 2)$, as observed in Fig. 8-5. Only the coordinates of that point satisfy both equations. You should verify that both equations in the given system are true when $x = 1$ and $y = 2$.

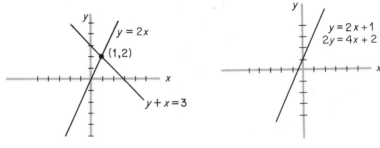

FIGURE 8-5   **FIGURE 8-6**

EXAMPLE
3

Consider the system

$$\begin{cases} 2y = 4x + 2 & (1) \\ y = 2x + 1. & (2) \end{cases}$$

Since we can obtain Eq. (1) by multiplying both members of Eq. (2) by 2, the equations are equivalent and they are represented by the same line in Fig. 8-6. Thus, every ordered pair satisfying one equation satisfies the other equation, which results in an infinite number of solutions. In short, every point on the line is a solution to this consistent system.

A consistent system of linear equations having more than one solution is called **dependent**. If exactly one solution exists, the system is **independent**. The system in Example 3 is then a consistent, dependent system while that in Example 2 is consistent and independent.

In the next sections we shall focus our attention on algebraic methods of solving a system of equations. The geometrical approach in Example 2 is not precise and should be used only when other methods are not feasible.

EXERCISE 8-1

In Problems 1–10, first write each equation in slope-intercept form and determine whether the given system is consistent and independent, consistent and dependent, or inconsistent. Then sketch the graph of the system and, if independent, approximate the common solution by inspection.

1. $\begin{cases} x - 2y = -3 \\ 3x + y = -2 \end{cases}$ CoмP

2. $\begin{cases} y = x - \frac{1}{2} \\ 2y = -1 + x \end{cases}$

3. $\begin{cases} y + 2x = 2 \\ y - 5 = -2x \end{cases}$

4. $\begin{cases} 2x - y = -11 \\ y + 5x = -7 \end{cases}$

5. $\begin{cases} 2x + y = 4 \\ 10y - 41 = -21x \end{cases}$

6. $\begin{cases} \frac{3}{2}y = 2x - 1 \\ y + \frac{2}{3} = \frac{4}{3}x \end{cases}$

7. $\begin{cases} y = 6x - 3 \\ 18x = 9 + 3y \end{cases}$ Suвst

8. $\begin{cases} 3x - 2y = 1 \\ 2x - 3y = -6 \end{cases}$

9. $\begin{cases} 2y = -x + 4 \\ y + 5 = 2x \end{cases}$ CoмP

10. $\begin{cases} 3(x + y) = 4(x - y) + 3 \\ 2(x - y) = 2y + 1 \end{cases}$

8-2
METHOD OF ELIMINATION BY ADDITION

When a system of two linear equations is given, it is possible by using suitable algebraic operations to obtain an equation in which only one variable appears. That is, one of the variables will be eliminated. After this equation has been solved, the value of the remaining variable can easily be determined.

To illustrate one procedure by which a variable can be eliminated, we shall solve the system

$$\begin{cases} 2x - 3y = -12 & \text{(1)} \\ 3x + y = -7 & \text{(2)} \end{cases}$$

by eliminating the variable y. If we multiply Eq. (2) by 3 [that is, multiply both members of Eq. (2) by 3], we obtain an equivalent system in which the coefficients of y in each equation differ only in sign:

$$\begin{cases} 2x - 3y = -12 & \text{(3)} \\ 9x + 3y = -21. & \text{(4)} \end{cases}$$

Since the left and right members of Eq. (3) are equal, each can be added to the corresponding member of Eq. (4). This yields an equation in one variable:

$$11x = -33$$

or

$$x = -3.$$

Replacing x in Eq. (1) by -3, we obtain

$$2(-3) - 3y = -12$$
$$-6 - 3y = -12$$
$$-3y = -6$$
$$y = 2.$$

Since we used Eq. (1) to obtain y, it is wise to choose Eq. (2) to check our solution. When $x = -3$ and $y = 2$, Eq. (2) is $3(-3) + 2 = -7$, which is true. Thus, the solution is indeed $x = -3$ and $y = 2$, or, equivalently, the solution set is $\{(-3, 2)\}$. Our procedure is referred to as **elimination by addition.**

Although we chose to eliminate y first, we can do the same for x. If we multiply Eq. (1) by 3 and multiply Eq. (2) by -2, in the resulting equivalent system the coefficients of x will differ only in sign:

$$\begin{cases} 6x - 9y = -36 & (5) \\ -6x - 2y = 14. & (6) \end{cases}$$

Adding Eq. (5) to Eq. (6) [that is, adding corresponding members of Eq. (5) to Eq. (6)], we obtain

$$-11y = -22$$

or

$$y = 2.$$

By replacing y in Eq. (1) by 2, we obtain $x = -3$, as expected. By either technique the solution set is $\{(-3, 2)\}$. Note, however, that the first approach is preferable, since it involved *one* multiplication operation rather than two. Clearly, the system is consistent and independent. Figure 8-7 indicates a geometrical representation of the system.

In the problem originally posed in Sec. 8-1, we were to solve the system

$$\begin{cases} 3x + 3y = 900 & (1) \\ \dfrac{18}{5}x - \dfrac{18}{5}y = 900. & (2) \end{cases}$$

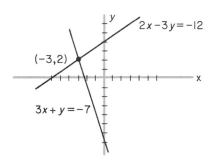

FIGURE 8-7

Multiplying Eq. (1) by $\frac{1}{3}$ and Eq. (2) by $\frac{5}{18}$, we obtain

$$\begin{cases} x + y = 300 & \text{(3)} \\ x - y = 250. & \text{(4)} \end{cases}$$

Adding Eq. (4) to Eq. (3) gives

$$2x = 550$$
$$x = 275.$$

Replacing x in Eq. (1) by 275 we find $y = 25$. Hence, the speed of the airplane in still air is 275 mi/hr and the speed of the wind is 25 mi/hr.

EXAMPLE 4   Solve the system

$$\begin{cases} y - 2x = 4 \\ 4x - 2y = -5. \end{cases}$$

Aligning the x and y terms,

$$\begin{cases} -2x + y = 4 & \text{(1)} \\ 4x - 2y = -5. & \text{(2)} \end{cases}$$

Multiplying Eq. (1) by 2 yields the equivalent system

$$\begin{cases} -4x + 2y = 8 & \text{(3)} \\ 4x - 2y = -5. & \text{(4)} \end{cases}$$

Adding Eq. (3) to Eq. (4) gives $0 = 3$ and we therefore have the equivalent system

$$\begin{cases} -4x + 2y = 8 & \text{(5)} \\ 0 = 3. & \text{(6)} \end{cases}$$

Since Eq. (6) is *never* true, the solution set of the original system is $\varnothing$. Observe that, using the slope-intercept form, we can write the original system as

$$\begin{cases} y = 2x + 4 \\ y = 2x + \dfrac{5}{2}. \end{cases}$$

These equations represent straight lines having slopes of 2 but with different y-intercepts, $(0, 4)$ and $(0, \frac{5}{2})$; that is, they determine distinct parallel lines (see Fig. 8-8). The system is inconsistent.

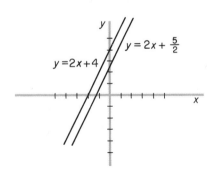

FIGURE 8-8

EXAMPLE
5

Solve the system

$$\begin{cases} 2x + y = 1 & (1) \\ 4x + 2y = 2. & (2) \end{cases}$$

Multiplying Eq. (1) by -2, we obtain the equivalent system

$$\begin{cases} -4x - 2y = -2 & (3) \\ 4x + 2y = 2. & (4) \end{cases}$$

Adding Eq. (4) to Eq. (3),

$$\begin{cases} 0 = 0 & (5) \\ 4x + 2y = 2. & (6) \end{cases}$$

Any solution of Eq. (6) is a solution of the system because Eq. (5) is always true. Looking at it another way, writing Eqs. (1) and (2) in their slope-intercept forms yields the equivalent system

$$\begin{cases} y = -2x + 1 \\ y = -2x + 1 \end{cases}$$

in which both equations represent the same line. Hence, the lines coincide (Fig. 8-9), Eqs. (1) and (2) are equivalent, and every point on the line is a solution. The given system is consistent and dependent and its solution set is

$$\{(x, y) \mid y = 1 - 2x\}.$$

Some solutions are: $x = 0, y = 1$; $x = 1, y = -1$; $x = -3, y = 7$, etc.

EXAMPLE
6

Solve

$$\begin{cases} \dfrac{1}{x} - \dfrac{2}{y} = 4 \\[2mm] \dfrac{3}{x} - \dfrac{4}{y} = 6. \end{cases}$$

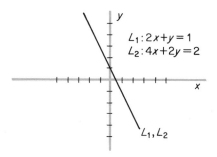

FIGURE 8-9

We may rewrite this *nonlinear* system as

$$\begin{cases} 1 \cdot \dfrac{1}{x} - 2 \cdot \dfrac{1}{y} = 4 \\[2mm] 3 \cdot \dfrac{1}{x} - 4 \cdot \dfrac{1}{y} = 6. \end{cases}$$

If we replace $1/x$ by w and $1/y$ by z, we obtain the *linear* system

$$\begin{cases} w - 2z = 4 \\ 3w - 4z = 6. \end{cases}$$

Solving for w and z gives $w = -2$ and $z = -3$. Hence,

$$x = \frac{1}{w} = -\frac{1}{2}$$

$$y = \frac{1}{z} = -\frac{1}{3}.$$

You should verify that these values satisfy the given system. This technique of *change of variable* will be useful for equations of this type when determinants are discussed in Sec. 8-5.

EXERCISE 8-2

Determine the solution sets of the following systems of equations by the method of elimination by addition.

1. $\begin{cases} x - 2y = -3 \\ 3x + y = -2 \end{cases}$ **2.** $\begin{cases} y = x - \frac{1}{2} \\ 2y = -1 + x \end{cases}$

3. $\begin{cases} 3x - 4y = 13 \\ 2x + 3y = 3 \end{cases}$ **4.** $\begin{cases} 2x - y = -11 \\ y + 5x = -7 \end{cases}$

5. $\begin{cases} 2x + y = 4 \\ 10y - 41 = -20x \end{cases}$ *ADD,TION*   **6.** $\begin{cases} \frac{3}{2}y = 2x - 1 \\ y + \frac{2}{3} = \frac{4}{3}x \end{cases}$

7. $\begin{cases} y = 6x - 3 \\ 18x = 9 + 3y \end{cases}$ **8.** $\begin{cases} 3x - 2y = 1 \\ 2x - 3y = -6 \end{cases}$

9. $\begin{cases} \frac{2}{3}x + \frac{1}{2}y = 2 \\ \frac{3}{8}x + \frac{5}{6}y = -\frac{11}{2} \end{cases}$

10. $\begin{cases} 3(x+y) = 4(x-y) + 3 \\ 2(x-y) = 2y + 1 \end{cases}$

11. $\begin{cases} 3v + 2w - 3 = v \\ w - 3v = -2 \end{cases}$

12. $\begin{cases} \dfrac{1}{x} - \dfrac{1}{y} = \dfrac{1}{12} \\ \dfrac{2}{x} + \dfrac{3}{y} = \dfrac{17}{12} \end{cases}$

13. $\begin{cases} \dfrac{1}{x} - \dfrac{3}{y} = \dfrac{13}{4} \\ \dfrac{1}{y} - \dfrac{1}{x} = -\dfrac{5}{4} \end{cases}$

14. $\begin{cases} \dfrac{x-y}{y+x} = \dfrac{2}{5} \\ y + \dfrac{x+1}{3} = 4 \end{cases}$

15. $\begin{cases} x - y = 4y \\ 0 = 2x - y \end{cases}$ SUBST. (16.) $\begin{cases} x + \dfrac{y - 2x}{y} = x - 7 \\ y - \dfrac{x + 3y}{x} = y - 2 \end{cases}$

ANr
METHOD

17. $\begin{cases} \dfrac{2}{x} + \dfrac{3}{y} = 18 \\ \dfrac{1}{x} - \dfrac{5}{y} = -30 \end{cases}$

18. $\begin{cases} \dfrac{7}{x+1} - \dfrac{6}{y-1} = 4 \\ \dfrac{3}{x+1} + \dfrac{4}{y-1} = -2 \end{cases}$

19. $\begin{cases} \dfrac{1}{a}(x-y) + \dfrac{1}{b}(x+y) = 1 \\ \dfrac{1}{a}(x+y) - \dfrac{1}{b}(x-y) = -1 \end{cases}$

20. $\begin{cases} a_1 x + b_1 y = c_1 \\ a_2 x + b_2 y = c_2 \end{cases}$

8-3
METHOD OF ELIMINATION BY SUBSTITUTION

In the **elimination-by-substitution** method of solving a system of two equations, we choose an equation and solve it for one unknown in the terms of the other, say y in terms of x. This in turn is substituted for y in the other equation. The resulting equation, in which only one variable may occur, is solved in the usual manner. The value of the remaining variable is then easily found.

EXAMPLE
7

By the elimination-by-substitution method, solve

$$\begin{cases} 2x + 3y = -10 & \quad (1) \\ 3x - 2y = -2. & \quad (2) \end{cases}$$

Solving Eq. (2) for y in terms of x,

$$y = \frac{3x + 2}{2}. \qquad (3)$$

By substitution, Eq. (1) is equivalent to

$$2x + 3\left(\frac{3x + 2}{2}\right) = -10,$$

which yields

$$4x + 9x + 6 = -20$$
$$13x = -26$$
$$x = -2.$$

Replacing x in Eq. (3) by -2 yields

$$y = \frac{3(-2) + 2}{2} = \frac{-4}{2} = -2.$$

You should verify that the solution set is $\{(-2, -2)\}$ by replacing x and y by -2 in Eqs. (1) and (2).

EXERCISE 8-3

Solve each of the systems in Problems 1–18 by the method of elimination by substitution.

1. $\begin{cases} x - 2y = -3 \\ 3x + y = -2 \end{cases}$

   2. $\begin{cases} y = x - \frac{1}{2} \\ 2y = -1 + x \end{cases}$

3. $\begin{cases} 2x - y = 1 \\ -x + 2y = 7 \end{cases}$

   4. $\begin{cases} 2x - y = -11 \\ y + 5x = -7 \end{cases}$

5. $\begin{cases} 2x + y = 4 \\ 10y - 41 = -20x \end{cases}$

   6. $\begin{cases} \frac{3}{2}y = 2x - 1 \\ y + \frac{2}{3} = \frac{4}{3}x \end{cases}$

7. $\begin{cases} 2y = -x + 4 \\ y + 5 = 2x \end{cases}$

   8. $\begin{cases} 3x - 2y = 1 \\ 2x - 3y = -6 \end{cases}$

9. $\begin{cases} 5x + 2y = 36 \\ 8x - 3y = -54 \end{cases}$

   10. $\begin{cases} 3(x + y) = 4(x - y) + 3 \\ 2(x - y) = 2y + 1 \end{cases}$

11. $\begin{cases} 3v + 2w - 3 = v \\ w - 3v = -2 \end{cases}$

    12. $\begin{cases} \dfrac{1}{x} - \dfrac{1}{y} = \dfrac{1}{12} \\ \dfrac{2}{x} + \dfrac{3}{y} = \dfrac{17}{12} \end{cases}$

13. $\begin{cases} \dfrac{1}{x} - \dfrac{3}{y} = \dfrac{13}{4} \\ \dfrac{1}{y} - \dfrac{1}{x} = \dfrac{-5}{4} \end{cases}$

    14. $\begin{cases} \dfrac{x - y}{y + x} = \dfrac{2}{5} \\ y + \dfrac{x + 1}{3} = 4 \end{cases}$

15. $\begin{cases} x - y = 4y \\ 0 = 2x - y \end{cases}$

    16. $\begin{cases} x + \dfrac{y - 2x}{y} = x - 7 \\ y - \dfrac{x + 3y}{x} = y - 2 \end{cases}$

17. $\begin{cases} \dfrac{2}{x} + \dfrac{3}{y} = 18 \\ \dfrac{1}{x} - \dfrac{5}{y} = -30 \end{cases}$

    18. $\begin{cases} \dfrac{7}{x + 1} - \dfrac{6}{y - 1} = 4 \\ \dfrac{3}{x + 1} + \dfrac{4}{y - 1} = -2 \end{cases}$

19. A 16-lb block is to be moved without acceleration along a horizontal surface by applying a force F parallel to the surface. The coefficient of kinetic friction between the block and the surface is .3. For this situation, Newton's laws lead to the system of equations

$$\begin{cases} F - .3N = 0 \\ N - 16 = 0, \end{cases}$$

where N is the normal force exerted on the block by the plane. Find F and N.

20. If the block in Problem 19 is to accelerate at 4 ft/sec², the appropriate system of equations is

$$\begin{cases} F - .3N = \dfrac{16}{32}(4) \\ N - 16 = 0. \end{cases}$$

Find F and N.

8-4

SYSTEMS OF LINEAR EQUATIONS IN THREE VARIABLES

Definition. The general linear equation in the three variables x, y, z is

$$Ax + By + Cz + D = 0,$$

where A, B, and C are not all zero.

Just as two equations were needed to obtain a unique solution to a linear system involving two variables, three equations are similarly needed for a linear system in three variables. The method of solution involves repetitive application of the techniques previously developed. One approach is to select two pairs of the given equations and eliminate the same variable from each. The resulting two equations, being in the same two variables, can then be solved by either elimination by addition or elimination by substitution. Substitution of these values in any of the original equations will yield the value of the remaining variable. This technique can be extended to a system of n linear equations in n variables by reducing the system to $n - 1$ equations in $n - 1$ variables, which is then reduced, etc. An example should clearly illustrate this procedure.

EXAMPLE
8

a. Solve the system

$$\begin{cases} 4x - y - 3z = 1 & (1) \\ 2x + y + 2z = 5 & (2) \\ 8x + y - z = 5. & (3) \end{cases}$$

Here we have a system of three linear equations in three variables. Selecting Eqs. (1) and (2), we have

$$\begin{cases} 4x - y - 3z = 1 \\ 2x + y + 2z = 5. \end{cases}$$

Adding these equations we obtain

$$6x - z = 6 \qquad\qquad (4)$$

in which y does not appear. Repeating this procedure, we select Eqs. (2) and (3) and multiply Eq. (3) by -1:

$$\begin{cases} 2x + y + 2z = 5 \\ -8x - y + z = -5. \end{cases}$$

Adding these equations, we obtain

$$-6x + 3z = 0 \qquad\qquad (5)$$

in which y does not appear. A new system is formed by Eqs. (4) and (5):

$$\begin{cases} 6x - z = 6 & (4) \\ -6x + 3z = 0. & (5) \end{cases}$$

Adding Eq. (4) to Eq. (5) and solving for z,

$$2z = 6$$
$$z = 3.$$

Substituting $z = 3$ in Eq. (4) gives $x = \frac{3}{2}$. Substituting $x = \frac{3}{2}$ and $z = 3$ in Eq. (1) gives $y = -4$. The solution set is $\{(\frac{3}{2}, -4, 3)\}$. You should verify this. We call $(\frac{3}{2}, -4, 3)$ an *ordered triple*.

b. Solve the system

$$\begin{cases} 2x + y - z = -2 & (1) \\ x - 2y = \dfrac{13}{2} & (2) \\ 3x + 2y - 2z = -\dfrac{9}{2}. & (3) \end{cases}$$

We shall solve this system by an approach that is an alternative of the previous example. From Eq. (2), $x = 2y + \frac{13}{2}$. By substituting for x in Eqs. (1) and (3) and simplifying, these two equations become

$$5y - z = -15 \qquad\qquad (4)$$
$$4y - z = -12 \qquad\qquad (5)$$

Solving the system formed by Eqs. (4) and (5) we find that $y = -3$ and $z = 0$. Since $x = 2y + \frac{13}{2} = \frac{1}{2}$, the solution of the given system is $x = \frac{1}{2}, y = -3, z = 0$, or, equivalently, the solution set is $\{(\frac{1}{2}, -3, 0)\}$.

EXERCISE 8-4

Find the solution sets.

1. $\begin{cases} x + y + z = 6 \\ x - y + z = 2 \\ 2x - y + 3z = 6 \end{cases}$

2. $\begin{cases} 2x - y + 3z = 12 \\ x + 2y - 3z = -10 \\ x + y - z = -3 \end{cases}$

ADITION 3. $\begin{cases} 2r - 3s + t = -2 \\ 3r + 3s - 2t = 2 \\ r - 6s + 3t = -2 \end{cases}$

4. $\begin{cases} x - z = 14 \\ y + z = 21 \\ x - y + z = -10 \end{cases}$

5. $\begin{cases} 2x + y + 6z = 3 \\ x - y + 4z = 1 \\ 3x + 2y - 2z = 2 \end{cases}$

6. $\begin{cases} x + y + z = -1 \\ 3x + y + z = 1 \\ 4x - 2y + 2z = 0 \end{cases}$

7. $\begin{cases} x + y = -6 \\ z = 4 \\ x - y + 2z = 16 \end{cases}$

8. $\begin{cases} \dfrac{x}{6} - \dfrac{y}{4} + \dfrac{z}{3} - 3 = -5 \\ x + y + z - 3 = 7 \\ \dfrac{x}{2} + \dfrac{3y}{4} - \dfrac{z}{12} = 13 \end{cases}$

9. $\begin{cases} a + b - c = 0 \\ a - 2b + 3c = 14 \\ \dfrac{3a}{2} + \dfrac{9b}{2} - c = 10 \end{cases}$

10. $\begin{cases} r - 2s = t + 4 \\ t - 3r = s + 6 \\ r + s - t = -10 \end{cases}$

11. $\begin{cases} x + y + 5z = 6 \\ x + 2y + w = 4 \\ 2y + z + w = 6 \\ 3x - 4z = 2 \end{cases}$

12. $\begin{cases} x - y + 3z + w = -14 \\ x + 2y - 3w = 12 \\ 2x + 3y + 6z + w = 1 \\ x + y + z + w = 6 \end{cases}$

8-5

DETERMINANTS OF ORDER 2

Consider a system of two linear equations in two variables of the general form $ax + by = c$, namely

$$\begin{cases} a_1 x + b_1 y = c_1 & \quad (1) \\ a_2 x + b_2 y = c_2. & \quad (2) \end{cases}$$

To solve for x, we can eliminate y by first multiplying Eq. (1) by b_2 and Eq. (2) by $-b_1$:

$$\begin{cases} a_1 b_2 x + b_1 b_2 y = b_2 c_1 \\ -a_2 b_1 x - b_1 b_2 y = -b_1 c_2. \end{cases}$$

Adding these equations,

$$a_1 b_2 x - a_2 b_1 x = b_2 c_1 - b_1 c_2.$$

Simplifying,

$$x(a_1b_2 - a_2b_1) = b_2c_1 - b_1c_2$$

$$x = \frac{b_2c_1 - b_1c_2}{a_1b_2 - a_2b_1}, \tag{3}$$

provided $a_1b_2 - a_2b_1 \neq 0$. In a similar fashion, we can show that

$$y = \frac{a_1c_2 - a_2c_1}{a_1b_2 - a_2b_1}. \tag{4}$$

Note that the denominators in Eqs. (3) and (4) are the same, $a_1b_2 - a_2b_1$. This number will be denoted by the square array of numbers

$$\begin{vmatrix} a_1 & b_1 \\ a_2 & b_2 \end{vmatrix}$$

enclosed by vertical bars. In this form, the number $a_1b_2 - a_2b_1$ is called a **determinant**. More precisely, it is called a **second-order determinant** since there are two entries in each row and column, as shown in the more general array:

$$\text{row 1} \longrightarrow \begin{vmatrix} a & b \\ c & d \end{vmatrix} = ad - bc.$$
$$\text{row 2} \longrightarrow$$
$$\uparrow \quad \uparrow$$
$$\text{col. 1 \quad col. 2}$$

In the future we shall find it convenient to simply call the above array a determinant, although what is meant is the number represented by the array. With this in mind, the value of a second-order determinant is simply the difference of the products of its diagonal entries. Thus,

$$\begin{vmatrix} 2 & 1 \\ 3 & -4 \end{vmatrix} = (2)(-4) - (1)(3) = -8 - 3 = -11$$

$$\begin{vmatrix} -3 & -2 \\ 0 & 1 \end{vmatrix} = (-3)(1) - (-2)(0) = -3 - 0 = -3$$

and

$$\begin{vmatrix} 2 & 0 \\ 0 & 2 \end{vmatrix} = (2)(2) - (0)(0) = 4.$$

Returning to our system of equations, we may now write Eqs. (3) and (4) as

$$x = \frac{\begin{vmatrix} c_1 & b_1 \\ c_2 & b_2 \end{vmatrix}}{\begin{vmatrix} a_1 & b_1 \\ a_2 & b_2 \end{vmatrix}}, \qquad y = \frac{\begin{vmatrix} a_1 & c_1 \\ a_2 & c_2 \end{vmatrix}}{\begin{vmatrix} a_1 & b_1 \\ a_2 & b_2 \end{vmatrix}}, \qquad a_1b_2 - a_2b_1 \neq 0.$$

Note that in both the equations for x and y above, we form the determinant in the denominator from the coefficients of the variables in the given equations, keeping in mind the proper order necessary. In finding x, the determinant in the numerator is that obtained by replacing the first column (or x-*column*) of the determinant in the denominator by the column of constants appearing to the right of the equality signs in the equations of the system. Similarly, for y the second column in the numerator would be replaced by the constant terms. The method described above is referred to as **Cramer's rule.**

EXAMPLE
9

Solve the following system by Cramer's rule:

$$\begin{cases} 2x + y + 5 = 0 & (1) \\ \quad 3y + x = 6. & (2) \end{cases}$$

We first write Eqs. (1) and (2) in the appropriate form:

$$\begin{cases} 2x + y = -5 \\ x + 3y = 6. \end{cases}$$

Using Cramer's rule we form the determinants for x and y according to the instructions given above. Thus,

$$x = \frac{\begin{vmatrix} -5 & 1 \\ 6 & 3 \end{vmatrix}}{\begin{vmatrix} 2 & 1 \\ 1 & 3 \end{vmatrix}} = \frac{(-5)(3) - (1)(6)}{(2)(3) - (1)(1)} = \frac{-15 - 6}{6 - 1} = -\frac{21}{5}$$

and

$$y = \frac{\begin{vmatrix} 2 & -5 \\ 1 & 6 \end{vmatrix}}{\begin{vmatrix} 2 & 1 \\ 1 & 3 \end{vmatrix}} = \frac{(2)(6) - (-5)(1)}{5} = \frac{12 + 5}{5} = \frac{17}{5}.$$

The solution set is $\{(-\frac{21}{5}, \frac{17}{5})\}$.

EXAMPLE
10

Solve the following system by Cramer's rule:

$$\begin{cases} \dfrac{1}{x} - \dfrac{2}{y} = 4 \\ \dfrac{3}{x} - \dfrac{4}{y} = 6. \end{cases}$$

Substituting w for $1/x$ and z for $1/y$ as in Example 6, we obtain

$$\begin{cases} w - 2z = 4 \\ 3w - 4z = 6. \end{cases}$$

Thus,

$$w = \frac{\begin{vmatrix} 4 & -2 \\ 6 & -4 \\ 1 & -2 \\ 3 & -4 \end{vmatrix}}{} = \frac{(-16) - (-12)}{(-4) - (-6)} = \frac{-4}{2} = -2$$

and

$$z = \frac{\begin{vmatrix} 1 & 4 \\ 3 & 6 \\ 1 & -2 \\ 3 & -4 \end{vmatrix}}{} = \frac{(6) - (12)}{2} = \frac{-6}{2} = -3.$$

Thus, $x = \dfrac{1}{w} = -\dfrac{1}{2}$ and $y = \dfrac{1}{z} = -\dfrac{1}{3}$. *Do not forget this final step.*

EXERCISE 8-5

1. Evaluate the following:

   a. $\begin{vmatrix} 2 & 1 \\ 3 & 2 \end{vmatrix}$
   b. $\begin{vmatrix} 3 & 2 \\ -5 & -4 \end{vmatrix}$
   c. $\begin{vmatrix} -2 & -1 \\ -2 & -1 \end{vmatrix}$

   d. $\begin{vmatrix} -3 & 1 \\ a & b \end{vmatrix}$
   e. $\begin{vmatrix} a & e \\ h & z \end{vmatrix}$
   f. $\begin{vmatrix} -2 & -a \\ -a & +2 \end{vmatrix}$

   g. $\begin{vmatrix} 1 & 2 \\ 3 & 4 \\ 2 & 1 \\ 5 & 6 \end{vmatrix}$
   h. $\begin{vmatrix} 6 & 2 \\ 1 & 5 \\ 2 & -6 \\ 5 & 3 \end{vmatrix}$

2. Solve for k if

   $$\begin{vmatrix} 2 & 3 \\ 4 & k \end{vmatrix} = 12.$$

Solve Problems 3–16 by Cramer's rule.

3. $\begin{cases} 2x - y = 4 \\ 3x + y = 5 \end{cases}$ DETER

4. $\begin{cases} 3x + y = 6 \\ 7x - 2y = 5 \end{cases}$

5. $\begin{cases} -2x = 4 - 3y \\ y = 6x - 1 \end{cases}$

6. $\begin{cases} x + 2y - 6 = 0 \\ y - 1 = 3x \end{cases}$

7. $\begin{cases} 2 - u = t \\ 3 + t = -u \end{cases}$

8. $\begin{cases} \frac{3}{2}w - \frac{1}{4}z = 1 \\ \frac{1}{3}w = 2 - \frac{1}{2}z \end{cases}$

9. $\begin{cases} s - \frac{1}{4}t = 1 \\ s + t = -4 \end{cases}$

10. $\begin{cases} \dfrac{1}{x} - \dfrac{1}{y} = \dfrac{2}{15} \\ \dfrac{3}{x} + \dfrac{4}{y} = \dfrac{27}{15} \end{cases}$

11.
$$\begin{cases} \dfrac{2}{x} - \dfrac{3}{y} = 10 \\ \dfrac{3}{y} - \dfrac{4}{x} = 15 \end{cases}$$

12.
$$\begin{cases} x - 2y = 4 \\ x - 6 = -y \end{cases}$$

13.
$$\begin{cases} x + 4 = 3x \\ 3x + 2y = 8 \end{cases}$$

14.
$$\begin{cases} 3(x + 2) = 5 \\ 6(x + y) = -8 \end{cases}$$

15.
$$\begin{cases} \frac{3}{2}x - \frac{1}{4}z = 1 \\ \frac{1}{3}x + \frac{1}{2}z = 2 \end{cases}$$

16.
$$\begin{cases} .6x - .7y = .33 \\ 2.1x - .9y = .69 \end{cases}$$

8-6

DETERMINANTS OF ORDER 3

The value of the **third-order determinant**

$$\begin{array}{c} \text{col. 1} \quad \text{col. 2} \quad \text{col. 3} \\ \downarrow \qquad \downarrow \qquad \downarrow \end{array}$$

$$\begin{array}{l} \text{row 1} \longrightarrow \\ \text{row 2} \longrightarrow \\ \text{row 3} \longrightarrow \end{array} \begin{vmatrix} a_1 & b_1 & c_1 \\ a_2 & b_2 & c_2 \\ a_3 & b_3 & c_3 \end{vmatrix}$$

is defined in the following manner. With each element in the array, we associate the second-order determinant obtained by crossing out the row and column in which the element lies. Hence, for a_2 we cross out the second row, which consists of the elements $a_2, b_2, c_2,$ and the first column, which consists of the elements $a_1, a_2, a_3,$

$$\begin{vmatrix} a_1 & b_1 & c_1 \\ a_2 & b_2 & c_2 \\ a_3 & b_3 & c_3 \end{vmatrix},$$

leaving the determinant

$$\begin{vmatrix} b_1 & c_1 \\ b_3 & c_3 \end{vmatrix},$$

called the **minor** of the element a_2. With each element and its minor is associated the number

$$(-1)^{i+j},$$

where i is the number of the row and j is the number of the column in which the element lies. Since a_2 lies in row 2 and column 1, we associate $(-1)^{2+1} = (-1)^3 = -1$. The **cofactor** of a_2 is the product of this number and the minor of a_2:

$$-1 \cdot \begin{vmatrix} b_1 & c_1 \\ b_3 & c_3 \end{vmatrix}.$$

To evaluate any third-order determinant, select **any** row (or column) and multiply each element in the row (column) by its cofactor. The sum of these values is defined to be the value of the determinant.

EXAMPLE
11

Evaluate the determinant

$$\begin{vmatrix} 2 & -1 & 3 \\ 3 & 0 & -5 \\ 2 & 1 & 1 \end{vmatrix}.$$

Evaluating along the first row,

$$2\left[(-1)^{1+1}\begin{vmatrix} 0 & -5 \\ 1 & 1 \end{vmatrix}\right] + (-1)\left[(-1)^{1+2}\begin{vmatrix} 3 & -5 \\ 2 & 1 \end{vmatrix}\right] + 3\left[(-1)^{1+3}\begin{vmatrix} 3 & 0 \\ 2 & 1 \end{vmatrix}\right]$$

$$= 2(1)(5) - 1(-1)(13) + 3(1)(3)$$
$$= 10 + 13 + 9$$
$$= 32,$$

which is the value of the determinant. Evaluating along the second column,

$$-1(-1)^{1+2}\begin{vmatrix} 3 & -5 \\ 2 & 1 \end{vmatrix} + 0 + 1(-1)^{3+2}\begin{vmatrix} 2 & 3 \\ 3 & -5 \end{vmatrix}$$

$$= 13 + 0 + 19$$
$$= 32$$

as before. The latter expansion is preferable since the zero in column 2 contributes nothing in the evaluation, simplifying the calculation. *It can be shown that the value of the determinant is unique and does not depend on the row or column chosen for its evaluation.*

Use of determinants in solving a system of three linear equations in three variables follows the same method as that for two equations.

EXAMPLE
12

Solve by determinants:

$$\begin{cases} 2x + y + z = 0 \\ 4x + 3y + 2z = 2 \\ 2x - y - 3z = 0. \end{cases}$$

We have

$$x = \dfrac{\begin{vmatrix} 0 & 1 & 1 \\ 2 & 3 & 2 \\ 0 & -1 & -3 \end{vmatrix}}{\begin{vmatrix} 2 & 1 & 1 \\ 4 & 3 & 2 \\ 2 & -1 & -3 \end{vmatrix}}.$$

The value of the denominator calculated along the first row is

$$2(-1)^{1+1}\begin{vmatrix} 3 & 2 \\ -1 & -3 \end{vmatrix} + 1(-1)^{1+2}\begin{vmatrix} 4 & 2 \\ 2 & -3 \end{vmatrix} + 1(-1)^{1+3}\begin{vmatrix} 4 & 3 \\ 2 & -1 \end{vmatrix}$$

$$= 2(+1)(-7) + 1(-1)(-16) + 1(+1)(-10)$$

$$= -8.$$

The numerator, calculated along the first column for convenience, is

$$0 + 2(-1)^{2+1}\begin{vmatrix} 1 & 1 \\ -1 & -3 \end{vmatrix} + 0 = 4.$$

Thus, $x = 4/-8 = -\frac{1}{2}$. Similarly,

$$y = \frac{\begin{vmatrix} 2 & 0 & 1 \\ 4 & 2 & 2 \\ 2 & 0 & -3 \end{vmatrix}}{-8} = \frac{2(-1)^{2+2}\begin{vmatrix} 2 & 1 \\ 2 & -3 \end{vmatrix}}{-8} = \frac{2(1)(-8)}{-8} = 2$$

$$z = \frac{\begin{vmatrix} 2 & 1 & 0 \\ 4 & 3 & 2 \\ 2 & -1 & 0 \end{vmatrix}}{-8} = \frac{2(-1)^{2+3}\begin{vmatrix} 2 & 1 \\ 2 & -1 \end{vmatrix}}{-8} = \frac{2(-1)(-4)}{-8} = -1.$$

The solution set is $\{(-\frac{1}{2}, 2, -1)\}$.

EXERCISE 8-6

1. Evaluate.

a. $\begin{vmatrix} 2 & 1 & 3 \\ 2 & 0 & 1 \\ -4 & 0 & 6 \end{vmatrix}$
   b. $\begin{vmatrix} 3 & 2 & 1 \\ 1 & -2 & 3 \\ -1 & 3 & 2 \end{vmatrix}$
   c. $\begin{vmatrix} 1 & 2 & -3 \\ 4 & 5 & 4 \\ 3 & -2 & 1 \end{vmatrix}$

d. $\begin{vmatrix} 1 & 0 & -1 \\ 0 & 1 & 0 \\ 1 & -1 & 1 \end{vmatrix}$
   e. $\begin{vmatrix} 2 & 1 & 5 \\ -3 & 4 & -1 \\ 0 & 6 & -1 \end{vmatrix}$
   f. $\begin{vmatrix} 1 & 2 & 3 \\ 4 & 5 & 4 \\ 3 & 2 & 1 \end{vmatrix}$

2. A *nomogram* is one type of graphical process that can be used to solve certain types of equations. One simple additive nomogram yields the basic (determinant) equation

$$\begin{vmatrix} \frac{1}{2}f_1 & \frac{1}{2} & 1 \\ -f_2 & 1 & 1 \\ -3f_1 & 0 & 1 \end{vmatrix} = 0.$$

What is the relation between f_1 and f_2?

Solve the systems of equations in Problems 3–12 by determinants.

3. $\begin{cases} x + y + z = 6 \\ x - y + z = 2 \\ 2x - y + 3z = 6 \end{cases}$

4. $\begin{cases} 2x - y + 3z = 12 \\ x + y - z = -3 \\ x + 2y - 3z = -10 \end{cases}$

5. $\begin{cases} 2x - 3y + 4z = 0 \\ x + y - 3z = 4 \\ 3x + 2y - z = 0 \end{cases}$

6. $\begin{cases} 3r - t = 7 \\ 4r - s + 3t = 9 \\ 3s + 2t = 15 \end{cases}$

7. $\begin{cases} 2x - 3y + z = -2 \\ x - 6y + 3z = -2 \\ 3x + 3y - 2z = 2 \end{cases}$

8. $\begin{cases} x - z = 14 \\ y + z = 21 \\ x - y + z = -10 \end{cases}$

9. $\begin{cases} 2x + y + 6z = 3 \\ x - y + 4z = 1 \\ 3x + 2y - 2z = 2 \end{cases}$

10. $\begin{cases} x + y + z = -1 \\ 3x + y + z = 1 \\ 4x - 2y + 2z = 0 \end{cases}$

DET 11. $\begin{cases} 5x - 7y + 4z = 2 \\ 3x + 2y - 2z = 3 \\ 2x - y + 3z = 4 \end{cases}$

12. $\begin{cases} 3x - 2y + z = 0 \\ -2x + y - 2z = 5 \\ \frac{3}{2}x + \frac{4}{5}y + 4z = 10 \end{cases}$

8-7
VERBAL PROBLEMS

Each of the problems in this section gives rise to a system of linear equations. After an appropriate system of equations has been determined, you may use any method of solution. The following examples will illustrate some basic techniques.

EXAMPLE

13

In a particular laboratory process, a 25-percent hydrogen peroxide solution (25 percent by volume is hydrogen peroxide) is to be combined with a 40-percent hydrogen peroxide solution to obtain 2 liters of a 30-percent solution. How many liters of each solution should be mixed?

Let x denote the number of liters of the 25-percent solution and y denote the number of liters of the 40-percent solution which should be combined. Then

$$x + y = 2.$$

In 2 liters of a 30-percent solution, there will be $.3(2) = .6$ liters of hydrogen peroxide. This hydrogen peroxide comes from two sources: $.25x$ liters of it come from the 25-percent solution, and $.40y$ liters come from the 40-percent solution. Hence

$$.25x + .40y = .6.$$

We now have a system of two linear equations in two unknowns:

$$\begin{cases} x + y = 2 & (1) \\ .25x + .40y = .6. & (2) \end{cases}$$

Multiplying Eq. (1) by 25 and Eq. (2) by -100 we have

$$\begin{cases} 25x + 25y = 50 & (3) \\ -25x - 40y = -60. & (4) \end{cases}$$

Adding Eq. (3) to Eq. (4) yields

$$-15y = -10$$

$$y = \frac{-10}{-15} = \frac{2}{3}.$$

Thus, from Eq. (1), $x = \frac{4}{3}$. Hence $\frac{4}{3}$ liters of the 25-percent solution and $\frac{2}{3}$ liter of the 40-percent solution must be combined.

EXAMPLE 14

A man has invested $5000, part at 5 percent and the remainder at 4 percent. His annual interest is $220. How much is invested at each rate?

Let $x =$ amount at 5 percent and $y =$ amount at 4 percent. Then,

$$\begin{cases} x + y = 5000 & (1) \\ .05x + .04y = 220. & (2) \end{cases}$$

Multiplying Eq. (1) by 5 and Eq. (2) by -100,

$$\begin{cases} 5x + 5y = 25{,}000 & (3) \\ -5x - 4y = -22{,}000. & (4) \end{cases}$$

Adding Eq. (3) to Eq. (4), we obtain $y = \$3000$. From Eq. (1), it follows that $x = \$2000$.

EXAMPLE 15

In a manufacturing process a large storage tank is filled in 3 hr by two pipes running simultaneously. If, after these pipes have run together for 2 hr, the large pipe is turned off and it takes an additional 5 hr to fill the tank, how long would it take each pipe to fill the tank alone?

Let x denote the number of hours it takes the large pipe to fill the tank alone, and y denote the number of hours required by the small pipe alone. Then the part of the tank that is filled in 1 hr by the large pipe is $1/x$, and by the small pipe it is $1/y$. Since the two pipes running simultaneously fill the tank in 3 hr, the part of the tank they fill together in 1 hr is $\frac{1}{3}$. Thus, $(1/x) + (1/y) = \frac{1}{3}$. Now, the part of the tank both pipes fill in 2 hr is $(2/x) + (2/y)$, and the part the small pipe fills in 5 hr is $5/y$. It follows that

$$\left(\frac{2}{x} + \frac{2}{y} \right) + \frac{5}{y} = 1$$

since the whole tank may be denoted by $\frac{1}{1} = 1$. Therefore, our system of equations is

$$\begin{cases} \dfrac{1}{x} + \dfrac{1}{y} = \dfrac{1}{3} \\ \dfrac{2}{x} + \dfrac{7}{y} = 1. \end{cases}$$

Using the techniques of Example 6 we obtain $x = \frac{15}{4}$ hr and $y = 15$ hr.

EXAMPLE
16

A man wishes to fence in a rectangular plot of ground. Two types of material are available, one costing \$5/ft and the other \$3/ft. If he uses the less expensive material, the total cost is \$450; if he uses the more expensive material for the front side, which is a long side, and the cheaper material for the remaining three sides, the total cost is \$550. What are the dimensions of the rectangular plot?

Let $x =$ length of the rectangular plot and $y =$ width of the rectangular plot. Then

$$\begin{cases} (3.00)[2x + 2y] = 450 \\ (5.00)[x] + (3.00)[x + 2y] = 550. \end{cases}$$

Simplifying, we have

$$\begin{cases} x + y = 75 \\ 4x + 3y = 275. \end{cases}$$

Solving this system we obtain $x = 50$ ft and $y = 25$ ft.

EXAMPLE
17

Suppose it is known that quantities x and y are related by an equation of the form

$$y = mx + b.$$

If measurements are made and it is found that $y = 4$ when $x = 3$, and $y = 13$ when $x = 6$, find m and b.

Substituting the data into the given equation we obtain the system

$$\begin{cases} 4 = 3m + b \\ 13 = 6m + b. \end{cases}$$

Solving gives $m = 3$ and $b = -5$.

EXERCISE 8-7

1. The sum of \$12,000 was invested, part at 5 percent and the remainder at $5\frac{1}{2}$ percent. The total interest at the end of 1 year was \$640. Find the amount invested at each rate.

2. A portion of a total investment of \$4000 yields 5 percent, the remainder yielding 4.7 percent. The interest the first year was \$195.50. How much was invested at each rate?

3. The perimeter of a rectangle is 26 ft. The length exceeds the width by 3 ft. Find the dimensions of the rectangle.

4. The length of a rectangle is twice the width. The perimeter is 60 ft. Find the dimensions.

5. How much of each of a 25-percent chemical solution and a 32-percent solution must be combined to make 75 cc of a 28-percent solution?

6. How many gallons of a 20-percent acid solution must be added to a 6-percent acid solution to make 40 gal of an 18-percent acid solution?

7. Len and Dick can do a job in 10 hr working together. After 6 hr, Dick takes 9 hr to finish. How long does it take each to do the job working alone?

8. One of two complementary angles is three-fifths the other one. Find the angles.

9. One of two supplementary angles is three-fifths the other one. Find the angles.

10. The graph of the equation $y = ax^2 + bx + c$ passes through the points $(2, 3), (-4, 1)$, and $(3, 8)$. Find a, b, and c and hence determine the equation.

11. The graph of the equation $y = ax^2 + bx + c$ passes through the points $(0, 0), (2, 8)$, and $(-1, 1)$. Find a, b, and c and hence determine the equation.

12. A boat went 10 mi downstream in 45 min but took 75 min for the return trip. Find the rate of the current.

13. The boom in Fig. 8-10 has weight W and is acted upon by the forces shown. T is the force exerted by the string and V and H are forces exerted

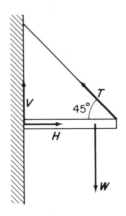

FIGURE 8-10

by the wall. From Newtonian mechanics it can be shown that for the boom in equilibrium,

$$\begin{cases} H - T\cos 45° = 0 \\ V + T\sin 45° - W = 0 \\ W(\tfrac{1}{2}) - T\sin 45° = 0. \end{cases}$$

If $W = 100$ lb, find the exact values for H, T, and V.

14. If the weight of the boom in Problem 13 is doubled, what are the values for H, T and V.

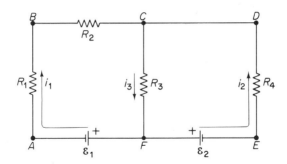

FIGURE 8-11

15. In the analysis of direct-current circuits as shown in Fig. 8-11, we use Kirchhoff's laws, which are essentially a special statement of the law of conservation of energy. The laws state

a.   The algebraic sum of the potential differences around any closed loop is equal to zero.

b.   The rate at which charge enters a junction—that is, the current—is equal to the rate at which charge leaves the same junction.

Application of these laws to the circuit of Fig. 8-11 leads to the system of linear equations

$$\begin{cases} R_1 i_1 + R_2 i_1 + R_3 i_3 + \varepsilon_1 = 0 & \text{(rule a, loop } ABCFA) \\ \qquad\qquad R_3 i_3 + R_4 i_2 + \varepsilon_2 = 0 & \text{(rule a, loop } FCDEF) \\ \qquad\qquad\qquad\quad i_1 + i_2 = i_3 & \text{(rule b at point } C). \end{cases}$$

If $R_1 = 2$, $R_2 = 3$, $R_3 = 5$, and $R_4 = 3$ ohms and $\varepsilon_1 = 15$ volts and $\varepsilon_2 = 2$ volts, solve the system of equations for i_1, i_2, and i_3 and, thus, find the current in amperes in each branch of the circuit.

16. In the circuit of Problem 15, if $R_1 = 1$, $R_2 = 1$, $R_3 = 4$, and $R_4 = 6$ ohms and $\varepsilon_1 = 2$ volts and $\varepsilon_2 = 60$ volts, find i_1, i_2, and i_3.

17. In the circuit of Problem 15, with $\varepsilon_1 = \varepsilon_2$ a student measured and found that $i_1 = i_2$. Show that, as a consequence,

$$R_1 + R_2 - R_4 = 0.$$

You may assume that $i_1 \neq 0$.

8-8
REVIEW

REVIEW QUESTIONS

Two methods by which a system of two linear equations in two variables can be solved are by elimination by _____(1)_____ and elimination by _____(2)_____ .

(1) **addition**   (2) **substitution**

If a system of two linear equations is geometrically represented by two parallel lines, what can be said about the solution of the system? _____ (3) _____

(3) **no solution or infinitely many solutions**

A system of two linear equations whose solution set is not empty is said to be
_____ (4) _____ .

(4) **consistent**

If there is one and only one solution for a system of two linear equations, the system is said to be (consistent)(inconsistent) and (dependent)(independent).
(5)   (6)

(5) **consistent**   (6) **independent**

The system

$$\begin{cases} x - y = 4 \\ 2x - 2y = 16 \end{cases}$$

can be classified as a(n) (consistent)(inconsistent) system.
(7)

(7) **inconsistent**

The system

$$\begin{cases} y = 2x - 4 \\ y = 6x \end{cases}$$

is said to be (consistent)(inconsistent) and (dependent)(independent).
(8)   (9)

(8) **consistent**   (9) **independent**

The system

$$\begin{cases} y - x = 1 \\ 2y - 2x = 2 \end{cases}$$

is said to be (consistent)(inconsistent) and (dependent)(independent).
(10)   (11)

(10) **consistent**   (11) **dependent**

To solve the system

$$\begin{cases} 3x - 4y = 10 \\ 8x + 5y = -40 \end{cases}$$

by the method of elimination of y, first multiply the top equation by
_____(12)_____ , the bottom equation by _____(13)_____ , and then
_____(14)_____ the resulting equations.

(12) **5**   (13) **4**   (14) **add**

The value of

$$\begin{vmatrix} 1 & 2 \\ 1 & -1 \end{vmatrix} \cdot \begin{vmatrix} 3 & 4 \\ -3 & 4 \end{vmatrix}$$

is _____(15)_____ .

(15) **−72**

The value of

$$\begin{vmatrix} 99 & 2 & 0 \\ 99 & 1 & 0 \\ 100 & 600 & 1 \end{vmatrix}$$

is _____(16)_____ .

(16) **−99**

The minor of entry 3 in

$$\begin{vmatrix} 0 & 1 & 3 \\ 2 & 4 & 0 \\ 1 & 5 & 2 \end{vmatrix}$$

is _____(17)_____ and its cofactor is _____(18)_____ .

(17) **6**   (18) **6**

REVIEW PROBLEMS

In Problems 1–22, solve the given system of equations by any method.

1. $\begin{cases} 2x - y = 6 \\ 3x + 2y = 5 \end{cases}$

2. $\begin{cases} 8x - 4y = 7 \\ y = 2x - 4 \end{cases}$

3. $\begin{cases} 3x + 5y = -6 \\ 2x - 6 = 5y \end{cases}$

4. $\begin{cases} 3y - 2x = 4 \\ 4x - 6y = 8 \end{cases}$

5. $\begin{cases} 5x + y = -2 \\ 20x + 2y = 1 \end{cases}$

6. $\begin{cases} 2x - y = 1 \\ -x + 2y = 7 \end{cases}$

7. $\begin{cases} 5x + 2y = 36 \\ 8x - 3y = -54 \end{cases}$

8. $\begin{cases} x + y = 3 \\ 3x + 2y = 19 \end{cases}$

9. $\begin{cases} \frac{2}{3}x + \frac{1}{2}y = 2 \\ \frac{3}{8}x + \frac{5}{6}y = -\frac{11}{2} \end{cases}$

10. $\begin{cases} \frac{1}{2}x - \frac{1}{4}y = \frac{1}{6} \\ x + \frac{1}{2}y = \frac{2}{3} \end{cases}$

11. $\begin{cases} \frac{2}{3}x + \frac{4}{9}y = 1 \\ 8x - 2y = 14 \end{cases}$

12. $\begin{cases} 3(x + y) - 2(x - y) = 4 \\ 2(x + y) + 5(x + y) = 16 \end{cases}$

13. $\begin{cases} 2(x - y) + 6(x + y) = 13 \\ 3(x + y) - 4(x - y) = 10 \end{cases}$

14. $\begin{cases} x + \dfrac{2y + x}{6} = 14 \\ y + \dfrac{3x + y}{4} = 20 \end{cases}$

15. $\begin{cases} 2x + 2x^2 - y = 2(x^2 - y) \\ 3y - 5x = 16 \end{cases}$

16. $\begin{cases} \frac{3}{2}x - \frac{4}{3}y = 16 \\ \frac{4}{5}y - \frac{3}{5}x = 1 \end{cases}$

17. $\begin{cases} \dfrac{2}{x} - \dfrac{3}{y} = 16 \\ \dfrac{5}{y} + \dfrac{6}{x} = 10 \end{cases}$

18. $\begin{cases} 5x + \dfrac{3}{y} = 15 \\ 3x - \dfrac{4}{y} = 8 \end{cases}$

19. $\begin{cases} x - 2y + 3z = 4 \\ 8x + y - 3z = 2 \\ x + y - 3 = z \end{cases}$

20. $\begin{cases} 2x + 3y - z = 2 \\ 5x + 2y + z = 5 \\ -x + 3y - z = 12 \end{cases}$

21. $\begin{cases} 3x - \frac{4}{5}y + \frac{3}{4}z = 6 \\ -z + \frac{2}{3}y = 5 - \frac{3}{2}x \\ x + y + z = 1 \end{cases}$

22. $\begin{cases} \frac{2}{3}x - \frac{3}{2}y + \frac{3}{4}z = 6 \\ 2x - 5y + \frac{3}{4}z = 10 \\ x - y + z = -6 \end{cases}$

In Problems 23–26, evaluate the given determinants.

23. $\begin{vmatrix} 2 & -1 \\ 4 & 7 \end{vmatrix}$

24. $\begin{vmatrix} 5 & 8 \\ 3 & 0 \end{vmatrix}$

25. $\begin{vmatrix} 1 & 2 & -1 \\ 0 & 1 & 4 \\ 1 & 2 & 2 \end{vmatrix}$

26. $\begin{vmatrix} 2 & 0 & 3 \\ 1 & 4 & 6 \\ -1 & 2 & -1 \end{vmatrix}$

Solve Problems 27 and 28 by Cramer's rule.

27. $\begin{cases} 3x - y = 1 \\ 2x + 3y = 8 \end{cases}$

28. $\begin{cases} x + 2y - z = 0 \\ y + 4z = 0 \\ x + 2y + 2z = 0 \end{cases}$

9

exponents and radicals

9-1

A REVIEW OF EXPONENTS

Although we introduced the laws of exponents and radicals in Chapter 1, a more detailed treatment is certainly necessary. Our purpose here is to consider the application of these rules to simplify various types of mathematical expressions. For your convenience, the basic rules of exponents are repeated here; they are valid for any exponent whatsoever.

Rule 1. $x^m \cdot x^n = x^{m+n}$.

Rule 2. $(x^m)^n = x^{mn}$.

Rule 3. $(xy)^n = x^n y^n$.

Rule 4. $\left(\dfrac{x}{y}\right)^n = \dfrac{x^n}{y^n}, \quad y \neq 0$.

Rule 5. $\dfrac{x^m}{x^n} = x^{m-n}, \quad m > n, x \neq 0$.

Rule 6. $\dfrac{x^m}{x^n} = \dfrac{1}{x^{n-m}},\quad m < n,\, x \neq 0.$

Rule 7. $\dfrac{x^n}{x^n} = 1,\quad x \neq 0.$

For consistency we also defined

$$x^0 = 1 \qquad\qquad \text{for } x \neq 0$$

$$x^{-n} = \frac{1}{x^n} \qquad\qquad \text{for } x \neq 0$$

$$x^{1/n} = \sqrt[n]{x} \qquad \text{where } \sqrt[n]{x} \text{ is the principal } n\text{th root of } x.$$

At times we shall require that you express a final result involving exponents in terms of positive exponents with no fractional exponents in the denominator. Remember that eliminating a fractional exponent from a denominator is called *rationalizing the denominator* and is accomplished by multiplying both the numerator and denominator of a fraction by the same appropriate quantity.

EXAMPLE
1

Simplify $(2x^{-1/2})^3$.

$$(2x^{-1/2})^3 = (2)^3(x^{-1/2})^3 = 8x^{-3/2} = \frac{8}{x^{3/2}}.$$

To rationalize the denominator of the final expression, we multiply both the numerator and the denominator by that power of x which will result in the smallest positive integral power of x in the denominator. In this case,

$$\frac{8}{x^{3/2}} = \frac{8}{x^{3/2}} \cdot \frac{x^{1/2}}{x^{1/2}} = \frac{8x^{1/2}}{x^2}.$$

Hence,

$$(2x^{-1/2})^3 = \frac{8x^{1/2}}{x^2}.$$

By studying the simplifications in Example 2, you will review many of the fundamental operations with exponents.

EXAMPLE
2

Simplify each of the following expressions.

a. $\left(\dfrac{1}{2}x^{-2}y^{1/3}\right)^2 = \dfrac{1}{4}x^{-4}y^{2/3} = \dfrac{y^{2/3}}{4x^4}.$

b. $\dfrac{(2x^{-3})^2}{x^{-4}} = \dfrac{4x^{-6}}{x^{-4}} = \dfrac{4}{x^2}.$

c. $\left(\dfrac{xy^{-1}}{y^2}\right)^{-2} = \left(\dfrac{x}{y^3}\right)^{-2} = \dfrac{x^{-2}}{y^{-6}} = \dfrac{y^6}{x^2}.$

d. $\dfrac{y^{2/3}}{x^{2/3}y^{1/3}} = \dfrac{y^{1/3}}{x^{2/3}} = \dfrac{y^{1/3}}{x^{2/3}} \cdot \dfrac{x^{1/3}}{x^{1/3}} = \dfrac{x^{1/3}y^{1/3}}{x}$.

e. $\dfrac{(xy)^2(x^{1/2})^6}{xy} = (xy)(x^3) = x^4y$.

f. $(64x^4)^{1/2} = (8^2)^{1/2}(x^4)^{1/2} = 8x^2$.

g. $\left(\dfrac{16}{81}\right)^{3/4} = \left(\dfrac{2^4}{3^4}\right)^{3/4} = \left[\left(\dfrac{2}{3}\right)^4\right]^{3/4} = \left(\dfrac{2}{3}\right)^3 = \dfrac{8}{27}$.

EXERCISE 9-1

Simplify the expressions in Problems 1–34. Express your answers with rationalized denominators and positive exponents only.

1. $x^{1/2} \cdot x^{3/2}$ 2. $x^{2/3} \cdot x^{4/3}$

3. $x^{4/3} \cdot x^{-1/3}$ 4. $x^{-1/2} \cdot x^{3/2}$

5. $(xy^{-1/2})y^{-1/2}$ 6. $x^{1/2}(x^{1/2}y)$

7. $x^{1/2} \cdot x^{1/3}$ 8. $x^{1/4} \cdot x^{1/3}$

9. $x^{-2} \cdot x^{7/2} \cdot x^{5/2}$ 10. $x^3 \cdot x^{-3/8} \cdot x^{-5/8}$

11. $(x^{1/2})^3$ 12. $(y^{-1/3})^6$

13. $(-x^{1/2})^3$ 14. $(-y^{1/3})^6$

15. $(2x^{-2}y^{1/3})^3$ 16. $(3x^{-3}y^{-1/2})^4$

17. $(x^{-1/2})^3$ 18. $(y^{-1/3})^2$

19. $(x^{-2/3})^2$ 20. $(x^{-1/5})^2$

21. $\left(\dfrac{x^{-4/3}}{y^{-2/3}}\right)^3$ 22. $\left(\dfrac{x^{-3/2}}{y^{-1}}\right)^4$

23. $\left(\dfrac{2^{1/3}x^{2/3}}{y^{1/3}}\right)^3$ 24. $\left(\dfrac{3^{1/4}x^{3/4}}{y^{1/2}}\right)^4$

25. $(27x^{15})^{-1/3}$ 26. $(8x^{-6})^{-1/3}$

27. $(27^{-1}x^{-15})^{-1/3}$ 28. $(8x^6)^{-1/3}$

29. $(-27x^{15})^{-1/3}$ 30. $(-8x^{-6})^{-1/3}$

31. $(-2)^{-3}\left(\dfrac{1}{x^{-1}}\right)^2$ 32. $\left(\dfrac{x^{-1}}{x^{1/3}}\right)^2$

33. $\left[6^{1/2} - \left(\dfrac{1}{6^{-4}}\right)^{3/8}\right]^0$ 34. $\left[6^{1/2} - \left(\dfrac{1}{6^{-4}}\right)^{1/8}\right]^0$

35. Experimental data indicate that the radius R of a nucleus of an isotope of mass number A is given by

$$R = r_0 A^{1/3}.$$

The value of r_0, the radius parameter, can be taken to be 1.5×10^{-13} cm. Determine the radius of a naturally occurring isotope of zinc whose mass number is 64.

36. The probability that an alpha particle will be scattered by a nucleus through an angle equal to or greater than some angle θ is given by the expression $\pi p^2 nt$, where n is the number of nuclei per unit volume, t is the thickness of the scattering material, and $p = (\sigma/\pi)^{1/2}$ is the impact parameter and depends on θ. Find an expression for the probability in terms of σ, n, and t.

9-2
FURTHER OPERATIONS WITH EXPONENTS

Although the title of this section may seem to imply that we are about to introduce a series of "new" operations involving exponents, in reality we shall introduce nothing new at all. We previously performed many algebraic manipulations such as the multiplication and division of multinomials, the addition and subtraction of fractions, and the simplification of complex fractions—but these were accomplished with little consideration of negative or fractional exponents. In essence, now we shall be employing all the techniques of the previous chapters in conjunction with the laws of exponents. We shall begin by illustrating some of the simplifications that can be accomplished with sums and differences of algebraic expressions containing exponents.

EXAMPLE
3

Simplify each of the following expressions.

a. $2x^{1/2} + (x^{1/6})^3$.

$$2x^{1/2} + (x^{1/6})^3 = 2x^{1/2} + x^{1/2} \qquad \text{(by Rule 2)}.$$

Combining similar terms we have

$$2x^{1/2} + (x^{1/6})^3 = 3x^{1/2}.$$

b. $(27x)^{1/3} - (2x^{1/3} - 4x^{1/3})$.

$$(27x)^{1/3} - (2x^{1/3} - 4x^{1/3}) = 3x^{1/3} - (-2x^{1/3})$$
$$= 5x^{1/3}.$$

c. $2x^{-1/2} + 3x^{-1/2}$.

$$2x^{-1/2} + 3x^{-1/2} = 5x^{-1/2}$$
$$= \frac{5}{x^{1/2}}$$
$$= \frac{5}{x^{1/2}} \cdot \frac{x^{1/2}}{x^{1/2}}$$
$$= \frac{5x^{1/2}}{x}.$$

EXAMPLE
4

Perform the indicated operations and simplify.

a. $x^{-2}(x^{-2} + y^3)$.

By the distributive axiom we have

$$x^{-2}(x^{-2} + y^3) = x^{-4} + x^{-2}y^3$$
$$= \frac{1}{x^4} + \frac{y^3}{x^2}$$
$$= \frac{1 + x^2y^3}{x^4}.$$

In the last step, the fractions were added in the usual manner.

b. $(x^{1/2} + y^{1/2})^2$.

This is merely the square of a binomial:

$$(x^{1/2} + y^{1/2})^2 = (x^{1/2})^2 + 2(x^{1/2}y^{1/2}) + (y^{1/2})^2$$
$$= x + 2x^{1/2}y^{1/2} + y.$$

c. $\dfrac{3x^4 - 2x^2}{x^{1/3}}$.

Here we can express the division of the multinomial by the monomial as

$$\frac{3x^4 - 2x^2}{x^{1/3}} = \frac{3x^4}{x^{1/3}} - \frac{2x^2}{x^{1/3}}$$
$$= 3x^{11/3} - 2x^{5/3}.$$

d. $\dfrac{(4x^4y)^{-1/2}(y^4)^{-1/8}}{3xy^{-3}}$.

Any expression involving only the product and/or quotient of monomials can be simplified by direct application of the rules of exponents:

$$\frac{(4x^4y)^{-1/2}(y^4)^{-1/8}}{3xy^{-3}} = \frac{(4^{-1/2}x^{-2}y^{-1/2})(y^{-1/2})}{3xy^{-3}}$$
$$= \frac{4^{-1/2}x^{-2}y^{-1}}{3xy^{-3}}$$
$$= \frac{y^2}{(4^{1/2})3x^3} = \frac{y^2}{2(3x^3)}$$
$$= \frac{y^2}{6x^3}.$$

e. $\dfrac{1}{x^{-1} + y^{-1}}$.

$$\frac{1}{x^{-1} + y^{-1}} = \frac{1}{\dfrac{1}{x} + \dfrac{1}{y}}$$
$$= \frac{1}{\dfrac{y + x}{xy}}$$
$$= 1 \cdot \frac{xy}{y + x}$$
$$= \frac{xy}{x + y}.$$

f. $\dfrac{3x^{1/5}y^{1/4}}{x^{2/5}y^{3/4}}$.

By the rules of exponents it should be clear that

$$\frac{3x^{1/5}y^{1/4}}{x^{2/5}y^{3/4}} = \frac{3}{x^{1/5}y^{1/2}} .$$

To rationalize the denominator, we employ the usual techniques:

$$\frac{3x^{1/5}y^{1/4}}{x^{2/5}y^{3/4}} = \frac{3}{x^{1/5}y^{1/2}} = \frac{3}{x^{1/5}y^{1/2}} \cdot \frac{x^{4/5}y^{1/2}}{x^{4/5}y^{1/2}}$$

$$= \frac{3x^{4/5}y^{1/2}}{xy} .$$

EXERCISE 9-2

In each of the following, perform the indicated operations and simplify. Express your answers with rationalized denominators and positive exponents only. Combine any resulting sum or difference of fractions into a single fraction.

1. $x^{-1} + y^{-1}$

2. $2x^{-1} - 1$

3. $2x^{1/2} - 5x^{1/2}$

4. $7x^{3/2} + 2x^{3/2} - 5x^{3/2}$

5. $2 + 2x^{-1}$

6. $3 - y^{-1}$

7. $4x^{-1/3} - 2(x^2)^{-1/6}$

8. $3x^{1/4} - (x^{-1/2})^{-1/2}$

9. $3x^{-1} + y$

10. $(x^{-1} + y^{-1})(4)$

11. $(x^{3/2} + 2)^2$

12. $(x^{1/3} - 1)^2$

13. $(x^{1/2} + 1)(x^{3/2} - 2)$

14. $(x^{2/3} + 3)(x^{1/3} - 1)$

15. $(x^{1/2} + x^{3/2})^2$

16. $(x^{1/3} - x^{2/3})^2$

17. $(x^{1/2} + y^{1/2})^2$

18. $(x^{1/3} + y^{1/3})^2$

19. $3x^{-2}(x^{-1} + y)$

20. $2x^{-3}(x^{-2} + y)$

21. $\dfrac{x^{-1} + y^{-1}}{y^{-1}}$

22. $\dfrac{y^{-1}}{x^{-1} + y^{-1}}$

23. $\dfrac{x^{4/3} + x^2 - x^4}{x^{1/3}}$

24. $\dfrac{y^{5/4} - y^3 - y^{1/4}}{y^{1/4}}$

25. $\dfrac{x^{1/2} - 3x^{1/3}}{x^{1/4}}$

26. $\dfrac{2x^{1/3} + 3x^{1/5}}{x^{1/2}}$

27. $\dfrac{(3x^{-2}y)^2}{xy}$

28. $\dfrac{(2x^{1/2}y)^3(4x)^{-1/2}}{x}$

29. $\dfrac{x^{-1/2}y^{-1/3}}{2}$

30. $\dfrac{x^{-2/3}y^{-1/2}}{3}$

31. $\dfrac{(3x^{-2}yw^{1/2})^2(2xy)}{(xyw)^{-1}}$

32. $\dfrac{(2xy)^{-2}(x^{1/2}w^2)(-x)^3}{x^{1/2}y^{-2}}$

33. $\dfrac{1^{-1} + 2^{-1} + 3^{-1}}{4^{-1}}$ **34.** $\dfrac{1}{8} \cdot 8^{2/3} - 2 \cdot 8^{-2/3}$

35. $\dfrac{x^{-1} - 4^{-1}}{x^{-1} + 4^{-1}}$ **36.** $(x^{-1} + b^{-1})(x^{-1} - b^{-1})$

37. $(2x^{3/4} + 1)(2x^{3/4} - 1)(4x^{3/2} + 1)$

38. $(y^{2/3} + 2)(y^{4/3} - 2y^{2/3} + 4)$

9-3
SOME BASIC OPERATIONS WITH RADICALS

The basic rules governing the manipulation of radicals, which follow logically from the rules of exponents, were stated in Chapter 1. We repeat them here for your convenience.

Rule 8. $x^{m/n} = \sqrt[n]{x^m} = (\sqrt[n]{x})^m$.

Rule 9. $\sqrt[n]{x}\,\sqrt[n]{y} = \sqrt[n]{xy}$.

Rule 10. $\dfrac{\sqrt[n]{x}}{\sqrt[n]{y}} = \sqrt[n]{\dfrac{x}{y}}$.

Rule 11. $\sqrt[m]{\sqrt[n]{x}} = \sqrt[mn]{x}$.

Remember that these properties apply only to real numbers. Do not try to apply them in any case involving an even root of a negative number. *Throughout this chapter we shall assume that all letters under consideration represent positive real numbers.*

EXAMPLE   a. $x^{2/3} = \sqrt[3]{x^2} = (\sqrt[3]{x})^2$ (Rule 8).

5   Note also that we could write

$$x^{2/3} = (x^2)^{1/3} \qquad \text{(rules of exponents)}$$
$$= \sqrt[3]{x^2} \qquad \text{(definition of fractional exponent)}$$

and equivalently

$$x^{2/3} = (x^{1/3})^2 \qquad \text{(rules of exponents)}$$
$$= (\sqrt[3]{x})^2 \qquad \text{(definition of fractional exponent).}$$

b. $\sqrt[3]{9}\,\sqrt[3]{3} = \sqrt[3]{9 \cdot 3} = \sqrt[3]{27} = 3$ (Rule 9).

c. $\dfrac{\sqrt[4]{64}}{\sqrt[4]{4}} = \sqrt[4]{\dfrac{64}{4}} = \sqrt[4]{16} = 2.$ (Rule 10).

d. $\sqrt[5]{\sqrt{x}} = \sqrt[10]{x}$ (Rule 11).

In conjunction with the above rules, let us indicate three very basic operations that can be performed with a radical.

A. REMOVING FACTORS FROM A RADICAL

Any factors in the radicand whose root can be found can be removed from the radicand by application of appropriate rules. The procedure is illustrated below with an indication made of the basic rule employed.

EXAMPLE
6

a. $\sqrt{8} = \sqrt{4\cdot 2} = \sqrt{4}\sqrt{2} = 2\sqrt{2}$ (Rule 9).

b. $\sqrt[3]{128} = \sqrt[3]{64\cdot 2} = \sqrt[3]{64}\sqrt[3]{2} = 4\sqrt[3]{2}$ (Rule 9).

c. $\sqrt{25x^5y^5} = \sqrt{25x^4y^4}\sqrt{xy} = 5x^2y^2\sqrt{xy}$ (Rule 9).

d. $\sqrt[3]{27x^4y^6z^7} = \sqrt[3]{27x^3y^6z^6}\sqrt[3]{xz} = 3xy^2z^2\sqrt[3]{xz}$ (Rule 9).

e. $\dfrac{\sqrt{8x^5}}{\sqrt{2x}} = \sqrt{\dfrac{8x^5}{2x}} = \sqrt{4x^4} = 2x^2$ (Rule 10).

f. $\sqrt{x^2 + 2x + 1} = \sqrt{(x+1)^2} = x + 1$, assuming $x + 1 \geq 0$. Otherwise we must write $\sqrt{(x+1)^2} = |x+1|$.

g. $\sqrt[3]{x^{-6}y^6} = \sqrt[3]{\dfrac{y^6}{x^6}} = \dfrac{\sqrt[3]{y^6}}{\sqrt[3]{x^6}} = \dfrac{y^{6/3}}{x^{6/3}} = \dfrac{y^2}{x^2}$ (Rules 10 and 8).

B. INTRODUCING FACTORS INTO THE RADICAND

The process of introducing a factor into the radicand is the reverse of the process of removing a factor. The technique should be clear from the illustrations below.

EXAMPLE
7

Introduce all coefficients into the radicand, assuming all literal symbols denote positive numbers.

a. $2b\sqrt{x} = \sqrt{4b^2}\sqrt{x} = \sqrt{4b^2x}$.

b. $x^2y\sqrt{xy} = \sqrt{x^4y^2}\sqrt{xy} = \sqrt{x^5y^3}$.

c. $2b\sqrt[3]{x} = \sqrt[3]{8b^3}\sqrt[3]{x} = \sqrt[3]{8b^3x}$.

d. $\dfrac{5x}{y}\sqrt{\dfrac{x}{y}} = \sqrt{\dfrac{25x^2}{y^2}}\sqrt{\dfrac{x}{y}} = \sqrt{\dfrac{25x^3}{y^3}}$.

e. $(x+y)\sqrt{\dfrac{1}{x^2-y^2}} = \sqrt{(x+y)^2}\sqrt{\dfrac{1}{x^2-y^2}}$

$= \sqrt{(x+y)^2\cdot\dfrac{1}{(x+y)(x-y)}}$

$= \sqrt{\dfrac{x+y}{x-y}}.$

C. REDUCING THE INDEX OF A RADICAL

Under certain conditions, basic manipulations can make it possible to reduce the index of a radical. Here again, the technique is merely an application of the basic rules, and the following example should clearly illustrate the process.

EXAMPLE
8

Reduce the index of each of the given radicals.

a. $\sqrt[4]{25} = \sqrt[4]{5^2} = 5^{2/4} = 5^{1/2} = \sqrt{5}$.

b. $\sqrt[6]{16x^2y^4} = \sqrt[6]{4^2x^2y^4} = 4^{2/6}x^{2/6}y^{4/6}$

$$= 4^{1/3}x^{1/3}y^{2/3} = (4xy^2)^{1/3}$$

$$= \sqrt[3]{4xy^2}.$$

c. $\sqrt[6]{\dfrac{x^3}{y^9}} = \left(\dfrac{x^3}{y^9}\right)^{1/6} = \dfrac{x^{1/2}}{y^{3/2}} = \left(\dfrac{x}{y^3}\right)^{1/2} = \sqrt{\dfrac{x}{y^3}} = \dfrac{1}{y}\sqrt{\dfrac{x}{y}}$.

It should be clear that the basic techniques used above can be viewed, in essence, as an application of Rule 11. For example, in Example 8a we could have written

$$\sqrt[4]{25} = \sqrt{\sqrt{25}} = \sqrt{5}.$$

Similarly, in Examples 8b and c

$$\sqrt[6]{16x^2y^4} = \sqrt[3]{\sqrt{16x^2y^4}} = \sqrt[3]{4xy^2}$$

and

$$\sqrt[6]{\dfrac{x^3}{y^9}} = \sqrt{\sqrt[3]{\dfrac{x^3}{y^9}}} = \sqrt{\dfrac{x}{y^3}} = \dfrac{1}{y}\sqrt{\dfrac{x}{y}}.$$

EXERCISE 9-3

In Problems 1–20, simplify the given expression by expressing it as a single radical where necessary and removing factors from the radicand whenever possible. Assume all such factors are positive.

1. $\sqrt{50}$

2. $\sqrt{48}$

3. $\sqrt{.01x^4}$

4. $\sqrt{.16x^{12}y^{12}}$

5. $\dfrac{\sqrt{36x^4y^6}}{\sqrt{81}}$

6. $\dfrac{\sqrt{25(x+y)^8}}{\sqrt{49(x-y)^4}}$

7. $\sqrt{8x^7y^4}$

8. $\sqrt{24(a+b)^3}$

9. $\sqrt[3]{8x^3y^4}$

10. $\sqrt[3]{24(a+b)^7}$

11. $\sqrt{x^2+4x+4}$

12. $\sqrt{x^2+6x+9}$

13. $\dfrac{\sqrt[3]{16x^7y^8}}{\sqrt[3]{8xy^5}}$

14. $\dfrac{\sqrt[4]{x^3y^3}}{\sqrt[4]{x^{-2}y^{-2}}}$

15. $\sqrt{2x^4 - 4x^2 + 2}$

16. $\sqrt{3x^2 - 12x + 12}$

17. $\sqrt[3]{\sqrt{x^{12}y^5w^{25}}}$

18. $\sqrt{\sqrt[3]{64x^{12}y^{11}w^7}}$

19. $\sqrt[4]{\dfrac{16x^5}{y^8}}$

20. $\sqrt[3]{\dfrac{1}{x^{-3}} \cdot y^6}$

In Problems 21–30, introduce the coefficients into the radicand. You may assume the coefficients are positive.

21. $3\sqrt{xy}$

22. $2\sqrt{x}$

23. $5a\sqrt{3x}$

24. $abc\sqrt{xy}$

25. $3x\sqrt[3]{xy}$

26. $2x\sqrt[4]{xy}$

27. $(x+2)\sqrt[4]{x-2}$

28. $(x-y)\sqrt[3]{x+y}$

29. $x^a\sqrt{y}$

30. $x^a\sqrt[3]{y}$

In Problems 31–42, reduce the indices of the given radicals and simplify where possible.

31. $\sqrt[4]{9}$

32. $\sqrt[4]{49}$

33. $\sqrt[4]{16x^4y^2}$

34. $\sqrt[6]{8}$

35. $\sqrt[6]{8x^9}$

36. $\sqrt[6]{27x^3y^3z^3}$

37. $\sqrt[4]{\dfrac{625x^2}{y^2}}$

38. $\sqrt[6]{\dfrac{625x^2}{y^2}}$

39. $\sqrt[12]{x^2y^4z^{10}}$

40. $\sqrt[15]{x^3y^{15}z^6}$

41. $\sqrt[2n]{x^n}$

42. $\sqrt[4n]{x^{2n}y^{4n}}$

43. The wavelength of the waves associated with a stream of molecules of a gas at temperature T can be written

$$\sqrt[6]{\frac{8\pi^3 h^6}{M^3 k^3 T^3}}.$$

Reduce this radical to a square root.

44. The grating space of a cubical crystal (the distance between ions in the crystal) can be written

$$\sqrt[4]{\frac{M^{4/3}}{8^{4/9}\rho^{4/3}N_0^{4/3}}}.$$

Express this grating space as a single radical of lower index.

9-4

ADDITION AND SUBTRACTION OF RADICALS

Recall from our discussion of the addition and subtraction of algebraic expressions that we were only able to combine *similar* or *like terms*. An analogous situation is true for the addition and subtraction of radicals.

Radicals are *similar* when they have the *same radicand* and the *same index*, and only radicals meeting this condition can be combined.

EXAMPLE

9

Simplify each of the given expressions.

a. $8\sqrt{xy} - 4\sqrt{xy} - 2\sqrt{xy}$.

Using the distributive axiom we can write

$$8\sqrt{xy} - 4\sqrt{xy} - 2\sqrt{xy} = (8 - 4 - 2)\sqrt{xy}$$
$$= 2\sqrt{xy}.$$

b. $(16\sqrt[3]{x} + 15 - \sqrt[3]{x}) - (15 - 3\sqrt[3]{x})$.

By removing grouping symbols and then combining similar terms we have

$$(16\sqrt[3]{x} + 15 - \sqrt[3]{x}) - (15 - 3\sqrt[3]{x})$$
$$= 16\sqrt[3]{x} + 15 - \sqrt[3]{x} - 15 + 3\sqrt[3]{x}$$
$$= (15 - 15) + (16 - 1 + 3)\sqrt[3]{x}$$
$$= 18\sqrt[3]{x}.$$

c. $\sqrt[3]{81} - \sqrt[3]{24}$.

Although the indices of the radicals are the same, we cannot perform the subtraction unless the radicands are also the same. They can, however, be made the same by appropriate manipulation. Note that

$$\sqrt[3]{81} = \sqrt[3]{27}\sqrt[3]{3} = 3\sqrt[3]{3}$$

and

$$\sqrt[3]{24} = \sqrt[3]{8}\sqrt[3]{3} = 2\sqrt[3]{3}.$$

Hence,

$$\sqrt[3]{81} - \sqrt[3]{24} = 3\sqrt[3]{3} - 2\sqrt[3]{3} = \sqrt[3]{3}.$$

d. $5\sqrt{75(x+y)} - 3\sqrt{12(x+y)}$.

Here we can write

$$5\sqrt{75(x+y)} = 5\sqrt{25 \cdot 3(x+y)} = 25\sqrt{3(x+y)}$$

and

$$3\sqrt{12(x+y)} = 3\sqrt{4 \cdot 3(x+y)} = 6\sqrt{3(x+y)}.$$

Hence,

$$5\sqrt{75(x+y)} - 3\sqrt{12(x+y)} = 25\sqrt{3(x+y)} - 6\sqrt{3(x+y)}$$
$$= 19\sqrt{3(x+y)}.$$

e. $\sqrt[3]{x^6 y} - \sqrt[9]{y^3}$.

$$\sqrt[3]{x^6 y} - \sqrt[9]{y^3} = x^2\sqrt[3]{y} - \sqrt[3]{y}.$$
$$= (x^2 - 1)\sqrt[3]{y}$$
$$= (x + 1)(x - 1)\sqrt[3]{y}.$$

f. $\sqrt[3]{\dfrac{3}{4}} + \sqrt[3]{\dfrac{2}{9}} - \sqrt[3]{\dfrac{1}{36}}.$

$$\sqrt[3]{\dfrac{3}{4}} + \sqrt[3]{\dfrac{2}{9}} - \sqrt[3]{\dfrac{1}{36}} = \sqrt[3]{\dfrac{6}{8}} + \sqrt[3]{\dfrac{6}{27}} - \sqrt[3]{\dfrac{6}{216}}$$

$$= \dfrac{1}{2}\sqrt[3]{6} + \dfrac{1}{3}\sqrt[3]{6} - \dfrac{1}{6}\sqrt[3]{6}$$

$$= \dfrac{2}{3}\sqrt[3]{6}.$$

EXERCISE 9-4

In Problems 1–26, perform the indicated operations and simplify.

1. $4\sqrt[3]{3} - 2\sqrt[3]{3} + \sqrt[3]{3}$ 2. $6\sqrt{7} - (2\sqrt{7} + 3\sqrt{7})$

3. $5\sqrt{6} + 6(2\sqrt{6} - 3\sqrt{6}) - \sqrt{6}$

4. $8\sqrt{10} - (-2\sqrt{10} + \sqrt{10})$

5. $x^2\sqrt[3]{y} + 4\sqrt[3]{y}$ 6. $4x^2\sqrt{2y} - y^2\sqrt{2y}$

7. $x^2\sqrt{6} + 3x\sqrt{6} + 3\sqrt{6}$ 8. $x^2\sqrt[4]{2x} + x\sqrt[4]{2x} - 2\sqrt[4]{2x}$

9. $3\sqrt{75} - 2\sqrt{12}$ 10. $\sqrt{75} - (\sqrt{27} - 2\sqrt{3})$

11. $5\sqrt{18} - (\sqrt{2} + 1)$ 12. $5\sqrt{8} - 2\sqrt{18} + 4$

13. $2y\sqrt{16x} - 3y\sqrt{9x}$ 14. $\sqrt[3]{16} - \sqrt[3]{54}$

15. $\sqrt[3]{128} - 6\sqrt[3]{16}$ 16. $2\sqrt[3]{54} - (\sqrt[3]{128} - 2\sqrt[3]{16})$

17. $\sqrt[3]{64} - \sqrt[6]{64} + 9$ 18. $\dfrac{1}{100}\sqrt{500} - (18\sqrt{5} + \sqrt{.05})$

19. $\sqrt{xy^3z} - \sqrt{xyz^3} + \sqrt{x^3yz}$ 20. $\sqrt{x^3 + x^2y} + \sqrt{y^3 + y^2x}$

21. $\sqrt{98x^2} + \sqrt[4]{4x^4} - 3\sqrt[6]{8x^6}$ 22. $\sqrt{98x} + \sqrt[4]{4x^2} - 3\sqrt[6]{8x^3}$

23. $\sqrt{4x^2 - 4y^2} - (x - y)\sqrt{\dfrac{x+y}{x-y}}$

24. $5x\sqrt[3]{3000y} - 2x\sqrt[3]{81y} - \sqrt[3]{24x^3y}$

25. $4x^2\sqrt{27y} - (\sqrt{75x^4y} - 2x^2\sqrt{3y}) + 3x^2y^{1/2}\sqrt{12}$

26. $2\sqrt{\dfrac{1}{15}} - 3\sqrt{\dfrac{5}{12}} + \sqrt{\dfrac{3}{5}}$

9-5
MULTIPLICATION OF RADICALS

The multiplication of radicals having the same index can be accomplished by employing Rule 9, namely

$$\sqrt[n]{x}\,\sqrt[n]{y} = \sqrt[n]{xy}.$$

This rule is valid only if the radicands are nonnegative. We shall assume throughout our discussion that the radicands satisfy this requirement.

EXAMPLE
10

Determine each of the indicated products and simplify.

a. $\sqrt{2x}\sqrt{8x}$.

$$\sqrt{2x}\sqrt{8x} = \sqrt{16x^2} = 4x.$$

b. $\sqrt{6x^2y}\sqrt{3xy}$.

$$\sqrt{6x^2y}\sqrt{3xy} = \sqrt{18x^3y^2} = \sqrt{9x^2y^2(2x)}$$
$$= 3xy\sqrt{2x}.$$

c. $\sqrt{3}(\sqrt{3} - \sqrt{6})$.

Employing the distributive axiom we have

$$\sqrt{3}(\sqrt{3} - \sqrt{6}) = \sqrt{3}(\sqrt{3}) - \sqrt{3}(\sqrt{6})$$
$$= \sqrt{9} - \sqrt{18}$$
$$= 3 - 3\sqrt{2}.$$

d. $(\sqrt{2} - \sqrt{3})^2$.

By the usual technique of squaring a binomial we can write

$$(\sqrt{2} - \sqrt{3})^2 = (\sqrt{2})^2 - 2(\sqrt{2})(\sqrt{3}) + (-\sqrt{3})^2$$
$$= 2 - 2\sqrt{6} + 3$$
$$= 5 - 2\sqrt{6}.$$

e. $(2\sqrt{2} - \sqrt{3})(3\sqrt{2} - 4\sqrt{3})$.

$$(2\sqrt{2} - \sqrt{3})(3\sqrt{2} - 4\sqrt{3})$$
$$= 2\sqrt{2}(3\sqrt{2} - 4\sqrt{3}) - \sqrt{3}(3\sqrt{2} - 4\sqrt{3})$$
$$= (2\sqrt{2})(3\sqrt{2}) - (2\sqrt{2})(4\sqrt{3}) - \sqrt{3}(3\sqrt{2}) + \sqrt{3}(4\sqrt{3})$$
$$= 6\sqrt{4} - 8\sqrt{6} - 3\sqrt{6} + 4\sqrt{9}$$
$$= 12 - 8\sqrt{6} - 3\sqrt{6} + 12$$
$$= 24 - 11\sqrt{6}.$$

f. $(a\sqrt[3]{a^2b^3})^4$.

$$(a\sqrt[3]{a^2b^3})^4 = a^4\sqrt[3]{a^8b^{12}} \qquad \text{(Rule 8)}$$
$$= a^4\sqrt[3]{(a^6)(b^{12})(a^2)}$$
$$= (a^4)(a^2)(b^4)\sqrt[3]{a^2}$$
$$= a^6b^4\sqrt[3]{a^2}.$$

In Example 10 the radicals that were multiplied had the same index. We can, however, multiply radicals whose indices differ by writing them as equivalent radicals with the same index. To do this, we appeal to fractional exponents as the following example shows.

EXAMPLE
11

Determine each of the indicated products and simplify.

a. $\sqrt{2}\cdot\sqrt[3]{3}$.

We first write the product in an exponential form using Rule 8; hence,

$$\sqrt{2}\cdot\sqrt[3]{3} = 2^{1/2}3^{1/3}.$$

Now we write the exponents as exponents *whose denominators are the same.* Since $1/2 = 3/6$ and $1/3 = 2/6$, we have

$$2^{1/2}\cdot3^{1/3} = 2^{3/6}\cdot3^{2/6}.$$

Now, by Rule 9 we have

$$2^{3/6}\cdot3^{2/6} = \sqrt[6]{2^3}\sqrt[6]{3^2} = \sqrt[6]{2^3\cdot3^2} = \sqrt[6]{72}.$$

Summarizing, we can write

$$\sqrt{2}\cdot\sqrt[3]{3} = 2^{1/2}\cdot3^{1/3} = 2^{3/6}\cdot3^{2/6} = \sqrt[6]{2^3\cdot3^2} = \sqrt[6]{72}.$$

b. $\sqrt[3]{9x^2}\sqrt{2x}$.

Using a procedure analogous to that used in part a we have

$$\sqrt[3]{9x^2}\sqrt{2x} = (9x^2)^{1/3}(2x)^{1/2}$$
$$= (9x^2)^{2/6}(2x)^{3/6}$$
$$= \sqrt[6]{(9x^2)^2(2x)^3}$$
$$= \sqrt[6]{(9^2x^4)(2^3x^3)} = \sqrt[6]{9^2\cdot2^3\cdot x^6\cdot x}$$
$$= x\sqrt[6]{648x}.$$

c. $\sqrt[4]{2x}\cdot\sqrt[3]{3x^2}$.

$$\sqrt[4]{2x}\cdot\sqrt[3]{3x^2} = (2x)^{1/4}(3x^2)^{1/3}$$
$$= (2^{1/4}\cdot x^{1/4})(3^{1/3}\cdot x^{2/3})$$
$$= (2^{3/12}\cdot x^{3/12})(3^{4/12}\cdot x^{8/12})$$
$$= (2^3\cdot3^4\cdot x^{11})^{1/12}$$
$$= \sqrt[12]{648x^{11}}.$$

EXERCISE 9-5

Determine each of the indicated products and simplify.

1. $\sqrt{2}\cdot\sqrt{8}$
2. $\sqrt{12}\cdot\sqrt{3}$
3. $\sqrt{3}\cdot\sqrt{4}$
4. $\sqrt{9}\cdot\sqrt{2}$
5. $2\sqrt{6}\cdot3\sqrt{3}$
6. $5\sqrt{9}\cdot2\sqrt{3}$
7. $\sqrt[3]{3}\cdot\sqrt[3]{9}\cdot\sqrt[3]{2}$
8. $\sqrt[3]{4}\cdot\sqrt[3]{16}\cdot\sqrt[3]{1}$
9. $(-\sqrt{3})^2$
19. $(-\sqrt{5})^3$
11. $\sqrt{2x}\cdot\sqrt{x}\cdot\sqrt{3x}$
12. $\sqrt{5y}\cdot\sqrt{2y}\cdot\sqrt{y}$
13. $\sqrt{3x}\sqrt[3]{x^2}$
14. $\sqrt[4]{x}\sqrt[3]{x^2}$

15. $\sqrt[3]{9x} \cdot \sqrt{3x}$ **16.** $\sqrt{8x^2} \cdot \sqrt[3]{4x^3}$

17. $(\frac{1}{2}\sqrt[3]{x})^4$ **18.** $(\frac{1}{3}\sqrt{x})^3$

19. $(\sqrt{3} + 2)^2$ **20.** $(1 - \sqrt{5})^2$

21. $(\sqrt{6} - 5)^2$ **22.** $(\sqrt{3} + 4)^2$

23. $(\sqrt{6} + 2)(\sqrt{2} - 1)$ **24.** $(\sqrt{3} - 1)(\sqrt{2} + 2)$

25. $(3 - \sqrt{5})(2 - \sqrt{3})$ **26.** $(1 - \sqrt{2})(1 + \sqrt{2})$

27. $\sqrt{3xy^2} \cdot \sqrt{2xy} \cdot \sqrt{3xy^3}$ **28.** $\sqrt{abc} \cdot \sqrt{a^2bc^2} \cdot \sqrt{abc^2}$

29. $2\sqrt[3]{xy}(3\sqrt[3]{y} - 2\sqrt[3]{xy})$ **30.** $5\sqrt[4]{ab}(1 - 2\sqrt[4]{ab})$

31. $3\sqrt[3]{x^2y} \cdot 2\sqrt{2x}$ **32.** $5\sqrt{ab^2} \cdot \sqrt[3]{ab}$

33. $(3\sqrt{8} + \sqrt{3})(8\sqrt{3} - \sqrt{8})$

34. $(5\sqrt{12} + \sqrt{5})(12\sqrt{5} - \sqrt{12})$

35. $(\sqrt{y^2 + 1} - 3)(\sqrt{y^2 + 1} + 3)$

36. $(\sqrt{x} - 1)(2\sqrt{x} + 5)$

37. $(\sqrt{2y} + 3)^2$ **38.** $(\sqrt{xy} - \sqrt{z})(2\sqrt{x} + \sqrt{y})$

39. $(\sqrt{2x^2y})^5$ **40.** $(\sqrt[4]{2x^3y^2z})^5$

9-6
DIVISION OF RADICALS

The division of radicals having the same index can be accomplished by employing Rule 10, namely

$$\frac{\sqrt[n]{x}}{\sqrt[n]{y}} = \sqrt[n]{\frac{x}{y}}.$$

In this section we shall require that the denominator of every fraction be rationalized.

EXAMPLE
12

Perform the indicated operations and simplify.

a. $\dfrac{\sqrt{98}}{\sqrt{2}} = \sqrt{\dfrac{98}{2}} = \sqrt{49} = 7.$

b. $\dfrac{\sqrt{7}}{\sqrt{3}} = \sqrt{\dfrac{7}{3}} = \sqrt{\dfrac{7}{3} \cdot \dfrac{3}{3}} = \dfrac{1}{3}\sqrt{21} = \dfrac{\sqrt{21}}{3}.$

Alternatively,

$$\frac{\sqrt{7}}{\sqrt{3}} = \frac{\sqrt{7}}{\sqrt{3}} \cdot \frac{\sqrt{3}}{\sqrt{3}} = \frac{\sqrt{21}}{3}.$$

c. $\dfrac{\sqrt{3x}}{\sqrt{2x}} = \sqrt{\dfrac{3x}{2x}} = \sqrt{\dfrac{3}{2}} = \sqrt{\dfrac{3}{2} \cdot \dfrac{2}{2}} = \dfrac{\sqrt{6}}{2}.$

d. $\dfrac{\sqrt[3]{x}}{\sqrt[3]{y}} = \dfrac{x^{1/3} \cdot y^{2/3}}{y^{1/3} \cdot y^{2/3}} = \dfrac{\sqrt[3]{xy^2}}{y}$.

e. $\dfrac{\sqrt[5]{x^3y^2}}{\sqrt[5]{xy}} = \sqrt[5]{\dfrac{x^3y^2}{xy}} = \sqrt[5]{x^2y}$.

Note that here the result is essentially obtained by dividing the radicands. An alternative procedure is as follows:

$$\dfrac{\sqrt[5]{x^3y^2}}{\sqrt[5]{xy}} = \dfrac{x^{3/5}y^{2/5}}{x^{1/5}y^{1/5}} \cdot \dfrac{x^{4/5}y^{4/5}}{x^{4/5}y^{4/5}} = \dfrac{\sqrt[5]{x^7y^6}}{xy}$$

$$= \dfrac{xy\sqrt[5]{x^2y}}{xy}$$

$$= \sqrt[5]{x^2y}.$$

Clearly, the original procedure is preferable.

In Example 12 we considered the division of radicals having the same index. When radicals having different indices are to be divided, we can accomplish the division by appealing, once again, to fractional exponents.

EXAMPLE 13

Perform the indicated operations and simplify.

a. $\dfrac{\sqrt{10}}{\sqrt[3]{10}}$.

In this instance the radicands are the same. Using fractional exponents we have

$$\dfrac{\sqrt{10}}{\sqrt[3]{10}} = \dfrac{10^{1/2}}{10^{1/3}} = 10^{(1/2)-(1/3)} = 10^{1/6} = \sqrt[6]{10}.$$

b. $\dfrac{\sqrt{x^2y}}{\sqrt[4]{x^2y}} = \dfrac{(x^2y)^{1/2}}{(x^2y)^{1/4}} = (x^2y)^{(1/2)-(1/4)}$

$$= (x^2y)^{1/4}$$

$$= \sqrt[4]{x^2y}.$$

c. $\dfrac{\sqrt{2}}{\sqrt[3]{3}}$.

In this case we are dividing radicals which have different indices and different radicands. We begin by appealing to fractional exponents, as before, in order to rationalize the denominator:

$$\dfrac{\sqrt{2}}{\sqrt[3]{3}} = \dfrac{2^{1/2}}{3^{1/3}} \cdot \dfrac{3^{2/3}}{3^{2/3}} = \dfrac{2^{1/2} \cdot 3^{2/3}}{3}.$$

To perform the multiplication in the numerator we proceed, as in the previous section, to express the fractional exponents with a common denominator. Hence,

$$\frac{2^{1/2}\cdot3^{2/3}}{3}=\frac{2^{3/6}\cdot3^{4/6}}{3}=\frac{\sqrt[6]{2^3\cdot3^4}}{3}=\frac{\sqrt[6]{648}}{3}.$$

d. $\dfrac{\sqrt[3]{2x}}{\sqrt{x}}.$

$$\frac{\sqrt[3]{2x}}{\sqrt{x}}=\frac{2^{1/3}\cdot x^{1/3}}{x^{1/2}}\cdot\frac{x^{1/2}}{x^{1/2}}=\frac{2^{1/3}x^{(1/3)+(1/2)}}{x}$$

$$=\frac{2^{1/3}x^{5/6}}{x}$$

$$=\frac{2^{2/6}x^{5/6}}{x}$$

$$=\frac{\sqrt[6]{4x^5}}{x}.$$

e. $\dfrac{\sqrt[4]{2}}{\sqrt[3]{xy^2}}=\dfrac{2^{1/4}}{x^{1/3}\cdot y^{2/3}}\cdot\dfrac{x^{2/3}y^{1/3}}{x^{2/3}y^{1/3}}$

$$=\frac{2^{3/12}x^{8/12}y^{4/12}}{xy}$$

$$=\frac{\sqrt[12]{8x^8y^4}}{xy}.$$

In the previous examples, the denominators of the fractions have been composed of a single term and the use of fractional exponents has been shown to be an effective approach to rationalizing such denominators. We sometimes encounter denominators, however, which themselves indicate an addition or subtraction process—denominators such as $\sqrt{3}-2$, $-2\sqrt{3}+\sqrt{5}$, and $3\sqrt{6}-4\sqrt{7}$. In general, if the denominator of a fraction is of the form $a\sqrt{b}+c\sqrt{d}$, the denominator can be rationalized by multiplying the numerator and denominator of the fraction by $a\sqrt{b}-c\sqrt{d}$. If the denominator is of the form $a\sqrt{b}-c\sqrt{d}$, then multiply by $a\sqrt{b}+c\sqrt{d}$. In each case you will find this technique eliminates all radicals from the denominator.

EXAMPLE
14

Perform the indicated operations and simplify.

a. $\dfrac{1}{\sqrt{2}-\sqrt{3}}.$

To rationalize the denominator we multiply the numerator and denominator of the fraction by $\sqrt{2}+\sqrt{3}$:

$$\frac{1}{\sqrt{2}-\sqrt{3}}=\frac{1}{\sqrt{2}-\sqrt{3}}\cdot\frac{\sqrt{2}+\sqrt{3}}{\sqrt{2}+\sqrt{3}}=\frac{\sqrt{2}+\sqrt{3}}{(\sqrt{2})^2-(\sqrt{3})^2}.$$

Note that, as indicated in the last step, the multiplication to be performed in the denominator is in the form of a special product. This is a characteristic occurrence in rationalizing a denominator which is of the type $a\sqrt{b} \pm c\sqrt{d}$. Now,

$$\frac{1}{\sqrt{2} - \sqrt{3}} = \frac{\sqrt{2} + \sqrt{3}}{(\sqrt{2})^2 - (\sqrt{3})^2} = \frac{\sqrt{2} + \sqrt{3}}{2 - 3}$$

$$= \frac{\sqrt{2} + \sqrt{3}}{-1}$$

$$= -(\sqrt{2} + \sqrt{3}).$$

b. $\dfrac{\sqrt{3}}{2\sqrt{3} + \sqrt{2}}$.

$$\frac{\sqrt{3}}{2\sqrt{3} + \sqrt{2}} = \frac{\sqrt{3}}{2\sqrt{3} + \sqrt{2}} \cdot \frac{(2\sqrt{3} - \sqrt{2})}{(2\sqrt{3} - \sqrt{2})}$$

$$= \frac{\sqrt{3}(2\sqrt{3} - \sqrt{2})}{(2\sqrt{3})^2 - (\sqrt{2})^2}$$

$$= \frac{\sqrt{3}(2\sqrt{3}) - \sqrt{3}(\sqrt{2})}{2^2(\sqrt{3})^2 - (2)}$$

$$= \frac{2(3) - \sqrt{6}}{(4)(3) - 2}$$

$$= \frac{6 - \sqrt{6}}{10}.$$

c. $\dfrac{\sqrt{5} - \sqrt{2}}{\sqrt{5} + \sqrt{2}}$.

$$\frac{\sqrt{5} - \sqrt{2}}{\sqrt{5} + \sqrt{2}} = \frac{\sqrt{5} - \sqrt{2}}{\sqrt{5} + \sqrt{2}} \cdot \frac{\sqrt{5} - \sqrt{2}}{\sqrt{5} - \sqrt{2}}$$

$$= \frac{(\sqrt{5} - \sqrt{2})^2}{(\sqrt{5})^2 - (\sqrt{2})^2}$$

$$= \frac{(\sqrt{5})^2 - 2\sqrt{5}\sqrt{2} + (\sqrt{2})^2}{5 - 2}$$

$$= \frac{5 - 2\sqrt{10} + 2}{3}$$

$$= \frac{7 - 2\sqrt{10}}{3}.$$

d. $\dfrac{x}{\sqrt{2} - 6}$.

$$\frac{x}{\sqrt{2} - 6} = \frac{x}{\sqrt{2} - 6} \cdot \frac{\sqrt{2} + 6}{\sqrt{2} + 6}$$

$$= \frac{x(\sqrt{2} + 6)}{2 - 36}$$

$$= -\frac{x(\sqrt{2} + 6)}{34}.$$

EXERCISE 9-6

Perform the indicated operations in Problems 1–34 and simplify. Express all answers with rationalized denominators and no fractional exponents.

1. $\dfrac{\sqrt{32}}{\sqrt{2}}$

2. $\dfrac{\sqrt{18}}{\sqrt{2}}$

3. $\dfrac{\sqrt{2a^3}}{\sqrt{a}}$

4. $\dfrac{\sqrt{3x^5}}{\sqrt{x}}$

5. $\dfrac{2x\sqrt{x^7}}{\sqrt{x}}$

6. $\dfrac{3a^2b\sqrt{a^3}}{2a\sqrt{a}}$

7. $\dfrac{\sqrt{3}}{\sqrt{7}}$

8. $\dfrac{\sqrt{8}}{\sqrt{3}}$

9. $\dfrac{\sqrt{18}}{\sqrt{3}}$

10. $\dfrac{\sqrt{15}}{\sqrt{5}}$

11. $\dfrac{\sqrt[3]{3}}{\sqrt{2}}$

12. $\dfrac{\sqrt[3]{2}}{\sqrt[4]{3}}$

13. $\dfrac{\sqrt[3]{6}}{\sqrt[3]{4}}$

14. $\dfrac{\sqrt{6}}{\sqrt[3]{4}}$

15. $\dfrac{\sqrt{2x}}{\sqrt[3]{x}}$

16. $\dfrac{\sqrt{y}}{\sqrt[3]{y}}$

17. $\dfrac{\sqrt{2x}}{\sqrt[3]{y}}$

18. $\dfrac{\sqrt[3]{3y}}{\sqrt[3]{x}} = \dfrac{\sqrt[3]{3yx^2}}{x}$

19. $\dfrac{2\sqrt{xy}}{\sqrt[4]{xy}}$

20. $\dfrac{2x\sqrt{x}}{\sqrt[3]{2xy}}$

21. $\dfrac{1}{2+\sqrt{3}}$

22. $\dfrac{1}{1-\sqrt{2}}$

23. $\dfrac{2}{\sqrt{3}-\sqrt{2}}$

24. $\dfrac{5}{\sqrt{6}-\sqrt{3}}$

25. $\dfrac{2\sqrt{2}}{\sqrt{2}-\sqrt{3}}$

26. $\dfrac{2\sqrt{3}}{\sqrt{5}-\sqrt{2}}$

27. $\dfrac{1+\sqrt{2}}{\sqrt{3}+\sqrt{6}}$

28. $\dfrac{3-\sqrt{5}}{\sqrt{2}+\sqrt{4}}$

29. $\dfrac{4\sqrt{3}-\sqrt{2}}{\sqrt{3}-\sqrt{2}}$

30. $\dfrac{\sqrt{3}-5\sqrt{2}}{\sqrt{2}+\sqrt{6}}$

31. $\dfrac{1}{x+\sqrt{5}}$

32. $\dfrac{x-3}{\sqrt{x}-1}+\dfrac{4}{\sqrt{x}-1}$

33. $\dfrac{5}{1+\sqrt{3}} - \dfrac{4}{2-\sqrt{2}}$

34. $\dfrac{4}{\sqrt{x}+2} \cdot \dfrac{x^2}{3}$

35. When the mass m shown in Fig. 9-1 is released from rest and falls through

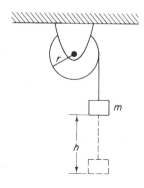

FIGURE 9-1

a distance h, the wheel can be shown to have an angular speed ω, given by

$$\omega = \sqrt{\frac{2mgh}{mr^2 + I}},$$

where I is the moment of inertia of the wheel and g is the acceleration due to gravity. Express ω with a rationalized denominator.

36. When a solid sphere of radius r and mass m is released from the top of an inclined plane of vertical height h, its speed v at the bottom, if the sphere rolls without slipping, can be shown to be

$$\sqrt{\frac{10gh}{7}}.$$

Express the speed in rationalized form.

37. If a block of mass m were released at the top of a frictionless inclined plane having the same vertical height h as in Problem 36, its speed v at the bottom of the incline can be shown to be

$$\sqrt{2gh}.$$

Find the ratio of the speed of the block to the speed of the sphere in Problem 36, and simplify.

38. If a circular ring of radius r has a charge per unit length of λ, it can be shown that the electric field intensity due to this ring at a point a distance y from the center of the ring on an axis perpendicular to the plane of the ring (Fig. 9-2), is given by the expression

$$\frac{\lambda r y}{2\epsilon_0 \sqrt{(r^2 + y^2)^3}},$$

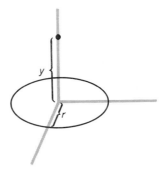

FIGURE 9-2

where ϵ_0 is a constant. Express the electric field intensity in rationalized form.

9-7
RADICAL EQUATIONS

Equations in which variables occur under radical signs are referred to as **radical equations**. A common procedure in the solution of such an equation is to raise both members to the same power so as to eliminate the radical. This is an operation which, you should recall, does not guarantee that the resulting equation is equivalent to the original one. When such an operation is performed it is imperative that you check all solutions for extraneous roots which may have been introduced. We are assuming, however, that the new equation contains all the roots of the original equation.

EXAMPLE
15

Solve the following equations.

a. $\sqrt{x - 7} = 4$.

$$\sqrt{x - 7} = 4$$
$$x - 7 = 16 \qquad \text{(squaring both members)}$$
$$x = 23.$$

Substituting $x = 23$ in the left member of the original equation we obtain $\sqrt{23 - 7} = \sqrt{16} = 4$, which is equal to the right member. The solution set is $\{23\}$.

b. $\sqrt{y - 3} - \sqrt{y} = -3$.

When an equation contains two radical expressions, it is advantageous to write the equation so that one radical is in each member:

$$\sqrt{y-3} = \sqrt{y} - 3$$
$$y - 3 = y - 6\sqrt{y} + 9 \qquad \text{(squaring both members)}$$
$$6\sqrt{y} = 12$$
$$\sqrt{y} = 2$$
$$y = 4. \qquad \text{(squaring both members)}$$

Substituting $y = 4$ into the left member of the original equation, we obtain $\sqrt{1} - \sqrt{4}$, which is equal to -1. Since this does not equal the right member, -3, we conclude that the solution set is $\varnothing$.

c. $\sqrt{(x-3)^3} - 64 = 0.$

This equation can be written as

$$(x - 3)^{3/2} = 64.$$

Raising each member to the two-thirds power,

$$[(x - 3)^{3/2}]^{2/3} = (64)^{2/3}$$
$$x - 3 = (64)^{2/3} = (4^3)^{2/3} = 4^2 = 16$$
$$x = 19.$$

The solution set is $\{19\}$, as you may verify.

EXERCISE 9-7

For Problems 1–16, find the solution set.

1. $\sqrt{x-2} = 5$

2. $\sqrt{x+7} = 9$

3. $\sqrt{2x-5} - 6 = 0$

4. $\sqrt{5x-6} - 16 = 0$

5. $2\sqrt{2x+1} = 3\sqrt{3x-9}$

6. $4\sqrt{8x+1} = 5\sqrt{3x+4}$

7. $\sqrt{x^2+33} = x + 3$

8. $\sqrt{\dfrac{x}{2}+1} = \dfrac{2}{3}$

9. $(y+6)^{1/2} = 7$

10. $\sqrt{4x-6} - \sqrt{x} = 0$

11. $\sqrt{7-2x} - \sqrt{x-1} = 0$

12. $(z-3)^{3/2} = 8$

13. $\sqrt{x} - \sqrt{x+1} = 1$

14. $\sqrt{y} + \sqrt{y+2} = 3$

15. $\sqrt[3]{\sqrt{x-5}} = 6$

16. $\sqrt{\dfrac{1}{x}} - \sqrt{\dfrac{2}{5x-2}} = 0$

17. The time T in seconds for one complete oscillation of a simple pendulum is

$$T = 2\pi\sqrt{\dfrac{L}{g}},$$

where L is the length of the pendulum and g is the acceleration due to gravity. Solve the equation for g.

18. The formula $f_r = \dfrac{1}{2\pi\sqrt{LC}}$ occurs in the study of alternating current. Here, f_r is resonant frequency, L is inductance, and C is capacitance. Solve for C.

9-8
REVIEW

REVIEW QUESTIONS

If it is true that $(x + 2y)^0 = 1$, then x is not equal to _____(1)_____ .

  (1) **−2y**

The statement $\left(\dfrac{x}{y}\right)^5 = \dfrac{x^5}{y^5}$ is true for all values of y except _____(2)_____ .

  (2) **zero**

$-2(xy)^2 = -2x^2y^2$, but $(-2xy)^2 =$ _____(3)_____ .

  (3) **4x²y²**

True or false: $\sqrt[3]{6}\,\sqrt[4]{6} = \sqrt[12]{6}$ _____(4)_____ .

  (4) **false**

To rationalize the denominator in the expression $1/\sqrt[3]{xy}$, we multiply both the numerator and denominator by the quantity _____(5)_____ .

  (5) **x²/³y²/³**

After the coefficient of the radical in $3x\sqrt[3]{2}$ is introduced into the radical, the radicand is _____(6)_____ .

  (6) **54x³**

When the denominator of $\sqrt{x^4}/\sqrt{y}$ is rationalized, the fraction can be written in simplified form as _____(7)_____ .

  (7) $\dfrac{x^2\sqrt{y}}{y}$

REVIEW PROBLEMS

In Problems 1–20, perform the indicated operations and simplify. Express all answers with rationalized denominators and no fractional exponents.

1. $(-3x^{1/2}y^2)^3$

2. $(2xy^3)^{1/2}(-3x^{3/2}y)^4$

3. $\dfrac{(2xy)^2(x^{1/2})^6}{8x^5y^5}$

4. $\dfrac{(3x^{1/2}y^{1/2})^3}{x^{1/2}}$

5. $x^{-3}(2x^4 + y^{-3})$

6. $(x^{1/2} + y^{1/2})(x^{1/2} - y^{1/2})$

7. $\sqrt{7}\cdot\sqrt{4}\cdot\sqrt{5}$

8. $\dfrac{\sqrt{3}\,\sqrt{6}}{\sqrt{2}}$

9. $\sqrt{2}\,(1 - \sqrt{6}\,)$

10. $(\sqrt{3} - \sqrt{2})(\sqrt{3} + 2\sqrt{2}\,)$

11. $\dfrac{1}{2\sqrt{3} + \sqrt{2}}$

12. $\dfrac{4\sqrt{6}}{\sqrt{6} + \sqrt{4}}$

13. $\sqrt[5]{\sqrt[3]{x^{10}}}$

14. $\sqrt[3]{\sqrt{\sqrt[3]{x^2}}}$

15. $(x - y)\sqrt{\dfrac{1}{x^2 - y^2}}$

16. $\sqrt[6]{\dfrac{x^6}{y^9}}$

17. $2\sqrt{8} - (5\sqrt{2} - \sqrt{2}\,)$

18. $\dfrac{\sqrt{2} + \sqrt{8}}{\sqrt{8} - \sqrt{6}}$

19. $\dfrac{\sqrt[3]{3x^2}}{\sqrt{2x}}$

20. $\dfrac{\sqrt[5]{x^2y}}{\sqrt[3]{xy}}$

In Problems 21–26, solve the given radical equation.

21. $\sqrt{2x + 5} = 5$

22. $\sqrt{3x - 4} = \sqrt{2x + 5}$

23. $2\sqrt{x - 2} = \sqrt{3}\,(\sqrt{x - 2}\,)$

24. $\sqrt[3]{11x + 9} = 4$

25. $\sqrt{x^2 + 5x + 25} = x + 4$

26. $\sqrt{z^2 + 2z} = 3 + z$

27. In Fig. 9-3 an arrangement of two charges q_1 and q_2 is indicated. Theory

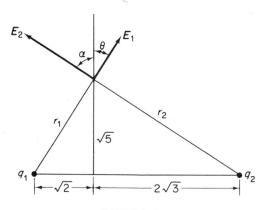

FIGURE 9-3

shows that the magnitudes of the electric field vectors E_1 and E_2 due to the charges q_1 and q_2, respectively, are given by

$$E_1 = \frac{kq_1}{r_1^2} \quad \text{and} \quad E_2 = \frac{kq_2}{r_2^2}.$$

Find expressions for the magnitudes of the *vertical components* of E_1 and E_2 and simplify each. *Hint:* For E_1 it is given by $E_1 \cos \theta$.

28. For the arrangement in Problem 27, find an expression for the magnitude of the *horizontal component* of E_1 and simplify.

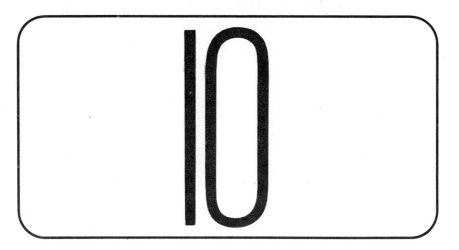

variation

10-1
DIRECT VARIATION

Given a square each side of which has length s, the formulas for the perimeter P and the area A of the square are, respectively,

$$P = 4s \quad \text{and} \quad A = s^2.$$

In each of these equations note that for all meaningful values of the variables employed, the ratio of the given powers of those variables is a constant. That is,

$$\frac{P}{s} = 4 \quad \text{and} \quad \frac{A}{s^2} = 1.$$

Such a constant ratio of variables characterizes the notion of *direct variation*.

Definition. The variable y is said to **vary directly** as (or be **directly proportional** to) the nth power of the variable x if $y = kx^n$, where k is called the **constant of variation** (or **constant of proportionality**).

In view of this definition, we could describe the relationships of the perimeter and area of a square to the length of a side by saying that "the perimeter of a square varies directly as the length of a side" and "the area of a square is directly proportional to the square of the length of a side." In the first case the constant of variation is 4; in the second it is 1. We shall see that many scientific laws are describable in terms of direct variation.

EXAMPLE 1

Hooke's law states that if the elastic limit is not exceeded, the force F exerted by a spring on an attached body varies directly as the displacement x of the body from its rest position. If a 10-lb force results from a displacement of 2 in., what displacement corresponds to a 3-lb force?

Since F varies directly as x, by definition $F = kx$. From the given data, if $x = 2$ in., then $F = 10$ lb; therefore,

$$F = kx \qquad\qquad (1)$$

$$10 \text{ lb} = k(2 \text{ in.})$$

$$k = \frac{10 \text{ lb}}{2 \text{ in.}} = 5 \text{ lb/in.}$$

This means that the ratio of F to x, namely the constant of variation k, is 5 lb/in. Now, replacing k in Eq. (1) by its value, we have

$$F = 5 \frac{\text{lb}}{\text{in.}}(x) \qquad \text{or} \qquad x = \frac{F}{5 \frac{\text{lb}}{\text{in.}}}.$$

Thus, if $F = 3$ lb,

$$x = \frac{F}{5 \frac{\text{lb}}{\text{in.}}} = \frac{3 \text{ lb}}{5 \frac{\text{lb}}{\text{in.}}} = \frac{3}{5} \text{ in.}$$

Note the algebraic treatment of the units involved. You should also realize that k can be determined if *any* pair of corresponding values of x and F are known (except $F = 0$ when $x = 0$).

EXAMPLE 2

If y varies directly as x^2, and $y = 8$ when $x = 4$, find y when $x = 10$.

By definition,

$$y = kx^2.$$

Substituting the given data into this equation,

$$8 = k(4)^2$$

$$k = \frac{1}{2}.$$

Thus,

$$y = \frac{1}{2}x^2.$$

When $x = 10$,

$$y = \frac{10^2}{2} = 50.$$

The basic definition of direct variation can be extended to the case where one variable varies directly as the *product* of powers of other variables. We call this *joint variation*.

Definition. The variable y is said to **vary jointly** as (or be directly proportional to the product of) x^m and z^n if $y = kx^m z^n$.

EXAMPLE
3

a. If y is directly proportional to the product of the square of x and the cube root of z, find y when $x = 2$ and $z = 16$ given that $y = 1$ when $x = 1$ and $z = 2$.

From the statement of the problem,

$$y = kx^2 \sqrt[3]{z}.$$

Substituting the given values ($y = 1$ when $x = 1$ and $z = 2$),

$$1 = k(1)^2 \sqrt[3]{2}$$

$$k = \frac{1}{\sqrt[3]{2}}.$$

Thus,

$$y = kx^2 \sqrt[3]{z} = \frac{1}{\sqrt[3]{2}} x^2 \sqrt[3]{z}.$$

When $x = 2$ and $z = 16$,

$$y = \frac{1}{\sqrt[3]{2}}(2)^2 \sqrt[3]{16} = 4\sqrt[3]{\frac{16}{2}} = 4\sqrt[3]{8} = 8.$$

b. If y varies jointly as x and w^2, how is y affected if x is increased by 20 percent and w is decreased by 10 percent?

$$y = kxw^2. \tag{1}$$

Let the initial values of y, x, and w be y_0, x_0, and w_0, respectively. Then

$$y_0 = kx_0 w_0^2. \tag{2}$$

Rather than solve for k, we proceed as follows. Let the new values of x and w be $1.2x_0$ and $.9w_0$. Then from Eq. (1), the corresponding value of y is

$$y = k(1.2x_0)(.9w_0)^2$$
$$= (1.2)(.81)(kx_0 w_0^2)$$
$$= (1.2)(.81)(y_0) \quad \text{[from Eq. (2)]}$$
$$\approx .97y_0.$$

Thus, y is decreased by approximately 3 percent.

EXAMPLE
4

The kinetic energy K of a particle varies jointly as the mass m of the particle and the square of the particle's speed v. When a particle of mass .1 kg has a speed of 20 m/sec, its kinetic energy is 20 kg m²/sec². What is the kinetic energy of the same particle when its speed is 10 m/sec?

$$K = kmv^2$$

$$20\frac{\text{kg m}^2}{\text{sec}^2} = k(.1 \text{ kg})\left(20\frac{\text{m}}{\text{sec}}\right)^2$$

$$20\frac{\text{kg m}^2}{\text{sec}^2} = 40k\frac{\text{kg m}^2}{\text{sec}^2}$$

$$k = \frac{1}{2}.$$

Note that here the constant of proportionality has no units associated with it. Now, when $v = 10$ m/sec,

$$K = \frac{1}{2}mv^2$$

$$K = \frac{1}{2}(.1 \text{ kg})\left(10\frac{\text{m}}{\text{sec}}\right)^2$$

$$K = 5\frac{\text{kg m}^2}{\text{sec}^2}.$$

EXERCISE 10-1

1. If y varies directly as x, and $y = 8$ when $x = 2$, find y when $x = 6$.

2. If y is directly proportional to $2x$, and $y = 8$ when $x = 2$, find y when $x = 3$.

3. If y is directly proportional to x^2, and $y = 2$ when $x = 4$, find y when $x = 6$.

4. If y is directly proportional to $\sqrt{x}$, and $y = 4$ when $x = 25$, find y when $x = 16$.

5. If y varies directly as x and z^2, and $y = 6$ when $x = 2$ and $z = \frac{1}{2}$, find y when $x = 4$ and $z = \sqrt{2}$.

6. If y varies directly as $\sqrt{x}$ and $\sqrt{z}$, find y when $x = 4$ and $z = 8$, if $y = 11$ when $x = 2$ and $z = 1$.

7. The weight W of an object is directly proportional to the product of the mass m of the object and the acceleration g due to gravity. If a mass of 2 slugs weighs 64 lb on the earth's surface where $g = 32$ ft/sec², what would the same mass weigh on the moon where $g = 5.3$ ft/sec²?

8. The elongation of a supporting cable varies directly as the load if the elastic limit is not exceeded. Find the elongation, in inches, when the load is 2 tons, if a load of 800 lb causes an elongation of 0.3 in.

9. When a rocket launches a space capsule from a planet (or satellite), it can be shown, using simplifying assumptions, that the capsule must be given an initial speed (or escape velocity) v if it is to escape the planet's gravitational field and that v is directly proportional to the square root of the radius r of the planet and the square root of the acceleration g due to gravity. The escape velocity on the earth is about 6.96 mi/sec, and to a first approximation the radius of the earth is four times that of the moon and the acceleration due to gravity on the earth is six times that of the moon. Find the escape velocity on the moon.

10. Experiments indicate that thin rods when heated expand principally in one dimension along their lengths. The change in length of a rod, ΔL, varies jointly as the original length of the rod, L_0, and the change in temperature, Δt. If a rod is initially 20 in. long at 50°C and then 20.0072 in. at 80°C, find its length at 110°C.

11. When a substance is heated without changing its state, the amount of heat Q added to the substance is directly proportional to the product of the mass m of the substance and the change in temperature, Δt. If it requires 60 calories of heat to raise the temperature of 100 gm of lead from 50° to 70°C, how many more calories are necessary to raise the temperature of the same mass an additional 10 Celsius degrees?

12. Experiments indicate that the potential difference V across an ohmic conductor varies directly as the current i and the resistance R. This is known as Ohm's law. If a conductor having a fixed resistance of x ohms has a potential difference of 30 volts across it when it carries a current of 2 amps, what is the potential difference across the same conductor when $i = 6$ amps?

13. The area of a triangle varies directly as the product of the length of the base and the length of the altitude. If the base is decreased by a factor of $\frac{1}{3}$, how must the altitude be changed if the area of the triangle is to be quadrupled?

14. Total internal reflection of a light ray takes place at the boundary of two surfaces if the angle of incidence is greater than the critical angle for those surfaces. The sine of the critical angle varies directly as the ratio of the indices of refraction of the two surfaces. If the index of refraction of air is 1.00 and that of glass is 1.50 and the critical angle for an air-glass boundary is 42°, find the critical angle for an air-water boundary if the index of refraction of water is 1.33. Is there more than one way to express the required ratio?

15. The potential V at a given point due to a number of distinct point charges varies directly as the algebraic sum of the ratios of the charges to the distances of the charges from the given point. Consider three point charges, two positive and one negative, all having magnitude q and equidistant

from a given point. How is the potential at the given point affected if the negative charge is doubled in magnitude and one of the positive charges is moved three times as far away from the given point?

16. In the equation $f = \dfrac{1}{2L}\sqrt{\dfrac{T}{m}}$, where m is a constant, the frequency f of a vibrating string of length L is directly proportional to the square root of the tension T in the string since

$$f = \frac{1}{2L\sqrt{m}}\sqrt{T} = k\sqrt{T}.$$

By squaring both members, show that T is directly proportional to the square of the frequency.

10-2
INVERSE AND COMBINED VARIATION

Many physical laws which are encountered in science and engineering express a particular functional relationship called *inverse variation*.

Definition. The variable y is said to **vary inversely** as (or be *inversely proportional* to) the nth power of x if $y = \dfrac{k}{x^n}$.

EXAMPLE 5

If y varies inversely as the cube root of z, and $y = 2$ when $z = 2$, find y when $z = 8$.

Since y varies inversely as $\sqrt[3]{z}$,

$$y = \frac{k}{\sqrt[3]{z}}.$$

When $z = 2$, then $y = 2$:

$$2 = \frac{k}{\sqrt[3]{2}}$$

$$k = 2\sqrt[3]{2}.$$

Thus,

$$y = \frac{2\sqrt[3]{2}}{\sqrt[3]{z}}.$$

When $z = 8$,

$$y = \frac{2\sqrt[3]{2}}{\sqrt[3]{8}} = \frac{2\sqrt[3]{2}}{2} = \sqrt[3]{2}.$$

EXAMPLE 6

The weight W of an object on or above the earth's surface varies inversely as the square of the distance d of the object from the center of the earth. Assume the radius of the earth to be 4000 mi. If an astronaut weighs 200 lb on earth, what does he weigh when at an altitude of 4000 mi above the earth's surface?

$$W = \frac{k}{d^2}.$$

On the surface we have $W = 200$ lb when $d = 4000$ mi:

$$200 \text{ lb} = \frac{k}{(4000 \text{ mi})^2}$$

$$k = 32 \times 10^8 \text{ lb mi}^2.$$

At an altitude of 4000 mi, we have $d = 8000$ mi and

$$W = \frac{k}{d^2} = \frac{32 \times 10^8 \text{ lb mi}^2}{(8000 \text{ mi})^2} = \frac{32 \times 10^8 \text{ lb mi}^2}{64 \times 10^6 \text{ mi}^2} = 50 \text{ lb.}$$

Sometimes a variable simultaneously varies directly as one quantity and inversely as another quantity. We speak of this situation as **combined variation**.

EXAMPLE
7

If y varies directly as x and inversely as z, find y when $x = 6$ and $z = 2$, if $y = 4$ when $x = 5$ and $z = 4$.

We express the fact that y varies directly as x and inversely as z by writing

$$y = \frac{kx}{z}.$$

Now, $y = 4$ when $x = 5$ and $z = 4$; therefore

$$4 = \frac{k \cdot 5}{4}$$

$$k = \frac{16}{5}.$$

Thus,

$$y = \frac{\frac{16}{5}x}{z} = \frac{16x}{5z}.$$

When $x = 6$ and $z = 2$ we have

$$y = \frac{16(6)}{5(2)} = \frac{48}{5}.$$

EXAMPLE
8

The resonant frequency f_r of a series alternating-current (a.c.) circuit containing an inductance L and a capacitance C varies inversely as the square root of the product of L and C. If $f_r = 10,000/\pi$ hertz (cycles per second) when $L = 5 \times 10^{-3}$ henries (H) and $C = 5 \times 10^{-7}$ farads (F), find f_r in terms of π when $L = 4 \times 10^{-3}$ H and $C = 9 \times 10^{-7}$ F.

Here

$$f_r = \frac{k}{\sqrt{LC}}$$

$$\frac{10{,}000}{\pi} = \frac{k}{\sqrt{(5 \times 10^{-3})(5 \times 10^{-7})}}$$

$$k = \frac{10{,}000}{\pi}\sqrt{(5 \times 10^{-3})(5 \times 10^{-7})} = \frac{10{,}000}{\pi}\sqrt{25 \times 10^{-10}}$$

$$= \frac{10{,}000}{\pi}(5)(10^{-5}) = \frac{.5}{\pi} = \frac{1}{2\pi}.$$

Thus,

$$f_r = \frac{1}{2\pi\sqrt{LC}}.$$

For the given data we have

$$f_r = \frac{1}{2\pi\sqrt{(4 \times 10^{-3})(9 \times 10^{-7})}} = \frac{1}{2\pi(6 \times 10^{-5})}$$

$$f_r = \frac{10^5}{12\pi} \text{ hertz.}$$

EXERCISE 10-2

1. If y varies inversely as x, and $y = 12$ when $x = 3$, find y when $x = 4$.

2. If y varies directly as x^2 and inversely as z, and $y = 8$ when $x = 2$ and $z = 2$, find y when $x = 3$ and $z = 4$.

3. If y varies directly as x^2 and z and inversely as $\sqrt{w}$, and if $y = 12$ when $x = 1$, $z = 3$, and $w = 4$, find y when $x = 4$, $z = 2$, and $w = \pi^2$.

4. If a varies directly as b^2 and c^3 and inversely as d^2, and if $a = 25$ when $b = 2$, $c = 2$, and $d = 2$, find a when $b = 3$, $c = 2$ and $d = 1$.

5. If y varies directly as x and inversely as z^2, find x when $y = 4$ and $z = 9$, if $y = 10$ when $x = 1$ and $z = 2$.

6. If y varies directly as x and inversely as $\sin z$, and $y = 7$ when $x = 4$ and $z = \pi/2$, find y when $x = \sqrt{2}$ and $z = \pi/4$.

7. Newton's second law of motion states that when a mass m is acted upon by a force F, the acceleration a of the mass is directly proportional to F and inversely proportional to m. If a mass of 50 gm initially at rest on a horizontal surface is acted upon by a force of 500 dynes, the acceleration is 10 cm/sec². What force would give a mass of 30 gm an acceleration of 75 cm/sec²?

8. The resistance of a copper wire varies directly as its length and inversely

as its cross-sectional area. If a copper wire of length 500 cm and radius .2 cm has a resistance of .025 ohms, what will the resistance be for the same copper wire of length 1000 cm and radius .1 cm?

9. Under certain conditions the illumination E of a surface varies directly as the intensity I of the light source and inversely as the square of the distance r of the surface from the source. What is the effect on the illumination if the intensity and distance are both doubled?

10. Coulomb's law states that the force F of attraction or repulsion between two electrostatic point charges of magnitudes q_1 and q_2 is proportional to the product of the magnitudes of the charges and inversely proportional to the square of the distance between them. If the magnitude of one charge is doubled and the distance between charges reduced by a factor of $\frac{1}{2}$, how is the force affected?

11. Referring to Example 6, to what altitude above the earth's surface must the astronaut be if his weight is to be 1 percent of his weight on the earth's surface?

12. The capacitance of a parallel-plate capacitor varies directly as the area of either of its plates and inversely as the separation of its plates. If the plates are 5 mm apart and 2 m² in area, the capacitance is 3.54×10^{-9} F. For square plates separated by 1 mm, what must be the length of a side for the capacitance to be 1 F? (Such capacitors are not found in practice.)

13. Newton's law of gravitation states that the attractive force F between masses m_1 and m_2 varies directly as the product of the masses and inversely as the square of the distance between their centers. If the centers of two 10-kg masses are 1 m apart and $F = 6.67 \times 10^{-9}$ newtons (N), find F when two 100-kg masses are 1 m apart.

14. Given a volume V containing n moles of gas at temperature T, the ideal gas law states that the pressure P of the gas is directly proportional to n and T and inversely proportional to V. If the volume of a particular sample of an ideal gas is reduced to one-third its original value during an isothermal process, one that takes place at constant temperature, what happens to the pressure of the gas?

15. Suppose in a laboratory experiment you took measurements of two related quantities, x and y, and obtained the following pairs of values:

x	3	6	12
y	240	60	15

Based on these data, would you conjecture that x varies inversely as y or inversely as $\sqrt{y}$? To support your conclusion, find an equation that relates x and y.

16. In studies of electric fields it is shown that the electric field intensity E at a distance r from an isolated point charge varies directly as the charge q and inversely as the square of the distance r. If a point charge of 2×10^{-9} coulombs sets up an electric field intensity of 2 N/C (newtons per coulomb) at a distance of 3 m from the charge, what is the electric field intensity for the same charge at a distance of 6 m from the charge?

10-3
REVIEW

REVIEW QUESTIONS

The statement that y varies directly as x means that the ratio of y to x is _____(1)_____.

(1) **constant**

If y varies directly as x and directly as w, we write $y = kxw$. For this statement we can also say that y varies _____(2)_____ as x and w.

(2) **jointly**

If y varies inversely as the reciprocal of x, then y varies directly as _____(3)_____.

(3) **x**

If $y = kL^2/\sqrt{w}$, then y varies _____(4)_____ as L^2 and _____(5)_____ as $\sqrt{w}$.

(4) **directly**   (5) **inversely**

Suppose y varies directly as $a^{3/2}$ and inversely as $b^{1/2}$. If a and b are both doubled, the value of y is multiplied by a factor of _____(6)_____.

(6) **2**

If y varies jointly as x^2 and w and inversely as the product of w^2 and x, then in more simple terms we can say y varies directly as _____(7)_____ and inversely as _____(8)_____.

(7) **x**   (8) **w**

REVIEW PROBLEMS

1. If y varies directly as x and inversely as w, and $y = 7$ when $x = 2$ and $w = 3$, find y when $x = 1$ and $w = 4$.

2. If y varies directly as x^2 and inversely as $\sqrt{w}$, and $y = 2$ when $x = 3$ and $w = 4$, find y when $x = \frac{1}{3}$ and $w = 9$.

3. If y varies jointly as x^2 and w and inversely as z and $y = 1$ when $x = 1$, $w = 2$, and $z = 3$, find y when $x = 3$, $w = 2$, and $z = 1$.

4. If y varies directly as $x^{3/2}$ and inversely as $\sqrt{z}$, and $y = 8$ when $x = 4$ and $z = 3$, find y when $x = 2$ and $z = 12$.

5. When a rod or cable is subjected to a stretching force, the stress is directly proportional to the strain produced. If for a particular steel wire a stress of 10^4 lb/in.^{2} produces a strain of 1/3000, what stress will produce a strain of 1/2000?

6. When an object is released from rest and falls freely, its speed varies directly as the square root of the product of the acceleration due to gravity and the distance fallen. Take the acceleration due to gravity to be 32 ft/sec^2. If a body acquires a speed of 8 ft/sec after falling a distance of 1 ft, what is its speed after falling 4 ft?

7. When a body travels in a circle at a constant speed, it is acted upon by a centripetal force which varies jointly as the mass of the body and the square of its speed and inversely as the radius of the circle. If a 2-kg mass traveling in a circle of radius 5 m with a speed of 10 m/sec is acted upon by a centripetal force of 40 newtons, what force is necessary to have a 3-kg mass travel in a circle of radius 6 m at a speed of 8 m/sec?

8. In some spectrographic instruments a charged particle is projected perpendicularly in a field of magnetic induction B. Under this condition the particle travels in a circle of radius R. The radius varies directly as the product of the mass m and speed v of the particle and inversely as the product of the charge q on the particle and the magnetic induction B. If a charge of mass m travels in a circle of radius 2 m in a magnetic field, what must be the mass of a charged particle with three times the charge if it is to travel in the same circle in the same field with the same speed?

9. The rate at which heat is transferred through a metal rod varies jointly as the cross-sectional area of the rod and the difference in temperature between its ends and inversely as the length of the rod. If 12,500 cal/sec (calories per second) is transmitted through an aluminum rod .5 cm long of cross-sectional area 500 cm^2 when the temperature difference between

its ends is 25°C, at what rate is heat transferred through a rod of the same material and length if its cross-sectional area is 600 cm² and the temperature difference is 80°C?

10. Under certain conditions the velocity of an object varies directly as its acceleration, which is constant, and the square of the time it has been in motion. (a) If the velocity is 144 ft/sec after 3 sec, what is the velocity after $\frac{1}{2}$ sec? (b) If the acceleration is 16 ft/sec², what is the constant of variation?

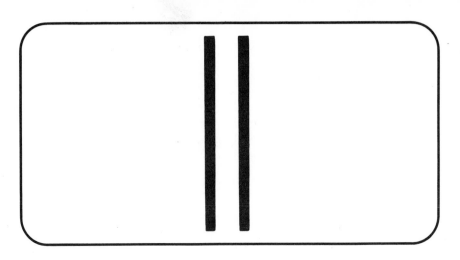

complex numbers

11-1
INTRODUCTION

Every problem encountered thus far in our study of technical mathematics has had a solution in the real number system. There are, however, many problems arising in applied mathematics which have *no* solution in that system. To satisfy the practical requirement that some type of solution to these problems indeed exists, the real number system can be extended to create the so-called *system of complex numbers*. This new system, which contains the real numbers as a subset, is sufficient to allow appropriate solutions to previously unsolvable problems.

Moreover, there are problems in science and engineering where, although a solution in the real number system does exist, the characteristics of that system make obtaining the solution tedious and laborious. In such problems, for example, the analysis of alternating-current circuits in electrical engineering, the complex number system permits a much simpler and more compact solution. We shall now see that the need for complex numbers arises quite naturally from consideration of solving certain equations.

THE J-OPERATOR AND COMPLEX NUMBERS

The solution set of the equation $x^2 = 1$ consists of all real numbers x whose square is 1. Since only 1 and -1 satisfy that condition, the solution set must be $\{1, -1\}$.

Now, suppose we consider the equation $x^2 = -1$. We know, of course, that the square of every real number is positive or zero. Thus, there is no real number x whose square is -1; that is, the symbol $\sqrt{-1}$ is not meaningful in the system of real numbers. Similarly, the real number system provides no solution of the equations $x^2 = -2$, $x^2 = -\frac{1}{2}$, and more generally $x^2 = -a$ where $a > 0$.

Because the real number system fails to satisfy our demands, we shall extend this system so that the result will indeed provide solutions of equations such as those above. Fortunately, since every negative real number is the product of -1 and a positive real number, it will be shown that we need consider only $\sqrt{-1}$ to accommodate the other situations. In fact, we shall also be able to obtain $\sqrt[4]{-16}$, $\sqrt[6]{-64}$, and other even nth roots of negative numbers.

To begin, we introduce (or invent, so to speak) a new type of number which, by its definition, is a solution of the equation $x^2 = -1$.

Definition. The **j-operator**, denoted j, is that number which when squared is equal to -1; that is,

$$j^2 = -1.$$

From the definition of j, it seems reasonable to express j as $\sqrt{-1}$; that is, j is a square root of -1. Since j is not a real number, we call it (according to custom) the **imaginary unit**, keeping in mind that the term *imaginary* does not mean that j is "impossible" or "does not exist", but only that it does not belong to **R**, the set of real numbers. Indeed, the practical applications of j to topics such as electrical theory are extensive.

By taking multiples of j, such as $2j$, $-5j$, and $6.4j$, we obtain what are referred to as *pure imaginary numbers*.

Definition. A **pure imaginary number** is a number that can be expressed in the form bj, where $b \in \mathbf{R}$.

To generalize the notion of $\sqrt{-1}$, we shall express a square root of *any* negative number as the product of a real number and j. The following definition gives the procedure.

Definition. If a is a negative number, then $\sqrt{a}$ is defined to be $j\sqrt{|a|}$.

Thus,

$$\sqrt{-3} = j\sqrt{|-3|} = j\sqrt{3}.$$

Loosely speaking, you can think of this result as that obtained by removing the factor $\sqrt{-1}$ from $\sqrt{-3}$. That is,

$$\sqrt{-3} = \sqrt{(-1)(3)} = \sqrt{-1}\sqrt{3} = j\sqrt{3}.$$

EXAMPLE
1

The following are pure imaginary numbers:

a. $\sqrt{-2} = j\sqrt{|-2|} = j\sqrt{2}.$

b. $-\sqrt{-9} = -(j\sqrt{9}) = -[j(3)] = -3j.$

c. $\sqrt{-32} = j\sqrt{32} = j(4\sqrt{2}) = 4j\sqrt{2}.$

d. $\sqrt{-\dfrac{1}{16}} = j\sqrt{\dfrac{1}{16}} = \dfrac{1}{4}j = \dfrac{j}{4}.$

e. $\sqrt{-2^{6}3^{4}} = 2^{3}3^{2}j = 72j.$

f. $\sqrt{-b}$, where $b > 0$, is $j\sqrt{b}.$

Assuming that arithmetic operations with j obey the same basic rules of multiplication as do the real numbers, the following example illustrates that any positive integral power of j has only one of four possible values: $j, -1, -j,$ or 1.

EXAMPLE
2

Powers of j:

a. $j^{1} = j.$

b. $j^{2} = -1.$

A note of caution. *The rule* $\sqrt{a}\sqrt{b} = \sqrt{ab}$ *which is true for nonnegative real numbers is **false** when **both** a and b are negative,* for by that rule we have

$$\sqrt{-1}\sqrt{-1} = \sqrt{(-1)(-1)} = \sqrt{1} = 1,$$

which is *incorrect*. By definition, however, the following *is* correct:

$$\sqrt{-1}\sqrt{-1} = j\cdot j = j^{2} = -1.$$

Thus, to help prevent any errors in computations, **always express square roots of negative numbers as multiples of j (that is, in the j-form) *before* performing any algebraic manipulations with them.**

c. $j^{3} = j^{2}\cdot j = (-1)j = -j.$

d. $j^{4} = j^{2}\cdot j^{2} = (-1)(-1) = 1.$

e. $j^{5} = j^{4}\cdot j = 1\cdot j = j.$

f. $j^{6} = j^{4}\cdot j^{2} = 1\cdot j^{2} = -1,$ etc.

We can infer from the above example that the values of j^{n}, for n a positive integer, can easily be obtained. First find the remainder when n

is divided by 4, and then raise j to that power which is equal to the remainder.

EXAMPLE
3

Evaluate each of the following:

a. j^{507}.

Since $507 \div 4$ is 126 with a remainder of 3,

$$j^{507} = j^3 = -j.$$

b. j^{26}.

Since $26 = (6)(4) + 2$,

$$j^{26} = j^2 = -1.$$

c. j^{1000}.

Since $1000 = (250)(4) + 0$,

$$j^{1000} = j^0 = 1.$$

The assertion that $j^0 = 1$ is consistent since

$$j^{1000} = (j^4)^{250} = 1^{250} = 1.$$

d. $j = j^5 = j^9 = j^{13} = j^{17} = j^{21}$, etc.

By combining the real number system with the concept of a pure imaginary number, we are led to the notion of a *complex number*, namely the sum of a real number and a pure imaginary number.

Definition.   A **complex number** is one of the form $a + bj$, where a and b are real numbers. The number a is called the **real part** of $a + bj$, and b is called the **imaginary part**.

Some examples of complex numbers are $-3 + 6j$, $4 + j$, $2 - \pi j$, and $-3 - 4j$. The real part of $-3 + 6j$ is -3 and the imaginary part is 6. If $a = 0$, then $a + bj$ reduces to bj, a pure imaginary number. If $b = 0$, then $a + bj$ reduces to a, a real number. Thus, *the set of complex numbers contains the real numbers as a subset.* For example, the real number 3 is a complex number since $3 = 3 + 0j$. Complex numbers which are not real numbers are frequently referred to simply as **imaginary numbers**. Thus the complex number 7 is real, but the complex number $7 + 2j$ is imaginary.

One final note on terminology. The complex numbers $a + bj$ and $a - bj$ are said to be (complex) **conjugates** of each other. Hence, the conjugate of $2 + 3j$ is $2 - 3j$, the conjugate of $-3 - 4j$ is $-3 + 4j$, the conjugate of $-5j$ is $5j$, and the conjugate of 6 is 6 because $6 = 6 + 0j$ and the conjugate of $6 + 0j$ is $6 - 0j$ or 6.

EXAMPLE
4

	Complex Number	Real Part	Imaginary Part	Conjugate
a.	$3 + 4j$	3	4	$3 - 4j$
b.	$-3 - j$	-3	-1	$-3 + j$
c.	-7	-7	0	-7
d.	$-j$	0	-1	j
e.	$\sqrt{-4}$	0	2	$-2j$
f.	$\sqrt[3]{-1}$	-1	0	-1
g.	0	0	0	0

Note in complex numbers c, f, and g that a real number is equal to its conjugate. Also observe in complex number f that $\sqrt[3]{-1}$ is a real number; $(-1)^3 = -1$ and thus $\sqrt[3]{-1} = -1$. Finally, in complex number g we note that zero is the only complex number which is both real and pure imaginary.

A proper definition of a complex number requires that we be precise as to when two complex numbers are equal. We say that $a + bj = c + dj$ if and only if $a = c$ and $b = d$. This means that **two complex numbers are equal if and only if they have the same real parts and the same imaginary parts.** Thus, $2 - 3j = -3j + 2$ and $.2 + j = j + \frac{1}{5}$, but $4 + 2j \neq 2 + 4j$ and $3 - j \neq j - 3$.

EXAMPLE
5

Solve each of the following equations assuming x and y are real numbers:

a. $2 - 5j = x + yj$.

By the definition of equality of complex numbers, we can equate the real parts of both members of the equation:

$$2 = x.$$

Equating imaginary parts,

$$-5 = y.$$

The solution is $x = 2$ and $y = -5$.

b. $x - y + xj = 3 + 2j$.

Equating real parts of both members of the equation,

$$x - y = 3. \tag{1}$$

Equating imaginary parts,

$$x = 2.$$

Substituting this value of x in Eq. (1) gives $2 - y = 3$ from which $y = -1$. The solution is $x = 2$ and $y = -1$.

EXERCISE 11-2

Simplify the expressions in Problems 1–8.

1. j^9 2. j^{14}

3. j^{15} 4. j^{488}

5. j^{66} 6. j^{243}

7. j^{324} 8. $j^2 \cdot j^3 \cdot j^4$

In Problems 9-20, write the given expression in bj-form.

9. $\sqrt{-81}$ 10. $\sqrt{-36}$

11. $-\sqrt{-25}$ 12. $\sqrt{-27}$

13. $\sqrt{-32}$ 14. $\sqrt{-144}$

15. $\sqrt{-\dfrac{1}{4}}$ 16. $-\sqrt{-\dfrac{1}{100}}$

17. $\sqrt{-.09}$ 18. $\sqrt{-.04}$

19. $\sqrt{-\pi^6}$ 20. $\sqrt{-4^{10}}$

For each of the complex numbers in Problems 21–36, state (a) the real part, (b) the imaginary part, and (c) the conjugate. Simplify first, if necessary.

21. $-6 + 5j$ 22. -2

23. $-31j - 2$ 24. $-3 + 5j$

25. $-3j$ 26. $2 - \sqrt{-9}$

27. $7 + \frac{3}{5}j$ 28. πj

29. j^{17} 30. j^2

31. 7 32. $-\sqrt[3]{-64x^6}$, where $x \in \mathbf{R}$

33. $\sqrt[3]{-8}$ 34. $6 + \sqrt{-3x^4y^4}$, where $x, y \in \mathbf{R}$

35. $-\sqrt{-49}$ 36. 0

Assuming x and y are real numbers, solve the equations in Problems 37–42.

37. $2x + yj = 4 - 6j$ 38. $x - 2j = 17 + yj$

39. $x + 4j = 17 + yj$ 40. $x + (x - y)j = 4j$

41. $(x + y) + (x - y)j = 18 - 14j$ 42. $x + yj = a - bj$

43. Two alternating-current voltages are given by the expressions $\frac{3}{2} + (y + 1)j$ and $(7 + x) - 3j$, respectively. If the voltages are equal, what are the values of x and y?

44. What can be said about a complex number which is equal to its conjugate?

45. State whether the following statements are true or false.
    a. $\frac{1}{2} + 2j = 2j + .5$
    b. $3 - \sqrt{-5} = 3 - j\sqrt{5}$
    c. $\sqrt[3]{-27} = 3j$
    d. $5 - 6j = (7 - 2) + (4 - 10)j$

46. Suppose $a + bj = 0$. To what must $a^2 + b^2$ be equal?

47. A Maxwell bridge is an alternating-current bridge circuit which permits the inductance L_x and resistance R_x of a coil to be measured. A theoretical analysis of the circuit yields the equation

$$R_x + j\omega L_x = R_2 R_3 \left(\frac{1}{R_1} + j\omega C_1 \right).$$

Determine the resistance and inductance of a coil in terms of the circuit constants R_1, R_2, R_3, C_1, and ω.

11-3
GEOMETRIC REPRESENTATION OF COMPLEX NUMBERS

Complex numbers can be geometrically represented in a plane with the use of a rectangular coordinate system. By one method, the complex number $a + bj$ is represented by the point whose coordinates are (a, b). The abscissa a corresponds to the real part of $a + bj$, and the ordinate b corresponds to the imaginary part of $a + bj$ [see Fig. 11-1(a)]. For example, in Fig. 11-1(b) the number $2 + 3j$ is represented by the point $(2, 3)$ in the first quadrant; the number $-3 - j$ is represented by the point $(-3, -1)$ in the third quadrant. The origin $(0, 0)$ corresponds to 0 or $0 + 0j$. Since the points in the plane are considered here to represent complex numbers,

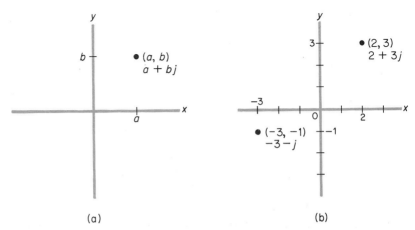

(a)   (b)

FIGURE 11-1

we refer to the plane as the **complex plane** (also called the *Argand plane*).

Observe in Fig. 11-2 that the point $(a, 0)$ on the x-axis corresponds to the real number $a = a + 0j$, while the point $(0, b)$ on the y-axis corre-

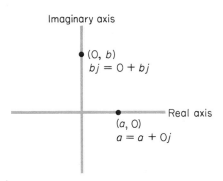

FIGURE 11-2

sponds to the pure imaginary number $bj = 0 + bj$. For these reasons, the x-axis is called the **real axis** and the y-axis is called the (pure) **imaginary axis**. The fact that $(0, 0)$ lies both on the real axis *and* the imaginary axis gives a geometric picture of a previous statement that 0 is both a real and a pure imaginary number.

Since the complex number $a + bj$ can be represented by the point (a, b), we can speak of the *point* $a + bj$ and no ambiguity should arise. Also, $a + bj$ is called the **rectangular form** of a complex number.

EXAMPLE
6

Plot the points corresponding to the complex numbers $2 + j$, $3 - 2j$, $-1 - 2j$, $-1 + 3j$, -3, 2, j, and $-2j$. (See Fig. 11-3).

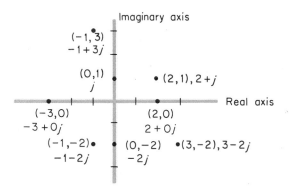

FIGURE 11-3

The complex number $a + bj$ can also be represented by the vector **OP** extending from the origin to the point $P(a, b)$, as indicated in Fig.

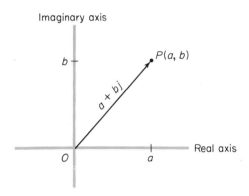

FIGURE 11-4

11-4. Hence, we can speak interchangeably of the *complex number* $a + bj$, the *point* $a + bj$, and the *vector* $a + bj$.

It is worthwhile to note that in practical uses, for instance in alternating-current circuit theory, complex numbers, and therefore vectors in the complex plane, are used to represent time-dependent currents and voltages. These are not, however, vector quantities in accordance with our discussion in Chapter 7, but are used as a convenience to indicate geometrically a *phase relationship* between these quantities. To eliminate any confusion with a vector in the usual sense, electrical engineers have adopted the more descriptive term *phasor* for these time vectors, and a diagram in which they appear is called a *phasor diagram*. Since no confusion will result in our discussions, we shall continue to use the word vector.

In Fig. 11-5 the angle θ which **OP** makes with the positive real axis is called the **amplitude** or **argument** of $a + bj$. The length r of **OP** is called the **absolute value** or **modulus** of $a + bj$ and is always positive or zero. From an observation of Fig. 11-5 we can conclude, using our

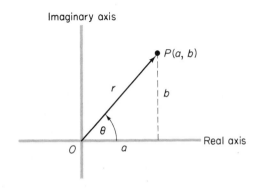

FIGURE 11-5

knowledge of trigonometry and the Pythagorean theorem, that

$$\cos \theta = \frac{a}{r} \qquad \text{and} \qquad a = r \cos \theta \qquad (1)$$

$$\sin \theta = \frac{b}{r} \qquad \text{and} \qquad b = r \sin \theta \qquad (2)$$

$$\tan \theta = \frac{b}{a} \qquad (3)$$

$$r = \sqrt{a^2 + b^2} \qquad (4)$$

By using (1) and (2) we can write

$$a + bj = r \cos \theta + (r \sin \theta)j$$
$$a + bj = r(\cos \theta + j \sin \theta)$$

We speak of $r(\cos \theta + j \sin \theta)$ as a **trigonometric** or **polar form** of a complex number.

To find a trigonometric form of the complex number $a + bj$, we use Eqs. (3) and (4):

$$\tan \theta = \frac{b}{a}, \qquad r = \sqrt{a^2 + b^2}.$$

Let us illustrate by expressing $2 + 2j$ in trigonometric form. We first note that $2 + 2j$ lies in the first quadrant. Second, since $a = 2$ and $b = 2$, we have

$$r = \sqrt{a^2 + b^2} = \sqrt{2^2 + 2^2} = \sqrt{8} = 2\sqrt{2}.$$

Also,

$$\tan \theta = \frac{b}{a} = \frac{2}{2} = 1.$$

Thus, θ is an angle whose tangent is 1. Clearly, $\theta = 45°$, $225°$, $45° + 360°$, $225° + 360°$, etc., satisfy this condition. It is customary, however, to choose θ to be the smallest nonnegative angle satisfying the condition. That is, $0° \leq \theta < 360°$. That means we limit the choice to $\theta = 45°$ or $\theta = 225°$. But since $2 + 2j$ is in the first quadrant, we must choose $\theta = 45°$. We can now write

$$2 + 2j = r(\cos \theta + j \sin \theta)$$
$$= 2\sqrt{2} (\cos 45° + j \sin 45°).$$

In Fig. 11-6, $2 + 2j$ is geometrically represented by both the point $P(2, 2)$ and the vector OP from the origin to $P(2, 2)$. We point out that since, in general, there are two choices for θ such that both $\tan \theta = b/a$

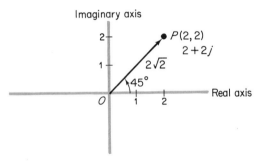

FIGURE 11-6

and $0° \le \theta < 360°$, the correct one will be evident by first determining the quadrant in which the point $a + bj$ lies.

EXAMPLE
7

a. Express $-3 + 3j\sqrt{3}$ in trigonometric form.

The point $-3 + 3j\sqrt{3}$ lies in the second quadrant. (See Fig. 11-7.)

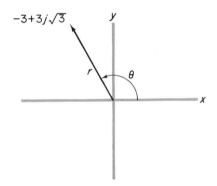

FIGURE 11-7

For $a = -3$ and $b = 3\sqrt{3}$ we have
$$r = \sqrt{a^2 + b^2} = \sqrt{(-3)^2 + (3\sqrt{3})^2} = \sqrt{9 + 27} = 6$$
and
$$\tan \theta = \frac{b}{a} = \frac{3\sqrt{3}}{-3} = -\sqrt{3}.$$

Since $\tan \theta = -\sqrt{3}$, θ may be either 120° or 300°. But $-3 + 3j\sqrt{3}$ is in the second quadrant and thus the correct choice is 120°. Therefore,
$$-3 + 3j\sqrt{3} = 6(\cos 120° + j \sin 120°).$$

b. Express $3 - 4j$ in trigonometric form.

The point $3 - 4j$ lies in the fourth quadrant. Also,
$$r = \sqrt{(3)^2 + (-4)^2} = \sqrt{25} = 5$$

and

$$\tan \theta = \frac{b}{a} = \frac{-4}{3} = -\frac{4}{3} = -1.3333.$$

Thus θ, being a fourth quadrant angle, must be (from tables) 306° 52′. We can therefore write

$$3 - 4j = 5(\cos 306° \ 52' + j \sin 306° \ 52').$$

The trigonometric form of a real number or a pure imaginary number is immediately evident, as the following example shows.

EXAMPLE
8

Express each of the following complex numbers in trigonometric form.

a. 8.

The point 8, or $8 + 0j$, lies on the positive x-axis and is eight units from the origin [Fig. 11-8(a)]. Therefore $\theta = 0°$, $r = 8$, and we have $8 = 8(\cos 0° + j \sin 0°)$.

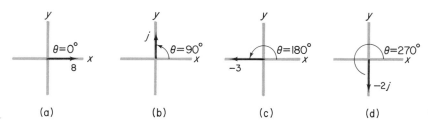

(a)   (b)   (c)   (d)

FIGURE 11-8

b. j.

The point j, or $0 + 1j$, lies on the positive y-axis and is one unit from the origin [Fig. 11-8(b)]. Thus, $\theta = 90°$ and $r = 1$. Therefore we have $j = 1(\cos 90° + j \sin 90°)$.

c. -3.

By inspection of Fig. 11-8(c), clearly $r = 3$ and $\theta = 180°$, and we conclude that $-3 = 3(\cos 180° + j \sin 180°)$.

d. $-2j$.

By inspection of Fig. 11-8(d), we have $r = 2$ and $\theta = 270°$. Hence $-2j = 2(\cos 270° + j \sin 270°)$.

EXAMPLE
9

Locate the number $2(\cos 300° + j \sin 300°)$ in the complex plane.

Since the absolute value is 2 and the argument is 300°, the number is represented by a vector of length 2 which makes an angle of 300° with the positive x-axis (Fig. 11-9).

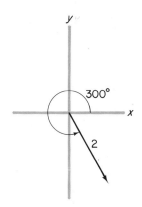

FIGURE 11-9

EXAMPLE
10

Express each of the following complex numbers in rectangular form.

a. $2(\cos 60° + j \sin 60°)$.

$$2(\cos 60° + j \sin 60°) = 2\left(\frac{1}{2} + j \cdot \frac{\sqrt{3}}{2}\right)$$
$$= 1 + j\sqrt{3}.$$

b. $8(\cos 90° + j \sin 90°)$.

$$8(\cos 90° + j \sin 90°) = 8(0 + j \cdot 1) = 8j.$$

c. $3(\cos 150° + j \sin 150°)$.

$$3(\cos 150° + j \sin 150°) = 3\left(-\frac{\sqrt{3}}{2} + j \cdot \frac{1}{2}\right)$$
$$= -\frac{3\sqrt{3}}{2} + \frac{3}{2}j.$$

d. $3(\cos 27° + j \sin 27°)$.

$$3(\cos 27° + j \sin 27°) = 3(.8910 + .4540j)$$
$$= 2.673 + 1.362j$$

EXERCISE 11-3

In Problems 1–18, (a) plot both the given number and its conjugate in the complex plane and (b) express the given number in trigonometric form.

1. $4 + 4j$

2. $-\sqrt{3} + j$

3. $2 - 2j\sqrt{3}$

4. -4

5. $5j$

6. $1 + j\sqrt{3}$

7. $-3 - 3j$

8. $-8j$

9. 1   **10.** $\sqrt{2} - j\sqrt{2}$

11. $-1 + j\sqrt{3}$ **12.** $-3\sqrt{3} + 3j$

13. $-\sqrt{3} - j$ **14.** $6j$

15. $-5j$ **16.** 64

17. $2 + 4j$ **18.** $3 - 4j$

In Problems 19–34, locate the given number in the complex plane and express it in rectangular form.

19. $2(\cos 30° + j \sin 30°)$ **20.** $3(\cos 240° + j \sin 240°)$

21. $2(\cos 120° + j \sin 120°)$ **22.** $4(\cos 210° + j \sin 210°)$

23. $2(\cos 315° + j \sin 315°)$ **24.** $3(\cos 150° + j \sin 150°)$

25. $3(\cos 270° + j \sin 270°)$ **26.** $2(\cos 180° + j \sin 180°)$

27. $4(\cos 0° + j \sin 0°)$ **28.** $4(\cos 240° + j \sin 240°)$

29. $3(\cos 330° + j \sin 330°)$ **30.** $\frac{1}{2}(\cos 90° + j \sin 90°)$

31. $2(\cos 15° + j \sin 15°)$ **32.** $3(\cos 40° + j \sin 40°)$

33. $2(\cos 245° + j \sin 245°)$ **34.** $4(\cos 340° + j \sin 340°)$

35. If all complex numbers with an absolute value of 1 were sketched in the plane, what geometric figure would they form?

36. If $3(\cos 220° + j \sin 220°) = a + bj$, find a and b.

37. If $2(\cos 0° + j \sin 0°) = a + \left(\dfrac{b}{c}\right)j$, find c.

38. If $4(\cos 500° + j \sin 500°) = a + bj$, find a and b.

In Problems 39–42, express the given number in rectangular form.

39. $3[\cos(-210°) + j \sin(-210°)]$

40. $8[\cos(-450°) + j \sin(-450°)]$

41. $\frac{3}{4}[\cos(-45°) + j \sin(-45°)]$

42. $\cos(-225°) + j \sin(-225°)$

11-4

OPERATIONS WITH COMPLEX NUMBERS IN RECTANGULAR FORM

Algebraic operations with complex numbers in the rectangular form $a + bj$ present no difficulty, since these operations are defined in such a way that the ordinary rules of arithmetic and algebra are preserved.

Definition.   *Addition of Complex Numbers:*

$$(a + bj) + (c + dj) = (a + c) + (b + d)j.$$

Subtraction of Complex Numbers:

$$(a + bj) - (c + dj) = (a - c) + (b - d)j.$$

Verbally, the sum (or difference) of two complex numbers is the complex number obtained by taking the sum (or difference) of their real parts and multiplying j by the sum (or difference) of their imaginary parts.

EXAMPLE
11

Perform the indicated operations and express the result in the form $a + bj$.

a. $(2 + 3j) + (4 + j) = (2 + 4) + (3 + 1)j = 6 + 4j.$

b. $2 + (8 - 3j) = (2 + 0j) + [8 + (-3)j] = (2 + 8) + [0 + (-3)]j$
$$= 10 - 3j.$$

c. $(6 + 3j) - (2 + 5j) = (6 - 2) + (3 - 5)j = 4 - 2j.$

d. $(-2 - 3j) + (2 - 4j) = (-2 + 2) + [-3 + (-4)]j$
$$= 0 - 7j = -7j.$$

e. $(5 - 2j) - (-6 + j) = [5 - (-6)] + (-2 - 1)j = 11 - 3j.$ More simply we could remove parentheses and write

$$(5 - 2j) - (-6 + j) = 5 - 2j + 6 - j = 11 - 3j.$$

f. $\sqrt{8} + \sqrt{-\dfrac{3}{2}} - \sqrt{-24} = \sqrt{8} + j\sqrt{\dfrac{3}{2}} - j\sqrt{24}$

$$= 2\sqrt{2} + \frac{1}{2}j\sqrt{6} - 2j\sqrt{6}$$

$$= 2\sqrt{2} - \frac{3}{2}j\sqrt{6} = 2\sqrt{2} - \frac{3j\sqrt{6}}{2}.$$

Definition.   *Multiplication of Complex Numbers:*

$$(a + bj)(c + dj) = a(c + dj) + bj(c + dj)$$
$$= ac + adj + bcj + bdj^2$$
$$= ac + (ad + bc)j - bd$$

$$(a + bj)(c + dj) = (ac - bd) + (ad + bc)j.$$

Rather than memorize this rule you need merely treat complex numbers as binomial expressions and multiply accordingly by the distributive law. Remember that all complex numbers should be expressed in the form $a + bj$ before multiplying them, and that whenever j^2 occurs, it must be replaced by -1.

EXAMPLE
12

a. $(3 + 2j)(2 - 5j) = 3(2 - 5j) + 2j(2 - 5j)$
$$= 6 - 15j + 4j - 10j^2 = 6 - 11j + 10$$
$$= 16 - 11j.$$

b. $j(-2 + 3j) = -2j + 3j^2 = -2j - 3 = -3 - 2j$.

c. $(8j)(4j) = 32j^2 = -32$.

d. $(-3 - 4j)(2 - j) = -3(2 - j) - 4j(-j) = -6 + 3j + 4j^2$
$$= -6 + 3j - 4 = -10 + 3j.$$

e. $(4 + 5j)^2$. We treat this as a square of a binomial:
$$(4 + 5j)^2 = (4)^2 + 2(4)(5j) + (5j)^2$$
$$= 16 + 40j - 25 = -9 + 40j.$$

f. $2\sqrt{-4}(3 - 2\sqrt{-3}) + j = 2(2j)(3 - 2j\sqrt{3}) + j$
$$= (4j)(3 - 2j\sqrt{3}) + j$$
$$= 12j - 8j^2\sqrt{3} + j$$
$$= 8\sqrt{3} + 13j.$$

g. $(a + bj)(a - bj) = a^2 - (bj)^2 = a^2 - b^2j^2 = a^2 + b^2$.

h. $(2 + 3j)(2 - 3j) = (2)^2 - (3j)^2 = 4 + 9 = 13$.

As seen in Example 12g and h, the product of a complex number and its conjugate is always a real number. Thus, to find $\dfrac{a + bj}{c + dj}$ it seems reasonable to multiply *both* the numerator and denominator by the conjugate of the denominator. The resulting denominator will be a real number, and the quotient therefore can easily be put in rectangular form. The procedure resembles that of rationalizing the denominator of an algebraic expression.

Definition.   *Division of Complex Numbers*:

$$\frac{a + bj}{c + dj} = \frac{a + bj}{c + dj} \cdot \frac{c - dj}{c - dj}, \qquad c + dj \neq 0$$

$$= \frac{ac - adj + bcj - bdj^2}{c^2 - d^2j^2}$$

$$= \frac{(ac + bd) + (bc - ad)j}{c^2 + d^2}$$

$$\frac{a + bj}{c + dj} = \frac{ac + bd}{c^2 + d^2} + \frac{bc - ad}{c^2 + d^2}.$$

To divide two complex numbers, multiply both numerator and denominator by the conjugate of the denominator. As a result, the denominator will then be a real number.

EXAMPLE

13

Perform each of the indicated operations.

a. $\dfrac{2 - j}{3 + 2j}$.

We multiply both the numerator and denominator by the conjugate of the denominator, namely $3 - 2j$:

$$\frac{2 - j}{3 + 2j} = \frac{2 - j}{3 + 2j} \cdot \frac{3 - 2j}{3 - 2j} = \frac{(2 - j)(3 - 2j)}{(3 + 2j)(3 - 2j)}$$

$$= \frac{6 - 4j - 3j + 2j^2}{9 - 4j^2} = \frac{4 - 7j}{13} = \frac{4}{13} - \frac{7}{13}j.$$

b. $\dfrac{2}{1 + j} = \dfrac{2}{1 + j} \cdot \dfrac{1 - j}{1 - j} = \dfrac{2(1 - j)}{(1 + j)(1 - j)} = \dfrac{2 - 2j}{1 - j^2}$

$$= \frac{2 - 2j}{2} = 1 - j.$$

c. $\dfrac{6}{5j} = \dfrac{6}{5j} \cdot \dfrac{-5j}{-5j} = \dfrac{-30j}{-25j^2} = \dfrac{-30j}{25} = -\dfrac{6}{5}j.$ If the denominator is a pure imaginary number, it suffices to multiply the given fraction by j/j. Thus,

$$\frac{6}{5j} = \frac{6}{5j} \cdot \frac{j}{j} = \frac{6j}{5j^2} = -\frac{6}{5}j.$$

d. $\dfrac{2 - \sqrt{-9}}{2 - \sqrt{-4}} = \dfrac{2 - 3j}{2 - 2j} = \dfrac{2 - 3j}{2 - 2j} \cdot \dfrac{2 + 2j}{2 + 2j}$

$$= \frac{4 + 4j - 6j - 6j^2}{4 - 4j^2} = \frac{10 - 2j}{8} = \frac{5}{4} - \frac{1}{4}j.$$

e. $(1 + j)^{-2} = \dfrac{1}{(1 + j)^2} = \dfrac{1}{1 + 2j + j^2} = \dfrac{1}{2j}$

$$= \frac{1}{2j} \cdot \frac{j}{j} = \frac{j}{-2} = -\frac{1}{2}j.$$

EXAMPLE 14

If $f(z) = z^2 + 6z + \dfrac{1}{z}$, find $f(3j)$.

$$f(3j) = (3j)^2 + 6(3j) + \frac{1}{3j}$$

$$= 9j^2 + 18j + \frac{1}{3j} \cdot \frac{j}{j}$$

$$= -9 + 18j + \frac{j}{-3} = -9 + \frac{53}{3}j.$$

EXERCISE 11-4

In Problems 1–54, perform the indicated operations and express the answer in rectangular form.

1. $2j + 3j - 3$

2. $2j - 6j + 4j$

3. $(3 - 5j) + (-6 + 4j)$

4. $(6j - 2) + (7 + 4j)$

5. $(7 - 3j) - (9 - 6j)$

6. $(4 + 3j) - (5j - 6)$

7. $\sqrt{-4} + 2\sqrt{-8} + 3\sqrt{12}$

8. $j^2 - 2j + j\sqrt{-16} + j^3\sqrt{-5}$

9. $(8 - \sqrt{-16}) - (\sqrt{-1} + 4j)$

10. $\sqrt{-5} - 2\sqrt{-16} + (3 - \sqrt{-2}) - (j\sqrt{2} + 1)$

11. $\sqrt{-5}\sqrt{-2}\sqrt{-20}$

12. $(8j)(2j) - (6j)(2j) + j^3(j^6)$

13. $3(2j)^3(3j)^2(j)^5$

14. $(-\sqrt{-4})(-\sqrt{-5})(-\sqrt{-20})$

15. $(2j)^4(-2j)^2(j)^6$

16. $j(2j)(3j)(4j)$

17. $(2 + j)(3 + 2j)$

18. $(1 + j)(6 - 2j)$

19. $(4 + 3j)(5 - 2j)$

20. $(-2 - j)(-3 + j)$

21. $2(2 - j)(3 + 2j)$

22. $2(-5 + j)(-5 - j)$

23. $(6 + 2j)(-7 - \sqrt{-4})$

24. $(2 - \sqrt{-3})(3 + \sqrt{-12})$

25. $(3 + 2j)^2$

26. $(8 - 4j)^2$

27. $(3 - \sqrt{-25})^2$

28. $(2 - \sqrt{-64})^2$

29. $(\frac{1}{3} - \frac{4}{5}j)(\frac{3}{8} + \frac{2}{7}j)$

30. $2(3 + \sqrt{-36})(2 + \sqrt{-72})(\sqrt{-4})$

31. $(3 - j) \div (4 + j)$

32. $(1 - j) \div (1 + j)$

33. $(2 - j) \div (3 - j)$

34. $(2 - j) \div (2 + j)$

35. $2 \div (1 - j)$

36. $\sqrt{-12} \div \sqrt{-9}$

37. $3 \div (2j)$

38. $(2 + 3j) \div (4 - 2j)$

39. $(6 + 4j) \div (2 - 3j)$

40. $3 \div [2(1 - j)]$

41. $6 \div (2 - \sqrt{-25})$

42. $1/[(2 + j)(2 - j)]$

43. $(1 + \sqrt{-5})^{-1}$

44. $(2 - \sqrt{-9})^{-1}$

45. $(2 - j)(3 + j)(j - 4)$

46. $\left(\frac{1}{3} + \frac{j}{4}\right) \div \left(\frac{2}{3} + \frac{j}{2}\right)$

47. $\left(\frac{2}{3} - \frac{j}{2}\right) \div \left(\frac{1}{3} + \frac{j}{4}\right)$

48. $\left(-\frac{1}{2} + \frac{\sqrt{3}}{2}j\right)^3$

49. $\left(\frac{1}{2} + \frac{\sqrt{3}}{2}j\right)^3$

50. $(2 - 3j)^{-2}$

51. $(1 + 2j)^{-2}$

52. $(6 - \sqrt{-4})^{-2}$

53. $(2 + \sqrt{-25})^2(2 + \sqrt{-16})^2$

54. $4 \div (2 + 3\sqrt{-9})^{-2}$

55. If $f(x) = x^2 - 2x + 1$, find $f(1 + j)$.

56. If $f(x) = 2x^2 - x + 1$, find $f(2 - j)$.

57. In electrical theory, the impedance Z (in ohms) for a certain alternating series circuit is given by

$$Z = \frac{E}{I},$$

where E is the voltage and I is the current. Find Z if $E = 5 + 5j$ (volts) and $I = 3 + 4j$ (amperes).

58. For the formula in Problem 57, find E if $Z = 2 + j$ and $I = 5 - j$.

59. It is shown in electrical theory that if two impedances Z_1 and Z_2 are connected in series, then the resulting impedance is Z, where

$$Z = Z_1 + Z_2.$$

Moreover, these impedances can be represented by complex numbers. Find Z if $Z_1 = 1 + j$ and $Z_2 = 2 + j$. If Z_1 and Z_2 are connected in parallel, then

$$\frac{1}{Z} = \frac{1}{Z_1} + \frac{1}{Z_2}.$$

Find Z for the values of Z_1 and Z_2 above.

60. The impedance Z of a certain circuit is given by

$$Z = Z_1 + \frac{Z_2 Z_3}{Z_2 + Z_3},$$

where $Z_1 = 1 + 3j$, $Z_2 = 1 + j$, and $Z_3 = 1 + 2j$. Find Z.

61. The sketch below indicates part of an electrical circuit. Kirchhoff's laws tell us that $I_1 + I_2 = I_3$. Find I_2 if $I_1 = 7 + 2j$ and $I_3 = 9 - 5j$.

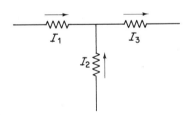

62. The Hay bridge is an alternating-current circuit used in electrical measurements. It is governed by the equation

$$\left(R_1 - j\frac{1}{\omega C_1} \right)(R_x + j\omega L_x) = R_2 R_3.$$

Find R_x and L_x in terms of the other circuit constants.

63. A student said that $-1/j = j^2/j = j$. Is his answer correct?

11-5

OPERATIONS WITH COMPLEX NUMBERS IN TRIGONOMETRIC FORM

Performing calculations such as $(1 + j)^{20}$ or $(2 + 2j)^{15}/(\sqrt{3} + j)^{21}$, which at first glance appear to be prohibitive, may often, in fact, be easily accomplished if the numbers involved are first expressed in trigonometric form. In the next section we shall indeed consider expressions of the type indicated above. But first, to lay some groundwork let us determine the product $z_1 z_2$, where

$$z_1 = r_1(\cos\theta_1 + j\sin\theta_1) \quad \text{and} \quad z_2 = r_2(\cos\theta_2 + j\sin\theta_2)$$

are two complex numbers in trigonometric form. We have

$$r_1(\cos\theta_1 + j\sin\theta_1)\cdot r_2(\cos\theta_2 + j\sin\theta_2)$$
$$= r_1 r_2(\cos\theta_1\cos\theta_2 + j\cos\theta_1\sin\theta_2 + j\sin\theta_1\cos\theta_2 + j^2\sin\theta_1\sin\theta_2)$$
$$= r_1 r_2[\cos\theta_1\cos\theta_2 - \sin\theta_1\sin\theta_2 + j(\sin\theta_1\cos\theta_2 + \cos\theta_1\sin\theta_2)]$$
$$= r_1 r_2[\cos(\theta_1 + \theta_2) + j\sin(\theta_1 + \theta_2)].$$

The last line follows by the application of the addition rules for the sine and cosine functions which we shall derive in Chapter 15. We can thus state the rule

> **The product of two complex numbers in trigonometric form is a complex number whose absolute value is the product of the given absolute values and whose amplitude is the sum of the given amplitudes.**

EXAMPLE
15

Multiply each of the following and express the answer in rectangular form.

a. $2(\cos 10° + j\sin 10°)\cdot 4(\cos 20° + j\sin 20°)$.

The product of the absolute values is $2\cdot 4 = 8$ and the sum of the amplitudes is $10° + 20° = 30°$. Thus

$$2(\cos 10° + j\sin 10°)\cdot 4(\cos 20° + j\sin 20°) = 8(\cos 30° + j\sin 30°)$$
$$= 8\left(\frac{\sqrt{3}}{2} + j\cdot\frac{1}{2}\right)$$
$$= 4\sqrt{3} + 4j.$$

b. $3(\cos 5° + j\sin 5°)\cdot 4(\cos 50° + j\sin 50°)\cdot 2(\cos 35° + j\sin 35°)$
$$= (3\cdot 4)[\cos(5° + 50°) + j\sin(5° + 50°)]\cdot 2(\cos 35° + j\sin 35°)$$
$$= (3\cdot 4\cdot 2)[\cos(5° + 50° + 35°) + j\sin(5° + 50° + 35°)]$$
$$= 24(\cos 90° + j\sin 90°) = 24(0 + j\cdot 1) = 24j.$$

In actual practice, we could multiply the three absolute values immediately as well as take the sum of the three amplitudes.

c. $(a + bj)j$.

If $a + bj = r(\cos\theta + j\sin\theta)$ and $j = 1(\cos 90° + j\sin 90°)$, then

$$(a + bj)j = r[\cos(\theta + 90°) + j\sin(\theta + 90°)],$$

which is the vector obtained by rotating $a + bj$ through an angle of $90°$. Thus, the effect of multiplying a vector by j is to rotate the vector through an angle of 90°. It is based on this rotational effect that we speak of j as an *operator*.

If we now turn our attention to division, suppose the quotient z_1/z_2 results in the complex number z_3. That is, let

$$\frac{r_1(\cos \theta_1 + j \sin \theta_1)}{r_2(\cos \theta_2 + j \sin \theta_2)} = r_3(\cos \theta_3 + j \sin \theta_3).$$

Then

$$r_1(\cos \theta_1 + j \sin \theta_1) = r_2(\cos \theta_2 + j \sin \theta_2) \cdot r_3(\cos \theta_3 + j \sin \theta_3)$$
$$= r_2 r_3 [\cos(\theta_2 + \theta_3) + j \sin(\theta_2 + \theta_3)].$$

Clearly $r_1 = r_2 r_3$ and $\theta_1 = \theta_2 + \theta_3$ will satisfy the equation. Equivalently, $r_3 = r_1/r_2$ and $\theta_3 = \theta_1 - \theta_2$ and so

$$\frac{r_1(\cos \theta_1 + j \sin \theta_1)}{r_2(\cos \theta_2 + j \sin \theta_2)} = \frac{r_1}{r_2}[(\cos(\theta_1 - \theta_2) + j \sin(\theta_1 - \theta_2)].$$

Therefore, we can say

The quotient of two complex numbers in trigonometric form is a complex number whose absolute value is the absolute value of the numerator divided by the absolute value of the denominator and whose amplitude is the result of subtracting the amplitude of the denominator from that of the numerator.

EXAMPLE
16

Perform the indicated operations and express the answer in rectangular form.

a. $\dfrac{3(\cos 75° + j \sin 75°)}{6(\cos 15° + j \sin 15°)} = \dfrac{3}{6}[\cos (75° - 15°) + j \sin (75° - 15°)]$

$$= \frac{1}{2}(\cos 60° + j \sin 60°)$$

$$= \frac{1}{2}\left(\frac{1}{2} + j\frac{\sqrt{3}}{2}\right) = \frac{1}{4} + j\frac{\sqrt{3}}{4}.$$

b. $\dfrac{3(\cos 10° + j \sin 10°) \cdot 4(\cos 150° + j \sin 150°)}{12(\cos 40° + j \sin 40°)}$

$$= \frac{(3 \cdot 4)[\cos (10° + 150°) + j \sin (10° + 150°)]}{12(\cos 40° + j \sin 40°)}$$

$$= \frac{12(\cos 160° + j \sin 160°)}{12(\cos 40° + j \sin 40°)}$$

$$= \frac{12}{12}[\cos (160° - 40°) + j \sin (160° - 40°)]$$

$$= \cos 120° + j \sin 120° = -\frac{1}{2} + j\frac{\sqrt{3}}{2}.$$

c. $\dfrac{2(\cos 15° + j \sin 15°)}{10(\cos 40° + j \sin 40°)} = \dfrac{2}{10}[\cos (15° - 40°) + j \sin (15° - 40°)]$

$$= \dfrac{1}{5}[\cos (-25°) + j \sin (-25°)].$$

Since $\cos (-25°) = \cos 25° = .9063$ and $\sin (-25°) = -\sin 25° = -.4226$, the previous expression is equal to

$$\tfrac{1}{5}(.9063 - .4226j) = .1813 - .0845j.$$

Addition and subtraction of complex numbers can be interpreted geometrically as addition and subtraction of vectors which have their initial points at the origin. The vector representing $a + bj$ has terminal point $P_1(a, b)$ and the vector for $c + dj$ has terminal point $P_2(c, d)$, as indicated in Fig. 11-10. OP_1 and OP_2 are vectors representing $a + bj$ and $c + dj$,

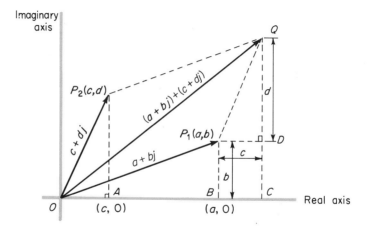

FIGURE 11-10

respectively. Vector OQ is the diagonal of the parallelogram having OP_1 and OP_2 as adjacent sides. We wish to show that OQ is the vector representing the sum

$$(a + bj) + (c + dj) = (a + c) + (b + d)j.$$

Since triangle OAP_2 is congruent to triangle P_1DQ, it follows, by corresponding sides, that $P_1D = OA = c$ and $DQ = AP_2 = d$. Thus, $OC = OB + BC = OB + P_1D = a + c$ and $CQ = CD + DQ = BP_1 + DQ = b + d$. As a result, OQ represents the vector $(a + c) + (b + d)j$, which is $(a + bj) + (c + dj)$.

EXAMPLE
17

Geometrically find the sum $(2 + 3j) + (3 + j)$.

The sum of $2 + 3j$ and $3 + j$ is represented geometrically by the diagonal

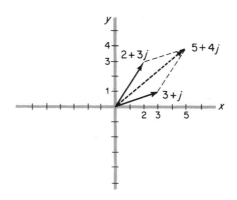

FIGURE 11-11

vector $5 + 4j$ of the parallelogram having the given vectors for adjacent sides. Refer to Fig. 11-11.

The subtraction $(a + bj) - (c + dj)$ can be stated in terms of the addition $(a + bj) + (-c - dj)$, where the vector representing $-c - dj$ is a vector equal in magnitude but opposite in direction to the vector $c + dj$.

EXAMPLE

18

Geometrically determine the difference $(2 + j) - (3 - 2j)$.

The difference $(2 + 3j) - (3 - 2j)$ is represented by the diagonal vector $-1 + 3j$ of the parallelogram whose adjacent sides are the vector for $2 + j$ and the vector which is equal in magnitude but opposite in direction to $3 - 2j$ (that is, $-3 + 2j$). Refer to Fig. 11-12.

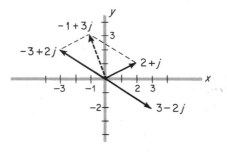

FIGURE 11-12

Just as complex numbers can be represented by vectors, liberal adaptation permits vector quantities, such as force and velocity, to be represented by complex numbers. As a result, problems which involve finding the resultant of a number of vector quantities may be readily solvable by means of algebra of complex numbers. That is, the resultant of two or more vectors can be obtained by first expressing each vector as a complex number and then adding these numbers in the usual manner. In the following example, we shall find the resultant of three forces acting at the same point (that is, *concurrent forces*).

EXAMPLE
19

Using complex algebra, find the resultant, in rectangular form, of the forces **A**, **B**, and **C** which act at the origin and are defined by

$$A = 30 \text{ lb}, \qquad \theta = 45°$$
$$B = 40 \text{ lb}, \qquad \theta = 120°$$
$$C = 20 \text{ lb}, \qquad \theta = 210°,$$

where θ is the angle between the vector and the positive *x*-axis. The system is indicated in Fig. 11-13.

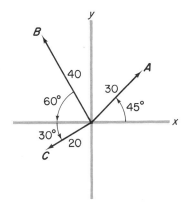

FIGURE 11-13

After first expressing each force in trigonometric form, we write them in rectangular form:

$$A = 30(\cos 45° + j \sin 45°) = 30\left(\frac{\sqrt{2}}{2} + j\frac{\sqrt{2}}{2}\right) = 15\sqrt{2} + 15j\sqrt{2}$$

$$B = 40(\cos 120° + j \sin 120°) = 40\left(-\frac{1}{2} + j\frac{\sqrt{3}}{2}\right) = -20 + 20j\sqrt{3}$$

$$C = 20(\cos 210° + j \sin 210°) = 20\left(-\frac{\sqrt{3}}{2} - j\cdot\frac{1}{2}\right) = -10\sqrt{3} - 10j.$$

We now add:

$$A + B + C = (15\sqrt{2} - 20 - 10\sqrt{3}) + (15\sqrt{2} + 20\sqrt{3} - 10)j$$
$$= -16.108 + 45.855j.$$

Thus, the resultant is $-16.108 + 45.855j$.

We close this section with a note concerning an application of complex numbers. The analysis of alternating-current (a.c.) circuits involves vector analysis, owing to the phase relationships that exist between the circuit currents and the potential differences across the circuit elements.

Figure 11-14 is an example of a simple a.c. series circuit. G is an a.c. generator, L an inductor, C a capacitor, and R a resistor. For the circuit shown, electrical theory shows that the applied voltage V_{app} is equal to the vector sum of V_L, V_C, and V_R—the voltages across the inductor, capacitor, and resistor, respectively. For example, Fig. 11-15 could represent these voltages and appropriate phase relationships. Since V_{app} is the vector sum of the vectors shown in Fig. 11-15, it should be clear that the problem can be handled by complex number notation.

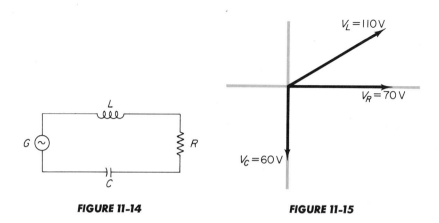

FIGURE 11-14   **FIGURE 11-15**

Most of the problems dealing with a.c. circuits are far more complex than this example indicates, and in those cases complex number notation is a distinct advantage. Certainly in this example the applied voltage could be found quite readily by the usual vector methods.

EXERCISE 11-5

In Problems 1–20, perform the indicated operations and express each answer in rectangular form.

1. $2(\cos 17° + j \sin 17°) \cdot 4(\cos 43° + j \sin 43°)$

2. $4(\cos 125° + j \sin 125°) \cdot 5(\cos 85° + j \sin 85°)$

3. $3(\cos 105° + j \sin 105°) \cdot 2(\cos 105° + j \sin 105°)$

4. $5(\cos 324° + j \sin 324°) \cdot 3(\cos 126° + j \sin 126°)$

5. $2(\cos 44° + j \sin 44°) \cdot 2(\cos 11° + j \sin 11°)$

6. $9(\cos 39° + j \sin 39°) \cdot 8(\cos 28° + j \sin 28°)$

7. $(\cos 12° + j \sin 12°) \cdot 3(\cos 17° + j \sin 17°) \cdot 7(\cos 61° + j \sin 61°)$

8. $3(\cos 20° + j \sin 20°) \cdot 4(\cos 167° + j \sin 167°) \cdot \frac{1}{6}(\cos 83° + j \sin 83°)$

9. $30(\cos 145° + j \sin 145°) \div 3(\cos 100° + j \sin 100°)$

10. $25(\cos 384° + j \sin 384°) \div 5(\cos 174° + j \sin 174°)$

11. $18(\cos 284° + j \sin 284°) \div 3(\cos 59° + j \sin 59°)$

12. $32(\cos 125° + j \sin 125°) \div 16(\cos 170° + j \sin 170°)$

13. $2(\cos 134° + j \sin 134°) \div 3(\cos 10° + j \sin 10°)$

14. $4(\cos 251° + j \sin 251°) \div 5(\cos 64° + j \sin 64°)$

15. $\dfrac{3(\cos 325° + j \sin 325°) \cdot 12(\cos 254° + j \sin 254°)}{18(\cos 9° + j \sin 9°)}$

16. $\dfrac{4(\cos 275° + j \sin 275°) \cdot 3(\cos 44° + j \sin 44°)}{6(\cos 19° + j \sin 19°)}$

17. $\dfrac{4(\cos 75° + j \sin 75°) \cdot 4(\cos 157° + j \sin 157°)}{3(\cos 50° + j \sin 50°) \cdot 2(\cos 2° + j \sin 2°)}$

18. $\dfrac{3(\cos 48° + j \sin 48°) \cdot 5(\cos 295° + j \sin 295°)}{4(\cos 200° + j \sin 200°) \cdot 3(\cos 23° + j \sin 23°)}$

19. $[3(\cos 15° + j \sin 15°)]^3$

20. $[2(\cos 75° + j \sin 75°)]^4$

Geometrically construct the sums and differences in Problems 21–24.

21. $(2 - 3j) + (-4 + 3j)$ 22. $(-8 - 4j) + (-6 - 2j)$

23. $(3 + 2j) - (2 + 4j)$ 24. $(3 + 5j) - (5 + 2j)$

Given the vectors **A**, **B**, *and* **C** *in Problems 25–28, find the resultant by using complex algebra.*

25. $A = 100, \theta = 330°$ 26. $A = 20, \theta = 225°$
   $B = 120, \theta = 45°$ $B = 30, \theta = 315°$
   $C = 50, \ \theta = 60°$ $C = 10, \theta = 60°$

27. $A = 120, \theta = 90°$ 28. $A = 200, \theta = 42°$
   $B = 49, \ \theta = 225°$ $B = 100, \theta = 85°$
   $C = 85, \ \theta = 315°$ $C = 50, \ \theta = 155°$

11-6
DEMOIVRE'S THEOREM

In evaluating an expression such as $(\sqrt{3} + j)^5$, it is obvious that expanding by repeated multiplication would be rather laborious. Fortunately, transformation to trigonometric form reduces the work considerably. Let us consider some various powers of the complex number $(\sqrt{3} + j)$ by using its trigonometric form $2(\cos 30° + j \sin 30°)$.

$$\sqrt{3} + j = 2(\cos 30° + j \sin 30°).$$

$$(\sqrt{3} + j)^2 = 2(\cos 30° + j \sin 30°) \cdot 2(\cos 30° + j \sin 30°)$$
$$= 2^2[\cos (30° + 30°) + j \sin (30° + 30°)]$$
$$= 2^2[\cos (2 \cdot 30°) + j \sin (2 \cdot 30°)].$$

$$(\sqrt{3} + j)^3 = (\sqrt{3} + j)^2(\sqrt{3} + j)$$
$$= 2^2[\cos (2 \cdot 30°) + j \sin (2 \cdot 30°)] \cdot 2[\cos 30° + j \sin 30°]$$
$$= 2^3[\cos (2 \cdot 30° + 30°) + j \sin (2 \cdot 30° + 30°)]$$
$$= 2^3[\cos (3 \cdot 30°) + j \sin (3 \cdot 30°)].$$

Similarly,

$$(\sqrt{3} + j)^4 = 2^4[\cos (4 \cdot 30°) + j \sin (4 \cdot 30°)].$$
$$(\sqrt{3} + j)^5 = 2^5[\cos (5 \cdot 30°) + j \sin (5 \cdot 30°)]$$
$$= 32(\cos 150° + j \sin 150°)$$
$$= 32\left(-\frac{\sqrt{3}}{2} + j \cdot \frac{1}{2}\right)$$
$$= -16\sqrt{3} + 16j.$$

Note that a procedure for raising a complex number to a power n has made itself clear. In effect, one merely raises its absolute value to the nth power and multiplies its amplitude by n. Thus $(\sqrt{3} + j)^{12}$ would be

$$2^{12}[\cos (12 \cdot 30°) + j \sin (12 \cdot 30°)] = 2^{12}[\cos (360°) + j \sin (360°)]$$
$$= 2^{12}[1 + j(0)]$$
$$= 2^{12}.$$

Since the choice of r and θ is immaterial to this procedure, we shall assume its application to **any** complex number $r(\cos \theta + j \sin \theta)$, and hence, we have the following important theorem.

DeMoivre's Theorem. $[r(\cos \theta + j \sin \theta)]^n = r^n(\cos n\theta + j \sin n\theta).$

The motivation for this theorem assumed n to be a positive integer, but it is true for any **integral** value of n.

EXAMPLE
20

DeMoivre's theorem:

a. $[2(\cos 10° + j \sin 10°)]^6 = 2^6[\cos (6 \cdot 10°) + j \sin (6 \cdot 10°)]$
$$= 64(\cos 60° + j \sin 60°)$$
$$= 64\left(\frac{1}{2} + j \frac{\sqrt{3}}{2}\right) = 32 + 32j\sqrt{3}.$$

b. $[3(\cos 30° + j \sin 30°)]^{-3} = 3^{-3}[\cos (-3 \cdot 30°) + j \sin (-3 \cdot 30°)]$
$$= \frac{1}{27}[\cos (-90°) + j \sin (-90°)]$$
$$= \frac{1}{27}[0 + j(-1)] = -\frac{1}{27}j.$$

EXAMPLE
21

DeMoivre's theorem:

a. $(1+j)^{20} = [\sqrt{2}\,(\cos 45° + j \sin 45°)]^{20}$

$\qquad = (\sqrt{2})^{20}[\cos (20 \cdot 45°) + j \sin (20 \cdot 45°)]$

$\qquad = 2^{10}(\cos 900° + j \sin 900°)$

$\qquad = 2^{10}(\cos 180° + j \sin 180°)$

$\qquad = -2^{10}.$

b. $\dfrac{(2+2j)^{15}}{(\sqrt{3}+j)^{21}} = \dfrac{[\sqrt{8}\,(\cos 45° + j \sin 45°)]^{15}}{[2\,(\cos 30° + j \sin 30°)]^{21}}$

$\qquad = \dfrac{(\sqrt{8})^{15}[\cos (15 \cdot 45°) + j \sin (15 \cdot 45°)]}{2^{21}[\cos (21 \cdot 30°) + j \sin (21 \cdot 30°)]}$

$\qquad = \dfrac{(2^{3/2})^{15}(\cos 675° + j \sin 675°)}{2^{21}(\cos 630° + j \sin 630°)}$

$\qquad = \dfrac{2^{45/2}}{2^{21}}[\cos(675° - 630°) + j \sin (675° - 630°)]$

$\qquad = 2^{3/2}(\cos 45° + j \sin 45°)$

$\qquad = 2\sqrt{2}\left(\dfrac{\sqrt{2}}{2} + j\dfrac{\sqrt{2}}{2}\right)$

$\qquad = 2 + 2j.$

If n is a positive integer, a number $w = r_1(\cos \varphi + j \sin \varphi)$ is said to be an nth root of the complex number $z = r(\cos \theta + j \sin \theta)$ if $w^n = z$. That is, w is an nth root of z if

$$w^n = [r_1(\cos \varphi + j \sin \varphi)]^n = r(\cos \theta + j \sin \theta),$$

or, by DeMoivre's theorem,

$$r_1^n(\cos n\varphi + j \sin n\varphi) = r(\cos \theta + j \sin \theta).$$

This will be true if $r_1^n = r$ and $n\varphi = \theta$; that is, for $r_1 = \sqrt[n]{r}$, the positive nth root of r, and $\varphi = \theta/n$. Hence, an nth root of z is given by

$$\sqrt[n]{r}\left[\cos \frac{\theta}{n} + j \sin \frac{\theta}{n}\right].$$

However, two numbers in trigonometric form represent the same number if, in addition to having the same absolute values, their amplitudes are coterminal angles. In other words, $n\varphi$ can differ from θ by any integral multiple of $360°$ (or 2π if θ is in radians). Symbolically, we can write

$$n\varphi = \theta + k \cdot 360°$$

$$\varphi = \frac{\theta}{n} + k \cdot \frac{360°}{n}, \qquad k = 0, 1, \ldots, n-1.$$

The reason that k takes only the n values from $k = 0$ to $k = n - 1$ is that for $k > n - 1$, repetition of the roots will occur.

In summary, there are exactly n distinct nth roots of a nonzero complex number $r(\cos\theta + j\sin\theta)$. These roots, having the same absolute value, are given by

$$\sqrt[n]{r}\left[\cos\left(\frac{\theta}{n} + k\cdot\frac{360°}{n}\right) + j\sin\left(\frac{\theta}{n} + k\cdot\frac{360°}{n}\right)\right],$$

$$k \in \{0, 1, \ldots, n-1\}.$$

EXAMPLE

22

Find the six sixth roots of -64.

We first express -64 in trigonometric form:

$$-64 = 64(\cos 180° + j\sin 180°).$$

If we denote the required roots by w_k, substituting in the above formula gives

$$w_k = \sqrt[6]{64}\left[\cos\left(\frac{180°}{6} + k\cdot\frac{360°}{6}\right) + j\sin\left(\frac{180°}{6} + k\cdot\frac{360°}{6}\right)\right]$$

$$= 2[\cos(30° + k\cdot60°) + j\sin(30° + k\cdot60°)], \qquad k \in \{0, 1, \ldots, 5\}.$$

Next we successively replace k by $0, 1, \ldots, 5$, resulting in the roots w_0, $w_1, \ldots, w_5$.

If $k = 0$,

$$w_0 = 2(\cos 30° + j\sin 30°) = 2\left(\frac{\sqrt{3}}{2} + j\cdot\frac{1}{2}\right) = \sqrt{3} + j.$$

If $k = 1$,

$$w_1 = 2(\cos 90° + j\sin 90°) = 2(0 + j) = 2j.$$

If $k = 2$,

$$w_2 = 2(\cos 150° + j\sin 150°) = 2\left(-\frac{\sqrt{3}}{2} + j\cdot\frac{1}{2}\right) = -\sqrt{3} + j.$$

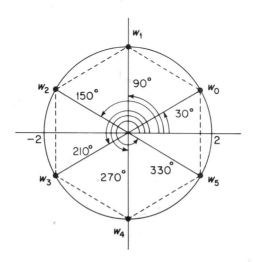

FIGURE 11-16

If $k = 3$,

$$w_3 = 2(\cos 210° + j \sin 210°) = 2\left(-\frac{\sqrt{3}}{2} + j \cdot -\frac{1}{2}\right) = -\sqrt{3} - j.$$

If $k = 4$,

$$w_4 = 2(\cos 270° + j \sin 270°) = 2(0 + j \cdot -1) = -2j.$$

If $k = 5$,

$$w_5 = 2(\cos 330° + j \sin 330°) = 2\left(\frac{\sqrt{3}}{2} + j \cdot -\frac{1}{2}\right) = \sqrt{3} - j.$$

You should verify that for $k > 5$, no new roots will be obtained. As shown in Fig. 11-16, the amplitudes of the roots are equally spaced and determine equal central angles of a circle of radius 2. The roots appear as vertices of a regular six-sided polygon in the complex plane.

EXAMPLE
23

Find the three cube roots of $-1 + j\sqrt{3}$.

$$-1 + j\sqrt{3} = 2(\cos 120° + j \sin 120°)$$

$$w_k = \sqrt[3]{2}\left[\cos\left(\frac{120°}{3} + k \cdot \frac{360°}{3}\right) + j \sin\left(\frac{120°}{3} + k \cdot \frac{360°}{3}\right)\right],$$

$$k \in \{0, 1, 2\}$$

$$= \sqrt[3]{2}\left[\cos(40° + k \cdot 120°) + j \sin(40° + k \cdot 120°)\right.$$

If

$$k = 0, \quad w_0 = \sqrt[3]{2}(\cos 40° + j \sin 40°)$$

$$k = 1, \quad w_1 = \sqrt[3]{2}(\cos 160° + j \sin 160°)$$

$$k = 2, \quad w_2 = \sqrt[3]{2}(\cos 280° + j \sin 280°).$$

See Fig. 11-17. Note that this problem could have been posed by saying, "solve the equation $x^3 = -1 + j\sqrt{3}$."

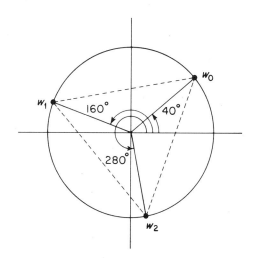

FIGURE 11-17

EXERCISE 11-6

Use DeMoivre's theorem to evaluate the expressions in Problems 1–20. Indicate your answer in both trigonometric and rectangular form.

1. $[2(\cos 10° + j \sin 10°)]^3$
2. $[3(\cos 25° + j \sin 25°)]^6$
3. $[\sqrt{2}(\cos 15° + j \sin 15°)]^8$
4. $[2(\cos 175° + j \sin 175°)]^4$
5. $(\cos 40° + j \sin 40°)^{-6}$
6. $[2(\cos 45° + j \sin 45°)]^{-3}$
7. $(\sqrt{2} + j\sqrt{2})^5$
8. $(-\sqrt{2} - j\sqrt{2})^4$
9. $(-\sqrt{3} + j)^4$
10. $\left(-\dfrac{\sqrt{3}}{2} - \dfrac{1}{2}j\right)^5$
11. $(\sqrt{3} + j)^4$
12. $\left(-\dfrac{1}{2} - j\dfrac{\sqrt{3}}{2}\right)^4$
13. $(-\sqrt{2} + j\sqrt{2})^6$
14. $(1 + j)^7$
15. $(1 - j)^{10}$
16. $(-\sqrt{2} + j\sqrt{2})^4$
17. $\dfrac{(\sqrt{2} + j\sqrt{2})^{10}}{(\sqrt{3} + j)^{10}}$
18. $\dfrac{(1 - j)^{15}}{(1 + j)^{12}}$
19. $(-2 + 2j)^{-8}$
20. $(1 - \sqrt{3}j)^{-4}$

Find the indicated roots in Problems 21–30. If you can do so without using tables, give the answer in rectangular form and indicate it geometrically.

21. Cube roots of 1
22. Fourth roots of 16
23. Square roots of $-8j$
24. Cube roots of -4
25. Square roots of j
26. Fourth roots of $\frac{1}{2}(1 - j\sqrt{3})$
27. Fourth roots of $2(-1 + j\sqrt{3})$
28. Fifth roots of $1 + j$
29. Sixth roots of $2 - 2j\sqrt{3}$
30. Sixth roots of $-1 - j$

In Problems 31–36 solve the given equations. Express your answers in rectangular form if convenient.

31. $x^3 = 1$
32. $x^4 = 1$
33. $x^5 - 1 = 0$
34. $x^6 - 1 = 0$
35. $x^3 = 1 + j$
36. $x^3 = -\sqrt{2} - j\sqrt{2}$

11-7
REVIEW

REVIEW QUESTIONS

The powers $j^2, j^3, j^4, j^5,$ and j^{-2} are equal, respectively, to _____(1)_____.

(1) **−1, −j, 1, j, −1**

The real part of $3 - 4j$ is _____(2)_____ and the imaginary part is
_____(3)_____ .

(2) **3**   (3) **−4**

Among the numbers $\sqrt[3]{-8}$, $\sqrt{-2}$, and $\sqrt{(-2)(-2)}$, which are pure imaginary?
_____(4)_____

(4) $\sqrt{-2}$

The number $a + bj$ is the conjugate of what number? _____(5)_____

(5) $a - bj$

The product of a complex number and its conjugate is (always)(sometimes) a real
number.   (6)

(6) **always**

In the representation of a complex number $a + bj$ in the complex plane, the
absolute value of $a + bj$ is the distance from $(0, 0)$ to the point _____(7)_____ .
This distance is given by the expression _____(8)_____ .

(7) **(a, b)**   (8) $\sqrt{a^2 + b^2}$

Representation of a complex number in polar form _____(is)(is not)_____ unique.
                                                        (9)

(9) **is not**

A trigonometric form for $z = 8$ is _____(10)_____ .

(10) **8(cos 0° + j sin 0°)**

In rectangular form, $3(\cos 270° + j \sin 270°)$ is equal to _____(11)_____ .

(11) **−3j**

The division $\dfrac{1}{1 - j}$ is performed by multiplying the numerator and denominator by _____(12)_____ , resulting in the answer _____(13)_____ .

(12) **1 + j**   (13) $\dfrac{1 + j}{2}$

$\sqrt{(-2)(-2)}$ equals _____(14)_____ but $\sqrt{-2}\sqrt{-2}$ equals _____(15)_____ .

(14) **2**   (15) **−2**

The nth roots of 1 lie on a circle of radius _____(16)_____ and determine the vertices of a regular polygon of _____(17)_____ sides.

(16) **1**   (17) **n**

The imaginary part of a complex number _____(is)(is not)_____ a real number.
                                              (18)

(18) **is**

If $1 + zj = 0$, then z equals _____(19)_____ .

(19) **j**

The square of a pure imaginary number _____(is)(is not)_____ a real number.
                                              (20

(20) **is**

The vector in the complex plane representing the number $-2 + j\sqrt{3}$ has its initial point at _____(21)_____ and terminates at the point _____(22)_____ .

(21) **(0, 0)**   (22) **$(-2, \sqrt{3})$**

If z is a sixth root of $-2 + j\sqrt{3}$, then z^6 equals _____(23)_____ .

(23) **$-2 + j\sqrt{3}$**

REVIEW PROBLEMS

In Problems 1–5, simplify the given expressions.

1. $(8 + j) - (-2 + \sqrt{-8}) + (\sqrt{5} + 2j) - 5j(j^2)$

2. $(2 + j)(3 - j) - (2 + j)(5j) + 6(j + 1)(j - 1)$

3. $\sqrt{-8}(6 + \sqrt{-5}) + 2j(-25 - \sqrt{-1}) + j^3(\sqrt{-9} + j\sqrt{-9})$

4. $\dfrac{(3 + j)(4 - j)}{(2 - j)(1 + j)}$

5. $\dfrac{(1 + j)^2(1 - 2j)^2}{(1 - j)^2(1 + 2j)^2}$

6. Solve for x and y if $(2x + y) + 7j = 8 - 6(x - 3y)j$

In Problems 7–10, transform to trigonometric form.

7. $4\sqrt{2} + 4j\sqrt{2}$

8. $\dfrac{3}{5} - \dfrac{3\sqrt{3}}{5}j$

9. $-\dfrac{\sqrt{3}}{2} + \dfrac{j}{2}$

10. $-27j$

In Problems 11–14, transform to rectangular form.

11. $\sqrt{3}(\cos 315° + j \sin 315°)$

12. $\sqrt{2}(\cos 135° + j \sin 135°)$

13. $\frac{1}{2}(\cos 210° + j \sin 210°)$

14. $\frac{2}{3}(\cos 60° + j \sin 60°)$

In Problems 15–17, evaluate the expressions.

15. $2(\cos 341° + j \sin 341°) \cdot 3(\cos 109° + j \sin 109°)$

16. $\dfrac{\sqrt{8}(\cos 284° + j \sin 284°)}{\sqrt{2}(\cos 59° + j \sin 59°)}$

17. $\dfrac{[2(\cos 22° + j \sin 22°)]^8 \cdot [3(\cos 125° + j \sin 125°)]^5}{[3(\cos 78° + j \sin 78°)]^4 \cdot [2(\cos 139° + j \sin 139°)]^2}$

In Problems 18–20, solve the given equation.

18. $x^8 - 1 = 0$ **19.** $x^2 - j\sqrt{3} + 1 = 0$

20. $x^6 = -\sqrt{2} - j\sqrt{2}$

quadratic equations

INTRODUCTION

When a ball is thrown vertically upward with an initial velocity of 56 ft/sec from a roof 176 ft above level ground, neglecting air resistance its vertical displacement s (in feet) *from the roof level* after t sec have elapsed is given by

$$s = 56t - 16t^2. \qquad (1)$$

Here, s is positive for displacement *above* roof level and negative for displacement *below* roof level. From physical considerations, we have $t \geq 0$. For example, if $t = 0$, then $s = 0$; if $t = 1$, then $s = 40$. Question: How long will it be before the ball strikes the ground?

We first note that when the ball strikes the ground it will be 176 ft below the roof level. At that time s will be -176 (remember that at roof level $s = 0$). Thus, setting $s = -176$ in Eq. (1),

$$-176 = 56t - 16t^2. \qquad (2)$$

Since the highest power of the variable which occurs is the second,

Eq. (2) is called a **second-degree equation**, or an **equation of degree 2**. It is most commonly referred to as a **quadratic equation**.

We saw in Chapter 5 that a linear equation has only one root. We now direct our attention to methods of solving quadratic equations, which, you will see, can have *two* distinct roots. A solution to Eq. (2) will be given in the next section.

12-2
SOLUTION OF QUADRATIC EQUATIONS BY FACTORING

Definition. A **quadratic equation** in the variable x is an equation which can be written in the form

$$ax^2 + bx + c = 0 \tag{12-1}$$

where a, b, and c are constants and $a \neq 0$.

A useful method of solving quadratic equations is based on factoring the expression $ax^2 + bx + c$. The method is a direct result of the following theorem.

Theorem. If the product of two (or more) factors is zero, then at least one of the factors must be zero. That is, if $ab = 0$, then $a = 0$, or $b = 0$, or both a and b are zero.

Proof: By assumption, $ab = 0$. If both a and b are zero, the theorem is true. If $a \neq 0$, we must show that $b = 0$. Since $1/a$ exists for $a \neq 0$, we can multiply both members of the equation $ab = 0$ by $1/a$:

$$\frac{1}{a}(ab) = \frac{1}{a}(0)$$

$$b = 0.$$

The proof for $b \neq 0$ is similar.

As mentioned above, the theorem can be extended to the case of more than two factors. From a practical standpoint, the theorem is applicable to the solution of $ax^2 + bx + c = 0$ when the left member is easily factorable. Once factored, it follows from the theorem that setting both factors equal to zero and solving each resulting equation for x will give the desired roots. The method is obviously applicable to equations of degree higher than 2, although difficulty in factoring often makes it prohibitive.

We shall now solve some equations by factoring. If necessary, you should review Sec. 3-2 on factoring.

EXAMPLE
1

Solve the quadratic equation $x^2 - 3x + 2 = 0$ by factoring.

The left member of the given equation factors easily:

$$x^2 - 3x + 2 = 0$$
$$(x - 2)(x - 1) = 0.$$

This can be interpreted as two quantities, $x - 2$ and $x - 1$, whose product is zero. By the theorem, at least one of the quantities *must* be zero. Thus, either

$$x - 2 = 0 \quad \text{or} \quad x - 1 = 0.$$
$$\text{If } x - 2 = 0, \quad \text{then } x = 2;$$
$$\text{If } x - 1 = 0, \quad \text{then } x = 1.$$

The solution set is $\{1, 2\}$ and you should verify by substitution that both elements satisfy the given equation. We point out that the equation $x^2 - 3x + 2 = 0$ may be solved graphically if you carefully sketch the equation $y = x^2 - 3x + 2$ and observe those values of x for which the curve meets the x-axis (that is, where $y = 0$). A sketch is given in Fig. 12-1.

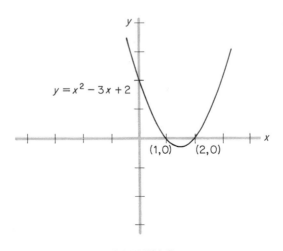

FIGURE 12-1

EXAMPLE
2

In the problem originally posed in the previous section, we obtained the equation

$$-176 = 56t - 16t^2,$$

which is equivalent to

$$16t^2 - 56t - 176 = 0.$$

Dividing both members by 8 and factoring,

$$2t^2 - 7t - 22 = 0$$
$$(2t - 11)(t + 2) = 0$$

If $2t - 11 = 0$, then $t = \frac{11}{2}$;

If $t + 2 = 0$, then $t = -2$.

Although both $t = \frac{11}{2}$ and $t = -2$ are roots of the given equation, a negative value of t has no physical significance here. Thus we choose $t = \frac{11}{2}$ sec as the answer to our problem. That is, the ball strikes the ground after $\frac{11}{2}$ sec.

EXAMPLE
3

a. Solve the equation $2x^2 + 4x = 0$ by factoring.

$$2x^2 + 4x = 0.$$

Factoring the left member,

$$2x(x + 2) = 0.$$

Thus,

$$2x = 0 \quad \text{or} \quad x + 2 = 0.$$

Equivalently,

$$x = 0 \quad \text{or} \quad x = -2 \quad \text{(solution set } \{0, -2\}\text{)}.$$

A common error would be to divide both members of $2x^2 + 4x = 0$ by $2x$, yielding $x + 2 = 0$ from which $x = -2$. Thus, a root is "lost," as previously discussed in Sec. 5-2.

b. Solve the equation $x^2 + 6x + 9 = 0$ by factoring.

$$x^2 + 6x + 9 = 0$$
$$(x + 3)(x + 3) = 0.$$

Thus, $x + 3 = 0$ or $x + 3 = 0$; that is, $x = -3$ or $x = -3$. The solution set is $\{-3, -3\} = \{-3\}$. Since each factor gave rise to the same root, -3, the root is called a **double root**, or a **root of multiplicity 2**. The graph of $y = x^2 + 6x + 9$ will touch, but not cross, the x-axis at $(-3, 0)$, as shown in Fig. 12-2.

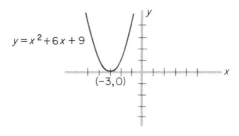

FIGURE 12-2

c. Solve the *cubic* equation $x^3 - 4x^2 - 5x = 0$ by factoring.

Although this equation is not quadratic (it is of the *third-degree*), the method involved is fundamental.

$$x^3 - 4x^2 - 5x = 0.$$

Factoring the left member,

$$x(x^2 - 4x - 5) = 0$$
$$x(x + 1)(x - 5) = 0.$$

Thus,

$$x = 0, \text{ or } x + 1 = 0, \text{ or } x - 5 = 0.$$

Hence, there are *three* roots:

$$x = 0 \text{ or } x = -1 \text{ or } x = 5$$

and the solution set is $\{0, -1, 5\}$

EXAMPLE
4

Solve the equation $(3x - 4)(x + 1) = -2$.

You should approach a problem such as this with caution. If the product of two quantities is equal to -2, it does *not* follow that at least one of the quantities must be -2. Why? To solve the equation, we first multiply the factors in the left member:

$$(3x - 4)(x + 1) = -2$$
$$3x^2 - x - 4 = -2.$$

Adding 2 to both members,

$$3x^2 - x - 2 = 0.$$

Factoring,

$$(3x + 2)(x - 1) = 0.$$

Thus,

$$3x + 2 = 0 \quad \text{or} \quad x - 1 = 0.$$

Equivalently,

$$x = -\tfrac{2}{3} \quad \text{or} \quad x = 1 \quad (\text{solution set } \{-\tfrac{2}{3}, 1\}).$$

From the examples presented thus far and the fact that a second-degree polynomial can be expressed as a product of two linear polynomials, it is reasonable to infer that **every quadratic equation has two roots, although they may not necessarily be distinct.**

EXAMPLE
5

Solve $w^2 = k$ by factoring.

$$w^2 = k$$
$$w^2 - k = 0$$
$$(w - \sqrt{k})(w + \sqrt{k}) = 0.$$

Thus,

$$w - \sqrt{k} = 0 \qquad \text{or} \qquad w + \sqrt{k} = 0$$
$$w = \sqrt{k} \qquad \text{or} \qquad w = -\sqrt{k}.$$

The solution set is $\{\sqrt{k}, -\sqrt{k}\} = \{\pm\sqrt{k}\}$. If k is negative, the roots are imaginary, being complex conjugates of each other. Otherwise, the roots are real. For example, the solution set of $x^2 = 4$ is $\{\pm\sqrt{4}\} = \{\pm 2\}$, but $x^2 = -4$ has solution set $\{\pm\sqrt{-4}\} = \{\pm 2j\}$. For the latter case, the graph of $y = x^2 + 4$ will not intersect the x-axis, indicating the absence of real roots (see Fig. 12-3).

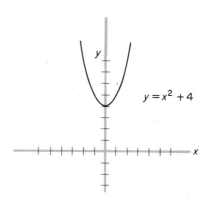

$y = x^2 + 4$

FIGURE 12-3

EXERCISE 12-2

In Problems 1–36, solve the given equation by factoring.

1. $x^2 + 3x + 2 = 0$ **2.** $x^2 + 5x + 6 = 0$

3. $x^2 + 9x + 14 = 0$ **4.** $x^2 + 8x + 15 = 0$

5. $y^2 - 7y + 12 = 0$ **6.** $t^2 - 4t + 4 = 0$

7. $x^2 - 2x - 3 = 0$ **8.** $x^2 + x - 12 = 0$

9. $x^2 - 12x + 36 = 0$ **10.** $x^2 - 1 = 0$

11. $x^2 - 4 = 0$ **12.** $x^2 - 16 = 0$

13. $x^2 - 8x = 0$ **14.** $2x^2 + 4x = 0$

15. $x^2 + 16 = 0$ **16.** $x^2 + 25 = 0$

17. $6x^2 + 12x + 6 = 0$ **18.** $3x^2 - 12x + 12 = 0$

19. $6x^2 + 7x - 3 = 0$ **20.** $10x^2 - x = 3$

21. $2z^2 + 7z = 4$ **22.** $-r^2 - r + 12 = 0$

23. $-t^2 + 3t + 10 = 0$ **24.** $x^2 - 2a + a^2 = 0$

25. $4x^2 + 1 = 4x$ 26. $8 + 2z - 3z^2 = 0$

27. $y(y + 4) = 5$ 28. $\frac{1}{7}y^2 = \frac{3}{7}y$

29. $x^2 - 5ax + 6a^2 = 0$ 30. $(x + 1)^2 - 5x + 1 = 0$

31. $(2x + 3)^2 - 3(x^2 - 9) = 0$

32. $(x + 3)^2 + 2(x - 2)^2 = 2(x^2 + 16)$

33. $x(x^2 - 64) = 0$ 34. $x^3 - 4x^2 - 5x = 0$

35. $(x + 3)(x + 1)(x - 2) = 0$ 36. $3(x^2 + 2x - 8)(x - 5) = 0$

In Problems 37–41, solve the given equation for the indicated letter.

37. $mgh + \frac{1}{2}mv^2 = c;\ v$ 38. $s = \frac{1}{2}at^2;\ t$

39. $T^2 = 4\pi^2\left(\dfrac{L}{g}\right);\ T$ 40. $P = \dfrac{E^2}{R + r} - \dfrac{E^2 r}{(R + r)^2};\ E$

41. $mgh = \frac{1}{2}mv^2 + \frac{1}{2}I\omega^2;\ \omega$

42. A rectangular field has an area of 2700 ft², and its length is 15 ft greater than its width. What are its length and width? Your work will be easier if you first convert units to yards and square yards.

43. The formula

$$S = 2\pi r^2 + 2\pi rh$$

gives the total surface area S of a cylinder of radius r and height h. Find r in order that a cylinder of height 2 in. will have an area of 48π in.²

44. The differences between the hypotenuse and the other two sides of a triangle are, respectively, 8 and 4 ft. Find the other two sides.

45. An open box is made from a square piece of tin by cutting out a 6-in. square from each corner and turning up the sides; the box contains 150 in.³ Find the area of the original square.

46. The length, width, and height of a bin are in the ratio of 3: 2: 1. Another bin has the same volume as the first, but it is 2 ft longer, 2 ft wider, and half as high as the first bin. Find the dimensions of the *second* bin.

12-3
COMPLETING THE SQUARE—THE QUADRATIC FORMULA

Solving quadratic equations by factoring can be quite tedious, as is evident by considering the equation $.3x^2 + x\sqrt{2} - \sqrt{6} = 0$. However, there is an approach that culminates in a formula for solving every quadratic equation.

In Example 5 of the previous section, the solution set of the equation $w^2 = k$ was shown to be $\{\pm\sqrt{k}\}$. We call $w^2 = k$ a **pure quadratic**

equation since it does not contain a first–degree term. We shall now show that every quadratic equation, whether easily factorable or not, can be transformed into the pure form by the method of *completing the square* and, as a result, easily solved.

In general, in the square of the binomial $x + k$, namely

$$(x + k)^2 = x^2 + 2kx + k^2,$$

the constant term k^2 is equal to the square of one-half the coefficient of the x-term, that is, the square of one-half the value of $2k$. Thus, the sum of an x^2-term and an x-term will become a perfect-square trinomial by **adding to it the square of one-half the coefficient of x.** Now *this is true only if the coefficient of the x^2-term is* 1.

For example, consider the equation

$$x^2 - 6x + 3 = 0.$$

This is not a pure quadratic equation due to the linear term $-6x$; the equation also cannot be easily factored. You should carefully study the following steps in the solution of the equation. We first write the equation in an equivalent form so that the left member consists only of an x^2-term and an x-term:

$$x^2 - 6x = -3.$$

If we then add to both members the square of one-half the coefficient of the x-term, namely $[\frac{1}{2}(-6)]^2$ or 9, the left member becomes a trinomial that is the square of a binomial:

$$x^2 - 6x + 9 = -3 + 9$$
$$(x - 3)^2 = 6.$$

If we think of $x - 3$ as a single quantity, say w, the equation is now in the pure form $w^2 = k$, which, by the discussion above, gives $w = \pm\sqrt{k}$. Hence,

$$x - 3 = \pm\sqrt{6}$$
$$x = 3 \pm \sqrt{6}.$$

Therefore the roots are $3 + \sqrt{6}$ and $3 - \sqrt{6}$, and the solution set is $\{3 \pm \sqrt{6}\}$.

The technique we have used to solve the quadratic equation is called **completing the square.**

EXAMPLE
6

Solve the equation $2x^2 + 3x - 4 = 0$ by completing the square.

$$2x^2 + 3x - 4 = 0$$
$$2x^2 + 3x = 4.$$

Dividing by 2 so that the coefficient of the x^2-term is 1,

$$x^2 + \frac{3}{2}x = 2.$$

Adding $[\frac{1}{2}(\frac{3}{2})]^2$ or $\frac{9}{16}$ to both members,

$$x^2 + \frac{3}{2}x + \frac{9}{16} = 2 + \frac{9}{16}$$

$$\left(x + \frac{3}{4}\right)^2 = \frac{41}{16}.$$

The equation is now in the pure form. Thus,

$$x + \frac{3}{4} = \pm\sqrt{\frac{41}{16}} = \pm\frac{\sqrt{41}}{4}$$

$$x = -\frac{3}{4} \pm \frac{\sqrt{41}}{4}$$

$$= \frac{-3 \pm \sqrt{41}}{4}.$$

The solution set is $\left\{\dfrac{-3 \pm \sqrt{41}}{4}\right\}$ and the roots are $\dfrac{-3 + \sqrt{41}}{4}$ and $\dfrac{-3 - \sqrt{41}}{4}$. Although these roots can be put in decimal form (a common practice in technology), in this text the simplest radical form is acceptable.

Applying the method of completing the square to the general quadratic equation $ax^2 + bx + c = 0$ culminates in a formula for solving every quadratic equation. Let

$$ax^2 + bx + c = 0, \qquad a \neq 0.$$

Then

$$ax^2 + bx = -c.$$

As before, we can divide both members by a since $a \neq 0$:

$$x^2 + \frac{b}{a}x = -\frac{c}{a}.$$

Since the coefficient of the x-term is $\dfrac{b}{a}$, we add the square of $\dfrac{1}{2}\left(\dfrac{b}{a}\right)$ to both members:

$$x^2 + \frac{b}{a}x + \frac{b^2}{4a^2} = -\frac{c}{a} + \frac{b^2}{4a^2}.$$

Factoring the left member and simplifying the right member,

$$\left(x + \frac{b}{2a}\right)^2 = \frac{b^2 - 4ac}{4a^2}.$$

This equation is of the pure form $w^2 = k$, where $w = x + \dfrac{b}{2a}$ and $k = \dfrac{b^2 - 4ac}{4a^2}$. Hence,

$$x + \frac{b}{2a} = \pm\sqrt{\frac{b^2 - 4ac}{4a^2}} = \pm\frac{\sqrt{b^2 - 4ac}}{2a}$$

$$x = -\frac{b}{2a} \pm \frac{\sqrt{b^2 - 4ac}}{2a}.$$

Thus, if $ax^2 + bx + c = 0$ and $a \neq 0$, then

$$x = \frac{-b \pm \sqrt{b^2 - 4ac}}{2a}. \tag{12-2}$$

The two values $\dfrac{-b + \sqrt{b^2 - 4ac}}{2a}$ and $\dfrac{-b - \sqrt{b^2 - 4ac}}{2a}$ can indeed be shown to satisfy $ax^2 + bx + c = 0$. Equation (12-2) is called the **quadratic formula** and, along with the hypothesis which precedes it, should be committed to memory. Be certain to use the quadratic formula correctly. It is *incorrect* to write $x = -b \pm \dfrac{\sqrt{b^2 - 4ac}}{2a}$.

If a, b, and c are real numbers, the expression $b^2 - 4ac$ under the radical sign in Eq. (12-2) is called the **discriminant** of the quadratic equation $ax^2 + bx + c = 0$. It is quite useful in determining the nature of the roots of the equation:

1. If $b^2 - 4ac > 0$, the two roots are *real* and *distinct*.
2. If $b^2 - 4ac = 0$, the two roots are *real* and *equal*—that is, a double root (or root of multiplicity 2) occurs.
3. If $b^2 - 4ac < 0$, then $\sqrt{b^2 - 4ac}$ is not a real number and the roots are *distinct imaginary numbers*—in fact, the roots are imaginary complex conjugates of each other.

Let us illustrate each of the three cases.

EXAMPLE
7

a. Solve $4x^2 - 17x + 15 = 0$ by the quadratic formula.

Here $a = 4$, $b = -17$, and $c = 15$. The discriminant is

$$b^2 - 4ac = (-17)^2 - 4(4)(15) = 49,$$

which is positive. Thus, there are two real, distinct roots of the equation:

$$x = \frac{-b \pm \sqrt{b^2 - 4ac}}{2a} = \frac{-(-17) \pm \sqrt{49}}{2(4)}$$

$$= \frac{17 \pm 7}{8}.$$

Therefore, $x = \dfrac{17+7}{8} = 3$ or $x = \dfrac{17-7}{8} = \dfrac{5}{4}$ and the solution set

is $\left\{3, \dfrac{5}{4}\right\}$. The graph of $y = 4x^2 - 17x + 15$ will cross the x-axis at

$(3, 0)$ and $\left(\dfrac{5}{4}, 0\right)$.

b. Solve $9t^2 + 12t + 4 = 0$ by the quadratic formula.

Here $a = 9$, $b = 12$, and $c = 4$. Since $b^2 - 4ac = 0$, the roots are real and equal:

$$t = \frac{-b \pm \sqrt{b^2 - 4ac}}{2a} = \frac{-12 \pm \sqrt{0}}{2(9)}.$$

Thus, $t = \dfrac{-12+0}{18} = -\dfrac{2}{3}$ or $t = \dfrac{-12-0}{18} = -\dfrac{2}{3}$. The solution set

is $\left\{-\dfrac{2}{3}, -\dfrac{2}{3}\right\} = \left\{-\dfrac{2}{3}\right\}$ and therefore $-\dfrac{2}{3}$ is a double root. The graph

of $y = f(t) = 9t^2 + 12t + 4$ will touch, but not cross, the t-axis at

$\left(-\dfrac{2}{3}, 0\right)$.

c. Solve $-x^2 = x + 1$ by the quadratic formula.

The equation is equivalent to $x^2 + x + 1 = 0$. Thus, $a = 1$, $b = 1$, and $c = 1$, and the discriminant is $b^2 - 4ac = 1^2 - 4(1)(1) = -3$. Thus, the roots are distinct, imaginary numbers:

$$x = \frac{-b \pm \sqrt{b^2 - 4ac}}{2a} = \frac{-1 \pm \sqrt{-3}}{2} = \frac{-1 \pm j\sqrt{3}}{2}.$$

The solution set is $\left\{\dfrac{-1+j\sqrt{3}}{2}, \dfrac{-1-j\sqrt{3}}{2}\right\}$. Note that the roots are

complex conjugates of each other. The graph of $y = f(x) = x^2 + x + 1$ will have no point in common with the x-axis, which indicates the absence of real roots. It is left as an exercise for you to sketch the graph.

EXERCISE 12-3

In Problems 1–6, solve the given equation by completing the square.

1. $x^2 + 4x - 3 = 0$
2. $x^2 - 6x + 2 = 0$
3. $x^2 + x - 1 = 0$
4. $x^2 - x + 1 = 0$
5. $2x^2 + 4x - 5 = 0$
6. $3x^2 - 9x + 4 = 0$

In Problems 7–40, solve the given equation by the quadratic formula. In Problems 7–16, first determine the nature of the roots by examination of the discriminant.

7. $x^2 + 2x - 15 = 0$
8. $x^2 - 2x - 24 = 0$
9. $x^2 - 8x + 16 = 0$
10. $x^2 + 18x + 81 = 0$

11. $x^2 - 4x + 5 = 0$

12. $x^2 - 2x + 2 = 0$

13. $x^2 - 5x + 3 = 0$

14. $2x^2 + 3x - 4 = 0$

15. $x^2 - 2x + 4 = 0$

16. $1 - 6x - 7x^2 = 0$

17. $6x^2 + 7x - 5 = 0$

18. $8x^2 - 18x + 9 = 0$

19. $2x^2 - 3x - 20 = 0$

20. $12x^2 + 4x - 5 = 0$

21. $x^2 + 4 = 0$

22. $x^2 - 3x + 1 = 0$

23. $x^2 + 2x = 0$

24. $x^2 - 12x = 0$

25. $x^2 - 8x + 5 = 0$

26. $2x^2 - 4x + 5 = 0$

27. $6r^2 + 8r + 3 = 0$

28. $3z^2 + 6z - 2 = 0$

29. $16x^2 + 4x = 5x^2$

30. $2x^2 + x - 5 = 0$

31. $-3x^2 + 5x = 4$

32. $-2x^2 - 6x = -5$

33. $\frac{2}{3}x^2 - \frac{1}{9}x + 2 = 0$

34. $\frac{3}{5}x^2 + \frac{1}{15}x - 3 = 0$

35. $.01x^2 + .2x - .6 = 0$

36. $.04x^2 - .1x - .09 = 0$

37. $2x^2 - \sqrt{3}\,x + 4 = 0$

38. $\sqrt{8}\,x^2 + \sqrt{5}\,x - \sqrt{2} = 0$

39. $x^2 = \dfrac{x + 5}{6}$

40. $\dfrac{x - 5}{7} = 2x^2$

In Problems 41 and 42, solve for x.

41. $x^2 - jx + 6 = 0$

42. $6x^3 + 8x^2 - 10x = 0$

43. If $3x^2 - 5xy - 4y^2 = 0$, solve for x by treating y as a constant.

44. If $2x^2 + 6xy - 9y^2 = 0$, solve for y by treating x as a constant.

45. Given the equation of motion

$$s = v_0 t + \tfrac{1}{2}at^2,$$

find t (in seconds) to two decimal places if $s = 15$ m, $v_0 = 18$ m/sec, and $a = -9.8$ m/sec².

46. When a projectile is fired so that its initial speed in the vertical direction is 152 ft/sec, it can be shown that the projectile will reach a height s above its starting level after t sec, where

$$s = 152t - 16t^2.$$

At what times will such a projectile reach a height of 280 ft above its starting level?

47. A chemist drew a quantity of acid from a full beaker containing 81 cc and then filled up the beaker with pure water. He then drew from the mixture the same amount that he drew the first time and found that there remained in the beaker 64 cc of pure acid. How many cubic centimeters did he draw each time?

48. A lot is 120 ft long and 80 ft wide. It is decided to double the area of the lot, still keeping it rectangular, by adding strips of equal width to one end and one side. Find the width of the strip.

49. In studying the dynamic behavior of galvanometers, one encounters the equation

$$Jm^2 + Dm + S = 0,$$

where J, D, and S are physical quantities that determine the characteristics of the motion of the galvanometer coil. Solve the equation for m.

50. In Problem 49, a galvanometer is said to be *overdamped* when $\dfrac{D^2}{4J^2} > \dfrac{S}{J}$, *underdamped* when $\dfrac{D^2}{4J^2} < \dfrac{S}{J}$, and *critically damped* when $\dfrac{D^2}{4J^2} = \dfrac{S}{J}$. For each case, what is the nature of the roots of the equation in Problem 49?

12-4
MAXIMUM AND MINIMUM OF QUADRATIC FUNCTIONS

You may have noticed that the curves in Figs. 12-1 to 12-3 have the same basic shape. Such curves, which are called *parabolas,* are discussed more fully in Chapter 18. Actually, the graph of any function of the form

$$y = f(x) = ax^2 + bx + c, \qquad a, b, c \in \mathbf{R}, a \neq 0,$$

called a **quadratic function**, is a parabola. If a is positive, the parabola opens upward; if a is negative, the parabola opens downward. For example, the graphs of $y = \frac{1}{2}x^2 - 2x + 3$ and $s = 56t - 16t^2$ in Fig. 12-4(a) and (b) illustrate this.

A parabola has a characteristic low point or high point, depending on

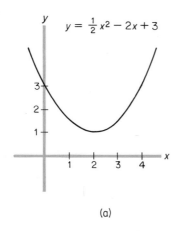

(a)

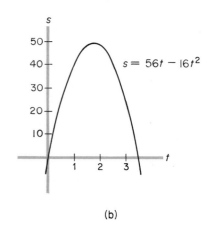

(b)

FIGURE 12-4

whether the curve opens upward or downward. Such a point corresponds to the minimum or maximum value of the function that the parabola represents. It will soon be shown that if $y = f(x) = ax^2 + bx + c$ is a quadratic function, then the abscissa of the low point (if $a > 0$) or high point (if $a < 0$) of the graph of f is $-\dfrac{b}{2a}$. That is, a quadratic function $y = f(x)$ achieves its minimum or maximum value when x is $-\dfrac{b}{2a}$. Thus, for $f(x) = \frac{1}{2}x^2 - 2x + 3$, the low point occurs when $x = -\dfrac{-2}{2(\frac{1}{2})} = 2$. At this point, $y = \frac{1}{2}(2)^2 - 2(2) + 3 = 1$ [see Fig. 12-4(a)].

Similarly, for the original problem in Sec. 12-1 concerning the ball, the maximum value of $f(t) = 56t - 16t^2$ occurs when $t = -\dfrac{56}{2(-16)} = \dfrac{7}{4}$. Thus, the maximum vertical displacement from the roof level that the ball attains occurs when $\frac{7}{4}$ sec have elapsed. This displacement is $56(\frac{7}{4}) - 16(\frac{7}{4})^2 = 49$ ft. See Fig. 12-4(b).

We conclude this section by showing that the quadratic function $f(x) = ax^2 + bx + c$ assumes its maximum or minimum value when $x = -\dfrac{b}{2a}$. Our technique parallels that used in the derivation of the quadratic formula. Recall that there we began by dividing the given quadratic equation by a. Here, we merely factor out a from each term. We then complete the square.

$$f(x) = ax^2 + bx + c$$
$$= a\left(x^2 + \frac{b}{a}x + \frac{c}{a}\right)$$
$$= a\left[\left(x^2 + \frac{b}{a}x\right) + \frac{c}{a}\right]$$
$$= a\left[\left(x^2 + \frac{b}{a}x + \frac{b^2}{4a^2}\right) + \frac{c}{a} - \frac{b^2}{4a^2}\right]$$
$$f(x) = a\left[\left(x + \frac{b}{2a}\right)^2 + \frac{4ac - b^2}{4a^2}\right]. \qquad (1)$$

Since the square of every real number is nonnegative, it follows that $\left(x + \dfrac{b}{2a}\right)^2 \geq 0$. This means that in Eq. 1 the factor

$$\left(x + \frac{b}{2a}\right)^2 + \frac{4ac - b^2}{4a^2}$$

will have its least value when $x + \dfrac{b}{2a} = 0$. This occurs when $x = -\dfrac{b}{2a}$.

Thus, if in Eq. (1) we have $a > 0$, then $f(x)$ takes on its least value when $x = -\dfrac{b}{2a}$. Similarly, if in Eq. (1) we have $a < 0$, then $f(x)$ will have its greatest value when $x = -\dfrac{b}{2a}$.

EXAMPLE
8

Find the dimensions of the rectangular region with perimeter 20 ft that has the largest area.

We first point out that both rectangles in Fig. 12-5 have a perimeter of 20 ft, but in (a) the area is 16 ft² and in (b) it is 24 ft². We want to obtain the

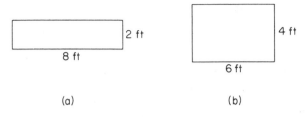

(a)   (b)

FIGURE 12-5

largest possible area. Suppose we denote one side of the rectangle by x. Then the opposite side also has length x. This leaves a total of $20 - 2x$ ft to be divided equally between the other two sides. Hence, each of these sides has length $(20 - 2x)/2$ or $10 - x$. An appropriate rectangle is shown in Fig. 12-6. You should verify that the perimeter is 20 ft. Now, let A denote the area of the rectangle. Then

$$A = x(10 - x)$$
$$A = 10x - x^2.$$

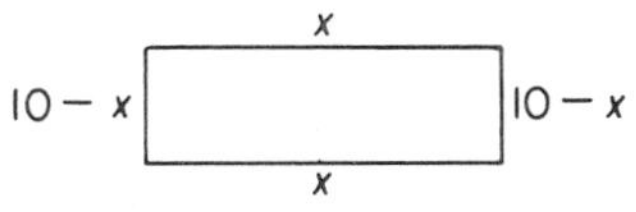

$$
\begin{array}{c}
x \\
10 - x \;\boxed{}\; 10 - x \\
x
\end{array}
$$

FIGURE 12-6

Thus the area is expressible as a quadratic function where $a = -1$ and $b = 10$. Thus A will have a maximum value when

$$x = -\frac{b}{2a} = -\frac{10}{2(-1)} = 5.$$

Thus, to obtain maximum area the dimensions should be 5 ft by $10 - 5$

or 5 ft. In short, the rectangle should be a square. In this case, the area is 25 ft².

EXERCISE 12-4

In Problems 1–6, find the value of x for which the given quadratic function has a maximum or minimum value. Specify as to whether it corresponds to a maximum or minimum. Then find the corresponding functional value.

1. $y = f(x) = 3x^2 - 6x$ **2.** $y = f(x) = -2x^2 - 16x + 3$

3. $y = f(x) = -4x^2 + 2x - 4$ **4.** $y = x^2$

5. $y = f(x) = 12 - x + x^2$ **6.** $y = x(x + 3) - 12$

7. For the rectangle shown below, express its area as a quadratic function of x. For what value of x will the area be a maximum?

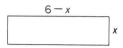

8. An object is thrown vertically upward from the ground with an initial velocity of 160 ft/sec. Its displacement s from its starting point after t sec have elapsed is given by

$$s = f(t) = -16t^2 + 160t.$$

After how many seconds will the ball reach its maximum height? What will this maximum height be? Sketch the graph of f.

9. A man wishes to fence in a rectangular plot adjacent to a building by using the building as one side of the enclosed area. If he has 2000 ft of fence, what should be the dimensions of the enclosed plot if the area is to be a maximum?

12-5
EQUATIONS LEADING TO QUADRATIC EQUATIONS

In some cases, an equation which is not of the quadratic type can lead to a quadratic equation by algebraic operations. We shall consider three such types: fractional equations, radical equations, and equations in *quadratic form*.

A. FRACTIONAL EQUATIONS

Recall that to solve a fractional equation, we first clear the equation of fractions by multiplying both members by the L.C.D. of the fractions involved. Since this does not guarantee that the resulting equation is equivalent to the original equation, it is important that any "roots" obtained in this manner be verified by substituting them in the given equation.

EXAMPLE
9

Solve

$$\frac{y+1}{y+3} + \frac{y+5}{y-2} = \frac{7(2y+1)}{y^2+y-6}. \tag{1}$$

First we multiply both members of Eq. (1) by the L.C.D., $(y+3)(y-2)$, of the three fractions involved:

$$(y+1)(y-2) + (y+5)(y+3) = 7(2y+1). \tag{2}$$

Due to this step, we must remember that Eq. (2) is not necessarily equivalent to Eq. (1). By simplifying Eq. (2) we have

$$y^2 - y - 2 + y^2 + 8y + 15 = 14y + 7$$
$$2y^2 - 7y + 6 = 0$$
$$(2y - 3)(y - 2) = 0$$
$$2y - 3 = 0 \quad \text{or} \quad y - 2 = 0$$
$$y = \tfrac{3}{2} \quad \text{or} \quad y = 2.$$

But y cannot equal 2 in the *original* equation, since this would result in division by zero. However, you should verify that $y = \tfrac{3}{2}$ does indeed satisfy Eq. (1). Hence, the solution set is $\{\tfrac{3}{2}\}$. In conclusion, although Eq. (1) leads to a quadratic equation, Eq. (2), the two equations are not equivalent.

B. RADICAL EQUATIONS

Recall that when solving a radical equation, we raise each member to the same power so as to eliminate the radical. Again, you must verify that any "roots" obtained satisfy the given equation.

EXAMPLE
10

Solve $\sqrt{x+2} - x + 4 = 0$.

Rewriting so as to avoid a radical after squaring,

$$\sqrt{x+2} = x - 4.$$

Squaring both members,

$$x + 2 = x^2 - 8x + 16.$$

Thus,

$$x^2 - 9x + 14 = 0$$
$$(x - 7)(x - 2) = 0$$
$$x = 7 \quad \text{or} \quad x = 2.$$

Checking, for $x = 7$ the left member of the *original* equation is $\sqrt{7+2} - 7 + 4$, which is 0, as is the right member. If $x = 2$, the left member is

$\sqrt{2+2} - 2 + 4$, which is 4, not agreeing with the right member. We conclude that the solution set is $\{7\}$.

EXAMPLE
11

Solve $\sqrt{x+6} - \sqrt{2x+5} = -1$.

It is best to rewrite the equation so that only one radical expression will appear in the left member:

$$\sqrt{x+6} = \sqrt{2x+5} - 1.$$

Squaring both members,

$$x + 6 = 2x + 5 - 2\sqrt{2x+5} + 1.$$

Simplifying,

$$2\sqrt{2x+5} = x.$$

Squaring both members again,

$$4(2x+5) = x^2$$
$$8x + 20 = x^2.$$

Thus,

$$x^2 - 8x - 20 = 0$$
$$(x - 10)(x + 2) = 0$$
$$x = 10, \qquad x = -2.$$

Substitution will show that the original equation will be satisfied only for $x = 10$. Thus, the solution set is $\{10\}$.

EXAMPLE
12

Solve $\sqrt[3]{x^3 + 19} = x + 1$.

Raising both members to the third power,

$$x^3 + 19 = x^3 + 3x^2 + 3x + 1$$
$$3x^2 + 3x - 18 = 0$$
$$x^2 + x - 6 = 0$$
$$(x + 3)(x - 2) = 0$$
$$x = -3, \qquad x = 2.$$

You may verify that both solutions satisfy the original equation. The solution set is $\{-3, 2\}$.

C. EQUATIONS IN QUADRATIC FORM

In some cases, an equation which is not quadratic can, by an appropriate substitution, be transformed into a quadratic equation. A few examples will illustrate how to solve such equations, which are said to be in **quadratic form**. In each case, a change of variable will be made to permit a solution.

EXAMPLE
13

Solve $\dfrac{1}{x^4} - \dfrac{8}{x^2} + 7 = 0$.

Rewriting the given equation,

$$\left(\frac{1}{x^2}\right)^2 - 8\left(\frac{1}{x^2}\right) + 7 = 0.$$

If we let $w = \dfrac{1}{x^2}$, then

$$w^2 - 8w + 7 = 0$$
$$(w - 7)(w - 1) = 0$$
$$w = 7, \qquad w = 1.$$

Returning to the variable x, this means that

$$\frac{1}{x^2} = 7 \qquad \text{or} \qquad \frac{1}{x^2} = 1.$$

Thus,

$$x^2 = \frac{1}{7} \qquad \text{or} \qquad x^2 = 1$$

$$x = \pm\frac{\sqrt{7}}{7} \qquad \text{or} \qquad x = \pm 1.$$

After checking we conclude that the solution set is

$$\left\{\pm\frac{\sqrt{7}}{7}, \pm 1\right\}.$$

EXAMPLE
14

Solve $x^6 + 9x^3 + 8 = 0$.

Rewriting the given equation,

$$(x^3)^2 + 9(x^3) + 8 = 0.$$

Suppose we replace x^3 by the variable w:

$$w^2 + 9w + 8 = 0$$
$$(w + 8)(w + 1) = 0$$
$$w = -8 \qquad \text{or} \qquad w = -1.$$

Returning to the variable x, we replace w by x^3. Thus, $x^3 = -8$ or $x^3 = -1$. The three roots of $x^3 = -8$ can be determined by going to polar form, or by factoring as follows:

$$x^3 = -8$$
$$x^3 + 8 = 0$$
$$x^3 + (2)^3 = 0.$$

By the formula for the sum of two cubes, we have

$$(x + 2)(x^2 - 2x + 4) = 0.$$

Setting the first factor equal to zero gives the root $x = -2$. Setting the second factor equal to zero and using the quadratic formula,

$$x^2 - 2x + 4 = 0$$

$$x = \frac{2 \pm \sqrt{-12}}{2}$$

$$= 1 \pm j\sqrt{3}.$$

Using the same technique for $x^3 = -1$ gives the roots -1 and $(1 \pm j\sqrt{3})/2$. Hence the solution set for the original equation contains six numbers:

$$\left\{ -1, -2, 1 \pm j\sqrt{3}, \frac{1 \pm j\sqrt{3}}{2} \right\}.$$

It should be noted that the given equation could have been factored immediately:

$$x^6 + 9x^3 + 8 = 0$$

$$(x^3 + 8)(x^3 + 1) = 0.$$

Setting each factor equal to zero and solving for x, we could then proceed in the same manner as given above.

EXAMPLE 15

Solve $x + \sqrt{x} - 2 = 0$.

Although this radical equation can be solved by the usual techniques, we shall show an alternative method. If $w = \sqrt{x}$, then the equation can be written as

$$w^2 + w - 2 = 0$$

$$(w - 1)(w + 2) = 0$$

$$w = 1 \quad \text{or} \quad w = -2.$$

Thus $\sqrt{x} = 1$ or $\sqrt{x} = -2$. The former gives the root $x = 1$ and the latter is unacceptable, since the principal square root of a number is never negative. The solution set is therefore $\{1\}$. We caution you that transformations, as in the instance above, may lead to extraneous roots; results should always be checked.

EXERCISE 12-5

In Problems 1–12, solve the indicated fractional equation.

1. $x^2 = \dfrac{x + 5}{6}$

2. $\dfrac{x}{3} = \dfrac{6}{x} - 1$

3. $\dfrac{3}{x-4} + \dfrac{x-3}{x} = 2$

4. $\dfrac{6x+7}{2x+1} - \dfrac{6x+1}{2x} = 1$

5. $\dfrac{2}{x-1} - \dfrac{6}{2x+1} = 5$

6. $\dfrac{6(x+1)}{2-x} + \dfrac{x}{x-1} = 3$

7. $\dfrac{2}{x-2} - \dfrac{x+1}{x+4} = 0$

8. $\dfrac{3}{x+1} + \dfrac{4}{x} - \dfrac{12}{x+2} = 0$

9. $\dfrac{2x-3}{2x+5} + \dfrac{2x}{3x+1} = 1$

10. $\dfrac{2}{x^2-9} - \dfrac{3x}{x+3} = \dfrac{1}{x-3}$

11. $\dfrac{3}{x^3-1} + \dfrac{x}{x^2+x+1} - \dfrac{4}{x-1} = 0$

12. $\dfrac{3}{x^2-4} + \dfrac{2}{x^2+4x+4} - \dfrac{4}{x+2} = 0$

In Problems 13–24, solve the given radical equation.

13. $3\sqrt{x+4} = x-6$

14. $x+2 = 2\sqrt{4x-7}$

15. $\sqrt{x+4} + \sqrt{x} - 6 = 0$

16. $\sqrt{3x} - \sqrt{5x+1} + 1 = 0$

17. $\sqrt{x+7} - \sqrt{2x} - 1 = 0$

18. $\sqrt{x} + \sqrt{2x+7} - 8 = 0$

19. $\sqrt{x} - \sqrt{2x+1} + 1 = 0$

20. $\sqrt{x-2} + 2 = \sqrt{2x+3}$

21. $\sqrt{x+5} + 1 = 2\sqrt{x}$

22. $\sqrt{\sqrt{x}+2} = \sqrt{2x-4}$

23. $\sqrt{\sqrt{x^2+2}} = 7$

24. $\sqrt{x+\sqrt{x-4}} = 2$

In Problems 25–42, solve the given quadratic form equation for x or y.

25. $x^4 - x^2 - 6 = 0$

26. $x^4 - 3x^2 - 10 = 0$

27. $\dfrac{1}{x^2} + \dfrac{6}{x} + 8 = 0$

28. $\dfrac{1}{x^2} + \dfrac{1}{x} - 12 = 0$

29. $\dfrac{1}{x^4} + \dfrac{9}{x^2} + 14 = 0$

30. $\dfrac{1}{x^4} - \dfrac{9}{x^2} + 8 = 0$

31. $x - 2\sqrt{x} - 3 = 0$

32. $6x - 5\sqrt{x} + 1 = 0$

33. $(x-3)^2 + 9(x-3) + 14 = 0$

34. $(x+5)^2 - 8(x+5) = 0$

35. $\dfrac{1}{(x-2)^2} - \dfrac{12}{x-2} + 35 = 0$

36. $\dfrac{2}{(x+4)^2} + \dfrac{7}{x+4} + 3 = 0$

37. $(x^2+2)^2 + 12(x^2+2) + 11 = 0$

38. $(x^2-5)^2 - 9(x^2-5) - 36 = 0$

39. $y^{2/3} + y^{1/3} - 2 = 0$

40. $2y^{-2/3} - 5y^{-1/3} - 3 = 0$

41. $y^6 + 8y^3 = 0$

42. $y^6 + 7y^3 - 8 = 0$

43. In a certain a.c. circuit, shown on the next page, the impedance Z (in ohms) is given by

$$Z = \sqrt{R^2 + X^2},$$

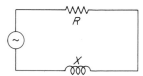

where R is the resistance (in ohms) and X is the reactance (in ohms). If $Z = 10$ ohms and $R = 8$ ohms, find X assuming that $X > 0$.

44. In a series a.c. circuit, resonance occurs when

$$2\pi f_r L = \frac{1}{2\pi f_r C},$$

where f_r is a resonant frequency, L is inductance, and C is capacitance. Solve for f_r assuming that $f_r > 0$.

45. An object is 120 in. from a wall. In order to focus the image of the object on the wall, a converging lens with a focal length of 24 in. is to be used. The lens is placed between the object and the wall at a distance of p in. from the object, where

$$\frac{1}{p} + \frac{1}{120 - p} = \frac{1}{24}.$$

Find p to one decimal place.

46. A tank can be filled by two pipes running together in $3\frac{3}{4}$ hr, but the larger one will fill it alone in 4 hr less time than the smaller one. What time is required for each pipe to fill it alone?

47. At 10 A.M. a small boat moving easterly out to sea with a speed of 20 mi/hr is 10 mi from the nearest point A of a straight shoreline, while a larger boat having a speed of 30 mi/hr starts from a point B, 10 mi south from A, to meet it. At what time, to the nearest minute, do the boats meet?

48. If $\omega L = \dfrac{1}{\omega C}$ and $2\pi f_r = \omega$, show that

$$f_r = \frac{1}{2\pi\sqrt{LC}}.$$

Assume that all quantities are positive.

49. A tank can be filled by three pipes in 1 hr. Three times as much water flows through the second pipe in a minute as through the first, and the third pipe alone would require 1 hr more to fill the tank than the second pipe alone. Find the time required by each pipe alone to fill the tank.

50. If a resistance R and reactance X are connected in parallel, the impedance Z of the circuit is given by

$$Z = \frac{RX}{\sqrt{R^2 + X^2}}.$$

Solve this equation for X.

12-6

SYSTEMS OF QUADRATIC EQUATIONS IN TWO VARIABLES

Definition. The **general quadratic equation in two variables** is an equation of the form

$$ax^2 + bxy + cy^2 + dx + ey + f = 0,$$

where a, b, and c are not all zero.

If $d = e = 0$, the equation is called a **pure quadratic equation.** As with systems of linear equations, the methods of solving systems of quadratic equations are primarily by elimination by addition or elimination by substitution, depending on the nature of the equations. Of course, the systems may be solved graphically, but one can only approximate the solution in that case, and determination of imaginary solutions is impossible. Thus, for the sake of accuracy and completeness of solution, we shall confine our attention to an algebraic approach to finding all ordered pairs (x, y) simultaneously satisfying the equations in the system.

EXAMPLE
16

One equation linear and one quadratic.

$$\begin{cases} x^2 + xy + y^2 = 3 & \text{(1)} \\ \qquad\quad x - y = 3. & \text{(2)} \end{cases}$$

The technique in this case is elimination by substitution. Solve the linear equation for one of the variables and substitute in the quadratic equation. Thus, from Eq. (2),

$$x = 3 + y. \qquad (3)$$

Substituting in Eq. (1),

$$(3 + y)^2 + (3 + y)y + y^2 = 3.$$

Simplifying,

$$3y^2 + 9y + 6 = 0$$
$$y^2 + 3y + 2 = 0$$
$$(y + 2)(y + 1) = 0$$
$$y = -2, \qquad y = -1.$$

Substitute these values in Eq. (3): When $y = -2$, then $x = 1$, and when $y = -1$, $x = 2$. The solution set is $\{(1, -2), (2, -1)\}$, which you should verify by substitution in *both* Eqs. (1) and (2). It is *imperative* that you realize that the solution cannot be written $x = 1$, $x = 2$, $y = -1$, $y = -2$. An appropriate manner could be $x = 1$ and $y = -2$, or $x = 2$ and $y = -1$.

EXAMPLE
17

A system of two pure quadratic equations with no xy-term.

$$\begin{cases} x^2 + y^2 = 4 & \text{(1)} \\ 9x^2 + y^2 = 9. & \text{(2)} \end{cases}$$

364

The technique in this case is elimination by addition, since both equations are linear in the variables x^2 and y^2. Multiplying each member of Eq. (1) by -1 and adding corresponding members of Eq. (1) to Eq. (2) gives

$$8x^2 = 5$$

$$x^2 = \frac{5}{8}$$

$$x = \pm\frac{\sqrt{10}}{4}.$$

Substituting $x = \sqrt{10}/4$ in Eq. (1) gives $y = \pm\frac{3}{4}\sqrt{6}$. Substituting $x = -\sqrt{10}/4$ in Eq. (1) gives $y = \pm\frac{3}{4}\sqrt{6}$. The solution set is therefore

$$\left\{ \left(\frac{\sqrt{10}}{4}, \frac{3\sqrt{6}}{4} \right), \left(\frac{\sqrt{10}}{4}, -\frac{3\sqrt{6}}{4} \right), \left(-\frac{\sqrt{10}}{4}, \frac{3\sqrt{6}}{4} \right), \left(-\frac{\sqrt{10}}{4}, -\frac{3\sqrt{6}}{4} \right) \right\}.$$

For illustrative purposes, the graphs of the given equations are indicated in Fig. 12-7. The four points of intersection correspond to the four solutions.

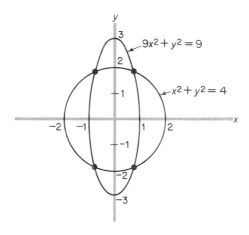

FIGURE 12-7

EXAMPLE
18

Solve the system

$$\begin{cases} 2x^2 + xy - y^2 = 20 & (1) \\ \qquad\qquad xy = 6. & (2) \end{cases}$$

From Eq. (2), if $x \neq 0$, then $y = 6/x$. Substituting in Eq. (1),

$$2x^2 + x\left(\frac{6}{x}\right) - \left(\frac{6}{x}\right)^2 = 20$$

$$2x^2 + 6 - \frac{36}{x^2} = 20$$

$$2x^4 + 6x^2 - 36 = 20x^2$$

$$2x^4 - 14x^2 - 36 = 0$$

$$x^4 - 7x^2 - 18 = 0$$
$$(x^2 - 9)(x^2 + 2) = 0.$$

Setting each factor equal to zero gives

$$x = \pm 3 \quad \text{or} \quad x = \pm j\sqrt{2}.$$

Using these values for x in the equation $y = 6/x$ gives the solution set:

$$\{(3, 2), (-3, -2), (j\sqrt{2}, -3j\sqrt{2}), (-j\sqrt{2}, 3j\sqrt{2})\}.$$

EXERCISE 12-6

Solve the systems of equations in Problems 1–14.

1. $\begin{cases} x^2 + y^2 = 25 \\ x^2 - y^2 = 7 \end{cases}$

2. $\begin{cases} y^2 - x^2 = 3 \\ 2y^2 + x^2 = 9 \end{cases}$

3. $\begin{cases} x^2 + 5y^2 = 4 \\ 3x^2 + 5y^2 = 12 \end{cases}$

4. $\begin{cases} 8y^2 + 16x^2 = 11 \\ 16y^2 - 16x^2 = -5 \end{cases}$

5. $\begin{cases} 2x^2 - 5y^2 = 10 \\ 3x^2 + 15y^2 = 195 \end{cases}$

6. $\begin{cases} 2x^2 - 3y^2 = 6 \\ 3x^2 - 2y^2 = 19 \end{cases}$

7. $\begin{cases} y^2 - x^2 = 28 \\ x - y = 14 \end{cases}$

8. $\begin{cases} x^2 - 2xy + 2y = 1 \\ y - x = 3 \end{cases}$

9. $\begin{cases} y^2 = 6x + 13 \\ 2x - y = 5 \end{cases}$

10. $\begin{cases} x^2 + 2xy + y^2 = 9 \\ 2x + 5y = 24 \end{cases}$

11. $\begin{cases} xy = 4 \\ 2x + y = 9 \end{cases}$

12. $\begin{cases} 3x^2 + 2y^2 = 21 \\ xy = 3 \end{cases}$

13. $\begin{cases} 3x^2 + 3y^2 - 6xy = 3 \\ 3x - y = 5 \end{cases}$

14. $\begin{cases} x^2 = y^2 + 14 \\ y = x^2 - 16 \end{cases}$

15. A man has three square flower beds all of one size and another square flower bed of a different size; the sum of the areas of the four beds is 156 ft² and the sum of their perimeters is 96 ft. Find the possible sizes of the flower beds.

16. A certain cloth when wet shrinks one-eighth in its length and one-sixteenth in its width. If the surface of a piece of cloth is dimished by $5\frac{3}{4}$ yd², and the length of the four sides by $4\frac{1}{4}$ yd, what were the length and width originally?

17. When a body A of mass 5 kg moving with a speed of 3 m/sec undergoes a completely elastic collision with a body B of mass 2 kg initially at rest, the principles of conservation of linear momentum and kinetic energy yield the system of equations

$$\begin{cases} 5(3) = 5V_A + 2V_B \\ \tfrac{1}{2}(5)(3^2) = \tfrac{1}{2}(5)V_A^2 + \tfrac{1}{2}(2)(V_B^2), \end{cases}$$

where V_A and V_B are the speeds of bodies A and B after the collision. Find V_A and V_B.

18. Two bodies A and B are moving at constant rates and in the same direction around a circle 36 ft in circumference. A makes one revolution in 3 sec less time than B, and A and B meet every 18 sec. What are their rates?

12-7
REVIEW

REVIEW QUESTIONS

A quadratic equation is of the _____(1)_____ degree and can be written in the form _____(2)_____ .

(1) **second**   (2) $ax^2 + bx + c = 0, a \neq 0$

The equation $(2x + 3)(x - 1) = 0$ has the solution set _____(3)_____ .

(3) $\left\{-\dfrac{3}{2}, 1\right\}$

If $ax^2 + bx + c = 0$, then $b^2 - 4ac$ is called the _____(4)_____ of the quadratic equation. If $b^2 - 4ac = 0$, then the roots are real and _____(5)_____ . If $b^2 - 4ac = 16$, the roots are real and _____(6)_____ .

(4) **discriminant**   (5) **equal**   (6) **distinct**

True or false: Every quadratic equation has two distinct roots. _____(7)_____

(7) **false**

In the equation $(x - 5)(x - 5) = 0$, we refer to 5 as being a root of multiplicity 2, or more simply a _____(8)_____ root.

(8) **double**

If the graph of $f(x) = ax^2 + bx + c = 0$ does not intersect the x-axis, then the equation $f(x) = 0$ has two _(real)(imaginary)_ roots.
                                                                                (9)

(9) **imaginary**

The roots of $x^2 + 100x = 0$ are _____(10)_____ .

(10) **0, −100**

The equation $x^{2/3} + 3x^{1/3} - 6 = 0$ is in quadratic form. We can transform the equation into a quadratic equation in w if we substitute w for _____(11)_____.

(11) $x^{1/3}$

To complete the square in the equation $x^2 - 7x = 0$, you would add to both members the number _____(12)_____.

(12) $\dfrac{49}{4}$

The discriminant of $2x^2 - 4x + 3 = 0$ is _____(13)_____.

(13) -8

Both roots of $2x^2 - 4x + 3 = 0$ are $\underset{(14)}{\underline{\text{(real)(imaginary)}}}$.

(14) **imaginary**

The system
$$\begin{cases} x^2 - 2xy - y^2 = 6 \\ \quad 2x - y = 3 \end{cases}$$
can be solved easily for x by solving the $\underset{(15)}{\underline{\text{(first)(second)}}}$ equation for y and substituting this value in the $\underset{(16)}{\underline{\text{(first)(second)}}}$ equation.

(15) **second**   (16) **first**

A system of two pure quadratic equations having no xy-term should be solved by the method of _____(17)_____.

(17) **elimination by addition**

The graph of $y = 2x^2 - 6x + 2$ opens $\underset{(18)}{\underline{\text{(upward)(downward)}}}$.

(18) **upward**

The graph of $y = -2x^2 + 4x - 3$ has a $\underset{(19)}{\underline{\text{(high)(low)}}}$ point when $x =$ _____(20)_____.

(19) **high**   (20) **1**

REVIEW PROBLEMS

Solve Problems 1–24 by any method.

1. $2x^2 - x = 0$

2. $x^2 - 10x + 25 = 0$

3. $x^2 + 10x - 25 = 0$

4. $2x^2 - 3x + 4 = 0$

5. $-2x^2 + x - 4 = 0$

6. $-3x^2 + 5x - 1 = 0$

7. $(8x - 5)(2x + 6) = 0$

8. $x(x - 9) = 0$

9. $-x^2 + 2x + 2 = 0$

10. $2(x^2 - 1) + 2x = x^2 - 6x + 1$

11. $5x^2 = 7x$

12. $3x^2 + 2x - 5 = 0$

13. $x^{2/3} - x^{1/3} - 12 = 0$

14. $x^4 + 5x^2 + 6 = 0$

15. $x^{-2} + 10x^{-1} + 25 = 0$

16. $x + 1 + 2\sqrt{x} = 0$

17. $\sqrt{\sqrt{x} + 1} = 5$

18. $\sqrt{x} - \sqrt{x + 1} = 6$

19. $x^6 - 16x^3 + 64 = 0$

20. $\begin{cases} x^2 + y^2 = 15 \\ x^2 - y^2 = 13 \end{cases}$

21. $\begin{cases} x^2 - 2xy + y^2 = 6 \\ 2y - 3x = 4 \end{cases}$

22. $\begin{cases} x^2 - 2x + y = 36 \\ x - y = -6 \end{cases}$

23. $\begin{cases} xy = 16 \\ 2x^2 - 4y^2 = -32 \end{cases}$

24. $jx^2 - 6x + j = 0.$

25. Find the maximum value of the function $y = -3x^2 + 6x - 4$.

26. Find the minimum value of the function $y = 4x^2 + 16x - 6$.

13

exponential and logarithmic functions

13-1
EXPONENTIAL AND LOGARITHMIC FUNCTIONS

In nearly all branches of science and engineering, there are pairs of physical quantities whose relationship to one another can be described mathematically by an equation of the form $y = b^x$, that is, a constant b raised to a variable power x. For example, as functions of time, the number of bacteria in a colony, the quantity of a radioactive substance, the charge on a discharging capacitor, and the temperature of a cooling body are all expressible in the basic form of a constant raised to a variable power.

Definition. The function f defined by

$$y = f(x) = b^x, \qquad b > 0, b \neq 1,$$

where the exponent x can be any real number, is called an **exponential function.**

The restriction $b \neq 1$ merely excludes from our discussion the rather trivial constant function $y = 1^x = 1$. The fact that the exponent can be any real number gives rise to the question of how to define an expression such as $6^{\sqrt{2}}$. Stated simply, we use an approximation method. We could first say that $6^{\sqrt{2}}$ is approximated by $6^{1.4} = 6^{7/5} = \sqrt[5]{6^7}$, which *is* defined. Furthermore, better approximations would be given by $6^{1.41} = \sqrt[100]{6^{141}}$ and $6^{1.414}$, etc. In this manner, a meaning of $6^{\sqrt{2}}$ becomes evident.

From the graphs of $y = 2^x$, $y = 3^x$, and $y = (\tfrac{1}{2})^x = 2^{-x}$, shown in Fig. 13-1, the following observations can be made. As indicated by the point of intersection $(0, 1)$ of the graphs, we have $b^0 = 1$ for every base b. Moreover, $y = b^x$ varies in two different ways, depending on whether

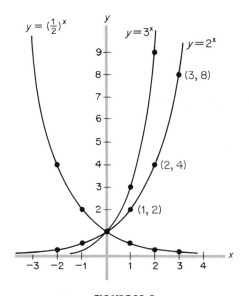

x	2^x	3^x	$(\tfrac{1}{2})^x$
-2	$\frac{1}{4}$	$\frac{1}{9}$	4
-1	$\frac{1}{2}$	$\frac{1}{3}$	2
0	1	1	1
1	2	3	$\frac{1}{2}$
2	4	9	$\frac{1}{4}$
3	8	27	$\frac{1}{8}$

FIGURE 13-1

$b > 1$ or $0 < b < 1$. If $b > 1$ as in $y = 2^x$ and $y = 3^x$, then as x increases, y also increases (without bound); we call such functions **increasing functions**. Also, y can take on values very close to zero. For example, if $x = -100$, then $2^x = 2^{-100} = 1/2^{100}$ is a number close to zero. Hence, the x-axis is an asymptote of $y = 2^x$. On the other hand, suppose $0 < b < 1$ as in $y = (\tfrac{1}{2})^x$. Then as x increases, y decreases, taking on values close to zero. Finally, in all cases note that the graph of an exponential function always lies above the x-axis.

Now, suppose we consider the graph of

$$x = 2^y.$$

You should note that this equation is *not* in the form of an exponential function *of* x according to our definition. Here the variables x and y are interchanged. Nevertheless, this equation does define a type of function which is essential to our future discussions. Let us locate some points on its graph. If $y = 0$, then $x = 1$, resulting in the point $(1, 0)$. When $y = 2$, then $x = 4$, and when $y = -1$, we have $x = \frac{1}{2}$, resulting in the points $(4, 2)$ and $(\frac{1}{2}, -1)$. Similarly, we obtain other points as indicated below. The graph shown in Fig. 13-2 is asymptotic to the y-axis. *Note that given any positive value of x, there is an exponent y such that $x = 2^y$. This exponent may be positive, negative, or zero.*

x	$\frac{1}{8}$	$\frac{1}{4}$	$\frac{1}{2}$	1	2	4	8
y	-3	-2	-1	0	1	2	3

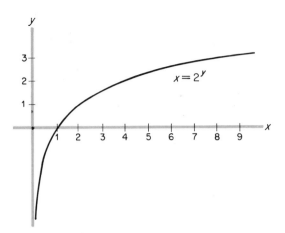

FIGURE 13-2

We stress that the equation $x = 2^y$ and all other functions of the form $x = b^y$, where $b > 0$ and $b \neq 1$, are defined for all positive values of x. Indeed, the equation

$$x = b^y, \qquad b > 0,\, b \neq 1, \tag{13-1}$$

means that every *positive* number x is expressible as a power y of the base b. The *exponent y* which satisfies $x = b^y$ is called the **logarithm (or log)** of x to the base b and we write

$$\log_b x = y. \tag{13-2}$$

That is, **to say that the log (or logarithm) to the base b of x is y means that b raised to the y power is x.** For example, since $8 = 2^3$, we can say that $\log_2 8 = 3$. It should therefore be clear that there is an equivalence between the exponential form of Eq. (13-1) and the logarithmic form of Eq. (13-2)—that is,

$$\log_b x = y \qquad \text{if and only if} \qquad x = b^y.$$

Therefore, the graph of $y = \log_2 x$ is identical to the graph of $x = 2^y$ in Fig. 13-2.

To avoid any confusion that may arise from the interchange of variables—for example, $x = b^y$ and $y = b^x$—it is customary to choose y as the dependent variable and, hence, $y = b^x$ and $y = \log_b x$ are considered to be the fundamental exponential and logarithmic functions, respectively. For illustrative purposes the general shapes of these functions for $b > 1$ are indicated in Fig. 13-3. Note that $y = \log_b x$ is an increasing function.

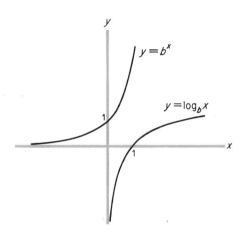

FIGURE 13-3

We remark that the functions $y = b^x$ and $y = \log_b x$ are said to be *inverses* of one another.

EXAMPLE   a. The statement $25 = 5^2$ is written in logarithmic form as $\log_5 25 = 2$; that is, the logarithm of 25 to the base 5 is 2.

1

b. Since $16 = 2^4$, then $\log_2 16 = 4$.

c. Since $4^{-2} = \frac{1}{16}$, $\log_4 \frac{1}{16} = -2$.

d. Since $10^0 = 1$, then $\log_{10} 1 = 0$.

e. Since $8^{2/3} = 4$, then $\log_8 4 = \frac{2}{3}$.

f. If $10^{2x} = 20$, then $\log_{10} 20 = 2x$ and $x = \dfrac{\log_{10} 20}{2}$.

EXAMPLE   a. The statement $\log_{10} 100 = 2$ can be written in exponential form as $10^2 = 100$.

2

b. $\log_{64} 8 = \frac{1}{2}$ means $64^{1/2} = 8$.

c. $\log_{10} \frac{1}{10} = -1$ means $10^{-1} = \frac{1}{10}$.

d. To find $\log_5 125$, we note that $5^3 = 125$. Thus $\log_5 125 = 3$.

e. $\log_b 0$ is undefined since $\log_b 0 = x$ means $b^x = 0$; but for $b > 0$ there is no value of x for which $b^x = 0$. **We may speak only of the log of a positive number.**

f. $\log_b 1 = 0$ since $b^0 = 1$, and $\log_b b = 1$ since $b^1 = b$. *You should be familiar with these results.*

EXAMPLE 3

a. If $\log_7 49 = x$, then in exponential form we have $7^x = 49$. By inspection, $x = 2$.

b. If $\log_2 (2^{6.2}) = x$, then $2^x = 2^{6.2}$ and, by inspection, $x = 6.2$.

c. Solve the equation $2 \log_x 49 = 4$ for x.

$$2 \log_x 49 = 4$$
$$\log_x 49 = 2,$$

which we can write in the equivalent exponential form as

$$x^2 = 49$$
$$x = 7.$$

We ignored $x = -7$ since $\log_{-7}$ is undefined.

d. Solve $\log_3 x = 4$ for x.

$$\log_3 x = 4$$

or equivalently

$$3^4 = x$$
$$81 = x.$$

e. Find x if $25^{x+2} = 5^{3x-4}$.

Since $25 = 5^2$, we can express both members of the given equation as powers of 5 and then equate exponents:

$$(5^2)^{x+2} = 5^{3x-4}$$
$$5^{2x+4} = 5^{3x-4}$$
$$2x + 4 = 3x - 4$$
$$8 = x.$$

f. Find x if $x = \log_7 \sqrt[9]{7^8}$.

$$x = \log_7 \sqrt[9]{7^8} = \log_7 7^{8/9}.$$

In exponential form this means

$$7^x = 7^{8/9}$$

and so

$$x = \tfrac{8}{9}.$$

g. Find x if $x = \log_{10} 10^c$.

Writing the given equation in exponential form, we obtain $10^x = 10^c$.

Thus, $x = c$. You should be familiar with this fact, namely

$$\log_{10} 10^c = c.$$

EXERCISE 13-1

In Problems 1–24, express the given logarithmic forms exponentially and the given exponential forms logarithmically.

1. $4^4 = 256$

2. $2^3 = 8$

3. $2^5 = 32$

4. $(\frac{1}{2})^3 = \frac{1}{8}$

5. $3^{-2} = \frac{1}{9}$

6. $1.52 = 10^{.1818}$

7. $\log_3 27 = 3$

8. $\log_5 125 = 3$

9. $\log_4 (\frac{1}{16}) = -2$

10. $\log_3 (\frac{1}{81}) = -4$

11. $\log_2 256 = 8$

12. $16^{1/2} = 4$

13. $\log_2 (\frac{1}{128}) = -7$

14. $\log_{10} (.001) = -3$

15. $\log_{.5} (\frac{1}{16}) = 4$

16. $8^{1/3} = 2$

17. $151.4 = 10^{2.1801}$

18. $-2 = \log_{1/3} 9$

19. $\log_{.5} 16 = -4$

20. $2^0 = 1$

21. $10^4 = 10,000$

22. $\log_{25} 625 = 2$

23. $4 = \log_{1/6} (\frac{1}{1296})$

24. $64^{1/3} = 4$

In Problems 25–32, sketch the graph of each of the given functions. If necessary, first express the equation in exponential form.

25. $y = f(x) = 3^x$

26. $y = f(x) = 3(2)^x$

27. $y = f(x) = 3^{-x}$

28. $y = f(x) = (\frac{1}{3})^x$

29. $y = f(x) = \log_4 x$

30. $y = f(x) = \log_3 x$

31. $y = f(x) = \log_2 (2x)$

32. $y = f(x) = \log_{10} x$

In Problems 33–48, find the value of x.

33. $\log_3 x = 4$

34. $\log_2 x = 4$

35. $\log_x 81 = 4$

36. $\log_x 16 = 2$

37. $\log_{10} 10^4 = x$

38. $\log_7 x = 2$

39. $\log_x (\frac{1}{81}) = -4$

40. $\log_{16} x = \frac{1}{2}$

41. $\log_{10} (x + 2)^2 = 2$

42. $\log_{16} x = \frac{1}{4}$

43. $\log_2 [4^2(8)] = x$

44. $\log_x 5 = 1$

45. $\dfrac{\log_3 27}{\log_3 81} = x$

46. $2 \log_x 49 = 4$

47. $\log_3 (\frac{27}{81}) = x$

48. $3 \log_x (\frac{1}{125}) = -9$

Since logarithms are exponents, their properties must be consistent with the rules of exponents previously developed. Indeed, the basic properties of logarithms are derived directly from the properties of exponents. In what follows let

$$\log_b x = w \quad \text{and} \quad \log_b y = z,$$

which in exponential form mean

$$b^w = x \quad \text{and} \quad b^z = y,$$

respectively.

a. Since $xy = b^w b^z = b^{w+z}$, in logarithmic form we have

$$\log_b (xy) = w + z.$$

Hence,

$$\log_b (xy) = \log_b x + \log_b y. \tag{13-3}$$

That is, *the logarithm of the product of two numbers is the sum of the logarithms of the numbers.* For example, $\log_{10} (3 \cdot 5) = \log_{10} 3 + \log_{10} 5$.

b. Since $x/y = b^w/b^z = b^{w-z}$, in logarithmic form we have

$$\log_b \left(\frac{x}{y} \right) = w - z.$$

Hence,

$$\log_b \left(\frac{x}{y} \right) = \log_b x - \log_b y. \tag{13-4}$$

That is, *the logarithm of a quotient is equal to the logarithm of the numerator minus the logarithm of the denominator.* For example, $\log_{10} \frac{2}{3} = \log_{10} 2 - \log_{10} 3$.

c. Since $x = b^w$, then $x^n = (b^w)^n = b^{nw}$, which means

$$\log_b (x^n) = nw.$$

Hence,

$$\log_b (x^n) = n \log_b x \tag{13-5}$$

That is, *the logarithm of the n–th power of a number is equal to n times the logarithm of the number.* For example, $\log_{10} 2^8 = 8 \log_{10} 2$. We can use this property to confirm a previous result:

$$\log_{10} 10^c = c \log_{10} 10 = c \cdot 1 = c.$$

EXAMPLE
4

The examples which follow illustrate the application of the basic properties of logarithms, Eqs. (13-3), (13-4), and (13-5).

a. $\log_4 [(17)(121)] = \log_4 17 + \log_4 121$ [Eq. (13-3)].

b. $\log_3 [(a)(b)(c)] = \log_3 [(ab)(c)]$

$= \log_3 (ab) + \log_3 c$ [Eq. (13-3)].

$= \log_3 a + \log_3 b + \log_3 c$ [Eq. (13-3)].

This technique can, of course, be extended to the logarithm of any number of factors.

c. $\log_2 \left(\dfrac{102}{37}\right) = \log_2 102 - \log_2 37$ [Eq. (13-4)].

d. $\log_4 \dfrac{(310)(26)}{(19)} = \log_4 [(310)(26)] - \log_4 19$ [Eq. (13-4)]

$= \log_4 310 + \log_4 26 - \log_4 19$ [Eq. (13-3)].

e. $\log_{10} (50.2)^6 = 6 \log_{10} 50.2$ [Eq. (13-5)].

f. $\log_7 \sqrt[5]{xy} = \log_7 (xy)^{1/5}$

$= \tfrac{1}{5} \log_7 (xy)$ [Eq. (13-5)]

$= \tfrac{1}{5} [\log_7 x + \log_7 y]$ [Eq. (13-3)].

g. $\log_3 \dfrac{(20)(\sqrt{13})}{3} = \log_3 [(20)(\sqrt{13})] - \log_3 3$ [Eq. (13-4)]

$= \log_3 20 + \log_3 13^{1/2} - \log_3 3$ [Eq. (13-3)]

$= \log_3 20 + \tfrac{1}{2} \log_3 13 - 1.$ [Eq. (13-5)]

h. $\log_5 [a(b + c)] = \log_5 a + \log_5 (b + c).$

Note, in particular, that $\log_5 (b + c) \neq \log_5 b + \log_5 c.$

EXAMPLE
5

a. Express $3 \log_2 10 + \log_2 15$ as a single logarithm.

$3 \log_2 10 + \log_2 15 = \log_2 10^3 + \log_2 15 = \log_2 [(10)^3(15)]$

$= \log_2 15,000.$

b. Express $2 \log_{10} 5 - \log_{10} 7$ as a single logarithm.

$2 \log_{10} 5 - \log_{10} 7 = \log_{10} 5^2 - \log_{10} 7$

$= \log_{10} \left(\dfrac{5^2}{7}\right) = \log_{10} \dfrac{25}{7}.$

c. Write $\log_{10} \sqrt[3]{\dfrac{x^5(x - 2)^8}{x - 3}}$ in terms of $\log_{10} x$, $\log_{10} (x - 2)$, and $\log_{10} (x - 3)$.

$\log_{10} \sqrt[3]{\dfrac{x^5(x - 2)^8}{x - 3}} = \log_{10} \left[\dfrac{x^5(x - 2)^8}{x - 3}\right]^{1/3} = \dfrac{1}{3} \log_{10} \dfrac{x^5(x - 2)^8}{x - 3}$

$= \dfrac{1}{3} \{\log_{10} [x^5(x - 2)^8] - \log_{10} (x - 3)\}$

$= \dfrac{1}{3} \{\log_{10} x^5 + \log_{10} (x - 2)^8 - \log_{10} (x - 3)\}$

$= \dfrac{1}{3} \{5 \log_{10} x + 8 \log_{10} (x - 2) - \log_{10} (x - 3)\}.$

EXERCISE 13-2

In Problems 1–10, write the given expression in terms of $\log_{10} x$, $\log_{10} y$, *and* $\log_{10} z$.

1. $\log_{10} (xy)$

2. $\log_{10} (x^2 yz)$

3. $\log_{10} \dfrac{x}{z^2}$

4. $\log_{10} x^6$

5. $\log_{10} \sqrt{x}$

6. $\log_{10} \dfrac{xy^2}{z^3}$

7. $\log_{10} (xy^2)^6$

8. $\log_{10} \dfrac{y\sqrt[3]{x}}{z^2}$

9. $\log_{10} \sqrt[6]{\dfrac{x^2 y^3}{z^5}}$

10. $\log_{10} \dfrac{1}{y\sqrt{z}}$

In Problems 11–22, use the basic properties of logarithms to express each of the given forms as a single logarithm.

11. $\log_2 7 + \log_2 3$

12. $\log_3 32 - \log_3 2$

13. $3 \log_{10} 7 - 5 \log_{10} 23$

14. $2 \log_{10} 39 - \frac{1}{2} \log_{10} 3$

15. $2 \log_3 4 + \log_3 2$

16. $4 \log_4 3 - \log_4 9$

17. $\log_4 4 - \log_4 200$

18. $3 \log_2 10 + \log_2 15$

19. $12 \log_4 3 + \log_4 3^{-2}$

20. $\log_{10} \sqrt[3]{x} - 3 \log_{10} x^2$

21. $\log_2 2x - \log_2 (x + 1)$

22. $2 \log_{10} x - \frac{1}{2} \log_{10} (x - 2)$

23. Write each of the following in terms of $\log_{10} x$, $\log_{10} (x + 2)$, and $\log_{10} (x - 3)$.

  a. $\log_{10} \sqrt{\dfrac{x^2(x - 3)^3}{x + 2}}$

  b. $\log_{10} [(x - 3)(x + 2)^2 \sqrt{x(x + 2)}]$

  c. $\log_{10} \dfrac{1}{x(x - 3)^2(x + 2)^3}$

13-3
COMMON LOGARITHMS

A. COMMON LOGARITHMS

For computational purposes, logarithms to the base 10, called **common logarithms**, are usually used. In this case, the subscript 10 denoting the base is usually omitted from the notation. Thus,

$$\log 100 \quad \text{means} \quad \log_{10} 100.$$

Since $\log 10 = 1$, the common logarithms of powers of 10 are readily obtainable:

$\log 1 = \log 10^0 = 0 \log 10 = 0$

$$\log 10 = \log 10^1 = 1 \qquad \log \frac{1}{10} = \log 10^{-1} = -1$$

$$\log 100 = \log 10^2 = 2 \qquad \log \frac{1}{100} = \log 10^{-2} = -2$$

$$\log 1000 = \log 10^3 = 3 \qquad \log \frac{1}{1000} = \log 10^{-3} = -3$$

$$\vdots \qquad\qquad\qquad \vdots$$

As pointed out in Chapter 1, every positive number can be written in scientific notation—that is, as the product of a number between 1 and 10 and some integral power of 10. Symbolically, for any positive number N,

$$N = m \cdot 10^c, \qquad 1 \leq m < 10, c \in \mathbf{Z}.$$

EXAMPLE
6

a.	$23 = (2.3)(10^1)$	$m = 2.3,$	$c = 1$
b.	$4 = (4)(10^0)$	$m = 4,$	$c = 0$
c.	$.0001212 = (1.212)(10^{-4})$	$m = 1.212,$	$c = -4$
d.	$121200 = (1.212)(10^5)$	$m = 1.212,$	$c = 5$

The determination of c is based on the number of decimal places that the decimal point in N is moved to obtain m. If moved to the left, c is positive; if moved to the right, then c is negative.

We shall now find the logarithm of a number written in scientific notation. We have

$$N = m \cdot 10^c = 10^c \cdot m$$
$$\log N = \log (10^c \cdot m) = \log 10^c + \log m$$
$$\log N = c + \log m. \tag{1}$$

Since $1 \leq m < 10$, $\log 1 = 0$, $\log 10 = 1$, and $y = \log x$ is an increasing function, it follows that $0 \leq \log m < 1$. Thus, from Eq. (1) the logarithm of a positive number N is composed of two parts:

a. An integer c, the exponent to which 10 must be raised to write N in scientific notation. We call c the **characteristic** of $\log N$.

b. A decimal, the value of $\log m$, called the **mantissa**, where $0 \leq$ mantissa < 1.

In Example 6, the characteristic of log 23 is 1, the characteristic of log 4 is 0, that of log .0001212 is −4, etc. Note that while the characteristic of a common logarithm may be negative, the mantissa cannot be.

EXAMPLE
7

a. $3.65 = 3.65 \times 10^0$; hence, log $3.65 = 0 + \log 3.65$.
b. $36{,}500 = 3.65 \times 10^4$; hence, log $36{,}500 = 4 + \log 3.65$.
c. $.0000365 = 3.65 \times 10^{-5}$; hence, log $.0000365 = -5 + \log 3.65$.
d. $365 = 3.65 \times 10^2$; hence, log $365 = 2 + \log 3.65$.

In Examples 7a–d the characteristics of the logarithms are 0, 4, −5, and 2, respectively. In each case, however, the mantissa is the value of log 3.65. Herein lies the major advantage of common logarithms—namely, that **the mantissas of the logarithms of numbers having the same sequence of digits are the same although the characteristics may be different.** Essentially, finding the logarithm of a number N involves finding the logarithm of a number between 1 and 10. A five-place table of approximate mantissas for numbers between 1 and 10 is provided in Appendix E. Decimal points have been omitted for ease of printing. We shall illustrate the use of the table.

EXAMPLE
8

a. Find log 36.56.

Since $36.56 = (3.656)(10^1)$, we have log $36.56 = 1 + \log 3.656$. We now look in the table for log 3.656. In the column labeled N we find 365 (actually 3.65) and move over horizontally to the column labeled 6. The entry 56301 (actually .56301) gives us the mantissa. Therefore, log $3.656 = .56301$ and

$$\log 36.56 = 1 + .56301 = 1.56301.$$

Note that in using the table our concern was with the sequence of digits 3656.

b. Find log 123.

Since $123 = (1.23)(10^2)$, we have log $123 = 2 + \log 1.23$. We now look in the table for log 1.23. In the column labeled N we find 123 (actually 1.23) and move over to the column labeled 0. Then the entry 08991 (actually .08991) gives us the mantissa. Thus

$$\log 123 = 2 + .089991 = 2.08991.$$

c. Find log 29.

Since $29 = (2.9)(10^1)$, we have log $29 = 1 + \log 2.9$. In the table we find log 2.900 to be .46240. Thus,

$$\log 29 = 1.46240.$$

d. Find log 6.215.

The characteristic is 0. From the table, log 6.215 is .79344. Therefore, log 6.215 = 0.79344.

e. Find log .0006273.

Since .0006273 = 6.273×10^{-4}, log .0006273 = $-4 + \log 6.273$. From the tables, log 6.273 = .79748. Hence,

$$\log .0006273 = -4 + .79748.$$

We caution you that this *cannot* be written as -4.79748. Actually, $-4 + .79748$ is a negative number and hence a negative mantissa, which is not permitted. To remedy the situation, we shall rewrite the characteristic -4 by adding and subtracting 10:

$$-4 = (10 - 4) - 10.$$

Thus

$$\log .0006273 = -4 + .79748$$
$$= (10 - 4) + .79748 - 10$$
$$= 6.79748 - 10.$$

The characteristic can be said to be $6 - 10$. This technique should be used whenever the characteristic is negative. The usefulness of this technique will become apparent later.

f. Find log .008515.

Since .008515 = $(8.515)(10^{-3})$, the characteristic is -3 or $7 - 10$. From the table, log 8.515 = .93018. Note that the asterisk in the body of the table indicates the mantissa has already undergone the change in its first two digits from 92 to 93. Hence, log .008515 = $7.93018 - 10$.

B. LINEAR INTERPOLATION—PROPORTIONAL PARTS

When it is necessary to find the logarithm of a number that lies between two given numbers in the table, we may use the method of linear interpolation explained in conjunction with the trigonometric tables in Chapter 7. The examples which follow should make the technique clear.

EXAMPLE
9

Find log 13.473.

We know that log 13.473 = $1 + \log 1.3473$. Although log 1.3473 is not in our table, log 1.3470 and log 1.3480 are. We now assume that log 1.3473 is a value between them. Looking them up we find log 1.3470 = .12937 and log 1.3480 = .12969. Thus, as a number increases from 1.3470 to 1.3480, the logarithm of the number increases by .00032 from .12937 to .12969. Our number 1.3473 is .0003/.0010 or 3/10 of the way from 1.3470 to 1.3480. We assume then that log 1.3473 is 3/10 of the way from .12937 to .12969. Therefore,

$$\tfrac{3}{10}(.00032) = .000096 \approx .00010.$$

Thus, $\log 1.3473 = .12937 + .00010 = .12947$ and our final answer is that $\log 13.473 = 1.12947$. The data obtained in this problem can be arranged in an orderly fashion as follows:

$$.0010 \left\{ .0003 \left\{ \begin{array}{l} \log 1.3470 = .12937 \\ \log 1.3473 = \quad ? \end{array} \right\} x \right\} .00032.$$
$$\log 1.3480 = .12969$$

$$\frac{.0003}{.0010} = \frac{x}{.00032}$$

$$x = \left(\frac{3}{10}\right)(.00032) = .000096 \approx .00010.$$

Hence,

$$\log 1.3473 = .12937 + x = .12937 + .00010 = .12947.$$

Finally,

$$\log 13.473 = 1.12947.$$

The technique used in Example 7 assumes that changes in $\log N$ are directly proportional to changes in N. That is, we assume the graph of $y = \log x$ is a straight line. Although this assumption is completely false, for small changes in N the error is small—well within reason for our purposes.

EXAMPLE
10

Find $\log 22064$.

Since $22064 = 2.2064 \times 10^4$, $\log 22064 = 4 + \log 2.2064$.

$$.0010 \left\{ .0004 \left\{ \begin{array}{l} \log 2.2060 = .34361 \\ \log 2.2064 = \quad ? \end{array} \right\} x \right\} .00019$$
$$\log 2.2070 = .34380$$

Now

$$\frac{.0004}{.0010} = \frac{x}{.00019} \text{ and } x = \frac{4}{10}(.00019) = .000076 \approx .00008.$$

Hence,

$$\log 2.2064 = .34361 + .00008 = .34369.$$

Finally, $\log 22064 = 4.34369$.

In the margins of many logarithm tables, small tables of **proportional parts** are available which greatly simplify the task of interpolation. We shall illustrate the use of these auxiliary tables by first repeating the problem of Example 10.

EXAMPLE
11

a. Find log 22064 using tables of proportional parts.

We have log 22064 = 4 + log 2.2064. As observed from the tables, log 2.2064 lies between the entries for 2.2060 and 2.2070, that is, between .34361 and .34380, respectively, which differ by .00019, or 19 units of the fifth decimal place. Clearly, log 2.2064 lies 4/10 of the way from log 2.2060 to log 2.2070. The proportional-parts table indicates a division of the number that heads it into tenths. Selecting the table labeled 19, $\frac{4}{10}(19)$ is seen to be $7.6 \approx 8$. Hence, we add eight units to the fifth-place digit of log 2.2060 = .34361. Hence, log 22064 = 4.34361 + .00008 = 4.34369, which of course agrees with the result of Example 10.

b. Find log 584.27 using tables of proportional parts.

We have log 584.27 = 2 + log 5.8427. We observe that log 5.8427 lies 7/10 of the way from log 5.8420 to log 5.8430, that is, 7/10 of the way from .76656 to .76664, numbers which differ by eight units of the fifth decimal place. Choosing the proportional-parts table labeled 8, we find that $(\frac{7}{10})(8)$ is $5.6 \approx 6$. Hence, we add six units to the fifth-place digit of log 5.8420. We then have log 584.27 = 2.76656 + .00006 = 2.76662.

C. ANTILOGARITHMS

When the logarithm of a number N is known and it is desired to find the number N, the process is referred to as finding the **antilogarithm**. The examples that follow illustrate the procedure.

EXAMPLE
12

Find N if log $N = 4.86052$; that is, find the antilogarithm of 4.86052.

The characteristic is 4 and the mantissa is .86052. Looking up the mantissa in the *body* of the table, we find it in the row headed by 725 and the column headed by 3. Thus,

$$.86052 = \log 7.253 \quad \text{and} \quad N = (7.253)(10^4) = 72530.$$

EXAMPLE
13

Find the antilogarithm of 8.68052 − 10.

The characteristic is 8 − 10 or −2 and the mantissa is .68052. We find that in the body of the table this mantissa corresponds to the number 4.792. Thus, .68052 = log 4.792 and our required number is $(4.792)(10^{-2}) = .04792$.

EXAMPLE
14

Find antilog 3.92026.

The characteristic is 3. The mantissa .92026 is not in the table, but it lies between .92023 and .92028, corresponding to the numbers 8.322 and 8.323, respectively. We assume the antilogarithm of .92026 lies between these two numbers. We proceed with our work as follows:

Number   Mantissa

$$.001\left\{x\left\{\begin{array}{cc} 8.322 & .92023 \\ ? & .92026 \\ 8.323 & .92028 \end{array}\right. \left.\begin{array}{c} \\ \end{array}\right\}.00003\right\}.00005$$

$$\frac{x}{.001} = \frac{.00003}{.00005}$$

$$x = \frac{3}{5}(.001) = .0006$$

$$8.322 + .0006 = 8.3226.$$

Thus antilog $.92026 = 8.3226$ and so antilog $3.92026 = 8.3226 \times 10^3 = 8322.6$.

To illustrate the use of the tables of proportional parts in finding antilogarithms, we repeat the problem in Example 14.

EXAMPLE
15

Find antilog 3.92026 by using tables of proportional parts.

By examining the body of the logarithm tables we find that the next smallest mantissa is .92023, corresponding to the number 8.322, which gives the first four figures of the required antilogarithm. Since the logarithm of the required number exceeds .92023 by .00003, or three units, and .92026 lies between mantissas in the table separated by five units, we use the proportional-parts table headed 5, find the number closest to 3 in the *body* of the table, and determine the corresponding number in the left column. From the table we see this number is 6, which becomes the fifth digit of our answer. Hence, antilog $3.92026 = 8.3226 \times 10^3 = 8322.6$, which of course agrees with the result of Example 14.

EXERCISE 13-3

In Problems 1–24, determine the logarithms of the given numbers. Use linear interpolation or tables of proportional parts where appropriate.

1.	$N = 12.62$	**2.**	$N = .241$
3.	$N = .0008621$	**4.**	$N = 22230$
5.	$N = 28.4$	**6.**	$N = .0007$
7.	$N = 92000$	**8.**	$N = 7128000$
9.	$N = 126.7$	**10.**	$N = .00006222$
11.	$N = .7766$	**12.**	$N = 806500$
13.	$N = 8214 \times 10^{-8}$	**14.**	$N = .001623$

15. $N = .0062454$ **16.** $N = 5.6307 \times 10^9$

17. $N = 135.64$ **18.** $N = 535.46$

19. $N = .064325$ **20.** $N = .000012345$

21. $N = 2.0034$ **22.** $N = 8630.4$

23. $N = 26347$ **24.** $N = 7.1351$

In Problems 25–44, find the antilogarithms of the given logarithms.

25. 0.98345   **26.** 0.99016

27. 4.82105   **28.** 6.83985

29. 6.45010 — 10   **30.** 7.42732 — 10

31. 2.10924   **32.** 3.17869

33. 8.56174 — 10   **34.** 4.59402 — 10

35. 1.74596   **36.** 2.76025

37. 7.63214 — 10   **38.** 1.74557

39. 6.66666 — 10   **40.** 7.69078 — 10

41. 6.52290 — 10   **42.** 2.53627 — 10

43. 3.80385   **44.** 1.81007

13-4
COMPUTATIONS WITH LOGARITHMS

In this section we shall show how the basic properties of logarithms, combined with an ability to find logarithms and antilogarithms, enable us to perform rather complex calculations quite easily.

EXAMPLE
16

Find (22.07)(.0624)(822).

Let

$$N = (22.07)(.0624)(822)$$
$$\log N = \log 22.07 + \log .0624 + \log 822$$

$$\left. \begin{array}{l} \log 22.07 = \ \ 1.34380 \\ \log .0624 = \ \ 8.79518 - 10 \\ \log 822 = \ \underline{\ \ 2.91487\ \ } \end{array} \right\} \ \text{add}$$

$$\log N = 13.05385 - 10$$
$$= 3.05385$$

The characteristic is 3 and the mantissa is .05385. Since antilog .05385 = 1.132,

$$N = (1.132)(10^3) = 1132.$$

EXAMPLE
17

Find $\dfrac{.01966}{.1238}$.

Let

$$N = \frac{.01966}{.1238}$$

$$\log N = \log .01966 - \log .1238$$

$$\left.\begin{array}{l} \log .01966 = 8.29358 - 10 \\ \log .1238 = 9.09272 - 10 \end{array}\right\} \text{ subtract}$$

We cannot perform the subtraction without getting a negative mantissa. To avoid this we write the characteristic $8 - 10$ in the equivalent form $18 - 20$:

$$\left.\begin{array}{l} \log .01966 = 18.29358 - 20 \\ \log .1238 = 9.09272 - 10 \end{array}\right\} \text{ subtract}$$

$$\log N = 9.20086 - 10$$

The characteristic is -1 and antilog $.20086 = 1.5880$. Hence,

$$N = (1.5880)(10^{-1}) = .15880.$$

EXAMPLE
18

Find $(.6201)^8$.

Let

$$N = (.6201)^8$$

$$\log N = 8 \log .6201$$

$$\left.\begin{array}{r} \log .6201 = 9.79246 - 10 \\ 8 \end{array}\right\} \text{ multiply}$$

$$\log N = 78.33968 - 80$$

$$= 8.33968 - 10$$

$$N = .021862$$

EXAMPLE
19

Find $\sqrt[4]{7637}$.

Let

$$N = \sqrt[4]{7637} = (7637)^{1/4}$$

$$\log N = \tfrac{1}{4} \log 7637$$

$$\left.\begin{array}{r} \log 7637 = 3.88292 \\ \tfrac{1}{4} \end{array}\right\} \text{ multiply}$$

$$\log N = 0.97073$$

$$N = 9.3482$$

EXAMPLE
20

Find $\sqrt[3]{.1974}$.

Let

$$N = \sqrt[3]{.1974} = (.1974)^{1/3}$$
$$\log N = \tfrac{1}{3} \log .1974$$

$$\log .1974 = 9.29535 - 10 \left.\rule{0pt}{22pt}\right\} \text{ multiply}$$
$$\tfrac{1}{3}$$

Since -10 is not divisible by 3 but -30 is, we rewrite the characteristic as $29 - 30$:

$$\log .1974 = 29.29535 - 30 \left.\rule{0pt}{22pt}\right\} \text{ multiply}$$
$$\tfrac{1}{3}$$
$$\log N = 9.76512 - 10$$

Interpolating, antilog $.76512 = 5.8226$ and, hence,

$$N = (5.8226)(10^{-1}) = .58226$$

EXAMPLE
21

Find N if $N = \dfrac{.00335(273)^4}{787\sqrt{.723}}$.

$$\log N = \log .00335 + 4(\log 273) - (\log 787 + \tfrac{1}{2} \log .723).$$

$$\log 273 = 2.43616 \left.\rule{0pt}{18pt}\right\} \text{ multiply}$$
$$4$$

$$\begin{array}{c} 9.74464 \\ \log .00335 = \underline{7.52504 - 10} \end{array} \left.\rule{0pt}{22pt}\right\} \text{ add}$$
$$\log \text{ numerator} = 17.26968 - 10 = 7.26968$$

$$\log .723 = 19.85914 - 20 \left.\rule{0pt}{22pt}\right\} \text{ multiply}$$
$$\tfrac{1}{2}$$

$$\begin{array}{c} 9.92957 - 10 \\ \log 787 = \underline{2.89597} \end{array} \left.\rule{0pt}{22pt}\right\} \text{ add}$$
$$\log \text{ denominator} = 12.82554 - 10$$
$$= 2.82554$$

$$\begin{array}{c} \log \text{ numerator} = 7.26968 \\ \log \text{ denominator} = \underline{2.82554} \end{array} \left.\rule{0pt}{18pt}\right\} \text{ subtract}$$
$$\log N = 4.44414$$
$$N = 27806.$$

EXERCISE 13-4

In Problems 1–30, use logarithms to evaluate the given expression.

1. $(1.26)(.00621)$

2. $(.00726)(86.4)$

3. $72.6 \div 2.71$

4. $851 \div 26.2$

5. $\sqrt[12]{462}$

6. $\sqrt[3]{8.66}$

7. $(.0746)^4$

8. $(5.2)^8$

9. $(2.07)^{1.3}$

10. $\dfrac{(5.236)(589.3)}{528}$

11. $\sqrt[3]{.0063572}$

12. $70063\sqrt{.002142}$

13. $\sqrt{\dfrac{74.4}{820}}$

14. $\sqrt{(82.6)(39.5)}$

15. $\dfrac{(819)(748)^2}{3670}$

16. $\dfrac{(.2166)\sqrt{812}}{(.0176)^3}$

17. $\dfrac{\sqrt{6407}}{\sqrt[3]{312.65}}$

18. $(2.71)(22.1)^{-2}$

19. $\dfrac{(.00614)\sqrt{2.19}}{\sqrt[3]{27.6}(.0626)^{.5}}$

20. $\dfrac{(3.15)^3}{\sqrt{(82.1)^2(.0762)^3}}$

21. $\dfrac{\sqrt{.007111}}{(8.3254)^2(-1.147)}$

22. $\dfrac{(-627.2)(348.4)^2}{\sqrt[3]{.007321}}$

23. $\sqrt{\dfrac{321.6(200)^2}{(.00473)}}$

24. $\sqrt[3]{\dfrac{82.54}{721.3}}$

25. $(2.004)^{-1/3}$

26. $\sqrt{\sqrt{(836.5)^2}}$

27. $\sqrt[3]{\dfrac{(817.4)(231.5)^2}{(100.2)(714.9)}}$

28. $\left(\sqrt{\dfrac{75.46}{800.3}}\right)^{2/3}$

29. $\left(\sqrt{\dfrac{(.002748)^{1/3}(321.5)^{.01}}{\sqrt[5]{528.6}}}\right)^{1/4}$

30. $\sqrt{\dfrac{(.6003)^2\sqrt{.00415}}{\sqrt[3]{321.6}}}$

31. The intensity level β of a sound wave of intensity I is defined by the equation

$$\beta = 10 \log \frac{I}{I_0},$$

where I_0 is a reference intensity taken to be 10^{-16} watt/cm^2, which corresponds, approximately, to the faintest audible sound. The intensity level is measured in decibels. If a second source has an intensity of 3×10^{-14} watt/cm^2, what is its intensity level to the nearest tenth of a decibel?

32. Referring to Problem 31, show that if a second sound source has an

intensity I, then its intensity level is equal to

$$\beta = 160 + 10 \log I.$$

33. Experimental data indicate that the radius R of a nucleus of an isotope of mass number A is

$$R = 1.2 \times 10^{-13} A^{1/3},$$

where R is in centimeters. The most abundant naturally occurring isotope of oxygen has a mass number of 16. Determine the radius of the nucleus.

34. When a gas undergoes an adiabatic expansion of compression (adiabatic means no heat is exchanged with the surroundings), the pressures and volumes are related by the equation

$$P_i V_i^\gamma = P_f V_f^\gamma,$$

where the subscripts i and f refer to initial and final values, respectively. If the gas is monatomic, then $\gamma = 1.67$. If such a gas has an initial volume of 10 liters at a pressure of 1 atmosphere, find the final pressure if the gas is compressed adiabatically to a volume of 5 liters. Determine the final pressure to the nearest tenth of an atmosphere.

35. When a charged capacitor of capacitance C is connected to an inductor of inductance L, an oscillation takes place in which energy is alternately exchanged between the magnetic field of the inductor and the electric field of the capacitor. The frequency of this oscillation, in hertz (1 Hz is 1 cycle/sec) is

$$f = \frac{1}{2\pi\sqrt{LC}}$$

If $L = 3 \times 10^{-3}$ H and $C = 20 \times 10^{-6}$ F, approximate π by 3.14 and use logarithms to determine the frequency of the oscillation to the nearest hertz.

13-5
NATURAL LOGARITHMS

The system of common logarithms, those to the base 10, is very useful for computation. Although any positive number other than 1 can be used as the base of a system of logarithms, there is only one other important system. The base of this system is the irrational number approximated by

$$2.718281828459\ldots,$$

which is simply denoted by e in honor of the great mathematician Euler. Logarithms to the base e are called **natural** or **Napierian logarithms**. Although e undoubtedly seems an unlikely number to be chosen as the base of a system of logarithms, it is to be noted that natural logarithms are of great importance in theoretical aspects of applied mathematics, science, and engineering technology. Furthermore, students of calculus find that many formulas are simpler when natural logarithms are used.

The natural logarithm of a number N can, of course, be designated $\log_e N$. We shall, however, adopt the customary abbreviation $\ln N$. Hence,

$$\ln N \quad \text{means} \quad \log_e N.$$

Since the properties of logarithms discussed in Sec. 13-2 were developed for any base, they are completely applicable to the system of natural logarithms. A table of natural logarithms will be found in Appendix F. Instructions accompanying the table extend its applicability to determining the natural logarithms of numbers not included in the table.

EXAMPLE
22

a. Find $\ln 516$.

Expressing 516 in scientific notation we have

$$\begin{aligned}
\ln 516 &= \ln (5.16 \times 10^2) \\
&= \ln 5.16 + \ln 10^2 \\
&= \ln 5.16 + 2 \ln 10 \\
&= 1.64094 + 4.60517 \\
\ln 516 &= 6.24611.
\end{aligned}$$

Note that accompanying the table of natural logarithms are values of $n \ln 10$ for some integral values of n.

b. Find $\ln .00462$.

$$\begin{aligned}
\ln .00462 &= \ln (4.62 \times 10^{-3}) \\
&= \ln 4.62 + \ln 10^{-3} \\
&= \ln 4.62 - 3 \ln 10 \\
&= 1.53039 - 6.90776 \\
\ln .00462 &= -5.37737.
\end{aligned}$$

Thus, $e^{-5.37737} = .00462$.

c. Find $\ln 8256$.

$$\begin{aligned}
\ln 8256 &= \ln (8.256 \times 10^3) \\
&= \ln 8.256 + 3 \ln 10 \\
&= \ln 8.256 + 6.90776.
\end{aligned}$$

We find $\ln 8.256$ by the usual interpolation techniques:

$$.01\left\{.006\left\{\begin{aligned} \ln 8.250 &= 2.11021 \\ \ln 8.256 &= \quad ? \\ \ln 8.260 &= 2.11142 \end{aligned}\right\}x\right\}.00121$$

$$\frac{.006}{.01} = \frac{x}{.00121} \quad \text{and} \quad x = \frac{(.006)(.00121)}{(.01)} \approx .00073.$$

Hence, $\ln 8.256 = 2.11021 + .00073 = 2.11094$ and

$$\ln 8256 = 2.11094 + 6.90776$$
$$= 9.01870.$$

d. Find $\ln e + \log \frac{1}{10}$.

$$\ln e + \log \tfrac{1}{10} = \log_e e + \log_{10} 10^{-1}$$
$$= 1 + (-1) = 0.$$

13-6
CHANGE OF BASE

Tables of logarithms to the bases 10 and e are those commonly available. Since at times it is desirable to change from one base to another, we shall derive a formula that will enable us to make such a transformation. Let

$$y = \log_a N.$$

Then in exponential form this means

$$a^y = N.$$

Taking logarithms to the base b of each member,

$$\log_b (a^y) = \log_b N$$
$$y \log_b a = \log_b N.$$

Hence,

$$y = \frac{\log_b N}{\log_b a}$$

and, finally,

$$\log_a N = \frac{\log_b N}{\log_b a} \tag{13-6}$$

Equation (13-6) can be used to convert logarithms from one base to another base. In particular, we note two special cases. If $a = 10$ and $b = e$, we have from Eq. (13-6),

$$\log N = \frac{\ln N}{\ln 10}$$

and

$$\ln N = (\ln 10)(\log N)$$

$$\ln N = 2.30259 \log N. \tag{13-7}$$

It also follows from Eq. (13-6) that

$$\log N = .43429 \ln N. \tag{13-8}$$

Equations (13-7) and (13-8) permit a most convenient method of converting logarithms from base 10 to base e and vice versa.

EXAMPLE
23

a. Find ln 124 by using tables of *common* logarithms.

From Eq. (13-7),

$$\ln 124 = 2.30259 \log 124$$
$$= (2.30259)(2.09342)$$
$$= 4.8203.$$

b. Find log 8.63 by using tables of *natural* logarithms.

From Eq. (13-8),

$$\log 8.63 = .43429 \ln 8.63$$
$$= (.43429)(2.15524)$$
$$= .93600.$$

EXAMPLE
24

Find $\log_5 100$.

By Eq. (13-6),

$$\log_5 100 = \frac{\log 100}{\log 5}$$
$$= \frac{2}{.69897} = 2.8614.$$

EXERCISE 13-6

In Problems 1–10, find the natural logarithm of the given number by direct use of the natural logarithm tables. Interpolate where necessary.

1. 122
2. .000435
3. 8640000
4. 7210×10^{-6}
5. .00321
6. 14100000
7. 32.65
8. 478.2
9. 62×10^5
10. 21.23×10^{15}

In Problems 11–15, find the given common logarithms by using tables of natural logarithms; find the given natural logarithms by using tables of common logarithms.

11. log 4800
12. ln 23.5
13. ln 1.62
14. log 46.35
15. ln .00132

In Problems 16–20, evaluate the given expressions.

16. $\log_5 27$
17. $\log_2 15$

$$Y = \frac{\log_b N}{\log_b a}$$

$$\frac{\log_{10} 15}{\log_{10} 2} = 3.9068$$

18. $\log_7 72$ **19.** $\log_6 .00032$

20. $\log_{100} 25$

21. The work, in joules, done by a 1-kg sample of nitrogen gas as its volume changes from an initial value V_i to a final value V_f during an isothermal (constant-temperature) process is given by the equation

$$W = 8.1 \times 10^4 \ln \frac{V_f}{V_i}.$$

If such a sample expands isothermally from a volume of 3 liters to a volume of 7 liters, determine the work done by the gas to the nearest hundred joules.

22. If the sample of gas in Problem 21 is compressed isothermally until its volume is reduced to one-half its original value, how much work is done by the gas to the nearest hundred joules?

23. In studies of electric fields, one often encounters a device called a cylindrical capacitor, two concentric cylindrical conductors. The capacitance per unit length of such a device, in microfarads per meter, can be shown to be inversely proportional to

$$\ln \frac{r_o}{r_i},$$

where r_o and r_i are the radii of the outer and inner cylinders, respectively. If the capacitance per unit length of a particular cylindrical capacitor is 10 μF/m when the outer radius is twice the inner radius, find the capacitance per unit length when it is three times the inner radius.

13-7
LOGARITHMS OF THE TRIGONOMETRIC FUNCTIONS

Sometimes trigonometric functions appear in computations which are suitable for using logarithms. In such cases it is, of course, necessary to be able to determine the logarithms of the trigonometric functions involved. This could be accomplished by simply applying the techniques we have already learned. For example, to find log sin 32° 10′, we could first determine the value of sin 32° 10′ and then find the logarithm of the resulting number. The task is simplified, however, by making use of the table of logarithms of the trigonometric functions which is found in Appendix D. It must be emphasized that in using these tables, −10 *must be affixed to every mantissa in the table*. Interpolation, of course, is accomplished in the usual manner. The following example illustrates both techniques of finding a logarithm of a trigonometric function.

EXAMPLE
25

Find log sin 46° 14'.

a. We shall first find sin 46° 14'.

$$10\left\{4\left\{\begin{array}{l}\sin 46° 10' = .7214\\ \sin 46° 14' = \quad ?\end{array}\right\}x \\ \sin 46° 20' = .7234\end{array}\right\}.0020$$

$$\frac{4}{10} = \frac{x}{.0020}; \qquad x = .0008.$$

Hence,

$$\sin 46° 14' = .7214 + .0008$$
$$= .7222.$$

Now, $.7222 = 7.222 \times 10^{-1}$ and by the usual techniques we find that $\log 7.222 = .85866$. Finally,

$$\log \sin 46° 14' = \log (.7222) = 9.85866 - 10.$$

b. To find log sin 46° 14' by using the table of logarithms of the trigonometric functions we write

$$10\left\{4\left\{\begin{array}{l}\log \sin 46° 10' = 9.8582 - 10\\ \log \sin 46° 14' = \qquad ?\end{array}\right\}x \\ \log \sin 46° 20' = 9.8594 - 10\end{array}\right\}.0012$$

$$\frac{4}{10} = \frac{x}{.0012}; \qquad x \approx .0005$$

Hence,

$$\log \sin 46° 14' = (9.8582 - 10) + (.0005)$$
$$= 9.8587 - 10.$$

You will note that there is a difference in the results of Examples 25a and b in the fourth decimal place. This difference is to be expected and is due, in part, to the accuracy of the tables involved—that is, the number of decimal places to which values in the tables are given. For our purposes it need cause no concern.

EXAMPLE
26

Find θ if $\sin \theta = \dfrac{2.296}{4.891}$

$$\log \sin \theta = \log 2.296 - \log 4.891$$
$$= 0.36097 - 0.68940$$
$$= (10.36097 - 10) - (0.68940)$$
$$\log \sin \theta = 9.67157 - 10$$
$$\approx 9.6716 - 10$$
$$\theta \approx 28°.$$

EXAMPLE
27

Find y if $y = \sin 24° \cos 86°$.

$$\log y = \log \sin 24° + \log \cos 86°$$
$$= (9.6093 - 10) + (8.8436 - 10)$$
$$= 18.4529 - 20$$
$$\log y = 8.4529 - 10$$
$$y \approx .02837.$$

In the last step we referred to the table of common logarithms.

EXERCISE 13-7

In Problems 1–10, use the table of logarithms of the trigonometric functions to find the value of the given expression.

1. $\log \sin 32° \, 20'$

2. $\log \tan 21° \, 10'$

3. $\log \cos 20° \, 12'$

4. $\log \sin 35° \, 18'$

5. $\log \tan 210° \, 13'$

6. $\log \cos 320° \, 44'$

7. $\log \cos 42° \, 04'$

8. $\log \sin 15° \, 27'$

9. $\log \sin (-194° \, 9')$

10. $\log \cos (-54° \, 52')$

11. The sine of the critical angle for the boundary of two given substances is given by the ratio of their indices of refraction:

$$\sin \theta_c = \frac{n_2}{n_1}.$$

Use logarithms to evaluate θ_c for a diamond-fluorite boundary if $n_1 = 2.4172$ and $n_2 = 1.4341$.

12. Snell's law, relating the indices of refraction to the angles of reflection and refraction, is given by

$$n_1 \sin \theta_1 = n_2 \sin \theta_2.$$

Use logarithms to find θ_2 given $n_1 = 2.264$, $n_2 = 1.3142$, $\theta_1 = 25° \, 20'$.

13. At intermediate angles of two light polaroid sheets, the amount of transmitted energy is given by

$$I = I_{max} \cos^2 \theta,$$

known as Malus' law. I_{max} represents the maximum amount of transmitted energy possible. Use logarithms to determine θ if $I = \frac{1}{2}I_{max}$.

14. For a closely wound circular coil of wire of N turns, cross-sectional area A, carrying a current i, in a magnetic field of induction B, the torque is given by

$$\tau = NiBA \sin \alpha.$$

If $N = 100$, $i = 3$ amp, $B = .6$ weber/m^2, $\alpha = 20°$, and the diameter of the coil is 8 cm, express the area A in units of square meters and use logarithms to evaluate τ.

15. Given the relation

$$y = \sqrt{4 \sin^3 \theta \sec^2 \theta},$$

determine the value of y when $\theta = 10°$.

13-8
EXPONENTIAL AND LOGARITHMIC EQUATIONS

We have seen that if a logarithmic equation is expressed in exponential form, in certain instances the value of an unknown quantity in the given equation is easily recognizable. For example, $\log_4 64 = x$ can be written $4^x = 64$, and clearly $x = 3$. If the unknown quantity cannot be recognized immediately, it may be determined by employing logarithms. In the basic logarithmic equation $\log_a b = c$, the unknown quantity may be a, b, or c, giving rise to the following three equations and their exponential equivalents:

a. $\log_x b = c$, $x^c = b$.

b. $\log_a x = c$, $a^c = x$.

c. $\log_a b = x$, $a^x = b$.

In general, to solve such logarithmic equations for x, express the equation in exponential form and take the common logarithm of each member.

EXAMPLE
28

Solve $\log_2 5 = x$.

Writing in exponential form,

$$2^x = 5.$$

Taking logarithms of each member yields

$$x \log 2 = \log 5$$

$$x = \frac{\log 5}{\log 2} = \frac{.69897}{.30103} = 2.3219.$$

The above line actually follows directly from the change of base formula, Eq. (13-6). It should be noted that the final step of the solution, that of dividing $\log 5$ by $\log 2$, can be accomplished by the use of logarithms or by usual long-division methods.

EXAMPLE
29

Solve $\log_{(4x+1)} 5 = 2.3219$.

Writing in exponential form,

$$(4x + 1)^{2.3219} = 5.$$

Taking logarithms of each member yields

$$2.3219 \log (4x + 1) = \log 5$$

$$\log (4x + 1) = \frac{\log 5}{2.3129} = \frac{.69897}{2.3219} = .30103.$$

Now, antilog $.30103 = 2$; hence,

$$4x + 1 = 2, \qquad x = \tfrac{1}{4}.$$

EXAMPLE
30

Solve $\log_4 (2x + 4) = 3$.

Writing in exponential form,

$$2x + 4 = 4^3 = 64.$$

Hence,

$$2x = 60$$

and

$$x = 30.$$

EXERCISE 13-8

Solve each of the following equations for x.

1. $10^{2x} = 10$		**2.** $4 = 2^{8x}$	
3. $\log_3 (2x) = 4$		**4.** $\log_5 125 = x$	
5. $12^x = 24$		**6.** $x = \log_6 3$	
7. $(1.02)^x = 3$		**8.** $\log_7 x = 3$	
9. $\log_{1.5} 18 = x$		**10.** $\log_4 x = 2.32$	
11. $\log_{2x} 16 = 1.4$		**12.** $\log_3 4 = 2x + 5$	
13. $\log_7 x = 3.1$		**14.** $\log_\pi 2 = x$	
15. $\log_x 4.3 = 1.3$		**16.** $\log_{3.1} x = 17.2$	
17. $3^{2x} \cdot 2^{3x} = 16$		**18.** $x = \log_7 \sqrt[3]{6}$	
19. $1.5^x \cdot 2^{2x} = 11.3$		**20.** $3.74^{(8x-1)} \cdot 2.61^{(4x+2)} = 6.31$	

13-9
LOGARITHMIC AND SEMILOGARITHMIC
GRAPH PAPER

Before considering the significance and use of logarithmic and semi-logarithmic graph papers, it is appropriate to consider the graph of one very important exponential function. We have spoken of the number e in relation to the system of natural logarithms, and it will be seen in further studies in engineering and technology that the functions $y = e^x$, $y = e^{-x}$, and their variations describe many physical phenomena. For example, the intensity of a radioactive sample after a time t is given by $I = I_0 e^{-\lambda t}$, the charge on a capacitor at a time t is given by $q = c\mathcal{E}(1 - e^{-t/rc})$, and the number of radioactive nuclei present after time t is given by $N = N_0 e^{-\lambda t}$.

The graph of $y = e^x$ is similar in shape to the graph of $y = 2^x$ which was given in Fig. 13-1. Values of e^x and e^{-x} are conveniently tabulated in Appendix B. The graphs of $y = e^x$ and $y = e^{-x}$ are shown in Fig. 13-4.

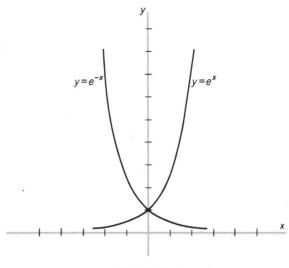

FIGURE 13-4

Often it is desirable to sketch a graph of a function $y = f(x)$ showing how $\log y$ varies with $\log x$. This is especially true when very large or very small values of x and/or y are involved. For example, if $10 \le x \le 1,000,000$, then $1 \le \log x \le 6$, an interval which is much more easily expressible graphically along an axis. If such a logarithmic graph is desirable, we can sketch it without having to determine the logarithms of any numbers by using a special **logarithmic paper**, as shown in Fig. 13-5. Such paper is ruled with logarithmic scales, rather than linear, both horizontally and vertically—that is, plotting a point (x, y) on it is equivalent to plotting the point $(\log x, \log y)$ on ordinary graph paper.

A similar type of graph paper called **semilogarithmic** graph paper, shown in Fig. 13-6, usually has the vertical axis ruled logarithmically and the horizontal axis ruled linearly. Thus, plotting a point (x, y) on it is equivalent to plotting $(x, \log y)$ on ordinary graph paper.

The most significant advantage of logarithmic and semilogarithmic graphs, aside from the ease of handling cumbersome numbers, is that certain functions that give rise to complex graphs on linear graph paper appear as straight lines when sketched on logarithmic or semilogarithmic paper. Let's consider some of these cases.

CASE A. The graph of $y = mx + b$.

We have seen in Chapter 6 that the graph of this linear equation is a straight line when sketched on ordinary graph paper.

CASE B. The graph of $y = ax^m$, where a and x are any positive numbers.

Taking the logarithms of both members yields

$$\log y = \log ax^m$$
$$\log y = \log a + \log x^m$$
$$\log y = m \log x + \log a. \tag{13-9}$$

We shall now make a change of variable in Eq. 13-9. Substituting y_L for $\log y$, x_L for $\log x$, and b for the constant $\log a$ yields

$$y_L = mx_L + b. \tag{13-10}$$

Equation (13-10) is a linear equation and, hence, equations of the form $y = ax^m$ appear as straight lines on logarithmic paper.

CASE C. The graph of $y = ae^{kx}$, where a is positive.

Taking logarithms of both members yields

$$\log y = \log a + \log e^{kx}$$
$$\log y = kx \log e + \log a. \tag{13-11}$$

Substituting y_L for $\log y$, b for the constant $\log a$, and m for the constant $k \log e$ yields

$$y_L = mx + b,$$

which is a linear equation. Hence, graphs of equations of the form $y = ae^{kx}$ are straight lines on semilogarithmic graph paper.

EXAMPLE
31

Sketch the graph of $y = 3x^2$ on logarithmic paper.

Since the equation is of the form $y = ax^m$, its graph on logarithmic paper must be a straight line. Hence, it is only necessary to locate two points. When $x = 1$, then $y = 3(1)^2 = 3(1) = 3$, and when $x = 10$, $y = 3(10)^2 = 300$. The graph is shown in Fig. 13-5.

EXAMPLE
32

Sketch the graph of $y = e^x$ on semilogarithmic graph paper.

Using values from the exponential tables in Appendix B, we obtain the

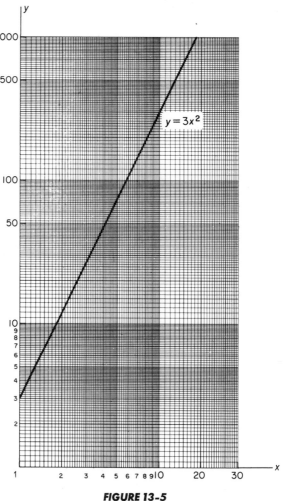

FIGURE 13-5

graph shown in Fig. 13-6. For example, if $x = 1$, then $y \approx 2.7$; when $x = 5$, then $y \approx 148.4$.

EXAMPLE
33

A student found that his data from an experiment resulted in a straight-line graph on semilogarithmic paper. If two of his data points are $x = 1$, $y = 2.81$, and $x = 4, y = 5.12$, find an equation which approximates the data.

We assume that the equation is of the form $y = ae^{kx}$. Substituting the given values,

$$2.81 = ae^{k} \tag{1}$$

$$5.12 = ae^{4k}. \tag{2}$$

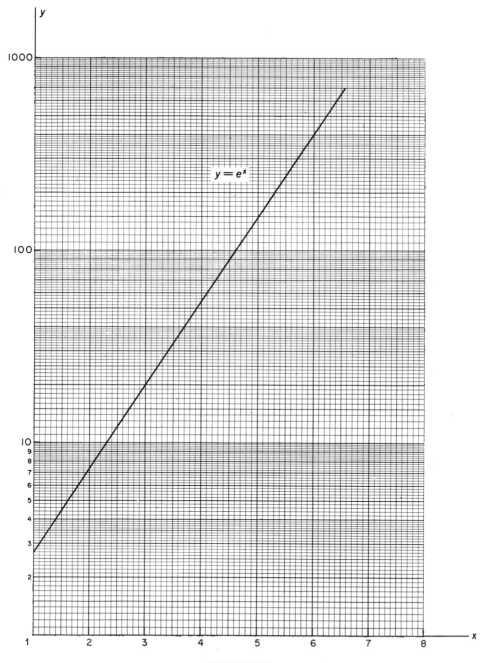

FIGURE 13-6

Dividing corresponding members of Eq. (2) by Eq. (1),

$$\frac{5.12}{2.81} = \frac{ae^{4k}}{ae^k} = e^{3k}.$$

Taking the natural logarithms of both members of this equation,

$$3k = \ln 5.12 - \ln 2.81$$
$$= 1.63315 - 1.03318 = .59997$$
$$k \approx .20.$$

Thus,

$$y = ae^{.20x}.$$

At (1, 2.81),

$$2.81 = ae^{.20} = a(1.2214)$$

$$a = \frac{2.81}{1.2214} \approx 2.3.$$

Thus,

$$y = 2.3e^{.20x}.$$

EXERCISE 13-9

1. Repeat Example 33 if the data points are (2, 3.61) and (4, 32.6).

2. Repeat Example 33 if the data points are (3, 3.46) and (8, 4.80) and if *logarithmic* paper were used.

13-10
A COMMENT ON EXPONENTIAL FORM OF A COMPLEX NUMBER

In the study of alternating-current circuits, many circuit quantities are expressible in the form of complex numbers. It is not uncommon for the engineer to encounter and, indeed, perform necessary computations using the exponential form of a complex number. The complex number $z = r(\cos \theta + j \sin \theta)$ can be expressed as

$$z = re^{j\theta},$$

where we define

$$e^{j\theta} = \cos \theta + j \sin \theta. \tag{13-12}$$

Equation (13-12), called *Euler's formula*, requires that the angle θ be expressed in radians.

EXAMPLE
34

Express $4 - 4j$ in exponential form.

We find that $r = \sqrt{(4)^2 + (-4)^2} = 4\sqrt{2}$ and that $\tan \theta = -4/4 = -1$. Since $4 - 4j$ lies in the fourth quadrant, $\theta = 7\pi/4$. Thus,

$$4 - 4j = re^{j\theta} = 4\sqrt{2}\, e^{(7\pi/4)j}.$$

EXAMPLE
35

Express $2e^{\pi j}$ in rectangular form.

Here $r = 2$ and $\theta = \pi$. Hence,

$$2e^{\pi j} = 2(\cos \pi + j \sin \pi) = 2(-1) = -2.$$

EXERCISE 13-10

In Problems 1–6, express the given complex numbers in exponential form.

1. $1 + j$ 2. -6

3. $-2\sqrt{3} + 2j$ 4. $-3 - 3j$

5. $-7j$ 6. $1 - j\sqrt{3}$

In Problems 7–12, express the given complex numbers in rectangular form.

7. $e^{\pi j/2}$ 8. $2e^{(4/3)\pi j}$

9. $3e^{3\pi j/4}$ 10. $4e^{\pi j}$

11. $2e^{(5\pi/6)j}$ 12. $e^{(3\pi/2)j}$

13-11
REVIEW

REVIEW QUESTIONS

The graph below is typical of a(n) <u>(exponential)(logarithmic)</u> function.
(1)

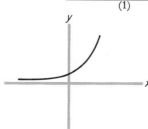

(1) **exponential**

The expression $10^{2 \log x}$ is equal to _____(2)_____.

(2) $\mathbf{x^2}$

If $\log x = 1.2222$, then $\log \sqrt{x} = $ _____(3)_____.

(3) **.6111**

$\ln e^{5x} = $ _____(4)_____.

(4)   **5x**

$$\log \frac{x^2 y^3}{z^4} = \underline{\hspace{1.5cm}(5)\hspace{1.5cm}} \log x + \underline{\hspace{1.5cm}(6)\hspace{1.5cm}} \log y - \underline{\hspace{1.5cm}(7)\hspace{1.5cm}} \log z.$$

(5)   **2**   (6)   **3**   (7)   **4**

$\log_N N = \underline{\hspace{2cm}(8)\hspace{2cm}}.$

(8)   **1**

$\log_N 0$ is $\underline{\hspace{2cm}(9)\hspace{2cm}}.$

(9)   **undefined**

If $\log x = .62148$, then $10^{.62148} = \underline{\hspace{2cm}(10)\hspace{2cm}}.$

(10)   **x**

In rectangular form, $6e^{\pi j} = \underline{\hspace{2cm}(11)\hspace{2cm}}.$

(11)   **−6**

The graph of $y = \log x$ is asymptotic to the $\underline{\hspace{2cm}(12)\hspace{2cm}}$ -axis.

(12)   **y**

If $\log N = 4$, then $\sqrt{N} = \underline{\hspace{2cm}(13)\hspace{2cm}}.$

(13)   **100**

REVIEW PROBLEMS

In Problems 1–6, determine the common logarithms of the given numbers.

1.   634.82   2.   .00064321

3. 8.62×10^{16} 4.   .084138

5.   6047.1   6. 9.72×10^{-3}

In Problems 7–12, determine the natural logarithms of the given numbers.

7.	.632	8.	468

7. .632   **8.** 468

9. 4.72   **10.** .000014

11. 32.47   **12.** 3247

In Problems 13–18, determine the value of N.

13. $\log N = 1.13465$ **14.** $\log N = 8.19893 - 10$

15. $\log N = .68078$ **16.** $\log N = 3.69034$

17. $\log N = 9.48004 - 10$ **18.** $\log N = 7.91100 - 10$

In Problems 19–28, determine the value of x.

19. $\log_6 x = 3.265$ **20.** $e^{(.3x+1)} = 6$

21. $7^{1.6x} = 3.429$ **22.** $\log_e \pi = x$

23. $4^{(.2x+1)} \cdot 3^{(16x-2)} = 362$ **24.** $\log \sin 34° \, 18' = x$

25. $\log \cos 34° \, 20' = x$ **26.** $\log_x 762.4 = 18.3$

27. $\left(\sqrt{\sqrt{321.6^2}}\right)^{1/3} = x$ **28.** $\dfrac{(18.629)(\sqrt{13}\sqrt{162.4})}{15.28(200.4)^{19}} = x$

29. Express $-2 - 2j$ in exponential form.

30. Express $4e^{(2\pi/3)j}$ in rectangular form.

14

graphs of the trigonometric functions

14-1
INTRODUCTION

In Chapter 7 the trigonometric functions were treated as mathematical entities and no consideration was given to their graphical characteristics. However, the graphs of the trigonometric functions have wide application in mathematically describing several occurrences in nature such as wave motion, vibrations and simple harmonic motion, and certain electrical phenomena. In this chapter we shall consider the graphs of the six trigonometric functions and some of their variations. We shall find that, in comparison to the graphs of algebraic functions, these graphs have strikingly unique features.

14-2
THE GRAPH OF THE SINE FUNCTION

To sketch the graph of the *basic sine function*

$$y = \sin x, \tag{14-1}$$

we shall use the techniques that were employed in graphing algebraic functions. That is, we shall assume arbitrary values for the independent variable x and determine the corresponding values of the dependent variable y. The resulting set of ordered pairs, (x, y), when plotted on a suitable coordinate system and joined by a smooth curve will be the graph of $y = \sin x$. Our assumed values for x and the corresponding values of y are given in Table 14-1. For the sake of completeness the angles are given in degrees as well as in radians. Note that the selected values of x are all special angles and the use of tables is not necessary to determine y. The graph of $y = \sin x$, where x is in radians, is shown in Fig. 14-1.

Table 14-1

x		$y = \sin x$	x		$y = \sin x$
Degrees	Radians		Degrees	Radians	
0	0	0	210	$\frac{7\pi}{6}$	$-\frac{1}{2} = -.5$
30	$\frac{\pi}{6}$	$\frac{1}{2} = .5$	240	$\frac{4\pi}{3}$	$-\frac{\sqrt{3}}{2} \approx -.87$
60	$\frac{\pi}{3}$	$\frac{\sqrt{3}}{2} \approx .87$	270	$\frac{3\pi}{2}$	-1
90	$\frac{\pi}{2}$	1	300	$\frac{5\pi}{3}$	$-\frac{\sqrt{3}}{2} \approx -.87$
120	$\frac{2\pi}{3}$	$\frac{\sqrt{3}}{2} \approx .87$	330	$\frac{11\pi}{6}$	$-\frac{1}{2} = -.5$
150	$\frac{5\pi}{6}$	$\frac{1}{2} = .5$	360	2π	0
180	π	0	390	$\frac{13\pi}{6}$	$\frac{1}{2} = .5$

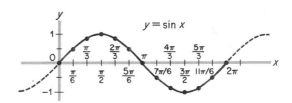

FIGURE 14-1

You should note the characteristic wavelike nature of this curve. Recall that the angles x and $x + 2\pi$ are coterminal angles and that trigonometric functions of coterminal angles are equal. For example, $13\pi/6$ ($= 390°$) is coterminal with $\pi/6$ ($= 30°$), and $\sin 13\pi/6 = \sin \pi/6 = 1/2$. More generally, $\sin(x + 2\pi) = \sin x$, which means the graph of $y = \sin x$ keeps repeating itself every 2π radians; that is, as x increases from 2π to

4π, 4π to 6π, 6π to 8π, and so forth, $\sin x$ changes exactly as it did in the interval from 0 to 2π. This repetition, or *periodicity*, is an inherent characteristic of the sine function. For this reason we say the sine function is **periodic** and has a **period** of 2π radians. Actually, since $\sin(x + 4\pi) = \sin(x + 6\pi) = \sin(x + 8\pi) = \sin x$, the numbers 4π, 6π, and 8π are also periods of $y = \sin x$. However, since 2π is the *least* positive number p such that $\sin(x + p) = \sin x$, we call 2π *the* period of $y = \sin x$.

It must be emphasized that Fig. 14-1 shows merely a portion of the graph of $y = \sin x$, for $y = \sin x$ is defined for all values of x and its graph extends indefinitely to the right and to the left. Indeed, although we have selected only positive values for x in Table 14-1, we could have chosen negative values with equal justification. We remark that the graph of the sine function over an interval of one complete period (2π radians) is called a **cycle** of the curve.

You should also observe in Fig. 14-1 that $y = \sin x$ ranges between values of $+1$ and -1 inclusive. In general, *one-half of the difference of the maximum and minimum values of a periodic function over one complete cycle is called the **amplitude** of the function*. Thus, the amplitude of $y = \sin x$ is 1, since $\frac{1}{2}[+1 - (-1)] = 1$.

In the following examples we shall show how to sketch variations of the sine function. We begin with functions of the form $y = a \sin x$, where a is a constant. You will see that although the function $y = a \sin x$ has the same period, 2π, as the basic sine function $y = \sin x$, its graph may differ in a vertical sense from that of $y = \sin x$. This difference in shape depends on a.

EXAMPLE
1

Sketch the graph of $y = 4 \sin x$.

As the values of $\sin x$ range between $+1$ and -1, the values of four times $\sin x$ range between $+4$ and -4. In other words, the effect is to multiply the y-coordinate of every point on the basic sine curve by 4, thus affecting the point's vertical distance from the x-axis. Hence, the amplitude of $y = 4 \sin x$ is $\frac{1}{2}[+4 - (-4)] = 4$. Furthermore, the period is still 2π since the 4 has no effect in a horizontal sense. Two cycles of the curve are shown in Fig. 14-2. Note that the independent variable is x, which represents an angle *in radians*.

EXAMPLE
2

Sketch the graph of $y = -2 \sin x$.

As $\sin x$ ranges from 1 to -1, $-2 \sin x$ ranges from -2 to 2. Thus, the amplitude is 2 and the period is 2π. Here, however, each value of $\sin x$ is multiplied by a negative number. The effect is to invert a sine curve; that is, wherever $y = 2 \sin x$ is positive, $y = -2 \sin x$ is negative, and wherever $y = 2 \sin x$ is negative, $y = -2 \sin x$ is positive. Two cycles of the graph of

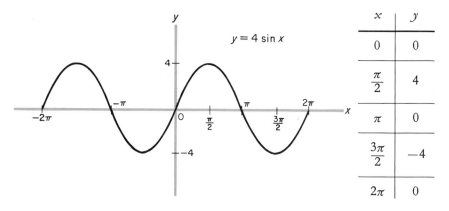

x	y
0	0
$\dfrac{\pi}{2}$	4
π	0
$\dfrac{3\pi}{2}$	−4
2π	0

FIGURE 14-2

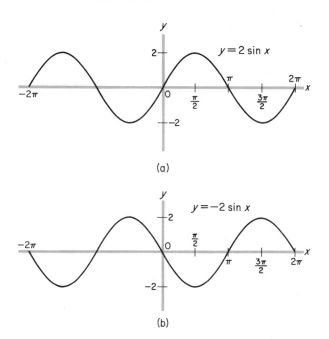

(a)

(b)

FIGURE 14-3

$y = 2 \sin x$ are shown in Fig. 14-3(a). The two cycles of $y = -2 \sin x$ shown in Fig. 14-3(b) are those of an inverted sine curve. We can also say that the graph of $y = -2 \sin x$ is the *reflection* in the x-axis of the graph of $y = 2 \sin x$. We can conclude from Examples 1 and 2 that **the amplitude of $y = a \sin x$ is $|a|$ and its period is 2π.**

We have considered graphs of sine functions of the form $y = a \sin x$. In each case, one cycle was completed as x varied from 0 to 2π radians, and we concluded that the period of such functions is 2π radians. The most general form of a sine function, however, can be written

$$y = a \sin(bx + c), \qquad b > 0. \tag{14-2}$$

The equation $y = a \sin x$ is a special case of Eq. (14-2), where $c = 0$ and $b = 1$. In Eq. (14-2) the quantity $bx + c$, which represents an angle, is called the **argument** of the sine function. As before, the amplitude of $y = a \sin(bx + c)$ is $|a|$, and the basic shape of the curve is that of a sine curve if a is positive or an inverted sine curve if a is negative.

The graph of $y = \sin x$ repeated itself whenever the argument x changed by 2π radians. That is, we could consider a cycle to begin when $x = 0$ and end when $x = 2\pi$. So it is that the graph of $y = a \sin(bx + c)$ repeats itself whenever the argument $bx + c$ changes by 2π radians. Hence, we can take a cycle to begin when

$$bx + c = 0 \qquad \text{or} \qquad x = -\frac{c}{b}$$

and end when

$$bx + c = 2\pi \qquad \text{or} \qquad x = \frac{2\pi - c}{b} = -\frac{c}{b} + \frac{2\pi}{b}.$$

For the most part the similarity ends there.

Although for one cycle of the graph of $y = a \sin(bx + c)$ the argument changes by 2π radians, the value of x changes by

$$\frac{2\pi - c}{b} - \left(-\frac{c}{b}\right) = \frac{2\pi}{b}.$$

That is, in terms of x the graph repeats itself whenever x changes by $2\pi/b$ radians. We conclude that the period of the function is $2\pi/b$.

Furthermore, depending on the values of b and c, the graph of $y = a \sin(bx + c)$ can be shifted horizontally to the right or left with respect to the graph of the basic sine function. The quantity $-c/b$ is called the **displacement** and represents the amount by which $y = a \sin(bx + c)$ is shifted in comparison to the basic sine curve. That is, a cycle of $y = \sin x$ always begins when $x = 0$, but a cycle of $y = a \sin(bx + c)$ begins when $x = -c/b$. If the displacement is positive, the shift is to the right; if the displacement is negative, the shift is to the left.

The discussion above leads to a procedure for sketching the graph of any equation of the form $y = a \sin(bx + c)$, where $b > 0$:

a. *Determine the amplitude of the curve. It will be $|a|$.*

b. *Determine the basic shape of the curve, that is, sine curve $(a > 0)$ or inverted sine curve $(a < 0)$.*

c. *Find x_b, a value of x representing the beginning of a cycle, by setting the argument of the function equal to zero and solving for x. For convenience you may replace x by x_b in this step.*

d. *Find x_e, the value of x representing the end of the cycle in step c, by setting the argument of the function equal to 2π and solving for x. For convenience you may replace x by x_e in this step.*

e. *Find the period p:*

$$p = x_e - x_b.$$

f. *Divide the interval of the cycle into four equal parts to determine the abscissas of significant points. By "significant" we mean those for which the function has maximum or minimum values, or is zero. These quarter-cycle points can be obtained by dividing p by 4 and successively adding the result to x_b.*

The important concept is that the graph of the general sine function $y = a \sin(bx + c)$ repeats itself every time the *argument* of the function changes by 2π radians, and so any arbitrary value of x could be chosen as the starting point of a cycle. However, for the sake of convenience, choose the starting point for which $bx + c$ is equal to zero as indicated in step c above.

EXAMPLE
3

Sketch the graph of $y = 3 \sin\left(2x + \dfrac{\pi}{2}\right)$.

This is Eq. (14-2) where $a = 3$, $b = 2$, and $c = \pi/2$.

a. The amplitude is 3.

b. Since $a > 0$, the basic shape of the curve is that of a sine curve.

c.
$$2x_b + \frac{\pi}{2} = 0$$

$$x_b = -\frac{\pi}{4}.$$

Thus, $x_b = -\pi/4$ is a point on the x-axis at which a cycle begins. Hence the curve is displaced $\pi/4$ units to the left.

d.
$$2x_e + \frac{\pi}{2} = 2\pi$$

$$x_e = \frac{3\pi}{4}.$$

That is, $x_e = 3\pi/4$ is the point on the x-axis at which the above cycle ends.

e. The period is

$$p = x_e - x_b = \frac{3\pi}{4} - \left(-\frac{\pi}{4}\right) = \pi \text{ radians.}$$

That is, the curve completes one cycle as x varies over an interval of π radians, for over that interval the argument $2x + \dfrac{\pi}{2}$ will vary over an interval of 2π radians.

   f. The interval from $-\pi/4$ to $3\pi/4$ (of length π) is divided into four equal subintervals by the points $-\pi/4$, 0, $\pi/4$, $\pi/2$, and $3\pi/4$. At $x = -\pi/4$, $y = 0$; when $x = 0$, $y = 3$; when $x = \pi/4$, $y = 0$; when $x = \pi/2$, $y = -3$; when $x = 3\pi/4, y = 0$.

One cycle of the graph is shown in Fig. 14-4.

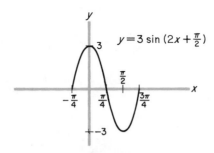

$$y = 3 \sin \left(2x + \tfrac{\pi}{2}\right)$$

FIGURE 14-4

EXAMPLE

4

Sketch the graph of $y = -2 \sin \left(\tfrac{1}{2}x\right)$.

   a. The amplitude is 2.

   b. The basic shape is an inverted sine curve.

   c. $\tfrac{1}{2}x_b = 0$, $x_b = 0$; the cycle starts at $x_b = 0$.

   d. $\tfrac{1}{2}x_e = 2\pi$, $x_e = 4\pi$; the cycle ends at $x_e = 4\pi$.

   e. The period is $4\pi - 0 = 4\pi$ radians.

   f. Significant points: when $x = 0$, π, 2π, 3π, and 4π.

One cycle of the graph is shown in Fig. 14-5.

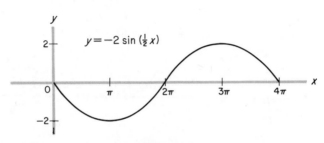

$$y = -2 \sin \left(\tfrac{1}{2}x\right)$$

FIGURE 14-5

EXAMPLE
5

Sketch the graph of the equation $y = \sin\left(\pi x - \dfrac{\pi}{4}\right)$.

a. The amplitude is 1.

b. The basic shape is a sine curve.

c. $\pi x_b - \dfrac{\pi}{4} = 0,\ x_b = \dfrac{1}{4}.$

d. $\pi x_e - \dfrac{\pi}{4} = 2\pi,\ x_e = \dfrac{9}{4}.$

e. $p = \dfrac{9}{4} - \dfrac{1}{4} = 2.$

f. Significant points: when $x = \frac{1}{4}, \frac{3}{4}, \frac{5}{4}, \frac{7}{4},$ and $\frac{9}{4}.$

One cycle of the graph is shown in Fig. 14-6.

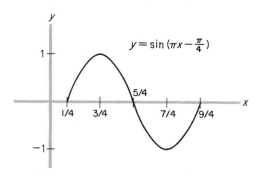

FIGURE 14-6

As a final note and summary, we point out that for the graph of $y = a \sin (bx + c),$

a.   *a* affects amplitude and reflection

b.   *b* affects the period

c.   *b* and *c* affect displacement.

EXERCISE 14-2

*In Problems 1–18, sketch one complete cycle of the graph of the given equation and label
the values of x for the quarter-cycle points. Do not plot needless points. State the amplitude
A and the period p.*

1. $y = 3 \sin x$ **2.** $y = -4 \sin x$

3. $y = \sin 2x$ **4.** $y = \sin \dfrac{x}{2}$

5. $y = 2 \sin \frac{1}{3}x$ **6.** $y = \frac{1}{2} \sin 3x$

7. $y = 4 \sin 2x$ **8.** $y = -3 \sin \frac{2}{3}x$

9. $y = -2 \sin 4x$

10. $y = 2 \sin 6x$

11. $y = \sin\left(x + \dfrac{\pi}{4}\right)$

12. $y = -\dfrac{1}{4} \sin\left(x + \dfrac{\pi}{2}\right)$

13. $y = 4 \sin\left(2x - \dfrac{\pi}{2}\right)$

14. $y = 2 \sin\left(x - \dfrac{2\pi}{3}\right)$

15. $y = -3 \sin\left(\dfrac{x}{2} + \dfrac{\pi}{2}\right)$

16. $y = 3 \sin\left(-\dfrac{3\pi}{2} + \dfrac{2}{3}\pi x\right)$

17. $y = \sin(1 + \pi x)$

18. $y = 1.2 \sin \dfrac{3\pi x}{2}$.

19. The transverse wave traveling on a stretched string is of the form $y = 2 \sin[\pi(2x - 20t)]$, where t is time in seconds. Sketch one cycle of this wave when $t = 2$ sec.

20. The velocity v of a particle undergoing simple harmonic motion is given by the relation $v = -2\pi f A \sin(2\pi f t)$, where f is the frequency in cycles per second, A is the maximum displacement from the equilibrium point, and t is time in seconds. Sketch the graph of this equation for a frequency of 50 cycles per second and $A = 10$ m.

21. The horizontal range R of a projectile having an initial speed v_0 and projected at an angle θ above the horizontal is given by

$$R = \frac{v_0^2}{g} \sin 2\theta.$$

a. Draw one cycle of the graph of this equation.

b. At what angle of projection, θ, does your graph indicate the horizontal range is maximum? Note the graph has physical significance for the first half-cycle only, for an object must be projected *above* the horizontal.

22. When the coil of a generator rotates in a magnetic field, the induced voltage ε in the coil as a function of time is given by the expression

$$\varepsilon = \varepsilon_m \sin(\omega t - \varphi),$$

where ε_m is the maximum induced voltage, ω is the angular velocity of the coil in radians per second, t is the time in seconds, and φ is the phase angle in radians. Let $\varepsilon_m = 20$ volts, $\omega = \pi/4$ radians/sec, and $\varphi = \pi/2$ radians. Sketch one cycle of the curve and determine, from your graph, the induced voltage when $t = 6$ sec.

14-3

THE GRAPH OF THE COSINE FUNCTION

To sketch the graph of the basic cosine function we shall employ the same techniques as we used for the sine function. Letting

$$y = \cos x, \tag{14-3}$$

the assumed values of x and the corresponding values of y are given in Table 14-2. For convenience we have again chosen the values of x to be special angles. The graph of $y = \cos x$ is shown in Fig. 14-7.

Table 14-2

x		$y = \cos x$	x		$y = \cos x$
Degrees	Radians		Degrees	Radians	
0	0	1	210	$\dfrac{7\pi}{6}$	$-\dfrac{\sqrt{3}}{2} \approx -.87$
30	$\dfrac{\pi}{6}$	$\dfrac{\sqrt{3}}{2} \approx .87$	240	$\dfrac{4\pi}{3}$	$-\dfrac{1}{2} = -.5$
60	$\dfrac{\pi}{3}$	$\dfrac{1}{2} = .5$	270	$\dfrac{3\pi}{2}$	0
90	$\dfrac{\pi}{2}$	0	300	$\dfrac{5\pi}{3}$	$\dfrac{1}{2} = .5$
120	$\dfrac{2\pi}{3}$	$-\dfrac{1}{2} = -.5$	330	$\dfrac{11\pi}{6}$	$\dfrac{\sqrt{3}}{2} \approx .87$
150	$\dfrac{5\pi}{6}$	$-\dfrac{\sqrt{3}}{2} \approx -.87$	360	2π	1
180	π	-1	390	$\dfrac{13\pi}{6}$	$\dfrac{\sqrt{3}}{2} \approx .87$

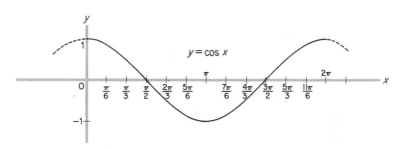

FIGURE 14-7

It should be clear to you that as with the function $y = \sin x$ the period of $y = \cos x$ is 2π radians and the amplitude is 1. Indeed, the complete discussion in Sec. 14-2 of amplitude, reflection, period, and displacement is directly applicable to variations of the cosine function. The following examples will illustrate.

EXAMPLE
6

Sketch the graph of the equation

$$y = 3 \cos 6x.$$

a. The amplitude is 3.

b. The basic shape is that of a cosine curve.

c. $6x_b = 0$, $x_b = 0$.

d. $6x_e = 2\pi$, $x_e = \dfrac{\pi}{3}$.

e. $p = \dfrac{\pi}{3} - 0 = \dfrac{\pi}{3}$.

f. Significant points: when $x = 0$, $\pi/12$, $\pi/6$, $\pi/4$, and $\pi/3$.

The graph is shown in Fig. 14-8.

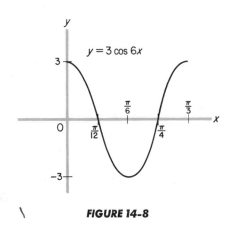

$y = 3 \cos 6x$

FIGURE 14-8

EXAMPLE 7

Sketch the graph of

$$y = -2 \cos (7x - \pi).$$

a. The amplitude is 2.

b. The basic shape is that of an inverted cosine curve.

c. $7x_b - \pi = 0$, $x_b = \pi/7$.

d. $7x_e - \pi = 2\pi$, $x_e = \dfrac{3\pi}{7}$.

e. $p = \dfrac{3\pi}{7} - \dfrac{\pi}{7} = \dfrac{2\pi}{7}$.

f. Significant points: when $x = \pi/7$, $3\pi/14$, $2\pi/7$, $5\pi/14$, and $3\pi/7$.

The graph is sketched in Fig. 14-9.

In electrical circuits it is often necessary to deal with the *phase relationship* between trigonometric graphs having the same period. For example, in a.c. circuits the voltage and current are both expressible as cosine or sine functions and the graphs of these functions do not necessarily pass through their zero (or maximum) values at the same instant of time. If they do not rise and fall together, a *phase difference* is said to exist between them.

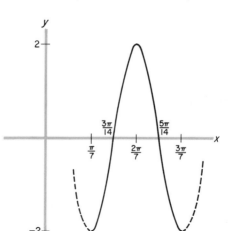

$$y = -2\cos(7x - \pi)$$

FIGURE 14-9

In Fig. 14-10 we have indicated three different phase relationships that exist between a voltage V and a current I in an a.c. circuit. Figures 14-10(a), (b), and (c), respectively, represent an a.c. generator connected in series to a pure resistor R, a pure capacitor C, and a pure inductor L. In (a), the current and voltage are said to be *in phase* because they pass through their zero and maximum values at the same instant of time. This relationship is characteristic of a purely resistive a.c. circuit. In (b), the current and voltage are said to be 90° *out of phase* since their respective maximum and zero values are displaced by $\frac{1}{4}$ cycle. Note that in discussing phase relation-

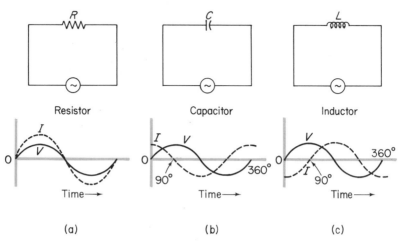

FIGURE 14-10

ships the engineer speaks in terms of *electrical degrees,* where 1 cycle represents 360 electrical degrees. More specifically we say that in (b) the current *leads* the voltage by 90° because it reaches its maximum (or zero) value $\frac{1}{4}$ cycle, or 90 electrical degrees, earlier than the voltage does as we view the curves from left to right. Equivalently we can say that the voltage *lags* the current by 90°. This is characteristic of a purely capacitive a.c. circuit. Finally, in (c) the voltage *leads* the current by 90°—characteristic of a purely inductive a.c. circuit.

EXERCISE 14-3

In Problems 1–12, sketch one complete cycle of each curve and label the values of x for the quarter-cycle points. State the amplitude A and the period p.

1. $y = \cos 2x$ **2.** $y = -3 \cos x$

3. $y = 4 \cos 4x$ **4.** $y = \frac{1}{2} \cos 5x$

5. $y = -4 \cos 3x$ **6.** $y = \cos\left(4x - \frac{\pi}{2}\right)$

7. $y = 2 \cos\left(x + \frac{\pi}{3}\right)$ **8.** $y = \cos\left(\frac{1}{4}x\right)$

9. $y = 2 \cos\left(\frac{x}{2} - \frac{2\pi}{3}\right)$ **10.** $y = 3 \cos\left(x - \frac{\pi}{6}\right)$

11. $y = -\cos\left(\pi x - \frac{\pi}{2}\right)$ **12.** $y = 2 \cos(3x + 1)$

13. The voltage across an inductor in an a.c. circuit can be described mathematically by the equation $y_1 = 2 \cos 3x$. The voltage across another component in the same circuit is given by $y_2 = \cos\left(3x - \frac{\pi}{2}\right)$. What is the phase relationship between these two voltage signals—that is, which of the curves leads the other and by how many electrical degrees?

14. If a uniform metal disk with its plane horizontal is suspended by a wire attached to its center, the device is called a torsional pendulum. If the disk is now rotated through an angle in the horizontal plane and released, it will execute simple angular harmonic motion. For such motion the angular displacement, θ, from the equilibrium position is given by the equation

$$\theta = \theta_m \cos (\omega t + \alpha),$$

where θ_m is the maximum displacement in radians, ω is the angular frequency in radians per second, t is time in seconds, and α is the phase angle in radians. Draw 1 cycle of the graph of this equation if $\omega = \pi$ and $\alpha = \pi/2$.

15. In a series a.c. circuit having a current $i = 2 \sin(120\,\pi t)$ amp, the voltage across a pure capacitor as a function of time is $v = -3 \cos(120\,\pi t)$. Sketch the graph of both of these equations on the same coordinate plane and determine the "phase difference" between i and v.

14-4
THE GRAPH OF THE TANGENT FUNCTION

While the graphs of the sine and cosine functions were both smooth curves exhibiting wavelike appearances, the graphs of the remaining trigonometric functions are strikingly different. Consider the function

$$y = \tan x. \tag{14-4}$$

In Table 14-3 are assumed values for x and the corresponding approximate values of $y = \tan x$. Figure 14-11 shows three cycles of the graph of $y = \tan x$.

Note that when $x = \pm\pi/2, \pm 3\pi/2$, etc., there is no point on the graph,

Table 14-3

x	0	$\dfrac{\pi}{6}$	$\dfrac{\pi}{3}$	$\dfrac{\pi}{2}$	$\dfrac{2\pi}{3}$	$\dfrac{5\pi}{6}$	π	$\dfrac{7\pi}{6}$	$\dfrac{4\pi}{3}$	$\dfrac{3\pi}{2}$
$\tan x$	0	.6	1.7	—	−1.7	−.6	0	.6	1.7	—

x	$-\dfrac{\pi}{6}$	$-\dfrac{\pi}{3}$	$-\dfrac{\pi}{2}$	$-\dfrac{2\pi}{3}$	$-\dfrac{5\pi}{6}$	$-\pi$	$-\dfrac{7\pi}{6}$	$-\dfrac{4\pi}{3}$	$-\dfrac{3\pi}{2}$
$\tan x$	−.6	−1.7	—	1.7	.6	0	−.6	−1.7	—

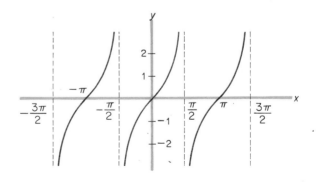

$y = \tan x$

FIGURE 14-11

for *near* these values of x the value of tan x increases or decreases without bound, depending on whether we approach these values of x from the left or right. The curve approaches, but does not intersect, the lines $x = \pm\pi/2, \pm3\pi/2$, etc. These lines are called **vertical asymptotes** of the curve; they are lines near which the function "blows up." It should be clear that $\tan(x + \pi) = \tan x$—that is, the curve repeats itself every π radians and the period of $y = \tan x$ is π radians. Moreover, the amplitude is undefined, as the graph shows.

14-5
GRAPHS OF THE COTANGENT, SECANT, AND COSECANT FUNCTIONS

The graphs of $y = \cot x$, $y = \sec x$, and $y = \csc x$ are conveniently sketched if we consider the reciprocal relationships

$$\cot x = \frac{1}{\tan x}, \qquad \sec x = \frac{1}{\cos x}, \qquad \csc x = \frac{1}{\sin x}.$$

Figure 14-12 shows the graph of $y = \sin x$ as a broken line and the graph of $y = \csc x$ as solid lines. We obtain the cosecant curve from the sine curve by using the reciprocal relationship in the following manner. First we sketch the graph of $y = \sin x$. For a point (x, y) on the graph of $y = \sin x$, we estimate the value of y. The reciprocal of this value is then the ordinate of the corresponding point on $y = \csc x$ with the same

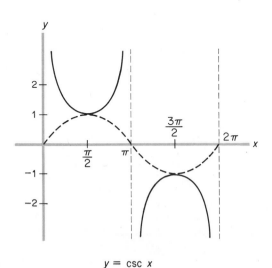

$y = \csc x$

FIGURE 14-12

abscissa. For example, when $\sin x = \frac{1}{2}$, then $\csc x = 2$; when $\sin x = 1$, then $\csc x = 1$; etc. Thus, when $\sin x = 0$, clearly $\csc x$ is undefined. By plotting a sufficient number of points, we can easily sketch the graph of $y = \csc x$, as shown in Fig. 14-12. Note that the lines $x = 0$, π, 2π, etc., are vertical asymptotes.

By similar reasoning we can also sketch the graphs of

$$y = \cot x = \frac{1}{\tan x} \quad \text{and} \quad y = \sec x = \frac{1}{\cos x}.$$

These graphs, with the graphs of the corresponding reciprocal functions shown in broken lines, are indicated in Figs. 14-13 and 14-14. Note that when $\tan x = 0$, $\cot x$ is undefined; when $\tan x$ is undefined, $\cot x = 0$. It should be clear that $\cot(x + \pi) = \cot x$, $\sec(x + 2\pi) = \sec x$, and $\csc(x + 2\pi) = \csc x$. That is, the period of $y = \cot x$ is π radians and the period of both $y = \sec x$ and $y = \csc x$ is 2π radians, as is the period of

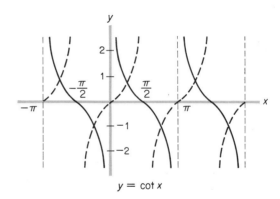

$$y = \cot x$$

FIGURE 14-13

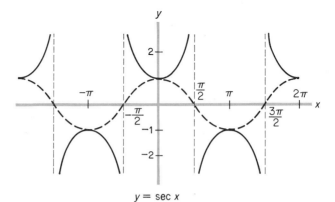

$$y = \sec x$$

FIGURE 14-14

their reciprocal functions. Furthermore, the amplitudes of $\cot x$, $\sec x$, and $\csc x$ are not defined.

EXERCISE 14-5

1. Sketch the tangent curve.

2. Sketch the cotangent curve.

3. Sketch the secant curve.

4. Sketch the cosecant curve.

14-6
COMBINATIONS OF TRIGONOMETRIC
FUNCTIONS—ADDITION OF ORDINATES

In certain applications, physical phenomena are representable mathematically as sums or differences of trigonometric functions. The graphs of such combinations of trigonometric functions may be sketched by a technique known as *addition of ordinates*. We shall illustrate this method by sketching the graph of $y = \sin x + \cos x$. This equation can be thought of as representing the sum of two functions:

$$y = y_1 + y_2, \qquad \text{where } y_1 = \sin x \text{ and } y_2 = \cos x.$$

We first sketch the graphs of $y_1 = \sin x$ and $y_2 = \cos x$ separately on the same coordinate plane, as shown in Fig. 14-15. These graphs are then combined by adding the ordinates corresponding to the same abscissa.

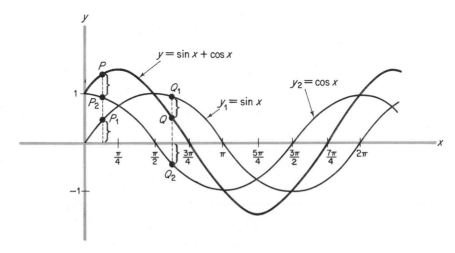

FIGURE 14-15

For example, point P is obtained by measuring the ordinate of P_1 (with a compass, for example) and adding it algebraically to the ordinate of P_2. To obtain point Q, we add the negative ordinate of point Q_2 to the ordinate of Q_1. After a suitable number of points are obtained in this manner, they are connected by a smooth curve. Thus we obtain the graph of $y = \sin x + \cos x$. We remark that engineers often speak of this addition of ordinates as the *principle of superposition*.

EXAMPLE
8

Sketch the graph of $y = \sin x + \frac{1}{3}\sin 3x$.

We first sketch the graphs of $y_1 = \sin x$ and $y_2 = \frac{1}{3}\sin 3x$ on the same coordinate plane (Fig. 14-16) using the facts that $\sin x$ has amplitude 1 and

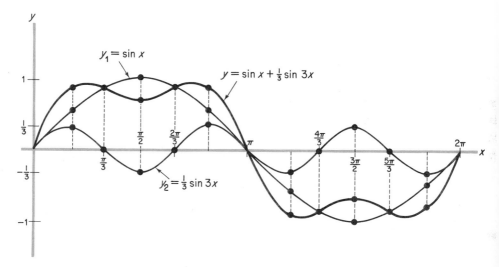

FIGURE 14-16

period 2π, but $\frac{1}{3}\sin 3x$ has amplitude $\frac{1}{3}$ and period $2\pi/3$. Next, at various values of x we "add" the "y-values" of the two graphs. We then connect the points by a smooth curve.

The French mathematician J. Fourier showed that we can analyze extremely complicated periodic waves as a combination of relatively simple waves. By his technique, well known in engineering as Fourier analysis, it can be shown that periodic waves can be represented as an infinite series—that is, an unending sum of terms—each term of which is a sine or cosine function.

EXAMPLE
9

The broken line of Fig. 14-17(a) is a sawtooth waveform, a periodic wave commonly encountered in the oscilloscope. Figure 14-17(b) indicates the graphs of the first six terms of the Fourier series for that periodic

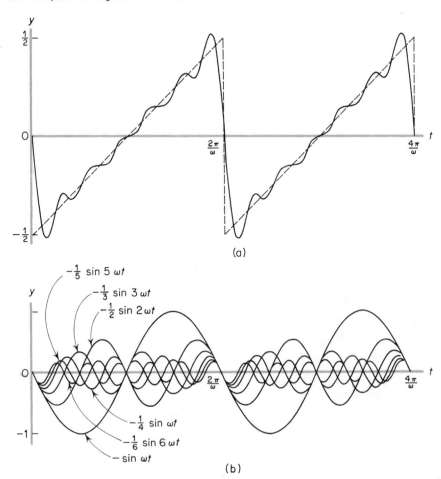

FIGURE 14-17 (Reprinted from R. Resnick and D. Halliday, *Physics*, 1st ed., by permission of John Wiley & Sons, Inc.)

wave. The solid line in Fig. 14-17(a) is the sum of the first six terms of the Fourier series found from the graphs in (b) by the technique of addition of ordinates. The solid line can be seen to be a fairly close approximation of the sawtooth waveform. As additional terms of the Fourier series are included, the approximation continually improves. Hence, the sawtooth waveform has been expressed as a sum of relatively simple sine curves.

EXERCISE 14-6

In Problems 1–8, sketch the graph of the given equation over the given interval by means of addition of ordinates.

1. $y = 2 + \sin x$; $[0, 2\pi]$ **2.** $y = 1 + \cos x$; $[0, 2\pi]$

3. $y = \sin x + \sin 2x$; $[0, 2\pi]$ **4.** $y = \sin 2x + 2 \sin x$; $[0, 2\pi]$

5. $y = \sin 2x + \cos 3x$; $[0, \pi]$ **6.** $y = 4 \sin x - 3 \cos x$; $[0, 2\pi]$

7. $y = x + \sin x$; $[0, 3\pi]$ **8.** $y = \cos 3x - \sin 2x$; $[0, \pi]$

9. Consider a string fixed at both ends in which a wavetrain is initiated by plucking the string. Waves will be reflected from both ends of the string with the resulting shape of the rope determined by the sum of the two waves, one traveling to the right and the other traveling to the left. If the wave traveling to the right is given by $y_1 = 2 \sin \pi(x - t)$ and the one to the left by $y_2 = 2 \sin \pi(x + t)$, find the shape of the rope as a function of position at $t = 3$ sec.

14-7
INVERSE TRIGONOMETRIC FUNCTIONS

If $x = \sin y$, then y is an *angle* whose sine is x and symbolically we write

$$y = \arcsin x,$$

which is read "y equals the arc sine of x," or equivalently, "y equals an angle whose sine is x." Another common notation for this is

$$y = \sin^{-1} x,$$

where the -1 is not an exponent but simply part of this new symbol. The symbol is read also as "y equals the arc sine of x." In a similar manner we define

$$\arccos x, \quad \cos^{-1} x$$
$$\arctan x, \quad \tan^{-1} x$$
$$\text{arccot } x, \quad \cot^{-1} x.$$

We have omitted arcsec x and arccsc x since they are expressible in terms of arcsin x and arccos x and, moreover, are rarely used.

EXAMPLE
10

a. Find y if $y = \arcsin \frac{1}{2}$.

Here, y is an angle whose sine is $\frac{1}{2}$. The sine function is positive in the first and second quadrants, and since the reference angle is 30° we have

$$y = \frac{\pi}{6}, \frac{5\pi}{6}, \frac{13\pi}{6}, \ldots$$

and

$$y = -\frac{7\pi}{6}, -\frac{11\pi}{6}, \ldots.$$

b. If $y = \cos^{-1}(-\frac{1}{2})$, by an analysis similar to part a we have

$$y = \pm\frac{2\pi}{3}, \pm\frac{4\pi}{3}, \pm\frac{8\pi}{3}, \ldots.$$

From Example 10 it should be clear that the equations $y = \arcsin x$ and $y = \arccos x$ do not define functions of x, since given a value of x there may correspond infinitely many values of y. To remedy this situation, we define so-called **principal values** whereby we restrict arcsin x, etc., to the following intervals:

$$-\frac{\pi}{2} \leq \operatorname{Arcsin} x \leq \frac{\pi}{2}, \qquad -1 \leq x \leq 1$$

$$0 \leq \operatorname{Arccos} x \leq \pi, \qquad -1 \leq x \leq 1$$

$$-\frac{\pi}{2} < \operatorname{Arctan} x < \frac{\pi}{2}, \qquad -\infty < x < \infty$$

$$0 < \operatorname{Arccot} x < \pi, \qquad -\infty < x < \infty.$$

Under these conditions, we call Arcsin x (spelled with a capital A), Arccos x, etc., **inverse trigonometric functions.**

EXAMPLE
11

a. If $y = \operatorname{Arcsin} \sqrt{2}/2$, then y is *the* angle between $-\pi/2$ and $\pi/2$ whose sine is $\sqrt{2}/2$. Clearly

$$y = \operatorname{Arcsin} \frac{\sqrt{2}}{2} = \frac{\pi}{4}.$$

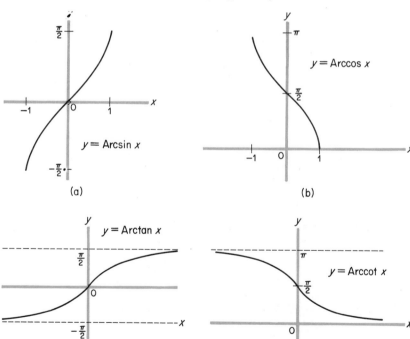

(a)

(b)

(c)

(d)

FIGURE 14-18

b. If $y = \text{Cos}^{-1} \frac{1}{2}$, then y is *the* angle between 0 and π whose cosine is $\frac{1}{2}$. Thus,

$$y = \text{Cos}^{-1}\frac{1}{2} = \frac{\pi}{3}.$$

The graphs of the inverse trigonometric functions are shown in Fig. 14-18.

EXERCISE 14-7

Evaluate each of the following.

1. $\text{Arcsin} \dfrac{\sqrt{3}}{2}$

2. $\text{Arcsin } 1$

3. $\text{Arccos } (-1)$

4. $\text{Cos}^{-1} \dfrac{\sqrt{3}}{2}$

5. $\text{Sin}^{-1} \left(-\frac{1}{2}\right)$

6. $\text{Arcsin} \left(-\frac{1}{2}\right)$

7. $\text{Arctan } 1$

8. $\text{Arctan} \left(-\sqrt{3}\right)$

9. $\sin \left(\text{Arcsin } \frac{1}{3}\right)$

10. $\cos \left[\text{Arccos} \left(-\frac{1}{5}\right)\right]$

11. $\cos \left[\text{Arcsin} \left(-\frac{1}{2}\right)\right]$

12. $\sin \left[\text{Arccos} (-1)\right]$

13. $\text{Arcsin} \left(\sin \dfrac{2\pi}{3}\right)$

14. $\text{Arcsin} \left(\sin \dfrac{\pi}{3}\right)$

14-8
THE OSCILLOSCOPE—LISSAJOUS FIGURES

We conclude this chapter on trigonometric graphs by considering the generation of *Lissajous figures* on the screen of a cathode-ray oscilloscope. We remark that the oscilloscope is one of the most useful and versatile instruments available to the engineer.

The heart of an oscilloscope is the cathode-ray tube in which a narrow beam of high-speed electrons strikes a special screen which fluoresces; that is, it gives off visible light at the point where the beam strikes the screen. The position at which the beam strikes the screen can be controlled by two pairs of *deflection plates*. A voltage signal applied to the horizontal deflection plates will cause the beam to move to the right or to the left depending on its polarity, while a signal applied to the vertical deflection plates will cause the beam to move up or down depending on its polarity.

When sinusoidal signals (that is, signals represented by sine or cosine curves) are applied to the two sets of deflecting plates, the path traced out by the flourescent dot formed by the impinging electron beam forms a pattern on the screen or face of the oscilloscope. Furthermore, the pattern will remain stationary as long as the amplitudes and phase relationships of the signals applied to the deflection plates do not change.

Let us consider an elementary situation in which a voltage signal in the shape of a sine curve is applied to the vertical deflection plates with *no signal* applied to the horizontal deflection plates. This situation is illustrated in Fig. 14-19 where, for convenience, we show rectangular coor-

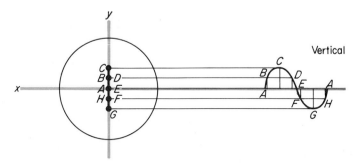

FIGURE 14-19

dinate axes on the screen of the oscilloscope. With no signal being applied to the horizontal deflection plates the dot will not move to the left or right and, hence, it will remain on the *y*-axis. Its motion on the *y*-axis is controlled by the signal applied to the vertical deflection plates as shown in Fig. 14-19. The dot is at the center of the screen when the voltage applied to the vertical deflection plates is at point *A* on the voltage signal. The dot moves upward along the *y*-axis and reaches a maximum height at point *C* as the voltage applied to the deflection plates reaches its maximum value at point *C*. Then, corresponding to a decreasing voltage signal between points *C* and *G*, the dot moves downward along the *y*-axis, reaching its lowest position at point *G*. Since the voltage signal increases to its original value in the last quarter of the cycle between points *G* and *A*, the dot will rise along the *y*-axis and return to its initial position. If the frequency of the signal to the vertical deflection plates were, for example, 60 cycles per second, the travels of the dot outlined above would be repeated 60 times each second. It is not surprising then that we would "see" a *stationary* vertical line on the oscilloscope screen. In practice oscilloscopes are designed to operate at a wide range of frequencies.

It should be clear that if the sinusoidal voltage shown in Fig. 14-19 were applied to the horizontal deflection plates with no signal applied to the vertical deflection plates, the dot would trace a path back and forth along the *x*-axis and we would "see" a horizontal line on the oscilloscope screen.

When sinusoidal signals are simultaneously applied to both sets of deflection plates, the pattern which results depends on the amplitudes, frequencies, and phase relationships of the two signals. The patterns observed on the screen when the ratio of the frequencies of the two signals can be expressed as a ratio of integers are called *Lissajous figures*.

As an example let us consider voltage signals that have the *same frequency*, have different amplitudes, and are 90° out of phase with one another. The mathematical form of such voltage signals applied to the horizontal and vertical deflection plates are given by Eqs. (14-5) and (14-6), respectively:

$$V_x = 4\sin(\omega t) \tag{14-5}$$

$$V_y = 2\cos(\omega t) \tag{14-6}$$

In each case we assume the value of ω is the same; it represents the constant quantity $2\pi f$, called the angular frequency. The physical situation

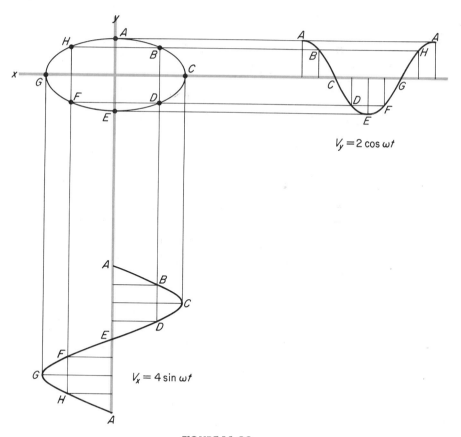

FIGURE 14-20

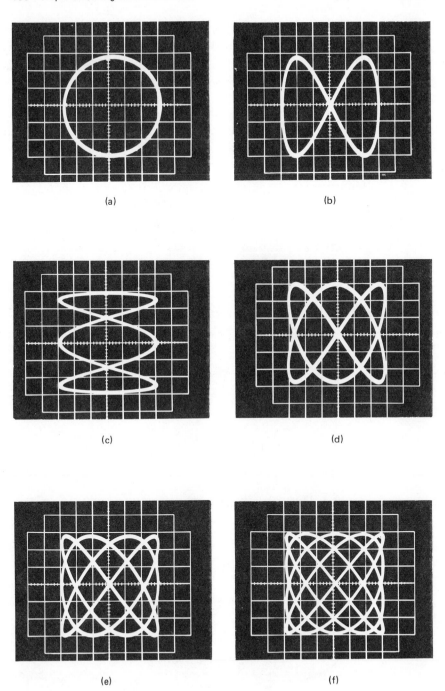

(a)

(b)

(c)

(d)

(e)

(f)

FIGURE 14-21 (Courtesy of Donald Lyons, Department of General Engineering, Pennsylvania State University.)

is illustrated in Fig. 14-20. By considering the horizontal and vertical deflections simultaneously, it should be clear that these voltage signals will result in the generation of the Lissajous figure shown, which is called an ellipse. Moreover, it should be apparent that if the amplitudes of the two signals were the same, the Lissajous figure would be a circle.

An important application of Lissajous figures is in the calibration of signal generators or, indeed, in any instance where a frequency comparison is desired. Photographs of actual Lissajous patterns for various values of the ratio f_y/f_x, where f_y and f_x are the frequencies of the vertical and horizontal signals, respectively, are shown in Fig. 14-21. Each of these patterns was generated by using signals of equal amplitudes to the vertical and horizontal deflection plates.

EXERCISE 14-8

1. Construct a diagram similar to Fig. 14-20 and determine the resulting pattern on the screen if $V_x = 2 \sin \omega t$ and $V_y = 2 \sin \omega t$.

2. In Problem 1, if the amplitude of the vertical deflection signal were increased while that of the horizontal deflection signal remained the same, how would the pattern on the screen change?

3. How would the pattern of Fig. 14-20 appear if the deflection voltages were switched, that is, if the vertical signal were applied to the horizontal deflection plates and the horizontal signal were applied to the vertical deflection plates?

4. As an example of the Lissajous patterns generated for ratios of vertical to horizontal frequencies which are expressible by integers, construct a diagram for

$$V_x = \sin 2\pi t$$

and

$$V_y = \sin 4\pi t,$$

which illustrates the ratio $f_y/f_x = 2/1$.

14-9
REVIEW

REVIEW QUESTIONS

The period of $y = 4 \sin 18x$ is equal to _____(1)_____. Its amplitude is _____(2)_____.

(1) $\pi/9$ (2) **4**

The values of $y = 3\sin\left(\frac{6}{97}x - \frac{2\pi}{63}\right)$ range from a minimum of
_____(3)_____ to a maximum of _____(4)_____.

(3) **−3** (4) **3**

In comparison to the graph of $y = \sin x$, the graph of $y = \sin\left(x + \frac{\pi}{2}\right)$
$\underset{(5)}{\underline{\text{(leads) (lags)}}}$ by $\pi/2$.

(5) **leads**

The periods of $y = \sin x$ and $y = \cos x$ are both equal to _____(6)_____,
but the periods of $y = \tan x$ and $y = \cot x$ are both equal to _____(7)_____.

(6) **2π** (7) **π**

The values of y where $y = \sec x$ for $-\infty < x < \infty$ can be expressed in set-builder notation as _____(8)_____.

(8) **$\{y \mid |y| > 1\}$**

The graph of $y = \sin x$ is the same as the graph of $y = -\cos x$ displaced $\pi/2$
radians to the $\underset{(9)}{\underline{\text{(left) (right)}}}$.

(9) **left**

Arcsin $\left(-\frac{1}{2}\right) =$ _____(10)_____ and Cos$^{-1}\left(-\frac{1}{2}\right) =$ _____(11)_____.

(10) **$-\pi/6$** (11) **$2\pi/3$**

REVIEW PROBLEMS

In Problems 1–12, sketch the graph of the given equation and state all particulars concerning amplitude and period where applicable.

1. $y = 2\sin\left(3x + \frac{\pi}{2}\right)$ 2. $y = -\cos 3x$

3. $y = \tan x$ 4. $y = \sin(7x - \pi)$

5. $y = 4\cos\left(\theta - \frac{\pi}{2}\right)$ 6. $y = 3\cos\left(2\theta + \frac{\pi}{6}\right)$

7. $y = \cos\left(\dfrac{x}{2} - \dfrac{\pi}{3}\right)$ **8.** $y = 3 \sin 8x$

9. $y = \sec x$ **10.** $y = -\sin\left(8\pi + \dfrac{\pi}{3}\right)$

11. $y = \sin 2x + 2 \cos x$ **12.** $y = 2 \sin x - \cos x$

13. Determine

     a. $\text{Arccos } 0 + \text{Arcsin } 0$

     b. $\text{Sin}^{-1}\left(\dfrac{\sqrt{2}}{2}\right) + \text{Cos}^{-1}\left(-\dfrac{\sqrt{2}}{2}\right)$

     c. $\text{Arcsin } \tfrac{1}{5} + \text{Arccos } \tfrac{1}{5}$

15

trigonometric formulas and equations

15-1

FUNDAMENTAL IDENTITIES

Each of the trigonometric reciprocal relationships which were given in Chapter 7 is an example of a **trigonometric identity**—that is, an equation involving trigonometric functions that is true for all values of the angles for which the equation is defined. For your convenience we repeat the identities here in their various forms:

$$\sin \theta = \frac{1}{\csc \theta} \qquad \csc \theta = \frac{1}{\sin \theta} \qquad \sin \theta \csc \theta = 1 \qquad (15\text{-}1)$$

$$\cos \theta = \frac{1}{\sec \theta} \qquad \sec \theta = \frac{1}{\cos \theta} \qquad \cos \theta \sec \theta = 1 \qquad (15\text{-}2)$$

$$\tan \theta = \frac{1}{\cot \theta} \qquad \cot \theta = \frac{1}{\tan \theta} \qquad \tan \theta \cot \theta = 1 \qquad (15\text{-}3)$$

With rather elementary algebraic manipulations it is possible to derive many other useful trigonometric·identities. For example, when an angle θ is drawn in standard position and a perpendicular is constructed from

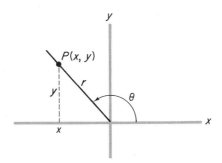

FIGURE 15-1

a point $P(x, y)$ on the terminal side of θ to the x-axis (see Fig. 15-1), we have $\sin \theta = y/r$, $\cos \theta = x/r$, and we can write

$$\frac{\sin \theta}{\cos \theta} = \frac{\dfrac{y}{r}}{\dfrac{x}{r}} = \frac{y}{r} \cdot \frac{r}{x} = \frac{y}{x} = \tan \theta.$$

Thus,

$$\frac{\sin \theta}{\cos \theta} = \tan \theta. \tag{15-4}$$

Using this result and the reciprocal relation of Eq. (15-3), we have

$$\frac{\cos \theta}{\sin \theta} = \cot \theta. \tag{15-5}$$

Moreover, by the Pythagorean theorem,

$$x^2 + y^2 = r^2.$$

Successively dividing each member of this equation by r^2, x^2, and y^2, respectively, we obtain three forms of this equation:

$$\frac{x^2}{r^2} + \frac{y^2}{r^2} = 1, \qquad 1 + \frac{y^2}{x^2} = \frac{r^2}{x^2}, \qquad \frac{x^2}{y^2} + 1 = \frac{r^2}{y^2}$$

or

$$\left(\frac{x}{r}\right)^2 + \left(\frac{y}{r}\right)^2 = 1, \qquad 1 + \left(\frac{y}{x}\right)^2 = \left(\frac{r}{x}\right)^2, \qquad \left(\frac{x}{y}\right)^2 + 1 = \left(\frac{r}{y}\right)^2.$$

Using the definitions of the trigonometric functions we see that these three equations are equivalent, respectively, to the identities

$$\sin^2 \theta + \cos^2 \theta = 1, \qquad \sin \theta = \pm \sqrt{1 - \cos^2 \theta}$$

$$\cos \theta = \pm \sqrt{1 - \sin^2 \theta} \tag{15-6}$$

$$1 + \tan^2 \theta = \sec^2 \theta, \qquad \tan \theta = \pm \sqrt{\sec^2 \theta - 1}$$
$$\sec \theta = \pm \sqrt{1 + \tan^2 \theta} \qquad \text{(15-7)}$$

$$1 + \cot^2 \theta = \csc^2 \theta, \qquad \cot \theta = \pm \sqrt{\csc^2 \theta - 1}$$
$$\csc \theta = \pm \sqrt{1 + \cot^2 \theta} \qquad \text{(15-8)}$$

In these equations note the use of the standard notation $\sin^2 \theta = (\sin \theta)^2$ and similarly for all trigonometric functions. The expression $\sin^2 \theta$ is read "sine squared theta" and means, explicitly, the square of the sine of the angle theta. Note that Eq. (15-6) implies $1 - \cos^2 \theta = \sin^2 \theta$, etc.

Equations (15-1)–(15-8) can be considered the fundamental trigonometric identities and their variations, and you should become totally familiar with them. These identities provide a means of proving other identities—that is, of changing a given expression to a particular equivalent form. In physical situations this can mean dealing with a simpler and, hence, more suitable form of a trigonometric expression.

EXAMPLE 1   When a ball of mass m is suspended from a string and pulled aside by a horizontal force F until the string makes an angle θ with the vertical, three forces are acting on the ball, as shown in Fig. 15-2. The downward gravita-

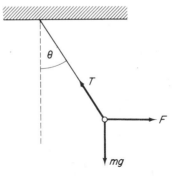

FIGURE 15-2

tional force is mg, the tension in the string is T, and F is the supporting force. The system, being stationary, is said to be in equilibrium and the relationships between the forces are determined by applying Newton's second law. For this case, by summing vertical and horizontal components we have

$$T \cos \theta - mg = 0$$

and

$$F - T \sin \theta = 0.$$

Solving the first of these equations for T, we obtain $T = mg/\cos \theta$; substituting this value into the second equation yields

$$F - \left(\frac{mg}{\cos\theta}\right)\sin\theta = 0$$

$$F = mg\left(\frac{\sin\theta}{\cos\theta}\right).$$

Finally, by Eq. (15-4),

$$F = mg\tan\theta.$$

In essence, we have used a trigonometric identity to show that

$$\left(\frac{mg}{\cos\theta}\right)\sin\theta = mg\tan\theta.$$

As with verbal problems there is no set procedure to follow to prove a given identity. In general, it is a question of showing, by algebraic manipulation and/or substitution using the fundamental identities, that one member of an identity is equal to the other member. When no hint of a procedure is evident, it is sometimes useful to express everything first in terms of sines and cosines and then attempt a simplification by valid operations. Also, it is usually best to reduce the more complex member to agree with the other member, although at times you may find it convenient to reduce both members (see Example 4b).

EXAMPLE
2

Prove the identity $\sec\theta - \tan\theta\sin\theta = \cos\theta$.

Expressing the left member in terms of sines and cosines, we shall show it is equal to the right member: by Eqs. (15-2) and (15-4),

$$\frac{1}{\cos\theta} - \frac{\sin\theta}{\cos\theta}\cdot\sin\theta \overset{?}{=} \cos\theta$$

$$\frac{1 - \sin^2\theta}{\cos\theta} \overset{?}{=} \cos\theta$$

$$\frac{\cos^2\theta}{\cos\theta} \overset{?}{=} \cos\theta \qquad [\text{Eq. (15-4)}]$$

$$\cos\theta = \cos\theta.$$

EXAMPLE
3

Prove the identity

$$\frac{\cos^2\theta}{1-\sin\theta} = 1 + \sin\theta.$$

By Eq. (15-6) we can write

$$\frac{1-\sin^2\theta}{1-\sin\theta} \overset{?}{=} 1 + \sin\theta$$

$$\frac{(1+\sin\theta)(1-\sin\theta)}{(1-\sin\theta)} \overset{?}{=} 1 + \sin\theta$$

$$1 + \sin\theta = 1 + \sin\theta.$$

EXAMPLE

4

a. Prove the identity $\sec^2 x - \dfrac{\sin^2 x}{\cos^2 x} = 1$.

$$\sec^2 x - \left(\dfrac{\sin x}{\cos x}\right)^2 \overset{?}{=} 1$$

From Eq. (15-4),

$$\sec^2 x - \tan^2 x \overset{?}{=} 1.$$

From Eq. (15-7),

$$1 = 1.$$

b. Prove the identity $\tan \theta + \cot \theta = \csc \theta \sec \theta$.

$$\tan \theta + \cot \theta \overset{?}{=} \csc \theta \sec \theta$$

$$\dfrac{\sin \theta}{\cos \theta} + \dfrac{\cos \theta}{\sin \theta} \overset{?}{=} \dfrac{1}{\sin \theta} \cdot \dfrac{1}{\cos \theta}$$

$$\dfrac{\sin^2 \theta + \cos^2 \theta}{\cos \theta \sin \theta} \overset{?}{=} \dfrac{1}{\sin \theta} \cdot \dfrac{1}{\cos \theta}$$

$$\dfrac{1}{\sin \theta \cos \theta} = \dfrac{1}{\sin \theta} \cdot \dfrac{1}{\cos \theta}.$$

EXERCISE 15-1

Prove the identities in Problems 1–38.

1. $\tan x \cos x = \sin x$

2. $\cot x \sin x = \cos x$

3. $\dfrac{\csc x}{\sec x} = \cot x$

4. $\dfrac{\sin^2 x}{1 - \sin^2 x} = \tan^2 x$

5. $\dfrac{1}{1 - \cos^2 x} = \csc^2 x$

6. $\dfrac{1}{1 + \tan^2 x} = \cos^2 x$

7. $\dfrac{1}{\csc^2 x - 1} = \tan^2 x$

8. $\dfrac{1 - \cos^2 x}{1 - \sin^2 x} = \tan^2 x$

9. $\dfrac{\cos^2 x}{1 - \sin^2 x} = 1$

10. $\tan x + \cot x = \sec x \csc x$

11. $\dfrac{1 + \tan^2 x}{\csc x} = \tan x \sec x$

12. $\dfrac{1 + \cot^2 x}{\sec x} = \cot x \csc x$

13. $\sin x \, (1 + \cot^2 x) = \csc x$

14. $(1 + \tan^2 x) \cos x = \sec x$

15. $\dfrac{1 - \sin x}{\cos x} = \sec x - \tan x$

16. $\cot x \, (\sec^2 x - 1) = \tan x$

17. $\dfrac{1}{\tan x + \cot x} = \cos x \sin x$

18. $\cot x + \tan x = \sec x \csc x$

19. $\csc^4 x - \cot^4 x = \csc^2 x + \cot^2 x$

20. $\dfrac{1 + \cot^2 x}{\cos^2 x \csc^2 x + 1} = 1$

21. $\sin^2 x - \cos^2 x + 1 - 2 \sin^2 x \cos^2 x = 2 \sin^4 x$

22. $\dfrac{2}{1 - \cos x} = \dfrac{2 \sec x}{\sec x - 1}$

23. $\dfrac{\cos x}{\tan x + \sec x} - \dfrac{\cos x}{\tan x - \sec x} = 2$

24. $\dfrac{\cos^2 x - \dfrac{\cos^2 x}{\csc^2 x}}{\sin^2 x - \dfrac{\cos^2 x}{\csc^2 x}} = \cot^4 x$

25. $\dfrac{\csc^4 x - 1}{\cot^2 x} = \csc^2 x + 1$

26. $\dfrac{\sin^2 x}{\cos x} (\tan x - \cos x \cot x) = \sin x \, (\tan^2 x - \cos x)$

27. $\sin x \left(\dfrac{\sin x + \tan x}{1 + \cos x} \right) = \sin x \tan x$

28. $\dfrac{\sin x \cos y}{\sin y \cos x} = \tan x \cot y$

29. $\sin x \cos x = \dfrac{\cot x}{1 + \cot^2 x}$

30. $\dfrac{1 + \tan x}{\sec x + \csc x} = \sin x$

31. $\dfrac{1}{\csc^2 x \, (\cos^4 x + \cos^2 x \sin^2 x)} = \tan^2 x$

32. $\dfrac{1}{\sec x - \tan x} - \dfrac{1}{\sec x + \tan x} = 2 \tan x$

33. $\dfrac{\sin^2 x + \cos^2 x}{\cos^2 x} = \sec^2 x$

34. $(1 + \cos x)\csc x + \dfrac{1}{\csc x \,(1 + \cos x)} = 2\csc x$

35. $\cos^4 x - \sin^4 x + 1 = 2\cos^2 x$

36. $\dfrac{\sin x \cos y - \sin y \cos x}{\cos x \cos y + \sin x \sin y} = \dfrac{\tan x - \tan y}{1 + \tan x \tan y}$

37. $\dfrac{\cos x \cos y - \sin x \sin y}{\sin x \cos y - \cos x \sin y} = \dfrac{1 - \cot x \cot y}{\cot x - \cot y}$

38. $\left(2 + \dfrac{2}{1 + \dfrac{\cos x}{\cos x + 1}}\right)\left(2 + \dfrac{2}{1 + \dfrac{1}{\cos x}}\right)$

$\qquad = -6\csc^2 x\,(3\cos x + 2)(\cos x - 1)$

39. When a weight W slides down an inclined plane of angle θ at constant speed, Newton's second law applied to the weight yields the equations

$$W \sin \theta - \mu N = 0$$

and

$$N - W \cos \theta = 0,$$

where N is the normal force exerted by the plane on the block and μ is the coefficient of friction. Solve the equations simultaneously and show that $\mu = \tan \theta$.

40. When a beam of circularly polarized light falls on a polarizing sheet, the resulting amplitude, E, of the electric field component is given by

$$E = \sqrt{E_x^2 + E_y^2},$$

where $E_x = E_m \sin(\omega t)$ and $E_y = E_m \cos(\omega t)$. Prove that for such a situation $E = E_m$. You may assume $E_m > 0$.

15-2
FUNCTIONS OF THE SUM AND DIFFERENCE OF ANGLES

In the theory of the interference and diffraction of electromagnetic waves, it is usual to assume that the electric field component of such a wave is of the form $E_0 \sin(\omega t + \varphi)$. In studies of reflection and refraction of light it is shown that the index of refraction of a prism is proportional to $\sin\left[\tfrac{1}{2}(\psi + \varphi)\right]$, where φ is the angle of a vertex of the prism and ψ is the angle of minimum deviation. In these and many other situations, it is useful to express a relation involving a function of a sum or difference of two angles in terms of functions of the individual angles. We shall derive such identities here and use them in the following section to develop the double- and half-angle formulas.

If α and β are two positive acute angles, then the sum $\alpha + \beta$ may be a first- or second-quadrant angle; these two cases are illustrated in Figs. 15-3 and 15-4. The discussion that follows refers to both figures.

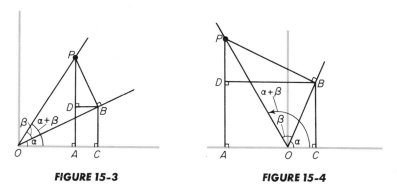

FIGURE 15-3   **FIGURE 15-4**

From any point P on the terminal side of $\alpha + \beta$, perpendiculars are constructed both to the x-axis at A and to the terminal side of α at B. From B, perpendiculars are constructed to the x-axis at C and to AP at D. Note that $AD = CB$. Furthermore, since

$$\sphericalangle BPD + \sphericalangle PBD = 90°$$

and

$$\sphericalangle DBO + \sphericalangle PBD = 90°,$$

it follows that

$$\sphericalangle BPD = \sphericalangle DBO.$$

However, $\sphericalangle DBO = \sphericalangle \alpha$ since they are alternate interior angles formed by parallel lines cut by a transversal. Thus, $BPD = \sphericalangle \alpha$.

Now, from $\triangle OCB$, $\sin \alpha = CB/OB$, that is, $CB = OB \sin \alpha$; from $\triangle PDB$, $\cos \alpha = DP/PB$, that is, $DP = PB \cos \alpha$. Therefore,

$$\sin(\alpha + \beta) = \frac{AP}{OP} = \frac{AD + DP}{OP} = \frac{CB + DP}{OP}$$

$$= \frac{OB \sin \alpha + PB \cos \alpha}{OP}$$

$$= \frac{OB}{OP} \sin \alpha + \frac{PB}{OP} \cos \alpha$$

$$= \cos \beta \sin \alpha + \sin \beta \cos \alpha$$

$$\mathbf{\sin(\alpha + \beta) = \sin \alpha \cos \beta + \cos \alpha \sin \beta.} \qquad (15\text{-}9)$$

Similarly,

$$\cos(\alpha + \beta) = \frac{OA}{OP} = \frac{OC - DB}{OP}.$$

From $\triangle OCB$, $\cos \alpha = OC/OB$, that is, $OC = OB \cos \alpha$; from $\triangle BPD$, $\sin \alpha = DB/BP$, that is, $DB = BP \sin \alpha$. Therefore,

$$\cos(\alpha + \beta) = \frac{OC - DB}{OP} = \frac{OB \cos \alpha - BP \sin \alpha}{OP}$$

$$= \frac{OB}{OP} \cos \alpha - \frac{BP}{OP} \sin \alpha$$

$$= \cos \beta \cos \alpha - \sin \beta \sin \alpha$$

$$\mathbf{\cos(\alpha + \beta) = \cos \alpha \cos \beta - \sin \alpha \sin \beta.} \qquad (15\text{-}10)$$

Although the proofs of Eqs. (15-9) and (15-10) have been demonstrated for positive acute angles α and β, it can be shown that these relations, and the ones that follow from them, are true for any values of α and β whatsoever. From Eq. (15-4),

$$\tan(\alpha + \beta) = \frac{\sin(\alpha + \beta)}{\cos(\alpha + \beta)}$$

$$= \frac{\sin \alpha \cos \beta + \cos \alpha \sin \beta}{\cos \alpha \cos \beta - \sin \alpha \sin \beta}.$$

Dividing the numerator and denominator by $\cos \alpha \cos \beta$,

$$\tan(\alpha + \beta) = \frac{\dfrac{\sin \alpha \cos \beta}{\cos \alpha \cos \beta} + \dfrac{\cos \alpha \sin \beta}{\cos \alpha \cos \beta}}{1 - \dfrac{\sin \alpha \sin \beta}{\cos \alpha \cos \beta}}$$

$$\mathbf{\tan(\alpha + \beta) = \frac{\tan \alpha + \tan \beta}{1 - \tan \alpha \tan \beta}.} \qquad (15\text{-}11)$$

Equations (15-9), (15-10), and (15-11) are called the **addition formulas**. Although we could derive formulas for $\cot(\alpha + \beta)$, $\sec(\alpha + \beta)$, and $\csc(\alpha + \beta)$, they are rarely used.

EXAMPLE 5

a. Find $\sin(75°)$ by using an addition formula.

Since $75° = 45° + 30°$, we use Eq. (15-9) and write

$$\sin 75° = \sin(45° + 30°)$$

$$= \sin 45° \cos 30° + \cos 45° \sin 30°$$

$$= \left(\frac{\sqrt{2}}{2}\right)\left(\frac{\sqrt{3}}{2}\right) + \left(\frac{\sqrt{2}}{2}\right)\left(\frac{1}{2}\right)$$

$$= \frac{\sqrt{6}}{4} + \frac{\sqrt{2}}{4} = \frac{\sqrt{6} + \sqrt{2}}{4}.$$

b. Given $\sin \alpha = \frac{1}{2}$, $\cos \beta = \frac{1}{3}$, and that α is a first-quadrant angle and β is a fourth-quadrant angle, find $\cos(\alpha + \beta)$.

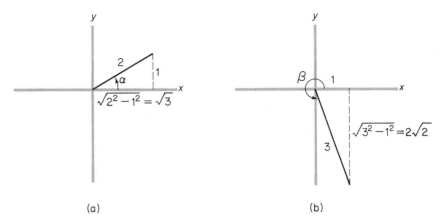

FIGURE 15-5

The angles α and β are shown in Fig. 15-5(a) and (b), respectively; the lengths of the sides indicated are found by the Pythagorean theorem. Hence,

$$\cos(\alpha + \beta) = \cos \alpha \cos \beta - \sin \alpha \sin \beta$$

$$= \left(\frac{\sqrt{3}}{2}\right)\left(\frac{1}{3}\right) - \left(\frac{1}{2}\right)\left(\frac{-2\sqrt{2}}{3}\right) = \frac{\sqrt{3} + 2\sqrt{2}}{6}.$$

The formulas for the functions of the difference of two angles can be most easily derived if we first express the trigonometric functions of a negative angle $-\theta$ in terms of the positive angle θ. In Fig. 15-6 we have indicated the angles θ and $-\theta$, both shown in standard position, along with points on their terminal sides. In each diagram, by construction the

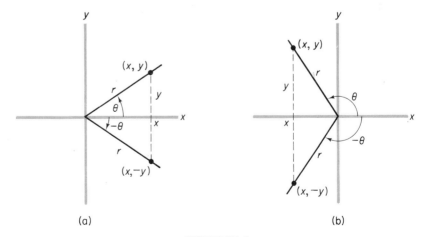

FIGURE 15-6

lengths of the radius vectors associated with both points are equal. Similarly, the abscissas are the same, but the ordinates differ in sign. By the definitions of the trigonometric functions,

$$\sin(-\theta) = \frac{-y}{r} = -\frac{y}{r} = -\sin\theta$$

$$\cos(-\theta) = \frac{x}{r} = \cos\theta$$

$$\tan(-\theta) = \frac{-y}{x} = -\frac{y}{x} = -\tan\theta.$$

These results, along with the results derived in a similar manner for the remaining trigonometric functions, are summarized by Eq. (15-12):

$$\begin{array}{ll} \sin(-\boldsymbol{\theta}) = -\sin\boldsymbol{\theta} & \cot(-\boldsymbol{\theta}) = -\cot\boldsymbol{\theta} \\ \cos(-\boldsymbol{\theta}) = \cos\boldsymbol{\theta} & \sec(-\boldsymbol{\theta}) = \sec\boldsymbol{\theta} \\ \tan(-\boldsymbol{\theta}) = -\tan\boldsymbol{\theta} & \csc(-\boldsymbol{\theta}) = -\csc\boldsymbol{\theta} \end{array} \qquad (15\text{-}12)$$

Thus, $\sin(-10°) = -\sin 10°$ and $\cos(-20°) = \cos 20°$. We remark that these results are valid for any angle θ, and you should verify these conclusions when θ is a third- or fourth-quadrant angle.

Now, if we substitute $-\beta$ for β in Eq. (15-9), we obtain

$$\sin[\alpha + (-\beta)] = \sin\alpha\cos(-\beta) + \cos\alpha\sin(-\beta)$$

and by Eq. (15-12) we conclude

$$\sin(\alpha - \beta) = \sin\alpha\cos\beta - \cos\alpha\sin\beta. \qquad (15\text{-}13)$$

The same substitution in Eqs. (15-10) and (15-11) yields, respectively,

$$\cos(\alpha - \beta) = \cos\alpha\cos\beta + \sin\alpha\sin\beta \qquad (15\text{-}14)$$

and

$$\tan(\alpha - \beta) = \frac{\tan\alpha - \tan\beta}{1 + \tan\alpha\tan\beta}. \qquad (15\text{-}15)$$

Equations (15-13), (15-14), and (15-15) are called the **subtraction formulas.**

EXAMPLE 6

Find $\sec 15°$ without the use of tables.

Since $\sec 15° = 1/\cos 15°$, we first find $\cos 15°$:

$$\cos 15° = \cos(45° - 30°)$$
$$= \cos 45° \cos 30° + \sin 45° \sin 30° \qquad [\text{Eq. (15-14)}]$$
$$= \frac{\sqrt{2}}{2} \cdot \frac{\sqrt{3}}{2} + \frac{\sqrt{2}}{2} \cdot \frac{1}{2}$$
$$= \frac{\sqrt{6} + \sqrt{2}}{4}.$$

Thus,

$$\sec 15° = \frac{4}{\sqrt{6} + \sqrt{2}}$$

$$= \frac{4}{\sqrt{6} + \sqrt{2}} \cdot \frac{\sqrt{6} - \sqrt{2}}{\sqrt{6} - \sqrt{2}}$$

$$= \frac{4(\sqrt{6} - \sqrt{2})}{4} = \sqrt{6} - \sqrt{2}.$$

EXAMPLE
7

Parallel and perpendicular lines: Section 6-4 made use of the facts that if two nonvertical lines L_1 and L_2 are parallel, their slopes m_1 and m_2 are equal, and if the two lines are perpendicular, the slope of one is the negative reciprocal of the slope of the other—that is, $m_2 = -1/m_1$. We are now able to prove both statements.

Figure 15-7(a) shows two parallel lines, L_1 and L_2, which are not horizontal. For each line we can determine the slope by selecting two arbitrary points

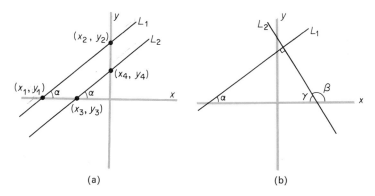

(a)   (b)

FIGURE 15-7

on the line. We choose for these points the intersection of the line with the x- and y-axes, and we define the angle α formed from the x-axis in a positive direction to the line as the **angle of inclination** of the line. Clearly, if the two lines are parallel, they must have the same angle of inclination. Furthermore,

$$m_2 = \frac{y_4 - y_3}{x_4 - x_3} = \tan \alpha$$

and

$$m_1 = \frac{y_2 - y_1}{x_2 - x_1} = \tan \alpha.$$

Hence, for parallel lines, $m_1 = m_2$. Note that if L_1 and L_2 are horizontal lines, they both have slopes of zero. Fig. 15-7(b) shows two perpendicular lines with angles of inclination α and β. Now,

$$\alpha + \gamma + 90° = 180°$$

and

$$\gamma + \beta = 180°.$$

Therefore

$$\alpha + \gamma + 90° = \gamma + \beta$$

and so

$$\alpha + 90° = \beta.$$

Thus,

$$\tan{(\alpha + 90°)} = \tan{\beta}.$$

Using Eqs. (15-9) and (15-10),

$$\tan(\alpha + 90°) = \frac{\sin(\alpha + 90°)}{\cos(\alpha + 90°)} = \frac{\sin \alpha \cos 90° + \cos \alpha \sin 90°}{\cos \alpha \cos 90° - \sin \alpha \sin 90°}$$

$$= \frac{0 + \cos \alpha}{0 - \sin \alpha} = -\cot \alpha.$$

Thus

$$-\cot \alpha = \tan \beta \qquad \text{or} \qquad -\frac{1}{\tan \alpha} = \tan \beta.$$

Since $m_1 = \tan \alpha$ and $m_2 = \tan \beta$,

$$m_2 = -\frac{1}{m_1}.$$

EXERCISE 15-2

In Problems 1–10, the trigonometric function of the given angle has been expressed in terms of a function of a sum or difference of special angles. Use the appropriate addition or subtraction formula to evaluate each. It should not be necessary to refer to trigonometric tables.

1. $\cos 75° = \cos{(30° + 45°)}$

2. $\sin 15° = \sin{(60° - 45°)}$

3. $\cos 105° = \cos{(60° + 45°)}$

4. $\cos 255° = \cos{(225° + 30°)}$

5. $\tan 15° = \tan{(60° - 45°)}$

6. $\tan 15° = \tan{(45° - 30°)}$

7. $\sin 165° = \sin{(120° + 45°)}$

8. $\sin 105° = \sin{(150° - 45°)}$

9. $\cos 195° = \cos{(225° - 30°)}$

10. $\cot 105° = 1/\tan{(60° + 45°)}$

In Problems 10–16, express the given angle as the sum or difference of special angles and evaluate.

11. $\sin(-150°)$ **12.** $\cos(-150°)$

13. $\cos(-15°)$ **14.** $\sin(-15°)$

15. $\tan 345°$ **16.** $\tan 255°$

In Problems 17–22, without performing any detailed calculations, determine whether the given statement is true or false.

17. $\sin(-85°) = -\sin 85°$

18. $\sin(-225°) = -\sin 225°$

19. $\tan(\alpha + \beta) = \tan \alpha + \tan \beta$

20. $\cos(-225°) = -\cos 45°$

21. $\cos(-225°) = \cos 225°$

22. $\sin(\alpha - \beta) = \sin \alpha - \sin \beta$

23. If α and β are second-quadrant angles and $\tan \alpha = -\frac{1}{2}$ and $\tan \beta = -\frac{2}{3}$, find values for (a) $\tan(\alpha + \beta)$ and (b) $\cos(\alpha + \beta)$.

24. Prove that $\dfrac{\sin(\alpha + \beta)}{\cos \alpha \cos \beta} = \tan \alpha + \tan \beta$.

In Problems 25–31, use trigonometric formulas to simplify the given equation, and then graph one cycle of the curve defined by the equation.

25. $y = 2 \sin(x + \pi)$

26. $y = 3 \cos(x + \pi)$

27. $y = 4 \cos\left(\dfrac{\pi}{2} - x\right)$

28. $y = 2 \sin\left(\dfrac{\pi}{2} - x\right)$

29. $y = \tan(x + \pi)$

30. $y = \sin\left(x + \dfrac{\pi}{2}\right)$

31. $y = -\cos(\pi - x)$

32. Express $\sin 18° \cos 10° - \cos 18° \sin 10°$ as a function of one angle only.

33. The displacement x of a particular object undergoing harmonic motion as a function of time t is

$$x = 2\sqrt{2} \, \cos\left(2t - \dfrac{\pi}{4}\right).$$

  a. By expanding the right member, show that such motion is effectively two different motions combined.

  b. For these two different motions, what is the contribution of each to the displacement when $t = \pi/4$?

34. The electric field components of two light waves vary with time at a given point as

$$E_1 = E_0 \sin(\omega t)$$
$$E_2 = E_0 \sin(\omega t + \varphi).$$

The electric field component is associated with the disturbance caused by the waves. Find the total disturbance $E_1 + E_2$.

35. For light passing symmetrically through a prism, the index of refraction of glass with respect to air is given by

$$n = \frac{\sin[\frac{1}{2}(\alpha + \beta)]}{\sin(\beta/2)},$$

where α is the deviation angle and β is the angle of the apex of the prism. If $\beta = 60°$, find an equivalent expression for n.

36. Suppose in a three-phase a.c. generator that the phases are expressed as $I \cos \theta$, $I \cos (\theta + 120°)$, and $I \cos (\theta + 240°)$. It is to be shown that each phase is numerically equal to the sum of the other phases but opposite in sign. To do this it suffices to show

$$I \cos \theta + I \cos (\theta + 120°) + I \cos (\theta + 240°) = 0.$$

Show that this is indeed the case.

15-3
DOUBLE- AND HALF-ANGLE FORMULAS

Sometimes it is quite useful and important to be able to express the trigonometric functions of twice an angle in terms of functions of the angle itself. By letting $\beta = \alpha$ in Eqs. (15-9), (15-10), and (15-11), we are able to derive these **double-angle formulas.**

From Eq. (15-9), this substitution yields

$$\sin(\alpha + \alpha) = \sin \alpha \cos \alpha + \cos \alpha \sin \alpha$$

$$\mathbf{\sin 2\alpha = 2 \sin \alpha \cos \alpha.} \qquad (15\text{-}16)$$

From Eq. (15-10), the same substitution yields

$$\cos(\alpha + \alpha) = \cos \alpha \cos \alpha - \sin \alpha \sin \alpha$$
$$= \cos^2 \alpha - \sin^2 \alpha.$$

By Eq. (15-6),

$$\cos^2 \alpha - \sin^2 \alpha = \cos^2 \alpha - (1 - \cos^2 \alpha)$$
$$= 2 \cos^2 \alpha - 1$$

or, alternatively,

$$\cos^2 \alpha - \sin^2 \alpha = (1 - \sin^2 \alpha) - \sin^2 \alpha$$
$$= 1 - 2 \sin^2 \alpha.$$

Thus,

$$\cos 2\alpha = \cos^2 \alpha - \sin^2 \alpha$$
$$= 2\cos^2 \alpha - 1 \qquad (15\text{-}17)$$
$$= 1 - 2\sin^2 \alpha.$$

Similarly, by letting $\beta = \alpha$ in Eq. (15-11) it follows that

$$\tan 2\alpha = \frac{2\tan \alpha}{1 - \tan^2 \alpha}. \qquad (15\text{-}18)$$

Expressions for $\cot 2\alpha$, $\sec 2\alpha$, and $\csc 2\alpha$ can be determined by use of the reciprocal relationships. *Remember that, in general, $\sin 2x \neq 2\sin x$, etc.*

EXAMPLE 8

Use a double-angle formula to evaluate $\sin 60°$.

If we let $2\alpha = 60°$, then $\alpha = 30°$ and

$$\sin 2\alpha = 2\sin \alpha \cos \alpha$$
$$\sin 60° = \sin (2 \cdot 30°) = 2\sin 30° \cos 30°$$
$$= 2\left(\frac{1}{2}\right)\left(\frac{\sqrt{3}}{2}\right) = \frac{\sqrt{3}}{2}.$$

EXAMPLE 9

Use a double-angle formula to evaluate $\cos 90°$.

If we let $2\alpha = 90°$, then $\alpha = 45°$ and

$$\cos 2\alpha = \cos^2 \alpha - \sin^2 \alpha$$
$$\cos 90° = \cos (2 \cdot 45°) = \cos^2 45° - \sin^2 45°$$
$$= \left(\frac{\sqrt{2}}{2}\right)^2 - \left(\frac{\sqrt{2}}{2}\right)^2 = 0.$$

Equivalently we could write

$$\cos 2\alpha = 2\cos^2 \alpha - 1$$
$$\cos 90° = 2\cos^2 45° - 1$$
$$= 2\left(\frac{\sqrt{2}}{2}\right)^2 - 1 = 1 - 1 = 0$$

or

$$\cos 2\alpha = 1 - 2\sin^2 \alpha$$
$$\cos 90° = 1 - 2\sin^2 45°$$
$$= 1 - 2\left(\frac{\sqrt{2}}{2}\right)^2 = 1 - 1 = 0.$$

EXAMPLE 10

If θ is a first-quadrant angle and $\sin \theta = \frac{3}{5}$, find $\tan 2\theta$. See Fig. 15-8.

By the Pythagorean theorem,

$$5^2 = x^2 + 3^2 \qquad \text{and} \qquad x = 4.$$

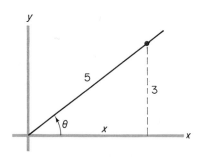

FIGURE 15-8

By Eq. (15-18),

$$\tan 2\theta = \frac{2 \tan \theta}{1 - \tan^2 \theta}$$

and since from Fig. 15-8, $\tan \theta = 3/x = 3/4$, we have

$$\tan 2\theta = \frac{2\left(\dfrac{3}{4}\right)}{1 - \left(\dfrac{3}{4}\right)^2} = \frac{\dfrac{3}{2}}{\dfrac{7}{16}} = \frac{24}{7}.$$

The **half-angle formulas** are used to express a function of half an angle in terms of the angle itself. To derive the half-angle formulas we first use the double-angle formula

$$\cos 2\alpha = 1 - 2 \sin^2 \alpha.$$

Letting $\alpha = \dfrac{\theta}{2}$ we have

$$\cos\left[2\left(\frac{\theta}{2}\right)\right] = 1 - 2 \sin^2 \frac{\theta}{2},$$

which can be written

$$2 \sin^2 \frac{\theta}{2} = 1 - \cos \theta.$$

Finally, dividing both members of the equation by 2 and then solving for $\sin \theta/2$ yields

$$\boldsymbol{\sin \frac{\theta}{2}} = \pm\sqrt{\frac{1 - \cos \theta}{2}}. \tag{15-19}$$

The choice of whether to use the plus or minus sign before the radical depends on $\sin \theta/2$; that is, it depends on the quadrant location of the given angle $\theta/2$. If $\theta/2$ is a first- or second-quadrant angle, the sign is positive; if $\theta/2$ is a third- or fourth-quadrant angle, the sign is negative.

Using the double-angle formula

$$\cos 2\alpha = 2 \cos^2 \alpha - 1$$

and again letting $\alpha = \dfrac{\theta}{2}$ yields

$$\cos\left[2\left(\frac{\theta}{2}\right)\right] = 2 \cos^2 \frac{\theta}{2} - 1.$$

Rearranging terms and dividing by 2 we have

$$\cos^2 \frac{\theta}{2} = \frac{1 + \cos \theta}{2}.\qquad(1)$$

Thus,

$$\cos \frac{\theta}{2} = \pm\sqrt{\frac{1 + \cos \theta}{2}}.\qquad\textbf{(15-20)}$$

Here we choose the positive sign if $\theta/2$ is a first- or fourth-quadrant angle and choose the negative sign if $\theta/2$ is a second- or third-quadrant angle. *Again, the sign depends on $\theta/2$, not θ.*

Since

$$\tan \frac{\theta}{2} = \frac{\sin \dfrac{\theta}{2}}{\cos \dfrac{\theta}{2}} = \frac{\sin \dfrac{\theta}{2}}{\cos \dfrac{\theta}{2}} \cdot \frac{2 \cos \dfrac{\theta}{2}}{2 \cos \dfrac{\theta}{2}}$$

$$= \frac{2 \sin \dfrac{\theta}{2} \cos \dfrac{\theta}{2}}{2 \cos^2 \dfrac{\theta}{2}},$$

we have from Eqs. (15-16) and (1),

$$\tan \frac{\theta}{2} = \frac{\sin \theta}{1 + \cos \theta}.\qquad\textbf{(15-21)}$$

EXAMPLE
11

Use a half-angle formula to determine $\sin 75°$.

We use Eq. (15-19) with $\theta = 150°$. Sin 75° is positive and so we use the positive sign before the radical:

$$\sin 75° = \sin \frac{150°}{2} = \sqrt{\frac{1 - \cos 150°}{2}}$$

$$= \sqrt{\frac{1 - \left(-\dfrac{\sqrt{3}}{2}\right)}{2}} = \sqrt{\frac{2 + \sqrt{3}}{4}}$$

$$= \frac{\sqrt{2 + \sqrt{3}}}{2}.$$

EXAMPLE

12

Use a half-angle formula to determine tan 105°.

We use Eq. (15-21) with $\theta = 210°$:

$$\tan 105° = \tan \frac{210°}{2} = \frac{\sin 210°}{1 + \cos 210°}$$

$$= \frac{-\dfrac{1}{2}}{1 + \left(-\dfrac{\sqrt{3}}{2}\right)}$$

$$= -\frac{1}{2 - \sqrt{3}}$$

$$= -\frac{1}{2 - \sqrt{3}} \cdot \frac{2 + \sqrt{3}}{2 + \sqrt{3}}$$

$$= -\frac{2 + \sqrt{3}}{4 - 3} = -2 - \sqrt{3}.$$

EXERCISE 15-3

In Problems 1–6, use a double-angle formula to evaluate the given expression. Trigonometric tables should not be needed.

1. sin 60°   2. cos 60°

3. cos 240°   4. sin 240°

5. tan 120°   6. tan 240°

In Problems 7–12, use a half-angle formula to evaluate the given expression. Trigonometric tables should not be needed.

7. sin 15°   8. cos 75°

9. cos 22.5°   10. sin 157.5°

11. tan 112.5°   12. sin 67.5°

In Problems 13–16, find sin x, cos x, tan x, sin 2x, cos 2x, and tan 2x from the given information.

13. $\cos x = 3/5$, sin x is positive

14. $\sec x = 5$

15. $\sin x = -1/3$, cot x is positive

16. $\cos(-x) = -1/4$, tan x is negative

In Problems 17–20, find sin x, cos x, tan x, sin (x/2), cos (x/2), and tan (x/2) from the given information. Use the facts that if $0 < x < \pi$, then $0 < \pi/2 < \pi/2$, and if $\pi < x < 2\pi$, then $\pi/2 < x/2 < \pi$.

17. $\cos x = 12/13$, sin x is positive

18. $\cot x = -8/15$, $\sin x$ is positive

19. $\sin x = -3/5$, $\tan x$ is positive

20. $\cos x = 5/13$, $\sin x$ is negative

21. If a projectile is fired from the ground at an angle θ with the horizontal with an initial speed V_0, the horizontal range of the projectile is given by

$$R = \frac{V_0^2 \sin 2\theta}{g}.$$

Determine another expression for R in terms of θ.

22. The index of refraction, n, of a prism whose apex angle is α and whose angle of minimum deviation is φ is given by

$$n = \frac{\sin[(\alpha + \varphi)/2]}{\sin(\alpha/2)}.$$

Show that

$$n = \sqrt{\frac{1 - \cos \alpha \cos \varphi + \sin \alpha \sin \varphi}{1 - \cos \alpha}}.$$

23. Express $\sin 3x$ in terms of $\sin x$. *Hint:* $\sin 3x = \sin(2x + x)$.

15-4
TRIGONOMETRIC EQUATIONS

A **trigonometric equation** is a conditional equation involving trigonometric functions of unknown angles. To solve such an equation means to find those angles satisfying the given equation. For our purposes, we shall consider as solutions only those angles x in the interval $0 \le x < 2\pi$, where x is in radians, or $0° \le \theta < 360°$ for θ given in degrees. Although no general method is available for solving trigonometric equations, use of algebraic methods and trigonometric identities affords us some means of determining the solution set. We shall illustrate by some examples.

EXAMPLE
13

Solve $2 \sin x = 1$.

$$2 \sin x = 1$$

$$\sin x = \frac{1}{2}.$$

Since $\sin x$ is positive in the first and second quadrants, $x = \pi/6$ or $x = 5\pi/6$. The solution set is $\{\pi/6, 5\pi/6\}$.

EXAMPLE
14

Solve $2 \sin^2 x - \sin x - 1 = 0$.

Factoring,

$$(2 \sin x + 1)(\sin x - 1) = 0.$$

Thus, $\sin x = -\frac{1}{2}$ or $\sin x = 1$. For $\sin x = -\frac{1}{2}$, $x = 7\pi/6$ or $x = 11\pi/6$. For $\sin x = 1$, $x = \pi/2$. Thus, the solution set is $\{\pi/2, 7\pi/6, 11\pi/6\}$.

EXAMPLE 15

Solve $\sin x + \cos x = 1$.

$$\sin x = 1 - \cos x$$
$$\sin^2 x = 1 - 2\cos x + \cos^2 x$$
$$1 - \cos^2 x = 1 - 2\cos x + \cos^2 x$$
$$2\cos^2 x - 2\cos x = 0$$
$$2\cos x (\cos x - 1) = 0$$
$$2\cos x = 0 \quad \text{or} \quad \cos x = 1$$
$$x = \frac{\pi}{2}, \frac{3\pi}{2} \quad \text{or} \quad x = 0.$$

Since we squared both members, we must check all solutions. For $x = 0$,

$$\sin 0 + \cos 0 = 0 + 1 = 1.$$

For $x = \frac{\pi}{2}$,

$$\sin \frac{\pi}{2} + \cos \frac{\pi}{2} = 1 + 0 = 1.$$

For $x = \frac{3\pi}{2}$,

$$\sin \frac{3\pi}{2} + \cos \frac{3\pi}{2} = -1 + 0 \neq 1.$$

Thus the solution set is $\{0, \pi/2\}$. An alternative method is to first square both sides of the given equation. Thus

$$\sin^2 x + 2\sin x \cos x + \cos^2 x = 1.$$

But $\sin^2 x + \cos^2 x = 1$. Thus the equation becomes $2\sin x \cos x = 0$. Factoring, $\sin x = 0$ or $\cos x = 0$. From the first equation, $x = 0$ or π. From the second one, $x = \pi/2$ or $3\pi/2$. Checking these values will give the above solution set.

EXAMPLE 16

Solve $2\cos x - \sec x = 1$.

$$2\cos x - \frac{1}{\cos x} = 1.$$

Multiplying both members by $\cos x$,

$$2\cos^2 x - 1 = \cos x$$
$$2\cos^2 x - \cos x - 1 = 0$$
$$(2\cos x + 1)(\cos x - 1) = 0.$$

Thus, $\cos x = -\frac{1}{2}$ or $\cos x = 1$. Hence, $x = 2\pi/3$, $4\pi/3$, or 0. Since we multiplied both members of the equation by a term involving the variable, you should verify that these solutions do indeed satisfy the original equation.

EXAMPLE
17

Solve $2 \sin x \cos x = 1$.

$$2 \sin x \cos x = 1$$

$$\sin 2x = 1.$$

Thus, if $2x = \pi/2$, then $x = \pi/4$. If $2x = 5\pi/2$, then $x = 5\pi/4$. Note that although $5\pi/2$ is beyond our usual considerations, it gives rise to an appropriate solution. The solution set is $\{\pi/4, 5\pi/4\}$. In fact, whenever you have an equation $f(n\theta) = k$, consider $n\theta$ on the interval $0 \le n\theta < 2n\pi$ so that you may be sure of finding all roots on the interval $0 \le \theta < 2\pi$.

EXERCISE 15-4

Solve for x where $0 \le x < 2\pi$.

1. $\sin x = \sqrt{2}/2$
2. $\cos x = -\sqrt{3}/2$
3. $\tan x = -1$
4. $\sec x = -1$
5. $2 \sin x \cos x = -\sqrt{2}/2$
6. $2 \sin x - \csc x = 1$
7. $\sin 2x = \sin x$
8. $2 \sin x \cos x - \sqrt{3} \sin x = 0$
9. $\sin^2 x - 2 \sin x + 1 = 0$
10. $\cos x - \cos 2x = 1$
11. $3 \cos x = \sqrt{3} \sin x$
12. $\sin 2x \cos x = 0$
13. $\sin x + \cos 2x = 4 \sin^2 x$
14. $\cos 2x = \sin x$
15. $2 \sin^2 x + \sin x = 1$
16. $\sin 2x = \sqrt{2} \cos x$
17. $\sin 2x + \cos x = 0$
18. $2 \sin x - \tan x = 0$
19. $\tan 2x + \sec 2x = 1$
20. $\sin x = \sin (x/2)$
21. $\sqrt{2 - \csc^2 x} = \csc x$
22. $\sin (x/2) = 1 - \cos x$

15-5
REVIEW

REVIEW QUESTIONS

Since $\sin \theta = 1/\csc \theta$ for all values of θ for which the equation is defined, that equation is an example of a(n) (conditional equation) (identity).
                                                        (1)

(1) **identity**

The mathematical statement $\tan \theta = -\sqrt{\sec^2 \theta - 1}$ implies that θ is in the
_____(2)_____ · or _____(3)_____ quadrant.

(2) **second**   (3) **fourth**

If $0 \leq x < 2\pi$, the equation $\sin x = 1$ has solution set _____(4)_____ and
the equation $\sin x = 1.1$ has solution set _____(5)_____.

(4) $\{\pi/2\}$ (5) $\varnothing$

Which of the following statements always are true?

a.   The graphs of $y = \cos 2x$ and $y = \sin\left(\dfrac{\pi}{2} - 2x\right)$ are identical.
b. $\cos(x + y) = \cos x + \cos y$
c. $\sin(x - y) = -\sin(y - x)$
d. $\cos(x - y) = \cos(y - x)$
e. $\tan\theta/\sec\theta = \csc\theta$
f. $\tan\theta/\sec\theta = 1/\csc\theta$ _____(6)_____

(6) **a, c, d, and f**

How many of the following expressions are equal?
a. $\sin^2\theta + \cos^2\theta$
b. $\csc^2\theta - \cot^2\theta$
c. $\sec^2\theta + \tan^2\theta$ _____(7)_____

(7) **two**

REVIEW PROBLEMS

In Problems 1–5, prove each of the given identities.

1. $\csc^2 x \tan^2 x - \sec x \cos x = \tan^2 x$

2. $\sec x \csc x = \tan x + \cot x$

3. $\dfrac{\sin^2 x}{1 - \cos x} = 1 + \cos x$

4. $\cot^2 x \sin^2 x + \tan^2 x \cos^2 x = 1$

5. $\dfrac{\sec^2 x}{\cot x} - \tan^3 x = \tan x$

In Problems 6–16, evaluate the given expression if $\alpha = 30°$, $\beta = 45°$, and $\theta = 60°$.

6. $\sin(\alpha + \beta)$ **7.** $\cos(\theta - \beta)$

8. $\tan\dfrac{\beta}{2}$ **9.** $\sin(\alpha - \beta)$

10. $\cos(\alpha + \beta)$ **11.** $\tan 2\beta$

12. $\sec(\alpha + \beta)$ **13.** $\cot(\alpha + \beta)$

14. $\csc(\alpha - \beta)$ **15.** $\cos \dfrac{\beta}{2}$

16. $\sin \dfrac{3\beta}{2}$

17. If $\tan x = 15/8$ and $\sin x$ is negative, find $\sin x$, $\cos x$, $\tan x$, $\sin 2x$, $\cos 2x$, $\tan 2x$, $\sin (x/2)$, $\cos (x/2)$, and $\tan (x/2)$.

18. If $\sin x = -\frac{3}{5}$ and $\sec x$ is positive, find all the trigonometric functions asked for in Problem 17.

19. Simplify and then sketch the graph of
$$ y = 2\sin\left(x + \frac{\pi}{4}\right) + 2\cos\left(x - \frac{\pi}{4}\right). $$

20. Solve $\sin x \cos x - \cot x = 0$.

21. Solve $\sin 2x - \sqrt{2}\,\sin x = 0$.

22. Solve $\dfrac{\sqrt{3}}{2} + \tan^2 x + \sin x - \sec^2 x = -1$.

16

oblique triangles and applications of angular measurement

16-1
THE LAW OF SINES

In Sec. 7-7 we found that the solution of a right triangle could be accomplished by applying the basic definitions of the trigonometric functions. If we turn our attention to the solution of *oblique triangles*—that is, triangles that do not contain a right angle—we must derive two very significant relations, both essential for a convenient solution of such triangles. The first is the *law of sines*.

Figure 16-1(a) and (b) indicates two oblique triangles *ABC* with

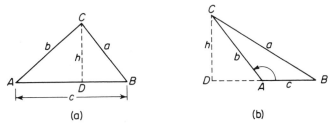

(a)   (b)

FIGURE 16-1

angles A, B, and C and sides a, b, and c. The first is an *acute triangle* and the second an *obtuse triangle,* so named because of the angles they contain. An acute triangle has three acute angles; an obtuse triangle has one obtuse angle, namely the one opposite the longest side. In each triangle the altitude CD has been constructed and is taken to be of length h. The discussion which follows is applicable to both parts of the figure. Keep in mind that in Fig. 16-1(b), $\sin A = \sin(180° - A) = \sin \angle CAD$. For each diagram we have

$$\sin A = \frac{h}{b} \qquad \text{or} \qquad h = b \sin A$$

and

$$\sin B = \frac{h}{a} \qquad \text{or} \qquad h = a \sin B.$$

Hence, equating these equal quantities,

$$a \sin B = b \sin A,$$

and dividing each member of this equation by the quantity $\sin A \sin B$ yields

$$\frac{a}{\sin A} = \frac{b}{\sin B}. \tag{16-1}$$

If for each triangle an altitude h were drawn from the vertex of angle A to side a or its extension, then in a similar manner we have

$$\sin B = \frac{h}{c} \qquad \text{or} \qquad h = c \sin B$$

and

$$\sin C = \frac{h}{b} \qquad \text{or} \qquad h = b \sin C.$$

Hence, equating these equal quantities,

$$b \sin C = c \sin B.$$

Dividing each member of this equation by the quantity $\sin B \sin C$ yields

$$\frac{b}{\sin B} = \frac{c}{\sin C}. \tag{16-2}$$

Finally, combining Eqs. (16-1) and (16-2) by the transitive axiom we have

$$\frac{a}{\sin A} = \frac{b}{\sin B} = \frac{c}{\sin C}. \tag{16-3}$$

Equation (16-3) is called the **law of sines** and states that *the lengths of the* *sides of any triangle are directly proportional to the sines of the angles opposite them.*

The law of sines (or *sine law*) is used to solve an oblique triangle if we are given two angles and any side or two sides and the angle opposite one of them. The solution of a problem of each of the above types will now be given.

CLASS A: GIVEN TWO ANGLES AND ONE SIDE

EXAMPLE
1

Solve the oblique triangle ABC given $A = 30°$, $B = 70°$, and $a = 4$. The triangle is illustrated in Fig. 16-2.

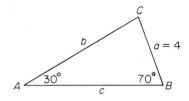

FIGURE 16-2

Since the sum of the angles in a triangle must equal 180°, we have $A + B + C = 180°$. Thus $C = 180° - 30° - 70° = 80°$. In Eq. (16-3), the sine law, to find b we pair the first and second expressions:

$$\frac{b}{\sin B} = \frac{a}{\sin A}$$

$$b = \frac{a \sin B}{\sin A} = \frac{(4)(\sin 70°)}{\sin 30°} = \frac{(4)(.9397)}{(.5000)} \approx 7.518.$$

Applying the law of sines once again to find c, we have

$$\frac{c}{\sin C} = \frac{a}{\sin A}$$

$$c = \frac{a \sin C}{\sin A} = \frac{(4)(\sin 80°)}{\sin 30°} = \frac{4(.9848)}{(.5000)} \approx 7.878.$$

Thus, the complete solution is $A = 30°$, $B = 70°$, $C = 80°$, $a = 4$, $b = 7.518$, and $c = 7.878$.

CLASS B: GIVEN TWO SIDES AND THE ANGLE
OPPOSITE ONE OF THEM

This class of problems presents a somewhat more difficult situation. When two sides of a triangle and the angle opposite one of them are given, we are not assured that these data determine only one triangle. Indeed, there may be one triangle or two triangles, or there may be no triangle at all. Due to the various possibilities, this class is called the **ambiguous case.** The following discussion is based on the assumption that, for this class of problems, the given parts of the triangle are a, b, and A. We can systematically deal with each possibility by considering the situations that arise when $A < 90°$ and those when $A \geq 90°$.

Angle A < 90°:

1. If $a < b \sin A$, that is, side a is smaller than the altitude $b \sin a$, there can be no triangle formed. Hence, there is **no solution** (Fig. 16-3).

FIGURE 16-3   **FIGURE 16-4**

2. If $a = b \sin A$, then a corresponds to the altitude and there is only **one solution**, which is the right triangle ABC (Fig. 16-4).

3. If $a > b$, then as a consequence $a > b \sin A$ and only **one triangle** exists, as shown in Fig. 16-5.

4. If $a > b \sin A$ **and** $a < b$, there are **two solutions**. As shown in Fig. 16-6, triangles ABC and $AB'C$ both satisfy the given conditions. Note that angles B' and B are supplementary, but B is acute and B' is obtuse.

5. If $a = b$, there is **one solution**, which is an isosceles triangle (Fig. 16-7).

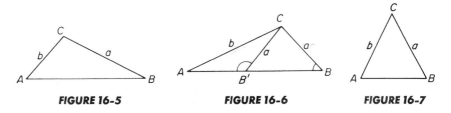

FIGURE 16-5   **FIGURE 16-6**   **FIGURE 16-7**

Angle A ≥ 90°: If A is either a right angle or an obtuse angle, there are only two possible situations:

1. If $a \leq b$, there is **no solution** (Fig. 16-8).

2. If $a > b$, there is **one solution** (Fig. 16-9).

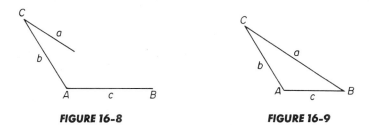

FIGURE 16-8   **FIGURE 16-9**

Notice that a fairly accurate sketch based on the given data will often make the number of solutions obvious.

EXAMPLE
2

Solve the oblique triangle ABC given that $a = 10$, $b = 5$, and $A = 60°$.

Since A is acute and $a > b$, there is exactly one triangle. An appropriate diagram is given in Fig. 16-10. By the law of sines we have $\dfrac{\sin B}{b} = \dfrac{\sin A}{a}$ and so

$$\sin B = \frac{b \sin A}{a} = \frac{(5)(.8660)}{(10)} = .4330.$$

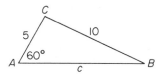

FIGURE 16-10

Since B is not opposite the longest side, we are assured that B is acute. Hence,

$$B \approx 25° \, 40'.$$

Since $A + B + C = 180°$, $C \approx 180° - 60° - 25° \, 40' = 94° \, 20'$. Also $\dfrac{c}{\sin C} = \dfrac{a}{\sin A}$ and so

$$c = \frac{a \sin C}{\sin A} = \frac{(10) \sin 94° \, 20'}{\sin 60°} = \frac{(10)(.9971)}{(.8660)} \approx 11.51.$$

Thus, the complete solution is $a = 10$, $b = 5$, $c = 11.51$, $A = 60°$, $B = 25° \, 40'$, and $C = 94° \, 20'$.

EXAMPLE
3

Solve the oblique triangle ABC given that $a = 6$, $b = 10$, and $A = 30°$.

Angle A is acute and $a < b$. Since $b \sin A = 10 \sin 30° = 10(.5) = 5$ and $a > b \sin A$, two solutions exist. The appropriate diagrams are shown in Fig. 16-11(a) and (b).

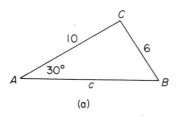
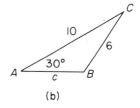

(a)   (b)

FIGURE 16-11

$$\sin B = \frac{b \sin A}{a} = \frac{(10)(.5000)}{6} \approx .8333.$$

Thus $B \approx 56° \, 26'$ or $B \approx 180° - 56° \, 26' = 123° \, 34'$.

CASE 1.   If $B = 56° \, 26'$ [Fig. 16-11(a)], then

$$C = 180° - A - B = 180° - 30° - 56° \, 26' = 93° \, 34'$$

and

$$c = \frac{a \sin C}{\sin A} = \frac{(6) \sin 93° \, 34'}{\sin 30°} = \frac{(6)(.9981)}{(.5000)} \approx 11.98.$$

Hence, the complete solution corresponding to Fig. 16-11(a) is $a = 6, b = 10, c = 11.98, A = 30°, B = 56° \, 26'$, and $C = 93° \, 34'$.

CASE 2.   If $B = 123° \, 34'$ [Fig. 16-11(b)], then

$$C = 180° - A - B = 180° - 30° - 123° \, 34' = 26° \, 26'$$

and

$$c = \frac{a \sin C}{\sin A} = \frac{6 \sin 26° \, 26'}{\sin 30°} = \frac{(6)(.4452)}{(.5000)} \approx 5.342.$$

Hence, a second complete solution, corresponding to Fig. 16-11(b), is $a = 6$, $b = 10$, $c = 5.342$, $A = 30°$, $B = 123° \, 34'$, and $C = 26° \, 26'$.

The solution of oblique triangles is applicable to many physical problems involving vector quantities. For example, if two forces acting in the directions AC and AD in Fig. 16-12 act simultaneously on an object located at A, the resultant force acting on the object is in the direction

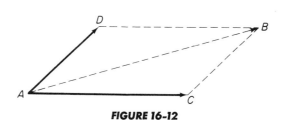

FIGURE 16-12

AB. That is, if the lengths of two sides of a parallelogram drawn from a common vertex are proportional to the magnitudes of the forces acting on an object located at that vertex, then the diagonal of the parallelogram drawn from that vertex is proportional to the resultant force acting on the object. Moreover, in addition to being representative of forces, the sides of the parallelogram can represent velocities, accelerations, displacements, or any vector quantities whatsoever. Hence, the problem of determining the resultant of two forces, velocities, accelerations, or displacements can be handed by solving an oblique triangle.

EXERCISE 16-1

Wherever possible, solve the oblique triangles ABC in Problems 1–22 from the given information.

1. $A = 50°, B = 100°, a = 20$
2. $A = 80°, C = 40°, c = 100$
3. $a = 50, b = 100, A = 60°$
4. $a = 28, b = 75, A = 30°$
5. $a = 55, b = 110, A = 30°$
6. $a = 36, b = 72, A = 30°$
7. $a = 116, b = 74, A = 81°$
8. $a = 18, b = 13, A = 55°$
9. $a = 7, b = 9, A = 20°$
10. $a = 67, b = 100, A = 25°$
11. $a = 46, b = 46, A = 27°$
12. $a = 70, b = 70, A = 70°$
13. $a = 27, b = 27, A = 96°$
14. $a = 113, b = 113, A = 135°$
15. $a = 85, b = 70, A = 172°$
16. $a = 15, b = 6, A = 135°$
17. $A = 60°, B = 72°, c = 80$
18. $B = 110°, C = 75°, b = 50$
19. $a = 72, b = 84, A = 104°$
20. $a = 67, b = 92, A = 110°$
21. $a = 20, b = 7, A = 140°$

22. $a = 110, b = 90, A = 110°$

23. Suppose forces $F_1 = 500$ lb and $F_2 = 700$ lb act on a body such that the resultant force F forms an angle of 42° with F_1. Find F and the angle that F_2 makes with F_1.

24. In one of the locks of the St. Lawrence Seaway, the angle of elevation α of the top C of a Canadian landmark from a ship at point A is $\alpha = 55°$ (Fig. 16-13). As the lock fills, the ship rises from A to B and the angle of elevation decreases to $\beta = 50°$. If the ship is 410 ft from the base of the landmark when at B, through what distance did the ship rise?

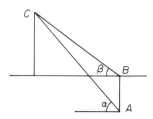

FIGURE 16-13

16-2
THE LAW OF COSINES

To develop the second relation which will be important in the solution of oblique triangles, consider the oblique triangle ABC indicated in Fig. 16-14. (Although angle A is acute, our results will also be true when A

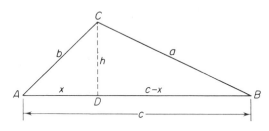

FIGURE 16-14

is obtuse.) If we let the length of the line segment AD be x, then the length of segment DB must be $c - x$. You will note that constructing the altitude CD has divided the oblique triangle into two right triangles. Applying the Pythagoream theorem, we have from $\triangle BCD$

$$a^2 = h^2 + (c - x)^2 \tag{1}$$

and from $\triangle ACD$

$$b^2 = h^2 + x^2. \tag{2}$$

Subtracting Eq. (2) from Eq. (1) yields

$$a^2 - b^2 = h^2 + (c - x)^2 - (h^2 + x^2)$$
$$a^2 - b^2 = c^2 - 2cx. \tag{3}$$

However, $\cos A = x/b$ and, therefore, $x = b \cos A$. Substituting this value for x in Eq. (3) and rearranging terms yields

$$a^2 = b^2 + c^2 - 2bc \cos A \tag{16-4}$$

In a similar manner it can be shown that

$$b^2 = a^2 + c^2 - 2\,ac \cos B \tag{16-5}$$

and

$$c^2 = a^2 + b^2 - 2\,ab \cos C \tag{16-6}$$

Each of Eqs. (16-4), (16-5), and (16-6), is a statement of the **law of cosines**. Verbally, the law of cosines states that *the square of any side of a triangle equals the sum of the squares of the other two sides minus twice their product times the cosine of the included angle.* If you learn and understand the *meaning* of the law of cosines it should be clear to you that there is little need to memorize all three equations. In fact, just knowing Eq. (16-4) will suffice since any angle in a triangle can be denoted by A.

The law of cosines (the *cosine law*) is initially used to solve an oblique triangle if you know either two sides and the included angle or three sides. A problem of each type will now be considered.

CLASS C: GIVEN TWO SIDES AND
THE INCLUDED ANGLE

EXAMPLE
4

Solve the oblique triangle ABC given that $a = 10$, $b = 40$, and $C = 120°$. The triangle is illustrated in Fig. 16-15.

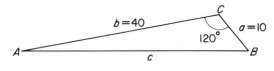

FIGURE 16-15

The initial step in the solution of a class C problem is to apply the law of cosines. By Eq. (16-6),

$$c^2 = a^2 + b^2 - 2ab \cos C$$
$$c^2 = (10)^2 + (40)^2 - 2(10)(40) \cos 120°$$
$$c^2 = 100 + 1600 - 800(-\tfrac{1}{2}) = 2100$$
$$c = \sqrt{2100} = 10\sqrt{21} \approx 45.83.$$

Although angles A and B can be found by repeated applications of the law of cosines [Eqs. (16-4) and (16-5)], the work is significantly simplified by using the law of sines. By Eq. (16-3),

$$\frac{a}{\sin A} = \frac{c}{\sin C}$$

$$\sin A = \frac{a \sin C}{c} = \frac{(10)(\sin 120°)}{(45.83)} = \frac{(10)(.8660)}{(45.83)} \approx .1890$$

$$A \approx 10° \, 54'.$$

Note that although there are two values of angle A—where $0° < A < 180°$—for which $\sin A \approx .1890$, since C is obtuse both A and B must be acute angles. Finally,

$$B = 180° - A - C = 180° - 10° \, 54' - 120°$$

$$B = 49° \, 06'.$$

Thus, the complete solution is $A = 10° \, 54'$, $B = 49° \, 06'$, $C = 120°$, $a = 10$, $b = 40$, and $c = 45.83$.

CLASS D: GIVEN THREE SIDES

EXAMPLE
5

Solve the oblique triangle ABC given that $a = 7$, $b = 6$, and $c = 8$. The triangle is illustrated in Fig. 16-16.

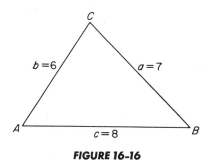

FIGURE 16-16

The initial step in the solution of a class D problem is to apply the law of cosines. By Eq. (16-6),

$$c^2 = a^2 + b^2 - 2ab \cos C$$

$$(8)^2 = (7)^2 + (6)^2 - 2(7)(6) \cos C$$

$$64 = 85 - 84 \cos C$$

$$84 \cos C = 21$$

$$\cos C = \tfrac{21}{84} = \tfrac{1}{4} = .2500.$$

Since $\cos C$ is positive, angle C is acute:

$$C \approx 75° \, 31'.$$

Again angles A and B can be found by repeated applications of the law of cosines [Eqs. (16-4) and (16-5)]. We choose here, however, to use the law of sines. From Eq. (16-3) we have

$$\frac{c}{\sin C} = \frac{b}{\sin B}$$

$$\sin B = \frac{b \sin C}{c} = \frac{(6) \sin 75° \, 31'}{(8)} = \frac{6(.9682)}{8} = .7262$$

$$B \approx 46° \, 34'.$$

Observe that there is no difficulty in concluding that B is an acute angle. At most there can be only one obtuse angle in a triangle and, if it exists, it must be opposite the longest side, namely c in our case. Hence, B must be an acute angle. Since we have already found the largest angle, angle C, and it is acute, the triangle ABC is clearly an acute triangle. As a general rule for class D problems such as this, you should make it a practice to first determine the largest angle as we have done here. Finally,

$$A = 180° - B - C = 180° - 46° \, 34' - 75° \, 31'$$
$$= 57° \, 55'.$$

The complete solution is $A = 57° \, 55'$, $B = 46° \, 34'$, $C = 75° \, 31'$, $a = 7$, $b = 6$, and $c = 8$.

EXAMPLE
6

Given the triangle ABC such that $a = 9$, $b = 8$, and $c = 2$, solve for A.

By the cosine law,

$$a^2 = b^2 + c^2 - 2bc \cos A$$
$$81 = 64 + 4 - 2(8)(2) \cos A$$
$$81 = 68 - 32 \cos A$$
$$32 \cos A = -13$$
$$\cos A = -\tfrac{13}{32} \approx -.4062.$$

Since $\cos A$ is negative, angle A is obtuse. From tables the related angle is $66° \, 02'$ and thus

$$A \approx 113° \, 58'.$$

EXERCISE 16-2

In Problems 1–10, if possible solve the oblique triangle from the given data.

1. $a = 20, b = 40, C = 28°$

2. $b = 7, c = 13, A = 135°$

3. $a = 16, b = 17, c = 18$

4. $a = 7, b = 4, c = 1$

5. $a = 15, c = 12, B = 115°$

6. $a = 5, b = 5, c = 10$

7. $a = 110, b = 85, c = 90$

8. $a = 10, c = 9, B = 60°$

9. $a = 5, b = 4, c = 8$

10. $a = 13, b = 15, c = 20$

11. Two forces act simultaneously on an object. $F_1 = 40$ lb and acts due west; $F_2 = 20$ lb and acts 62° east of north. Find the magnitude and direction of the resultant force acting on the object.

12. Forces acting on an object tend to give it simultaneous velocities in two directions. If $v_1 = 12$ ft/sec directed due east, and $v_2 = 15$ ft/sec directed 25° east of north, what is the magnitude and direction of the resultant velocity?

13. A student claims that he underwent two separate displacements of 5 ft and 8 ft, respectively, and ended up exactly 10 ft from his starting point. Find the angle between the 5-ft displacement and the resultant displacement of 10 ft. What is the angle between the two displacements?

14. Two engines and separate steering mechanisms of a speedboat effectively create two velocities: one, 22 mi/hr, 65° east of north, and a second, 50 mi/hr, 52° east of north. Determine the magnitude and direction of the resultant velocity.

15. Figure 16-17 shows a cross-sectional view of a roadside sign. The sign is 7 ft tall and three sections of supporting beams are 2.5, 3.5, and 4.5 ft, respectively. Find the distance x from the base of the sign to the bottom of the back leg.

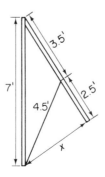

FIGURE 16-17

USE OF LOGARITHMS IN SOLVING TRIANGLES

EXAMPLE 7

Solve the oblique triangle ABC given $a = 116.32$, $b = 82.642$, and $A = 81° 20'$. The triangle is shown below.

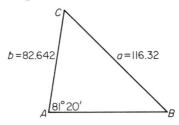

By the law of sines

$$\sin B = \frac{b \sin A}{a}$$

$$\log \sin B = \log b + \log \sin A - \log a$$
$$= \log 82.642 + \log \sin 81° 20' - \log 116.32$$
$$= 1.91720 + (9.9950 - 10) - 2.06565$$
$$= 9.84655 - 10,$$

which when rounded off to four decimal places gives

$$\log \sin B = 9.8466 - 10.$$

From the body of the table of logarithms of the trigonometric functions we find

$$10\left\{ {}^{x}\left\{\begin{matrix} \log \sin 44° 30' = 9.8457 - 10 \\ \log \sin \quad ? \quad = 9.8466 - 10 \end{matrix}\right\}^{.0009} \\ \log \sin 44° 40' = 9.8469 - 10 \end{matrix}\right\}.0012$$

$$x = \frac{.0009}{.0012}(10) \approx 8.$$

Thus,

$$B = 44° 38'.$$

Then $C = 180° - 44° 38' - 81° 20' = 54° 02'$. From the law of sines

$$c = \frac{a \sin C}{\sin A}$$

$$\log c = \log a + \log \sin C - \log \sin A$$
$$= \log 116.32 + \log \sin 54° 02' - \log \sin 81° 20'.$$

Interpolating to find $\log \sin 54° 02'$,

$$10\left\{ 2\left\{\begin{matrix} \log \sin 54° 00' = 9.9080 - 10 \\ \log \sin 54° 02' = \quad ? \end{matrix}\right\}^{x} \\ \log \sin 54° 10' = 9.9089 - 10 \end{matrix}\right\}.0009$$

$$x = \tfrac{2}{10}(.0009) \approx .0002$$

and

$$\log \sin 54° 02' = 9.9082 - 10.$$

Hence,

$$\log c = 2.06565 + (9.9082 - 10) - (9.9950 - 10)$$
$$\approx 1.9788.$$

Now antilog .9788 $\approx$ 9.524 and thus

$$c \approx 9.524 \times 10^1 = 95.24.$$

The complete solution of the triangle is $A = 81° 20'$, $B = 44° 38'$, $C = 54° 02'$, $a = 116.32$, $b = 82.642$, and $c = 95.24$. To obtain the same degree of accuracy by usual arithmetic techniques would be an extremely tedious task.

EXERCISE 16-3

In Problems 1–5, solve each of the oblique triangles logarithmically.

1. $a = 110.2$, $A = 64° 42'$, $C = 15° 33'$
2. $a = 643.7$, $b = 327.8$, $A = 40° 20'$
3. $A = 50° 14'$, $B = 100° 34'$, $a = 22.62$
4. $A = 64° 26'$, $a = 4.768$, $b = 2.832$
5. $A = 142° 31'$, $a = 16.73$, $b = 12.29$

16-4
APPLICATIONS OF ANGLES AND ANGULAR MEASUREMENT

A. AREA OF A TRIANGLE

Recall from geometry that the area of any triangle is equal to one-half the product of the lengths of its base and altitude. For each triangle ABC in Fig. 16-18, c is the base and h is the altitude. Moreover, in each case

$$\sin A = \frac{h}{b}$$

or

$$h = b \sin A.$$

Thus,

$$\text{Area} = \tfrac{1}{2}(\text{Base})(\text{Altitude})$$
$$= \tfrac{1}{2}c(b \sin A)$$
$$\textbf{Area} = \tfrac{1}{2}\textbf{\textit{bc}} \, \textbf{sin} \, \textbf{\textit{A}}. \tag{16-7}$$

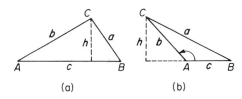

FIGURE 16-18

By appropriate labeling we can also show area $= \frac{1}{2}ac \sin B = \frac{1}{2}ab \sin C$. That is, the **area of a triangle is equal to one-half the product of any two sides and the sine of the included angle.**

EXAMPLE
8

Find the area of triangle ABC given that $b = 10$, $c = 15$, and $A = 35°$.

$$\text{Area} = \frac{1}{2}bc \sin A$$
$$= \frac{1}{2}(10)(15) \sin 35°$$
$$= 75(.5736)$$
$$= 43.02 \text{ square units.}$$

EXAMPLE
9

Find the area of the triangle in Fig. 16-19.

The area is given by $\frac{1}{2}ab \sin C$. To find C we first note that by the sine law,

$$\sin B = \frac{b \sin A}{a} = \frac{10(.3420)}{5} = .6840.$$

$$B \approx 43° \ 10'.$$

Thus, $C = 180° - 20° - 43° \ 10' = 116° \ 50'$ and

$$\text{Area} = \frac{1}{2}(10)(5) \sin 116° \ 50'$$
$$= 25(.8923) \approx 22.3 \text{ square units.}$$

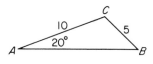

FIGURE 16-19

B. CIRCULAR ARC LENGTH

From geometry it is established that the length s of an arc on a circle is proportional to the central angle θ subtended by the arc—that is, s varies directly with θ (Fig. 16-20). Symbolically, $s = k\theta$, where the constant of variation k can be evaluated if corresponding values of s and θ are known. From Sec. 7-2 recall that associated with a central angle of

FIGURE 16-20

2π radians is an arc length equal to the circumference of the circle, $2\pi r$. Hence,

$$s = k\theta$$
$$2\pi r = k(2\pi)$$
$$k = r.$$

Thus,

$$s = r\theta, \qquad \theta \text{ in radians.} \tag{16-8}$$

EXAMPLE
10

Find the radius of a circle on which a central angle of 210° subtends an arc of length 14.65 m.

$$210° = (210°)\frac{\pi}{180°} = \frac{7\pi}{6} \text{ radians}$$

$$s = r\theta$$

$$r = \frac{s}{\theta} = \frac{14.65}{7\pi/6} \approx 4.0 \text{ m.}$$

C. AREA OF A SECTOR OF A CIRCLE

From geometry, the area A of a sector of a circle (Fig. 16-21) varies directly as the size θ of its central angle. Symbolically,

$$A = k\theta.$$

FIGURE 16-21

But when $\theta = 2\pi$, then $A = \pi r^2$ and

$$\pi r^2 = k(2\pi)$$
$$k = \tfrac{1}{2}r^2$$
$$A = \tfrac{1}{2}r^2\theta, \qquad \theta \text{ in radians.} \tag{16-9}$$

EXAMPLE
11
For a circle of radius 2 ft, find the area of a sector whose central angle is 30°.

$$30° = 30°\left(\frac{\pi}{180°}\right) = \frac{\pi}{6} \text{ radians}$$

$$A = \frac{1}{2}r^2\theta = \frac{1}{2}(2^2)\left(\frac{\pi}{6}\right) = \frac{\pi}{3} \text{ ft}^2.$$

D. ANGULAR SPEED

From physics it is known that the average linear speed, $\bar{v}$, of an object is the average rate of change of distance with respect to time—that is, the average speed is the ratio of distance s to time t:

$$\bar{v} = \frac{s}{t}. \tag{16-10}$$

Let Fig. 16-22 represent a circular disk of radius r rotating about an axis through O, perpendicular to the plane of the paper. The segment

FIGURE 16-22

OP serves as a reference. This is a situation describable by angular quantities. The **angular displacement** is the angle θ, in radians, through which the body rotates, and the **average angular speed**, $\bar{\omega}$, is, as in Eq. (16-10), the ratio of the angular displacement to time—that is,

$$\bar{\omega} = \frac{\theta}{t}$$

or

$$\theta = \bar{\omega}t. \tag{16-11}$$

While the body rotates through an angle θ, in radians, the point P on the rim moves a distance s given by

$$s = r\theta.$$

Thus, using Eq. (16-11) to substitute for θ,

$$s = r(\bar{\omega}t)$$

$$\boldsymbol{s = r\bar{\omega}t} \qquad \text{($\bar{\omega}$ in radians per unit time).} \tag{16-12}$$

To relate angular speed $\bar{\omega}$ to linear speed $\bar{v}$ we need only substitute for s in Eq. (16-12) the value from Eq. (16-10):

$$\bar{v}t = r\bar{\omega}t$$

$$\bar{v} = r\bar{\omega}, \qquad\qquad (16\text{-}13)$$

where $\bar{\omega}$ is in radians per unit time.

EXAMPLE
12

A rotating circular disk of radius 2 ft makes 10 revolutions/sec. Find (a) the angular speed of the body, (b) the linear speed of a point on the rim, and (c) the distance moved by that point in 10 sec.

a. $\bar{\omega} = \dfrac{10 \text{ rev}}{\text{sec}} = 10\dfrac{\text{rev}}{\text{sec}}\left(2\pi\dfrac{\text{rad}}{\text{rev}}\right) = 62.8$ radians/sec.

b. $\bar{v} = r\bar{\omega} = (2)(62.8) = 125.6$ ft/sec.

c. $s = r\bar{\omega}t = (2)(62.8)(10) = 1256$ ft.

EXERCISE 16-4

In Problems 1–6, find the area of the triangle ABC having the given parts.

1. $b = 6, c = 8, A = 20°$

2. $a = 10, b = 5, C = 14°$

3. $a = 7, c = 12, B = 130°$

4. $a = 20, b = 13, A = 55°$

5. $a = 40, A = 70°, b = 40$

6. $A = 100°, a = 100, b = 80$

In Problems 7–10, the radius r (in inches) and central angle θ of a circle are given. Determine the length of the arc subtended by θ and the area of the sector determined by θ. You may express your answer in terms of π.

7. $r = 12, \theta = \pi/6$

8. $r = 25, \theta = \pi/3$

9. $r = 18, \theta = 40°$

10. $r = 20, \theta = 171°$

11. Suppose a pendulum of length 20 in. swings through an arc of length 12 in. Through what angle, in degrees, does the pendulum swing?

12. The minute hand of a clock is 6 in. long. In a time period of 50 min, through what distance does the tip of the minute hand move?

13. What is the angular speed in radians per second of a wheel which makes 900 revolutions/min?

14. Three concentric circles have radii of 1, 1.5, and 3 ft. What is the area of the segment between the inner and middle circles bounded by radii forming a central angle of 87°?

15. A semicircular traffic rotary is to be built so that an automobile traveling at a constant speed of 30 mi/hr can traverse the semicircle in 25 sec. What is the required radius in feet?

16. In the traffic rotary of Problem 15, what is the average angular speed of the car?

17. The wheel of an automobile has a radius of 15 in. If the automobile is traveling at a constant speed of 30 mi/hr, what is the angular speed of the wheel?

18. A drive belt runs around two pulleys whose diameters are 6 in. and 2 ft. If the larger pulley makes 3 revolutions every second, determine
    a. The angular speed of the small pulley in radians per second.
    b. The linear speed of a point on the rim of each pulley in feet per second.
    c. The area, in square feet, of a sector swept out by a radius line on each pulley in 2 sec.

19. The propeller of a fan rotates at a constant angular speed of 1200 revolutions/min. The radius of the propeller is 5 in. Determine (a) the angular speed in radians per second, (b) the angular displacement after 28 sec, (c) the linear and angular speeds of a point on the propeller 3 in. from the center, and (d) the linear and angular speeds of a point on the rim of the propeller.

16-5
REVIEW

REVIEW QUESTIONS

True or false: Given two sides and the included angle of an oblique triangle, two solutions may occur. _____(1)_____

(1) **false**

The law of _____(2)_____ is the first relation employed in the solution of an oblique triangle if three sides are given.

(2) **cosines**

The law of sines is a statement of proportionality between the _____(3)_____ of a triangle and the _____(4)_____.

(3) **sides**   (4) **sines of the opposite angles**

When solving an oblique triangle where the lengths of three sides are given, it is advisable to solve first for the ___(largest)(smallest)___ angle.
                                                                                                                          (5)

(5) **largest**

The largest angle of a triangle is found opposite the side of ___(greatest)(least)___
                                                                                                                          (6)
length.

(6) **greatest**

For a circle of radius 1 ft, a central angle of 30° subtends an arc of length
_____(7)_____.

(7) **$\pi/6$ ft**

The area of the triangle ABC where $b = 6, c = 8$, and $A = 30°$ is
_____(8)_____.

(8) **12 square units**

State in the order given, that is, a, b, and c, how many triangles are possible in each case.
a. $a = 30, b = 25, A = 45°$
b. $a = 20, b = 30, A = 45°$
c. $a = 2\sqrt{3}, b = 4, A = 60°$ (9)

(9) **1, 0, 1**

REVIEW PROBLEMS

1. Solve the oblique triangles by employing the laws of sines and cosines and determine the area of each triangle.
   a. $A = 30°, B = 70°, a = 3$
   b. $C = 40°, B = 110°, b = 9$
   c. $A = 60°, a = 20, b = 10$
   d. $a = 6, b = 8, A = 10°$
   e. $A = 130°, b = 20, a = 10$

2. It is required that the area of a circular track in the shape of a washer have an area of 25,000 ft². Determine the inside and outside radii of the track

if the circumference of the inner edge is 200 ft. If a man runs along the very center of the track, how many revolutions must he make to run 2 mi?

3. A phonograph turntable rotates at $33\frac{1}{3}$ revolutions/min. What is the linear speed of a point 3 in. from the center of the turntable?

4. Find the area of triangle ABC given $a = 25$, $b = 12$, and $C = 120°$.

inequalities

17-1
LINEAR INEQUALITIES IN ONE VARIABLE

In Sec. 1-2, the inequality symbols $<$, $>$, $\leq$, and $\geq$ were introduced along with the notion of an interval. In defining inequalities we shall use the greater than relation ($>$) but the others ($<$, $\geq$, $\leq$) also apply.

Definition. An **inequality** is a statement that one number is greater than another number.

Inequalities are, of course, represented by means of inequality symbols. If two inequalities have their inequality symbols pointing in the same direction, then the inequalities are said to have the *same sense*. However, if the symbols point in opposite directions, the inequalities are said to be *opposite in sense* or one is said to have the *reverse sense* of the other. Thus, the inequalities $a < b$ and $c < d$ have the same sense, but $a < b$ and $c > d$ are opposite in sense.

As with equations, there are various types of inequalities. An **absolute inequality** is one which is true for *all* allowable values of the symbols employed. A **conditional inequality** is one which is true for *some*, but not all, of the possible values.

EXAMPLE
1

a. $x^2 \geq 0$ is an absolute inequality since the square of every real number x is positive or zero.

b. $x^2 > 0$ is a conditional inequality since it is true if and only if x is different from zero.

The solution set of an inequality is the set of those values of its variables for which the inequality is a true statement. To solve an inequality we apply algebraic operations so that the solution set is clearly evident. As with equations, there are some rules governing inequalities that must be noted.

Rule 1. If the same number is added to or subtracted from both members of an inequality, the resulting inequality has the same sense as the original inequality. Symbolically, if $a < b$, then $a + c < b + c$ and $a - c < b - c$.

EXAMPLE
2

a. $4 < 9$ and $4 + 3 < 9 + 3$.

b. If $x - 7 > 0$, then

$$x - 7 + 7 > 0 + 7$$

or

$$x > 7.$$

c. If $x + 6 > -4$, then

$$x + 6 - 6 > -4 - 6$$
$$x > -10.$$

Rule 2. If both members of an inequality are multiplied or divided by the same *positive* number, the resulting inequality has the same sense as the original inequality. Symbolically, if $a < b$ and $c > 0$, then $ac < bc$ and $\dfrac{a}{c} < \dfrac{b}{c}$.

EXAMPLE
3

a. $7 < 10$ and $7(3) < 10(3)$. Also, $\dfrac{7}{3} < \dfrac{10}{3}$

b. If $\dfrac{x}{3} < 4$, then

$$\frac{x}{3}(3) < 4(3)$$
$$x < 12.$$

c. If $8x - 16 > 24$, then

$$\frac{8x - 16}{8} > \frac{24}{8}$$

$$x - 2 > 3.$$

Now, using Rule 1,

$$x - 2 + 2 > 3 + 2$$

$$x > 5.$$

Rule 3. If both members of an inequality are multiplied or divided by the same *negative* number, then the resulting inequality has the *reverse* sense of the original inequality. Symbolically, if $a < b$ and $c < 0$, then $ac > bc$ and $\dfrac{a}{c} > \dfrac{b}{c}$.

EXAMPLE 4

a. $15 > 7$ but

$$15(-2) < 7(-2)$$

$$-30 < -14.$$

Also,

$$15 > 7$$

but

$$\frac{15}{-2} < \frac{7}{-2}$$

$$-\frac{15}{2} < -\frac{7}{2}.$$

b. If $-x < -7$, then

$$(-x)(-1) > (-7)(-1)$$

$$x > 7.$$

Rule 4. If both members of an inequality are different from zero and are of the same sign, then their respective reciprocals are unequal in the opposite sense.

EXAMPLE 5

a. $6 < 9$ but $\dfrac{1}{6} > \dfrac{1}{9}$.

b. $-3 < -2$ but $\dfrac{1}{-3} > \dfrac{1}{-2}$.

Rule 5. If both members of an inequality are positive and are raised to the same positive power, then the resulting inequality has the same sense as the original inequality. Symbolically, if $a > 0, b > 0, n > 0$, and $a > b$, then

$$a^n > b^n$$

and

$$\sqrt[n]{a} > \sqrt[n]{b}.$$

EXAMPLE 6

a. $5 > 4$ and

$$(5)^2 > (4)^2$$
$$25 > 16.$$

b. $4 < 9$ and

$$\sqrt{4} < \sqrt{9}$$
$$2 < 3.$$

c. $1 > -4$ but $(1)^2 > (-4)^2$ is false. Why doesn't Rule 5 apply here?

If Rules 1–3 are applied to an inequality, the resulting inequality has the same solution set as the original inequality and we say the resulting inequality is *equivalent* to the original inequality.

Definition. A **linear inequality** in the variable x is an inequality which is equivalent to

$$ax + b > 0 \qquad (\text{or } <, \leq, \geq),$$

where a and b are constants and $a \neq 0$.

We shall now give some examples of solving inequalities, primarily of the linear type. In each case the given inequality will be replaced by an equivalent inequality until the solution set is evident. Whereas a linear *equation* gave rise to exactly one solution, a linear *inequality* gives rise to an infinite number of solutions which can be depicted as an interval on the real number line.

EXAMPLE 7

a. Solve $2(x - 3) > -10$.

$$2(x - 3) > -10$$

$$2x - 6 > -10 \qquad \text{(distributive axiom)}$$

$$2x - 6 + 6 > -10 + 6 \qquad \text{(adding 6 to both members)}$$

$$2x > -4 \qquad \text{(simplifying)}$$

$$\frac{2x}{2} > \frac{-4}{2} \qquad \text{(dividing both members by 2)}$$

$$x > -2 \qquad \text{(simplifying)}.$$

The solution set consists of all real numbers to the right of -2, denoted $\{x \mid x > -2\}$ or $(-2, \infty)$. This set is shown in Fig. 17-1.

$$\{x \mid x > -2\} = (-2, \infty)$$

FIGURE 17-1

b. Solve $5 - 2x \geq 4$.

$$5 - 2x \geq 4$$

$\quad -2x \geq -1$ (subtracting 5 from both members)

$\quad \dfrac{-2x}{-2} \leq \dfrac{-1}{-2}$ (dividing both members by -2 and changing the sense of the inequality)

$\quad x \leq \dfrac{1}{2}.$

The solution set consists of all real numbers to the left of and including $\frac{1}{2}$, denoted $(-\infty, \frac{1}{2}]$ or $\{x \mid x \leq \frac{1}{2}\}$, as shown in Fig. 17-2.

$$\{x \mid x \leq \tfrac{1}{2}\} = (-\infty, \tfrac{1}{2}]$$

FIGURE 17-2

c. Solve $\frac{1}{2}(x - 3) \leq 3(2x + 3) + 6$.

$$\tfrac{1}{2}(x - 3) \leq 3(2x + 3) + 6.$$

Multiplying both members by 2, the L.C.D., gives

$$x - 3 \leq 6(2x + 3) + 12 \qquad \text{(Rule 2)}$$
$$x - 3 \leq 12x + 18 + 12 \qquad \text{(distributive axiom)}$$
$$-11x \leq 33 \qquad \text{(Rule 1)}$$
$$x \geq -3 \qquad \text{(Rule 3).}$$

The solution set is $\{x \mid x \geq -3\} = [-3, \infty)$, as shown in Fig. 17-3.

$$\{x \mid x \geq -3\} = [-3, \infty)$$

FIGURE 17-3

d. Solve $\dfrac{x}{2} - \dfrac{5}{4} < \dfrac{x}{4} + \dfrac{3}{2}$.

$$\dfrac{x}{2} - \dfrac{5}{4} < \dfrac{x}{4} + \dfrac{3}{2}.$$

Multiplying both members by 4, the L.C.D.,

$$2x - 5 < x + 6 \qquad \text{(Rule 2)}$$
$$x < 11 \qquad \text{(Rule 1).}$$

The solution set is $(-\infty, 11) = \{x \mid x < 11\}$, as shown in Fig. 17-4.

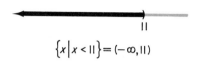

$$\left\{x \mid x < 11\right\} = (-\infty, 11)$$

FIGURE 17-4

EXAMPLE
8

Solve $3(4 - x) + 2 < 4(x - 2) - 7x$.

$$3(4 - x) + 2 < 4(x - 2) - 7x$$
$$12 - 3x + 2 < 4x - 8 - 7x$$
$$14 - 3x < -3x - 8$$
$$14 < -8 \qquad \text{(Rule 1).}$$

Since the inequality $14 < -8$ is never true, the solution set is $\varnothing$.

EXAMPLE
9

Solve $2(x - 6) > 2x - 15$.

$$2(x - 6) > 2x - 15$$
$$2x - 12 > 2x - 15$$
$$-12 > -15 \qquad \text{(Rule 1).}$$

Since the inequality $-12 > -15$ is true for all real numbers x, the solution set is $(-\infty, \infty)$.

EXERCISE 17-1

In Problems 1–24, find the solution set of the given inequality and indicate your answer geometrically on the real number line.

1. $2x < 4$

2. $3x + 7 > 0$

3. $3x - 4 \leq 2$

4. $3 - 5x \geq 5$

5. $-1 - x < 2x + 5$

6. $3 + x > 1 + x$

7. $3(2 - 3x) \geq 4(1 - 4x)$

8. $8(x + 1) + 1 < 3(2x) + 1$

9. $\frac{x}{2} - 4 \leq 3x + 2$

10. $3 + x < 1 + x$

11. $\frac{2}{3}(x + 4) \geq \frac{3}{2}(2x - 1)$

12. $x - \frac{1}{2}x \leq \frac{x}{3} + \frac{x}{4} + 2$

13. $2(3 - x) + 4(1 - 3x) < -5 - 14x$

14. $-\dfrac{x-3}{2}+4 < -\dfrac{3-x}{2}+\dfrac{x}{3}$

15. $0x \le 0$ **16.** $4x-1 \le 4(x-2)+7$

17. $\frac{2}{3}x > \frac{5}{6}x$ **18.** $\frac{7}{4}x > -\frac{2}{3}x$

19. $.1(.03x+4) \ge .02x+.434$ **20.** $\dfrac{.5x-1}{-3} > \dfrac{7(x+1)}{-2}$

21. $3x+2 < \sqrt{3}-x$

22. $\sqrt{2}(x+\sqrt{2}) > \sqrt{5}(\sqrt{2}-\sqrt{5}x)$

23. $\dfrac{x}{2}+\dfrac{x}{3} > \dfrac{x}{4}+\dfrac{x}{5}$ **24.** $\dfrac{3(2x-2)}{2} \le \dfrac{6x-3}{5}+\dfrac{x}{10}$

25. The amount of current needed for a certain electrical appliance is more than 7 amp but less than 9.5 amp. If 12 such appliances are to be used, and *I* represents the total number of amperes needed, using inequalities what can be said about *I*?

26. Using inequalities, symbolize the statement: The number of man-hours x to produce a blueprint is not less than 15 hr or more than 18 hr.

27. If a block slides down a rough inclined plane, it can be deduced from the principle of conservation of energy that the gravitational potential energy U of the block at the top of the inclined plane is greater than the kinetic energy K at the bottom of the plane. If $K = 36$ (joules), geometrically indicate the possible values of U on the real number line.

17-2
NONLINEAR INEQUALITIES

In this section we shall learn how to solve more complicated inequalities than those of the linear type. A concept that will be useful in writing solution sets of such inequalities is that of the *union* of two sets.

Definition. If A and B are sets, then the **union** of A and B, written $A \cup B$, is the set of all elements that are in A or B or in both A and B. That is,

$$A \cup B = \{x \mid x \in A \text{ or } x \in B\}.$$

EXAMPLE
10

a. If $A = \{1, 2, 3\}$ and $B = \{1, 5, 6\}$, then

$$A \cup B = \{1, 2, 3, 5, 6\}.$$

b. If $A = \{1, 2, 3, 4\}$ and $B = \{6\}$, then

$$A \cup B = \{1, 2, 3, 4, 6\}.$$

c. If A and B are the intervals $[-2, 3]$ and $[2, 4]$, respectively, then $A \cup B$

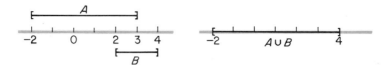

FIGURE 17-5

is the set of all points between and including -2 and 4 (Fig. 17-5).
Thus, $A \cup B = [-2, 4]$.

d. If $A = (-\infty, 3)$ and $B = (5, \infty)$, then
$$A \cup B = (-\infty, 3) \cup (5, \infty).$$

e. If $A = (-\infty, 0]$ and $B = (0, \infty)$, then
$$A \cup B = (-\infty, \infty) = \mathbf{R}.$$

f. Suppose that

$A = [-1, 5],$ $\qquad B = \{x \mid -2 < x < 2\},$ $\qquad C = (-\infty, \infty)$

$D = (-\infty, 3),$ $\qquad$ and $\qquad E = \varnothing.$

Then

$$
\begin{array}{ll}
A \cup B = (-2, 5] & A \cup D = (-\infty, 5] \\
D \cup E = (-\infty, 3) & B \cup E = B \\
A \cup C = C & B \cup D = (-\infty, 3)
\end{array}
$$

$$A \cup A = A$$

In the last section we discussed a method for solving a linear inequality such as $2x + 6 > 0$. It involved performing operations on the inequality until the solution set was evident. For example, we would write the inequality $2x + 6 > 0$ in the equivalent form $x > -3$ and hence conclude that the solution set is the interval $(-3, \infty)$.

But suppose we were asked to solve the inequality

$$x^2 - x - 2 > 0,$$

which is not linear. How could we proceed? One method is to graph the equation $y = f(x) = x^2 - x - 2$ and observe when y is positive. From Fig. 17-6 we see that $y > 0$ if $x < -1$ or $x > 2$. That is, the solution set of $x^2 - x - 2 > 0$ is $(-\infty, -1) \cup (2, \infty)$.

Since the accuracy of the graphical technique is quite limited, we shall develop another method. Consider again Fig. 17-6. Observe that the roots of the equation $f(x) = 0$ are $x = -1$ and $x = 2$. Moreover, these values of x determine three intervals on the real number line:

$$(-\infty, -1), \qquad (-1, 2), \qquad \text{and} \qquad (2, \infty).$$

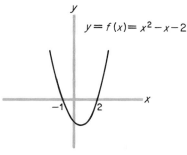

FIGURE 17-6

Given any of these intervals, note that $f(x)$ does not change sign throughout the interval. That is, if $x \in (-\infty, -1)$, then $f(x) > 0$. If $x \in (-1, 2)$, then $f(x) < 0$, etc. In short, $f(x)$ must be strictly positive or strictly negative on $(-\infty, -1)$ as well as on the other intervals. This means that to determine the sign of $f(x)$ on each interval, it is sufficient to determine its sign at an arbitrary point in each interval. For instance, $-2 \in (-\infty, -1)$ and $f(-2) = 4$. Thus, $f(x) > 0$ on $(-\infty, -1)$. Since $0 \in (-1, 2)$ and $f(0) = -2$, then $f(x) < 0$ on $(-1, 2)$. Similarly, $3 \in (2, \infty)$ and $f(3) = 4$; thus, $f(x) > 0$ on $(2, \infty)$. See Fig. 17-7. We conclude that the solution set of $x^2 - x - 2 > 0$ is, as before, $(-\infty, -1) \cup (2, \infty)$.

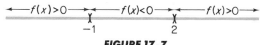

FIGURE 17-7

EXAMPLE
11

Solve the inequality $x^2 - 3x - 10 < 0$.

Let $f(x) = x^2 - 3x - 10 = (x + 2)(x - 5)$. The roots of the equation $f(x) = 0$ are clearly -2 and 5. These roots determine three intervals:

$$(-\infty, -2), \qquad (-2, 5), \qquad \text{and} \qquad (5, \infty).$$

Since $-3 \in (-\infty, -2)$, the sign of $f(x)$ on $(-\infty, -2)$ is the same as that of $f(-3)$. However, when $x = -3$, then $x + 2 < 0$ and $x - 5 < 0$ and so the product $(x + 2)(x - 5) > 0$. That is, $f(-3) > 0$ and therefore $f(x) > 0$ on $(-\infty, -2)$. Note that it is not necessary to actually evaluate $f(-3)$. We summarize this argument by using the following notations:

$$\text{since } f(-3) = (-)(-) = (+), \ f(x) > 0 \text{ on } (-\infty, -2).$$

For the other intervals we find that

$$\text{since } f(0) = (+)(-) = (-), \ f(x) < 0 \text{ on } (-2, 5)$$

and

$$\text{since } f(6) = (+)(+) = (+), \ f(x) > 0 \text{ on } (5, \infty).$$

The solution set of $(x + 2)(x - 5) < 0$ is $(-2, 5)$. See Fig. 17-8.

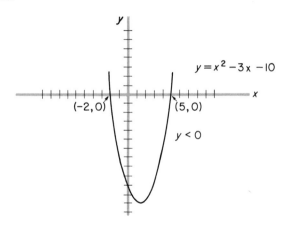

FIGURE 17-8

EXAMPLE
12

Solve the inequality $x(x - 1)(x + 4) \le 0$.

If $f(x) = x(x - 1)(x + 4)$, then the roots of the equation $f(x) = 0$ are 0, 1, and -4 and they determine four intervals:

$$(-\infty, -4), \quad (-4, 0), \quad (0, 1), \quad \text{and} \quad (1, \infty).$$

Determining the sign of $f(x)$ at a point on each interval we find that

since $f(-5) = (-)(-)(-) = (-)$, $f(x) < 0$ on $(-\infty, -4)$

since $f(-2) = (-)(-)(+) = (+)$, $f(x) > 0$ on $(-4, 0)$

since $f(\frac{1}{2}) = (+)(-)(+) = (-)$, $f(x) < 0$ on $(0, 1)$

and

since $f(2) = (+)(+)(+) = (+)$, $f(x) > 0$ on $(1, \infty)$.

The solution set of $x(x - 1)(x + 4) \le 0$ is

$$(-\infty, -4] \cup [0, 1].$$

Note that -4, 0, and 1 are included in the solution set since the stated inequality involved an $\le$ relationship and at these three values of x, $f(x) = 0$.

EXAMPLE
13

Solve $\dfrac{x^2 - 6x + 5}{x} < 0$.

Let

$$f(x) = \frac{x^2 - 6x + 5}{x} = \frac{(x - 1)(x - 5)}{x}.$$

For the case of a quotient, we solve the inequality by considering the intervals determined by the roots of $f(x) = 0$, namely 1 and 5, and those values of x for which f is undefined. Here, f is undefined when $x = 0$. Thus, we consider the intervals

$$(-\infty, 0), \quad (0, 1), \quad (1, 5), \quad \text{and} \quad (5, \infty).$$

We find that

$$\text{since } f(-1) = (-)(-)/(-) = (-), \ f(x) < 0 \text{ on } (-\infty, 0)$$
$$\text{since } f(\tfrac{1}{2}) = (-)(-)/(+) = (+), \ f(x) > 0 \text{ on } (0, 1)$$
$$\text{since } f(2) = (+)(-)/(+) = (-), \ f(x) < 0 \text{ on } (1, 5)$$

and

$$\text{since } f(6) = (+)(+)/(+) = (+), \ f(x) > 0 \text{ on } (5, \infty).$$

Therefore the solution set is

$$(-\infty, 0) \cup (1, 5).$$

EXAMPLE 14

Solve the inequality $x^2 + x + 1 > 0$.

By the use of the quadratic formula, we find that the roots of $x^2 + x + 1 = 0$ are imaginary. We therefore turn to the graph of $y = f(x) = x^2 + x + 1$ for any information it may yield (Fig. 17-9). Since for all x

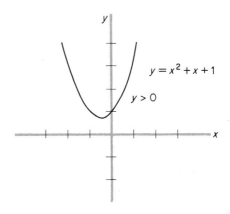

FIGURE 17-9

we have $y > 0$, the solution set is **R**—that is, the interval $(-\infty, \infty)$. It is worth noting that since

$$x^2 + x + 1 = (x + \tfrac{1}{2})^2 + \tfrac{3}{4},$$

$f(x)$ is always at least $\tfrac{3}{4}$ and hence the given inequality is true for all x.

EXERCISE 17-2

In Problems 1–8, find the union of the given sets.

1. $\{1, 2, 3\}; \{-1, 3, 6\}$

2. $\{2, 4, 6, 8, \ldots\}; \{0, 4, 8, 12, \ldots\}$

3. $\varnothing; \{0\}$ 4. $[-10, 2]; [-1, 10]$

5. $(-\infty, 4); [4, \infty)$ 6. $(-3, 1/2); [0, 1/2]$

7. $\{x \,|\, x > 2\}; \{x \,|\, x > 1\}$ 8. $\{x \,|\, x > 0\}; \{x \,|\, x \leq 0\}$

In Problems 9–40, solve the given inequality and, where appropriate, indicate the solution set on the real number line.

9. $x^2 + 3x + 2 < 0$ 10. $x^2 - 5x + 6 < 0$

11. $2x^2 - 4x > 0$ 12. $6x^2 + 12x > 0$

13. $x^2 + 7x + 10 \geq 0$ 14. $x^2 + 7x + 6 \geq 0$

15. $6x^2 + 7x - 3 \leq 0$ 16. $10x^2 - x - 3 \leq 0$

17. $x^2 - 2x + 1 > 0$ 18. $x^2 - 6x + 9 < 0$

19. $9x^2 + 5x + 2 < 0$ 20. $5 + 2x - 3x^2 < 0$

21. $x^2 - 16 > 0$ 22. $x^2 - 25 < 0$

23. $x^2 + 6 > 0$ 24. $x^2 - 6 < 0$

25. $4x^2 + 4x < 3$ 26. $2x^2 - 3x > -1$

27. $(x - 2)(x + 1)(x - 4) > 0$ 28. $x^3 - 2x^2 - 8x < 0$

29. $x^2(x^2 - 1) < 0$ 30. $(x + 5)(3 - x)(x + 1) < 0$

31. $\dfrac{x - 3}{x + 1} > 0$ 32. $\dfrac{x - 4}{x + 2} < 0$

33. $\dfrac{x + 3}{x - 2} < 0$ 34. $\dfrac{x + 4}{2x - 1} > 0$

35. $\dfrac{(x - 3)(x + 1)}{x} < 0$ 36. $\dfrac{(x + 5)(x)}{x - 3} > 0$

37. $\dfrac{x + 2}{(x - 1)(x - 6)} > 0$ 38. $\dfrac{x}{(x + 1)(x - 1)} < 0$

39. $\dfrac{x}{x^2 + 3x + 2} < 0$ 40. $\dfrac{2x}{x^2 + 6x + 9} > 0$

17-3
ABSOLUTE VALUE INEQUALITIES

Recall from Sec. 1-5 that the absolute value of a real number is its distance from the origin, disregarding direction. Aside from its geometrical interpretation, absolute value can formally be defined as follows:

Definition. The **absolute value** of a real number x, written $|x|$, is

$$|x| = \begin{cases} x, & \text{if } x > 0 \\ 0, & \text{if } x = 0 \\ -x, & \text{if } x < 0. \end{cases}$$

Applying the definition, $|5| = 5$, $|-4| = -(-4) = 4$, $|\frac{1}{2}| = \frac{1}{2}$, $-|2| = -2$, and $-|-6| = -6$. Also, $|\pi - 4| = 4 - \pi$. Why?

EXAMPLE 15

a. Solve $|x - 4| = 3$.

This equation asserts that $x - 4$ is a number three units from the origin. Thus,

$$x - 4 = 3 \quad \text{or} \quad x - 4 = -3.$$

If $x - 4 = 3$, then $x = 7$; if $x - 4 = -3$, then $x = 1$. The solution set is $\{1, 7\}$.

b. Solve $|x - 5| = -2$.

Since the absolute value of a number is never negative, the solution set is $\varnothing$.

c. Solve $|8 - 5x| = 6$.

The equation is true if $8 - 5x = 6$ or $8 - 5x = -6$. The first case yields $x = 2/5$ and the second $x = 14/5$. The solution set is $\{2/5, 14/5\}$.

It should be pointed out that if a and b are two points on the real number line, then $|a - b|$ represents the distance between a and b. For example, to solve the equation $|x - 4| = 3$ in Example 15a means to find all points x which are exactly three units from 4. Thus, x can be 1 or 7. We also remark that $|a - b| = |b - a|$.

If $|x| < 2$, then the distance of x from the origin is less than two units. Thus, x must lie between the points -2 and 2; equivalently, $-2 < x < 2$ [Fig. 17-10(a)]. However, if $|x| > 2$, then x must be greater than two units from the origin; hence, $x > 2$ or $x < -2$ [Fig. 17-10(b)]. Of course, if $|x| \leq 2$, then $-2 \leq x \leq 2$, and if $|x| \geq 2$, then $x \geq 2$ or $x \leq -2$.

(a)   (b)

FIGURE 17-10

EXAMPLE 16

a. Solve $|x - 5| < 3$.

The number $x - 5$ must be less than three units from the origin and so $-3 < x - 5 < 3$: we must simultaneously have

$$x - 5 > -3 \quad \textit{and} \quad x - 5 < 3.$$

Thus $x > 2$ and $x < 8$ simultaneously and the solution set is the interval $(2, 8)$. The solution procedure may be "set up" as follows:

$$-3 < x - 5 < 3$$
$$-3 + 5 < x < 3 + 5 \quad \text{(adding 5)}$$
$$2 < x < 8,$$

and the solution set is $(2, 8)$. This set can be characterized as the set of all points which are less than three units from 5.

b. Solve $|3 - 2x| \le 5$.

$$-5 \le 3 - 2x \le 5$$
$$-5 - 3 \le -2x \le 5 - 3 \quad \text{(subtracting 3)}$$
$$-8 \le -2x \le 2$$
$$4 \ge x \ge -1 \quad \text{(dividing by } -2\text{)}.$$

The solution set is $[-1, 4]$. Note that the sense of the original inequality is reversed in the last step due to division by a negative number.

EXAMPLE
17

a. Solve $|x + 6| \ge 4$.

The number $x + 6$ must be at least four units from the origin:

$$x + 6 \le -4 \quad or \quad x + 6 \ge 4.$$

Thus,

$$x \le -10 \quad or \quad x \ge -2.$$

Since $\{x \mid x \le -10\} = (-\infty, -10]$ and $\{x \mid x \ge -2\} = [-2, \infty)$, the solution set is the union of these intervals:

$$(-\infty, -10] \cup [-2, \infty).$$

b. Solve $|3x - 4| > 1$.

Let us use an approach which is different from that in part a. Since $|a - b|$ can be interpreted as the distance between a and b, then clearly $|3x - 4| > 1$ means that the distance between $3x$ and 4 is greater than one unit. Consequently, we must have

$$3x < 3 \quad or \quad 3x > 5.$$

Thus,

$$x < 1 \quad or \quad x > \tfrac{5}{3}.$$

The solution set is

$$(-\infty, 1) \cup (\tfrac{5}{3}, \infty).$$

EXAMPLE
18

Using absolute value notation, express the fact that

a. x is less than four units from 7:

$$|x - 7| < 4.$$

b. x differs from 2 by at least 3:
$$|x - 2| \geq 3.$$

c. $x < 9$ and $x > -9$ simultaneously:
$$|x| < 9.$$

d. x is strictly within two units of -3:
$$|x - (-3)| < 2$$
$$|x + 3| < 2.$$

e. x is strictly within σ (sigma) units of μ (mu):
$$|x - \mu| < \sigma.$$

EXERCISE 17-3

In Problems 1–8, write an equivalent form without the absolute value symbol.

1. $|-6|$

2. $|\pi^{-2}|$

3. $|7 - 3|$

4. $|(-6 - 4)/2|$

5. $|4(-7/3)|$

6. $|3 - 8| - |8 - 3|$

7. $|x| < 4$

8. $|x| < 8$

9. Using the absolute value symbol, express the fact that
   a. x is strictly within three units of 7.
   b. x differs from 2 by less than 3.
   c. x is no more than five units from 7.
   d. The distance between 7 and x is 4.
   e. $x + 4$ is strictly within two units of the origin.
   f. x is strictly between -3 and 3.
   g. $x < -6$ or $x > 6$
   h. $x - 6 > 4$ or $x - 6 < -4$
   i. the number of hours, x, that a machine will operate efficiently differs from 105 by less than 3.

10. If $|x - \mu| \leq 2\sigma$, in what interval does x lie?

In Problems 11–34, find the solution set of the given equation or inequality.

11. $|x| = 6$

12. $|-x| = 3$

13. $\left|\dfrac{x}{3}\right| = 2$

14. $\left|\dfrac{4}{x}\right| = 8$

15. $|x - 5| = 8$

16. $|4 + 3x| = 2$

17. $|5x - 2| = 0$

18. $|7x + 3| = x$

19. $|7 - 4x| = 5$

20. $|1 - 2x| = 1$

21. $|x| < 4$

22. $|x| < 7$

23. $|x - 4| < 16$ **24.** $|x + 5| < 6$

25. $|x| > -6$ **26.** $|x| < -2$

27. $|x + 1| > 6$ **28.** $|x - 2| > 4$

29. $\left|\dfrac{x}{2} + 6\right| \geq 2$ **30.** $\left|\dfrac{x}{3} - 5\right| \leq 4$

31. $|5 - 2x| \leq 1$ **32.** $|4x - 1| \geq 0$

33. $\left|\dfrac{3x - 8}{2}\right| \geq 4$ **34.** $\left|\dfrac{x - 8}{4}\right| \leq 2$

35. In the manufacture of a certain machine, the average dimension of a part is .01. Using the absolute value symbol, express the fact that an individual measurement x of a part does not differ from the average by more than .005.

17-4
LINEAR INEQUALITIES IN TWO VARIABLES

Suppose each week a chemical manufacturer has available 60 kg of a certain chemical element and uses it all to manufacture x units of com-

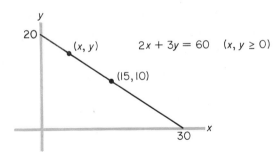

FIGURE 17-11

pound A and y units of compound B. If each unit of A requires 2 kg of the element and each unit of B requires 3 kg, then the possible combinations of A and B which can be produced satisfy the equation

$$2x + 3y = 60, \quad \text{where} \quad x, y \geq 0.$$

The solution set is represented by the line segment in Fig. 17-11. For example, if 15 units of A are made requiring 30 kg of the element, then, since all 60 kg of the element must be used, 10 kg of B must be made requiring 30 units of the element.

On the other hand, suppose the manufacturer does not necessarily wish to use his total supply of the element. The possible combinations now could be described by the inequality

$$2x + 3y \leq 60, \qquad \text{where } x, y \geq 0. \qquad (1)$$

When linear inequalities in *one* variable were discussed, their solution sets were represented geometrically on the real number line by intervals. However, for an inequality in *two* variables, as in (1), the solution set is usually a *region* in the coordinate plane. We shall find the region corresponding to (1) after considering inequalities in general.

Definition. A **linear inequality** in the variables x and y is an inequality which can be written in the form

$$ax + by + c < 0 \qquad (\text{or} \leq 0, \geq 0, > 0)$$

where a, b and c are constants and a and b are not both zero.

Geometrically, the solution set of an inequality in x and y is the set of all points (x, y) in the plane whose coordinates satisfy the inequality. In particular, the graph of the line $y = mx + b$ separates the plane into three distinct parts (Fig. 17-12). First, there are the points (x, y) whose coor-

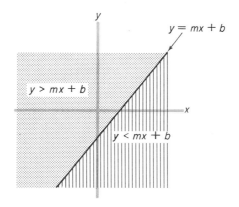

FIGURE 17-12

dinates satisfy $y = mx + b$, that is, the line itself. Second, there are the points (x, y) which lie above the line and satisfy $y > mx + b$. Third, there is the region below the line consisting of points (x, y) satisfying $y < mx + b$.

To apply these facts, we shall first find the solution set of $x + y < 4$. The line $x + y = 4$ is first sketched by choosing two points on it, for instance, the intercepts $(4, 0)$ and $(0, 4)$ (see Fig. 17-13). By writing the inequality in the equivalent form $y < -x + 4$, we conclude that the solution set consists of all points lying below this line. Part of this region has

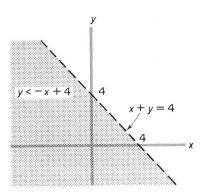

FIGURE 17-13

been shaded in the diagram. Thus, if (x_0, y_0) is *any* point in this region, then its ordinate y_0 is less than the quantity $-x_0 + 4$ (Fig. 17-14). If we had required that $y \leq -x + 4$, the line $y = -x + 4$ would also have been included in the solution set, as indicated by the solid line in Fig. 17-15. *We shall adopt the conventions that a solid line is included and that a broken line is not included in the solution set.*

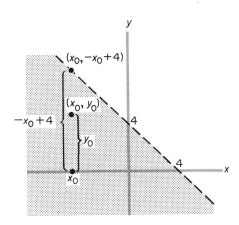

FIGURE 17-14

EXAMPLE
19

Find the region described by $2(2x - y) < 2(x + y) + 4$.

We first rewrite the inequality in an equivalent form so that y alone appears to the left of the inequality symbol:

$$2(2x - y) < 2(x + y) + 4$$
$$4x - 2y < 2x + 2y + 4$$
$$-4y < -2x + 4$$

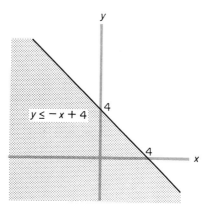

FIGURE 17-15

$$-y < -\frac{x}{2} + 1$$

$$y > \frac{x}{2} - 1.$$

Note that the sense of the original inequality has been changed since we multiplied both members of the inequality by -1, a negative number. We next sketch the line $y = (x/2) - 1$ by noting its intercepts are $(2, 0)$ and $(0, -1)$, and then shade the region above the line (see Fig. 17-16).

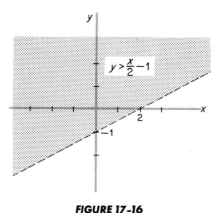

FIGURE 17-16

EXAMPLE

20

Find the solution set of $y \leq 3$.

Since x does not appear, the inequality is assumed to be true for all values of x. Thus, the solution set consists of the line $y = 3$ *and* the region below it since the y-coordinate of each point in that region is less than 3 (see Fig. 17-17).

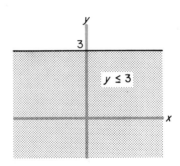

FIGURE 17-17

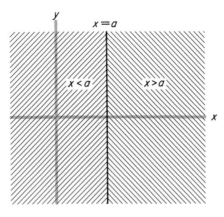

FIGURE 17-18

For a vertical line, $x = a$, we speak of regions to the right ($x > a$)
or to the left ($x < a$) of the line (Fig. 17-18).

EXAMPLE
21

Find the region described by $x \leq 4$.

Since y does not appear, the inequality is assumed to be true for all
values of y. The solution set consists of the line $x = 4$ and the region to
the left of the line (Fig. 17-19).

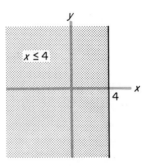

FIGURE 17-19

The solution set of a system of inequalities is the set of all points whose coordinates simultaneously satisfy all the given inequalities. Geometrically, it is the region which is common to all the regions determined by the given inequalities. For example, let us find the solution set of

$$\begin{cases} y - 3x < 6 \\ x - y \le -3. \end{cases}$$

The system is equivalent to

$$\begin{cases} y < 3x + 6 \\ y \ge x + 3. \end{cases}$$

Note that each inequality has been written so that y is isolated. Thus, the appropriate regions with respect to the corresponding lines will become apparent. We now sketch the lines $y = 3x + 6$ and $y = x + 3$ and then shade the region which is simultaneously *below* the first line and *on or above* the second line (see Fig. 17-20). When sketching the lines, it

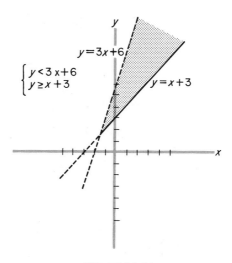

FIGURE 17-20

is best to draw broken lines everywhere until it is clear which portions of the lines are to be included in the solution set.

EXAMPLE
22

Find the region described by

$$\begin{cases} 2x + 3y \le 60 \\ x \ge 0 \\ y \ge 0. \end{cases}$$

This system relates to the discussion at the beginning of this section. The last two inequalities restrict the solution set to points on or to the right

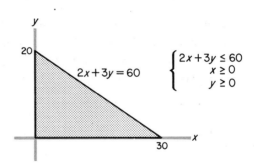

FIGURE 17-21

of the y-axis *and* on or above the x-axis. The desired region is indicated in Fig. 17-21.

EXAMPLE
23

Find the region described by

$$\begin{cases} 2x + y > 3 \\ x \geq y \\ 2y - 1 > 0. \end{cases}$$

The system is equivalent to

$$\begin{cases} y > -2x + 3 \\ y \leq x \\ y > \tfrac{1}{2}. \end{cases}$$

We sketch the lines $y = -2x + 3$, $y = x$, and $y = \frac{1}{2}$ and then shade the region which is simultaneously *above* the first line, *on or below* the second line, and *above* the third line (Fig. 17-22).

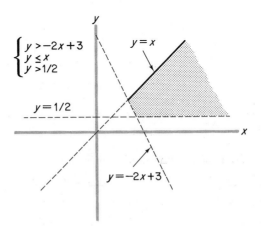

FIGURE 17-22

EXERCISE 17-4

Sketch the region described by the following inequalities.

1. $y > 2x$ **2.** $y < 3x - 4$

3. $2x + 3y \leq 6$ **4.** $x + 3y > 12$

5. $x + 5y < -5$ **6.** $3x + y \leq 0$

7. $-x \geq 2y - 4$ **8.** $2x + y \geq 10$

9. $2(x + y) > 2(2x - y)$ **10.** $-x < 2$

11. $\frac{3}{2}x + \frac{4}{3}y > \frac{3}{2}(x - y)$

12. $2(x^2 + 4x + y) > 4(x^2 - y + 1) - 2x^2$

13. $\begin{cases} y \leq 2x \\ x > 2y \end{cases}$ **14.** $\begin{cases} x - y < 1 \\ y - x \leq 1 \end{cases}$

15. $\begin{cases} 3x - 2y < 6 \\ x - 3y > 9 \end{cases}$ **16.** $\begin{cases} 2x + 3y > -6 \\ 3x - y < 6 \end{cases}$

17. $\begin{cases} 2x + 3y \leq 6 \\ x \geq 0 \end{cases}$ **18.** $\begin{cases} 2y - 3x < 6 \\ x < 0 \end{cases}$

19. $\begin{cases} 2x - 2 \geq y \\ 2x \leq 3 - 2y \end{cases}$ **20.** $\begin{cases} \frac{3}{2}x - \frac{3}{4}y \geq 1 \\ x(x + 1) - 5 \leq xy - x(y - x) \end{cases}$

21. $\begin{cases} x - y > 4 \\ x < 2 \\ y > 1 \end{cases}$ **22.** $\begin{cases} 2x + y < -1 \\ y > -x \\ 2x + 4 < 0 \end{cases}$

23. $\begin{cases} y < 2x + 1 \\ y > 1 \\ x > \frac{1}{2} \end{cases}$ **24.** $\begin{cases} 4x + 3y \geq 12 \\ y \geq x \\ 2y \leq 3x + 6 \end{cases}$

25. $\begin{cases} 5y - 2x \leq 10 \\ 4x - 6y \leq 12 \\ y \geq 2 \end{cases}$ **26.** $\begin{cases} 3x + y > -6 \\ x - y > -5 \\ x \geq 0 \end{cases}$

17-5
REVIEW

REVIEW QUESTIONS

The solution set of $-2x > 4$ consists of all numbers to the ___(right)(left)___
                                                                    (1)
of the number ____(2)____ on the real number line.

(1) **left**   (2) **−2**

True or false: The inequality $x > 2$ is equivalent to $x^2 > 4$. ____(3)____

(3) **false**

The solution set of $|x - 5| = 0$ is _____(4)_____.

(4) **{5}**

The solution set of $|x| < 0$ is _____(5)_____.

(5) ∅

The solution set of $x^2 > 9$ is _____(6)_____.

(6) **(−∞, −3) ∪ (3, ∞)**

The solution set of $|x - 4| \leq 3$ is _____(7)_____.

(7) **[1, 7]**

Using absolute value notation, the fact that $2x$ is strictly within seven units of 4 would be written _____(8)_____.

(8) **$|2x - 4| < 7$**

If $y_1 \geq m_1 x + b_1$ and $y_2 \leq m_2 x + b_2$ is a system of inequalities, which of the regions 1, 2, 3, or 4 in the diagram below would correspond to the solution? _____(9)_____

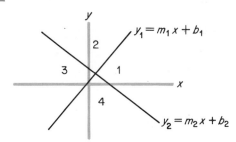

(9) **3**

The inequality $x \geq 4$ is satisfied by all points to the _(right)(left)_ of the line
$\qquad\qquad\qquad\qquad\qquad\qquad\qquad\qquad\qquad\qquad$ (10)
$x = 4$ and _(includes)(excludes)_ the line itself.
$\qquad\quad$ (11)

(10) **right**   (11) **includes**

The solution set of the system $\begin{cases} y > 4 \\ y \le -2 \end{cases}$ is _____(12)_____.

(12) $\varnothing$

If $A = \{x \mid x \ge 2\}$ and $B = \{x \mid x \le 2\}$, then $A \cup B = $ _____(13)_____.

(13) $(-\infty, \infty)$

The solution set of $x(x - 3) > 0$ is _____(14)_____ and the solution set of $x(x - 3) < 0$ is _____(15)_____.

(14) $(-\infty, \mathbf{0}) \cup (\mathbf{3}, \infty)$ (15) $(\mathbf{0}, \mathbf{3})$

REVIEW PROBLEMS

In Problems 1–16, find the solution set.

1. $2 - x < 3 + x$

2. $2(4 - \frac{3}{5}x) \le 5$

3. $x \ge 2x - (7 + x)$

4. $3x - 8 \ge 4(x - 2)$

5. $x^2 - x - 20 < 0$

6. $x^2 + 4x - 12 \ge 0$

7. $\dfrac{x - 4}{x + 3} < 0$

8. $x^2 + x + 2 < 0$

9. $x(x + 2)(x + 4) \ge 0$

10. $\dfrac{x^2 + 3x}{x^2 + 2x - 8} \le 0$

11. $|x - 5| > 0$

12. $|2x - 6| < 0$

13. $|8x + 3| > 6$

14. $|\frac{2}{3}x - 5| \le 2$

15. $|3 - 2x| = 7$

16. $\left|\dfrac{5x - 8}{13}\right| = 0$

In Problems 17–22, sketch the region defined by the given system.

17. $\begin{cases} 2x + y < 4 \\ -y + 2x > 5 \end{cases}$

18. $\begin{cases} 3x + 2y > 5 \\ -3y + 5x < 7 \end{cases}$

19. $\begin{cases} 3x + y < 4 \\ 8x - y > -2 \end{cases}$

20. $\begin{cases} y \ge 5 \\ 2x - y < -2 \end{cases}$

21. $\begin{cases} x + y > 1 \\ 3x - 5 \ge y \\ y > 2x \end{cases}$

22. $\begin{cases} 3x + y > -4 \\ x - y < -5 \\ x > 0 \end{cases}$

analytic geometry

18-1
INTRODUCTION: THE CONIC SECTIONS

In Chapter 6 the Cartesian coordinate system was introduced and the graphs of equations and functions were considered. In particular, the various forms of an equation of a straight line were discussed as well as the relationship between the slopes of parallel and perpendicular lines. In the following sections we extend the concepts of analytic geometry developed thus far and concern ourselves with **conic sections** or **conics**, those curves that result from the intersection of a plane and a cone. The visual interpretation of these important curves as cuts of a cone made by a plane, and a significant investigation into their mathematical characteristics, were first accomplished by the Greek mathematician Apollonius, a contemporary of Archimedes.

The four conics—the circle, parabola, ellipse, and hyperbola—are shown as cuts of a cone in Fig. 18-1(a), (b), (c), and (d), respectively. The **ellipse** in (c) occurs when the intersecting plane cuts the cone obliquely, and the **circle** in (a) occurs when the intersecting plane is perpendicular to the axis of the cone. In a sense, the circle may be considered

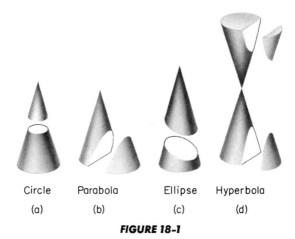

Circle   Parabola   Ellipse   Hyperbola

(a)   (b)   (c)   (d)

FIGURE 18-1

a special case of the ellipse. The **parabola** in (b) occurs when the intersecting plane cuts the cone obliquely while passing through its base, and the **hyperbola** in (d) occurs when an intersecting plane, not containing the axis of the cone, cuts both **nappes**—that is, the parts of the cone on either side of the vertex. Note that the cone is thought of as extending indefinitely on both sides of its vertex.

In comparison to these **regular conics**, certain special cuts in which the intersecting plane passes through the vertex give rise to the **irregular** or **degenerate conics**. If, in the case of Fig. 18-1(a), the intersecting plane passes through the vertex without cutting a nappe, the resulting intersection is merely a point. The degenerate case of the hyperbola, namely two intersecting straight lines, occurs when the intersecting plane contains the axis of the cone, as shown in Fig. 18-2(a). Finally, the parabola degenerates to a straight line if the intersecting plane is tangent to the surface of the cone and passes through the vertex, as shown in Fig. 18-2(b).

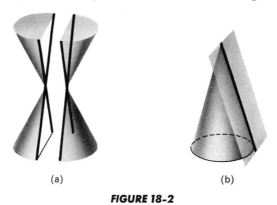

(a)   (b)

FIGURE 18-2

18-2
THE DISTANCE FORMULA

If $P_1(x_1, y_1)$ and $P_2(x_2, y_2)$ are two points in the xy-plane, a formula can be derived for the distance between them—that is, for the length of the line segment joining the points. Refer to Fig. 18-3. Through P_1 and P_2

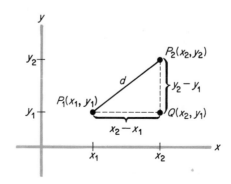

FIGURE 18-3

we construct horizontal and vertical segments, respectively, which will intersect at $Q(x_2, y_1)$. Thus, a right triangle P_1QP_2 is formed whose hypotenuse has length d and whose sides have lengths $x_2 - x_1$ and $y_2 - y_1$. By the Pythagorean theorem,

$$d^2 = (x_2 - x_1)^2 + (y_2 - y_1)^2.$$

Since d cannot be negative,

$$\boxed{d = \sqrt{(x_2 - x_1)^2 + (y_2 - y_1)^2}}$$

which is called the **distance formula**. Since $(x_2 - x_1)^2 = (x_1 - x_2)^2$ and $(y_2 - y_1)^2 = (y_1 - y_2)^2$, the order of the subscripts in the distance formula is immaterial. Thus, given any two points, the distance between these points can be determined by choosing either point as (x_1, y_1) in the formula.

EXAMPLE
1

a. Find the distance d between the points $(-3, 4)$ and $(5, -2)$.

Letting $(-3, 4)$ be the point (x_1, y_1) and $(5, -2)$ the point (x_2, y_2), by the distance formula we have

$$d = \sqrt{[5 - (-3)]^2 + (-2 - 4)^2} = \sqrt{64 + 36} = \sqrt{100} = 10.$$

You should verify that choosing $(5, -2)$ to be (x_1, y_1) produces the same result.

b. Find the distance d of $(2, -3)$ from the origin.

The distance of $(2, -3)$ from $(0, 0)$ is

$$d = \sqrt{(2-0)^2 + (-3-0)^2} = \sqrt{13}.$$

EXERCISE 18-2

1. Determine the distance between the given points.
   a. $(0, 5), (2, -2)$ b. $(1, 3), (1, 4)$
   c. $(1, -\tfrac{7}{2}), (-7, -\tfrac{5}{2})$ d. $(8, 6)$, origin

2. Determine the distance between the given points.
   a. $(5, 11), (1, 8)$ b. $(-7, -4), (-9, -1)$
   c. $(4, 0), (-1, -\sqrt{11})$ d. $(-4, 5)$, origin

18-3
THE CIRCLE

Definition. A **circle** is the set of all points in a plane which are at a given distance from a fixed point in the plane.

The fixed point is called the **center** and the given distance is called the **radius** of the circle. Let the radius of a circle be r and the coordinates of the center be (h, k), as shown in Fig. 18-4. By the distance formula,

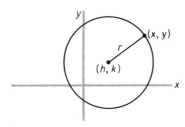

FIGURE 18-4

a typical point (x, y) on the circle most satisfy

$$\sqrt{(x-h)^2 + (y-k)^2} = r$$

or, after squaring both members,

$$(x - h)^2 + (y - k)^2 = r^2. \tag{18-1}$$

It can also be shown that all points on the graph of Eq. (18-1) lie on the given circle. Thus, Eq. (18-1) is an equation of the circle. More precisely, it is the **standard form** of an equation of the circle with center at (h, k) and radius r.

Expanding Eq. (18-1) we have

$$x^2 - 2hx + h^2 + y^2 - 2ky + k^2 = r^2$$
$$x^2 + y^2 - 2hx - 2ky + (h^2 + k^2 - r^2) = 0.$$

In general, then, the equation of any circle can be expressed in the form

$$x^2 + y^2 + Dx + Ey + F = 0, \tag{18-2}$$

where D, E, and F are constants, and Eq. (18-2) is called the **general form** of an equation of a circle. If an equation cannot be put in this form, its graph is not a circle.

By Eq. (18-1), an equation of the circle whose center is at the origin is, since $h = k = 0$,

$$x^2 + y^2 = r^2. \tag{18-3}$$

EXAMPLE
2

a. The standard form of an equation of the circle of radius four units with center at $(-2, 1)$ is, by Eq. (18-1) with $h = -2$, $k = 1$, and $r = 4$,

$$[x - (-2)]^2 + (y - 1)^2 = (4)^2$$

or

$$(x + 2)^2 + (y - 1)^2 = 16.$$

Expanding and simplifying will result in the general form

$$x^2 + y^2 + 4x - 2y - 11 = 0.$$

b. An equation of the circle of radius $\sqrt{5}$ with center at the origin is, by Eq. (18-3),

$$x^2 + y^2 = (\sqrt{5})^2$$
$$x^2 + y^2 = 5.$$

c. The equation $(x - 4)^2 + (y + 2)^2 = 36$ has the form of Eq. (18-1) with $h = 4$, $k = -2$, and $r = 6$ and, hence, is an equation of the circle with center at $(4, -2)$ and radius 6.

d. We can determine the center and radius of a circle whose equation is in general form by the method of completing the square. This will allow the equation to be expressed in standard form. For example, given the equation

$$x^2 + y^2 - 2x - 6y + 6 = 0$$

we write

$$(x^2 - 2x) + (y^2 - 6y) = -6.$$

Completing the square in x and in y, we have

$$(x^2 - 2x + 1) + (y^2 - 6y + 9) = -6 + 1 + 9,$$

which, when factored, yields

$$(x - 1)^2 + (y - 3)^2 = 4.$$

Hence, the given equation defines the circle with center at $(1, 3)$ and radius two units.

e. Although every equation of a circle in standard form can be expressed in the general form, an equation of the form of Eq. (18-2) need not necessarily be that of a circle. For example,

$$x^2 + y^2 - 2x - 6y + 10 = 0$$

is equivalent to

$$(x - 1)^2 + (y - 3)^2 = 0,$$

which defines a circle of radius zero, that is, a point. Also, consider

$$x^2 + y^2 - 2x - 6y + 11 = 0$$
$$(x - 1)^2 + (y - 3)^2 = -1.$$

Since the sum of two squares is always positive or zero, there are no points in the plane satisfying this equation.

EXERCISE 18-3

In Problems 1–10, find the standard and general forms of an equation of the circle determined by the given conditions. Sketch each circle.

1. Center at $(0, 0)$, radius two units

2. Center at $(0, 0)$, radius five units

3. Center at $(0, 2)$, radius $\sqrt{2}$ units

4. Center at $(3, 0)$, radius $\sqrt{3}$ units

5. Center at $(-2, 6)$, radius four units

6. Center at $(0, 1)$, radius $\sqrt{3}$ units

7. Center at $(-4, -7)$, radius k units

8. Center at $(1, -1)$, radius one unit

9. Center at $(\sqrt{2}, \sqrt{3})$, radius two units

10. Center at $(a, -b)$, radius c units

In Problems 11–24, describe the graph of the given equation.

11. $x^2 + y^2 = 16$

12. $x^2 + y^2 = 64$

13. $(x - 2)^2 + y^2 = 9$

14. $x^2 + (y - 3)^2 = 0$

15. $(x + 3)^2 + (y - 3)^2 = 3$

16. $(x + 6)^2 + (y + 1) = 4^2$

17. $x^2 - 2x + y^2 - 4y + 4 = 0$

18. $x^2 + 4x + y^2 - 6y + 9 = 0$

19. $x^2 + y^2 + 6y + 5 = 0$

20. $x^2 + y^2 - 12x + 27 = 0$

21. $x^2 - 8x + y^2 + 4y + 21 = 0$

22. $x^2 + 2x + y^2 - 10y + 26 = 0$

23. $y^2 - 2y + x^2 + 4x - 2 = 0$

24. $4x^2 + 4x + 4y^2 - 12y + 5 = 0$

Recall in Chapter 12 we considered systems of quadratic equations. In Problems 25–28, you are given a system of two equations. Use a crude sketch to predict the number of real distinct solutions of each system.

25. $\begin{cases} x^2 + y^4 = 4 \\ x^2 - 2x + y^2 - 2y = 2 \end{cases}$ **26.** $\begin{cases} x^2 - 4x + y^2 = -3 \\ x^2 - 10x + y^2 = -21 \end{cases}$

27. $\begin{cases} x^2 + y^2 = 1 \\ x^2 - 6x + y^2 = -8 \end{cases}$ **28.** $\begin{cases} x^2 + y^2 = 81 \\ x - y = 1 \end{cases}$

29. Which of the following equations are *not* equations of circles?

    a. $x^2 + y^2 = 4$ b. $x^2 + 2y^2 = 4$

    c. $x^2 - (y - 2)^2 = 4$ d. $x^2 + 2xy + y^2 = 4$

    e. $2x^2 + 2y^2 = 4$ f. $2x + 2y = 4$

    g. $\dfrac{1}{x^2} + \dfrac{1}{y^2} = \dfrac{1}{4}$

30. A particle moves in a plane so that it is always twice as far from $(4, 0)$ as from $(1, 0)$. Determine whether its path is a circle.

31. A racing automobile travels around a circular track at a speed of 157 ft/sec and makes one complete revolution every 40 sec. Taking the origin of a coordinate system at the center of the track and approximating π by 3.14, determine an equation of the path of the automobile.

32. In Problem 31, if you were standing on the track and considered your position as the origin of a coordinate system with the positive y-axis passing through the center of the track, what would be an equation of the path of the automobile?

33. When a particle having mass m and charge q enters a magnetic field of induction B with a velocity v at right angles to B, it can be shown that the particle will travel in a circle of radius r, where

$$r = \frac{mv}{qB}.$$

This is one of the basic operating principles of a mass spectrograph and an important consideration in the design of cyclotrons. If a singly charged lithium ion enters a magnetic field of induction $B = .4$ webers/m² with a speed of 1.17×10^5 m/sec, determine an equation of its circular path. For this ion, $q = 1.60 \times 10^{-19}$ C and $m = 1.16 \times 10^{-26}$ kg.

18-4
THE PARABOLA

Definition. A **parabola** is the set of all points in a plane that are equidistant from a given straight line and a given point not on the line.

The given point is called the **focus** and the given line the **directrix.** Let $2p$, where $p > 0$, be the distance from the focus to the directrix. For convenience we choose the focus at $F(0, p)$ and the directrix whose equation is $y = -p$ as shown in Fig. 18-5. The line through the focus perpen-

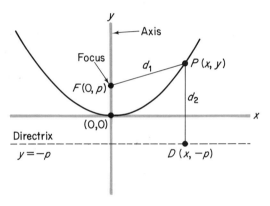

FIGURE 18-5

dicular to the directrix is called the **axis** of the parabola. Thus, the origin is the midpoint of the segment of the axis which connects the focus and the directrix. To derive an equation of the parabola we select an arbitrary point $P(x, y)$ on the parabola. It should be clear that the point of intersection of the directrix and the perpendicular to the directrix from P is $D(x, -p)$ as shown. From the definition of a parabola,

$$d_1 = d_2$$
$$\sqrt{(x - 0)^2 + (y - p)^2} = y + p.$$

Squaring both members yields

$$x^2 + (y - p)^2 = (y + p)^2,$$

which, after simplification, becomes

$$x^2 = 4py.$$

This is the **standard form** of an equation of the parabola with focus at $(0, p)$ and directrix $y = -p$. In addition, it can be shown that if the coordinates of any point satisfy the equation $x^2 = 4py$, then the point lies on the given parabola.

The point at which the parabola intersects its axis is called the **vertex** of the parabola. It is the midpoint of the segment of the axis which joins the focus and the directrix. Clearly, the vertex of the parabola above is at the origin.

More generally, a parabola with vertex at the origin and a coordinate axis for its axis has four possible orientations, the equations of which can be derived in the manner above. The results are indicated in Fig. 18-6. There is no reason for you to attempt to memorize which

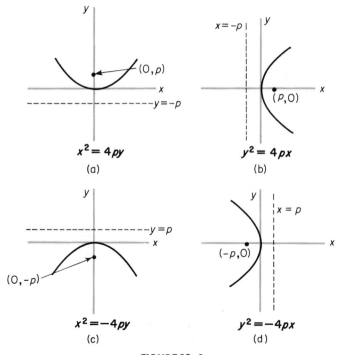

FIGURE 18-6

type of parabola corresponds to each equation, for the equation itself reveals the orientation of the parabola it represents. Consider, for example, Fig. 18-6(b) and the equation $y^2 = 4px$. Since p is considered positive and y^2 is never negative, it follows that $x\ [= y^2/(4p)]$ is never negative. Thus the parabola opens toward the right, as the graph indicates. You are urged to interpret mentally the remaining graphs and equations of Fig. 18-6 in

a similar manner. In each case, p *is the distance from the vertex to the focus.*
We remark that *only* Figs. 18-6(a) and (c) are graphs of functions of x.

EXAMPLE   a. Determine the vertex, focus, and equation of the directrix of the parabola
3              $y^2 = -16x$. Sketch the graph.

The vertex of the parabola is at the origin. Since $x = y^2/-16$, x is never
positive and the parabola opens to the left as shown in Fig. 18–7. Also,

$$4p = 16$$
$$p = 4.$$

Thus the distance from the vertex to the focus is 4 and so the focus is
at $(-4, 0)$. The equation of the directrix is $x = 4$. To aid you in sketching the graph we point out that **the "width" of a parabola at its focus
is 4p.** This means the parabola $y^2 = -16x$ is 16 units wide along a
line perpendicular to its axis and passing through $(-4, 0)$. From this
information we plot the points $(-4, 8)$ and $(-4, -8)$. The graph is
given in Fig. 18-7.

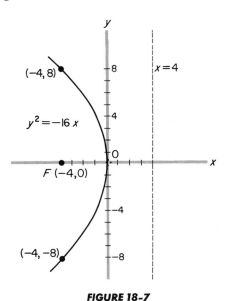

FIGURE 18-7

b. Determine an equation of the parabola with vertex at $V(0, 0)$ and focus
at $F(0, -\frac{3}{2})$.

Refer to Fig. 18-8. From the locations of the focus and vertex, the parabola must open downward and, as a result, y is never positive. Hence,
an equation is of the form $x^2 = -4py$. Since p, the distance from the
focus to the vertex, is $\frac{3}{2}$, the required equation is

$$x^2 = -4(\tfrac{3}{2})y = -6y.$$

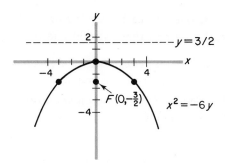

FIGURE 18-8

Consider the parabola whose axis is parallel to the x–axis and whose vertex is the point $V(h, k)$ as shown in Fig. 18-9, where $2p$, $p > 0$, is the

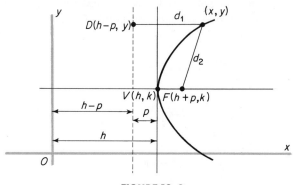

FIGURE 18-9

distance from the focus to the directrix. With the vertex at (h, k), the focus must be at $F(h + p, k)$ and the directrix must be the line $x = h - p$. From the dimensions shown in the diagram, we can deduce that the point of intersection of the directrix and a perpendicular to the directrix from an arbitrary point (x, y) on the parabola is $D(h - p, y)$. From the definition of a parabola,

$$d_2 = d_1$$

$$\sqrt{[x - (h + p)]^2 + (y - k)^2} = x - (h - p).$$

Squaring and simplifying yields

$$y^2 - 2yk + k^2 = 4xp - 4hp,$$

which can be factored and written

$$(y - k)^2 = 4p(x - h).$$

In a similar manner the standard forms of equations of other parabolas

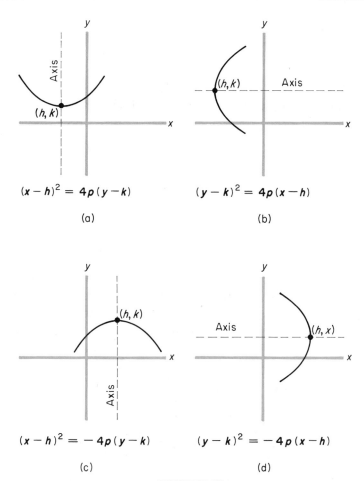

$(x - h)^2 = 4p(y - k)$

(a)

$(y - k)^2 = 4p(x - h)$

(b)

$(x - h)^2 = -4p(y - k)$

(c)

$(y - k)^2 = -4p(x - h)$

(d)

FIGURE 18-10

with vertex at (h, k) and axis parallel to a coordinate axis can be derived. The results are shown in Fig. 18-10.

You should carefully compare the graphs and equations in Figs. 18-6 and 18-10. It is left as an exercise for you to show by expanding the standard form that a general form of an equation of a parabola can be written

$$Ax^2 + Dx + Ey + F = 0, \qquad A, E \neq 0 \qquad (18\text{-}4)$$

for a parabola whose axis is parallel to the y-axis, and

$$Cy^2 + Dx + Ey + F = 0, \qquad C, D \neq 0 \qquad (18\text{-}5)$$

for a parabola whose axis is parallel to the x-axis. Conversely, for A, C, D, and E unequal to zero, Eqs. (18-4) and (18-5) satisfy the requirements of the locus for a parabola and can be transformed into standard form. In the case that $E = 0$ in (18-4) or $D = 0$ in (18-5), three degenerate cases may arise.

For example, if $E = 0$ in Eq. (18-4), then

$$Ax^2 + Dx + F = 0.$$

Let us consider different values of A, D, and F.

a. For $x^2 + 4x + 4 = 0$,

$$(x + 2)(x + 2) = 0$$
$$x = -2$$

and, hence, the set of all points satisfying the given equation is the straight line $x = -2$.

b. For $x^2 - 2x - 8 = 0$,

$$(x - 4)(x + 2) = 0$$
$$x = 4 \qquad \text{and} \qquad x = -2$$

and two distinct parallel straight lines occur. We remark that this is the only degenerate case that cannot be illustrated by a conic section.

c. For $x^2 + 4x + 5 = 0$,

$$x^2 + 4x + 4 = -1$$
$$(x + 2)^2 = -1$$

and, hence, no parabola exists.

EXAMPLE
4

Find an equation of the parabola with its vertex at $V(-1, 4)$ and its focus at $F(-1, 1)$.

By indicating the data in the plane (see Fig. 18-11), we conclude that the parabola must open downward; the equation is of the form $(x - h)^2 = -4p(y - k)$, and $p = 4 - 1 = 3$. Thus an equation is

$$(x + 1)^2 = -4(3)(y - 4)$$

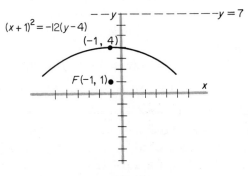

FIGURE 18-11

or

$$(x + 1)^2 = -12(y - 4).$$

The graph is shown in Fig. 18-11. Note that the equation of the directrix is $y = 7$.

EXAMPLE
5

Find the coordinates of the focus and vertex and the equation of the directrix of the parabola

$$y^2 - 8y - 8x + 24 = 0.$$

Rearranging terms and completing the square in y we have

$$(y^2 - 8y + 16) = 8x - 24 + 16 = 8x - 8$$
$$(y - 4)^2 = 8(x - 1).$$

Hence, the vertex is the point $(1, 4)$ and the parabola opens to the right. Also, since $4p = 8$, $p = 2$ and the focus lies two units to the right of the vertex at the point $(3, 4)$. A point on the directrix lies two units to the left of the vertex and so the equation of the directrix is $x = -1$.

EXAMPLE
6

When an object is thrown straight upward with a speed of 64 ft/sec, its height h above its starting point, in feet, as a function of time t, in seconds, is given by the quadratic function $h = 64t - 16t^2$. How high does the ball rise above its starting point?

Since

$$h = 64t - 16t^2$$
$$h = -16(t^2 - 4t)$$
$$h - 64 = -16(t^2 - 4t + 4)$$
$$(t - 2)^2 = -\tfrac{1}{16}(h - 64),$$

the graph of the given quadratic function is a parabola, opening downward, with vertex at $(2, 64)$ as shown in the figure below. Hence, the maximum height reached is 64 ft.

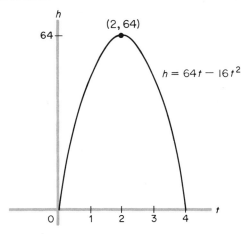

In general, if the graph of a quadratic function $y = f(x)$ crosses the x-axis at the points $(x_1, 0)$ and $(x_2, 0)$, then the variable y has its maximum (or minimum) value when

$$x = \frac{x_1 + x_2}{2}. \tag{18-6}$$

EXAMPLE
7

Find the maximum value of the function $y = 20x - x^2$.

To determine the values of x for which the graph crosses the x-axis, we set $y = 0$. Then

$$0 = 20x - x^2$$
$$0 = x(20 - x)$$

and

$$x = 0 \quad \text{or} \quad x = 20.$$

By Eq. (18-6), the maximum value is attained when

$$x = \frac{0 + 20}{2} = 10.$$

Hence, the maximum value is

$$y = 20x - x^2$$
$$= 20(10) - (10)^2$$
$$= 100.$$

You should use this technique to verify the result in Example 6.

EXERCISE 18-4

In Problems 1–12, find the coordinates of the focus and vertex, an equation of the directrix, and sketch each parabola.

1. $y^2 = 4x$
2. $y^2 = -6x$
3. $x^2 = -8y$
4. $y^2 = -x$
5. $x^2 = 2y$
6. $(x + 2)^2 = 4(y - 7)$
7. $(y + 2)^2 = \frac{1}{2}x$
8. $4x^2 = 3(y - 1)$
9. $(x - 4)^2 = 8(y + 3)$
10. $(y - 7)^2 = 12(x - 4)$
11. $(y + 2)^2 = -\frac{1}{4}(x - 1)$
12. $y = x^2$

*In Problems 13 and 14, apply the **definition** to derive the standard form of an equation of the parabola described.*

13. Focus $(-5, 2)$, directrix $x = 1$
14. Vertex $(3, 4)$, focus $(3, 6)$

In Problems 15–24, determine the standard form of an equation of the parabola from the given information without the use of the definition.

15. Focus $(0, 3)$, directrix $y = -5$

16. Vertex $(2, 4)$, directrix $y = 6$

17. Focus $(1, 4)$, vertex $(3, 4)$

18. Focus $(0, 5)$, directrix $x = -10$

19. Vertex $(-3, 2)$, directrix $y = 4$

20. Focus $(-2, -2)$, vertex $(-2, -6)$

21. Focus $(0, \frac{3}{2})$, directrix $y = -\frac{3}{2}$

22. Vertex $(3, 1)$, directrix $x = 6$

23. Focus $(-3, -2)$, vertex $(-4, -2)$

24. Focus $(0, -4)$, directrix $y = 4$

In Problems 25–29, transform the given equation into standard form, determine the coordinates of the focus and vertex, and find an equation of the directrix.

25. $y^2 - 6y + 4x + 1 = 0$

26. $2y^2 + 4y - x - 4 = 0$

27. $3x^2 - 12x - y + 12 = 0$

28. $x^2 + 3y - 8x + 19 = 0$

29. $y^2 + 4y - x + 5 = 0$

In Problems 30–33, use a crude graph of the given system to predict how many distinct real solutions exist.

30. $\begin{cases} y^2 = x - 1 \\ y = x^2 - 2x + 1 \end{cases}$ 　　　　**31.** $\begin{cases} y = x^2 - 3x \\ x + y = -4 \end{cases}$

32. $\begin{cases} x = (y - 1)^2 \\ x - y = 2 \end{cases}$ 　　　　**33.** $\begin{cases} y^2 = 2x + 1 \\ x^2 + y^2 = 9 \end{cases}$

34. The power P developed in a resistor of resistance R ohms carrying a current of i amp is $P = i^2R$. If a resistor has a resistance of 10 ohms, sketch a graph of power against current.

35. When an object is thrown straight upward with an initial velocity of 20 ft/sec, its height h, in feet, as a function of time t, in seconds, is given by $h = 20t - 16t^2$. At what time does the object reach its maximum height and what is that height?

36. The displacement s of an object from a reference point is given by $s = 3t^2 - 24t + 103$, where s is in feet and t is in seconds. What is the minimum displacement of the particle from the reference point?

37. If a light source is placed between the focus and vertex of a parabolic mirror, the rays of light diverge after reflection, while if the light source is placed outside the focus, the rays converge after reflection. This is how automobile headlights can give a broad beam of light (high beams) from one filament and a narrow beam of light (low beams) from a second filament. If the equation of a cross-sectional view of a parabolic mirror is $y^2 = 20x$, where x and y are in centimeters, what may the distance be between the vertex of the mirror and a filament if the light rays are to diverge after reflection from the mirror?

38. When a 4-lb object moves in a straight line on a horizontal surface 3 ft above the ground, its energy E, neglecting any rotation, can be written $E = 12 + \frac{1}{16}v^2$, where v is the speed of the object in feet per second, and E is measured in foot-pounds. The value of v is taken to be positive for one direction of motion and negative for the opposite direction. Sketch a graph of this energy function for the values of v from $v = -4$ ft/sec to $v = +4$ ft/sec.

18-5
THE ELLIPSE

Definition. An **ellipse** is the set of all points in a plane such that for each point the sum of its distances from two fixed points is a constant.

The two fixed points are called **foci** (the plural of focus) of the ellipse. Let $2c$, $c > 0$, be the distance between the foci. For convenience we shall locate the foci on the x-axis at $F_1(c, 0)$ and $F_2(-c, 0)$ [see Fig. 18-12].

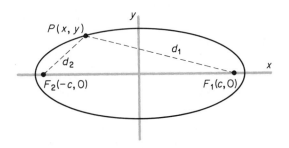

FIGURE 18-12

Furthermore, let the constant sum referred to in the definition be $2a$. To eliminate from our consideration those points on the segment joining the foci, we must require that $2a > 2c$, or $a > c$. Selecting an arbitrary point $P(x, y)$ on the ellipse and applying the definition, we have

$$d_1 + d_2 = \text{constant}$$
$$\sqrt{(x - c)^2 + (y - 0)^2} + \sqrt{(x + c)^2 + (y - 0)^2} = 2a.$$

Rearranging,

$$\sqrt{(x-c)^2 + y^2} = 2a - \sqrt{(x+c)^2 + y^2}.$$

Squaring both members,

$$(x-c)^2 + y^2 = 4a^2 - 4a\sqrt{(x+c)^2 + y^2} + (x+c)^2 + y^2,$$

which simplifies to

$$a^2 + cx = a\sqrt{(x+c)^2 + y^2}.$$

Squaring both members and rearranging,

$$a^4 - a^2c^2 + c^2x^2 - a^2x^2 - a^2y^2 = 0$$

$$a^2(a^2 - c^2) + x^2(c^2 - a^2) - a^2y^2 = 0.$$

Dividing both members by $a^2(a^2 - c^2)$,

$$1 - \frac{x^2}{a^2} - \frac{y^2}{a^2 - c^2} = 0$$

or, when rearranged,

$$\frac{x^2}{a^2} + \frac{y^2}{a^2 - c^2} = 1.$$

Since $a > c$, $a^2 - c^2$ is a positive quantity. If we let

$$a^2 - c^2 = b^2, \tag{18-7}$$

then we can write

$$\frac{x^2}{a^2} + \frac{y^2}{b^2} = 1. \tag{18-8}$$

Thus, every point on the given ellipse satisfies Eq. (18-8). Conversely, any point whose coordinates satisfy Eq. (18-8) is a point on the ellipse. Equation (18-8) is the **standard form** of the ellipse with **center**, the midpoint of the line segment joining the foci, at the origin. The foci are $F_1(c, 0)$ and $F_2(-c, 0)$ with $a^2 = b^2 + c^2$, as is sketched in Fig. 18-12.

The relationship of a, b, and c is easily seen from the triangle in Fig. 18-13. From Eq. (18-8), if $y = 0$, then $x = \pm a$; the x-intercepts are $(\pm a, 0)$. If $x = 0$, then $y = \pm b$; the y-intercepts are $(0, \pm b)$. Segment V_1V_2, which has length $2a$ and passes through the foci, is called the **major axis**. Segment P_1P_2, which has length $2b$, is called the **minor axis**. We speak of a as the **semimajor axis**; *it is the distance from the center of the ellipse to the end of the major axis.* Similarly, we speak of b as the length of the **semiminor axis**. The end points of the major axis, V_1 and V_2, are called the **vertices** of the ellipse.

In the case of an ellipse with center at the origin and whose foci lie on

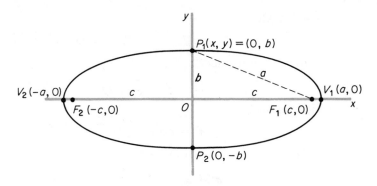

FIGURE 18-13

the y-axis [at $(0, c)$ and $(0, -c)$], the standard form is

$$\frac{y^2}{a^2} + \frac{x^2}{b^2} = 1. \tag{18-9}$$

Here, the major axis lies on the y-axis.

In Eqs. (18-8) and (18-9), remember that since $a > b$, **the larger denominator is always a^2.** Thus, if the larger denominator is in the x^2-term, the major axis is horizontal. If it is in the y^2-term, the major axis is vertical.

It can be shown that the standard form of an equation of an ellipse with center (h, k) and major axis horizontal is given by

$$\frac{(x - h)^2}{a^2} + \frac{(y - k)^2}{b^2} = 1, \tag{18-10}$$

and for one whose major axis is vertical the equation is

$$\frac{(y - k)^2}{a^2} + \frac{(x - h)^2}{b^2} = 1. \tag{18-11}$$

EXAMPLE
8

Sketch the graph of $4x^2 + 9y^2 = 36$ indicating the center, vertices, foci, and semimajor and semiminor axes.

To obtain the standard form we divide both members by 36:

$$\frac{x^2}{9} + \frac{y^2}{4} = 1.$$

The center is at the origin. Since the larger denominator (9) is under x^2, the major axis lies on the x-axis. Since $a^2 = 9$ and $b^2 = 4$, then $a = 3$ and $b = 2$. Thus, the semimajor axis has length three units and the semiminor axis has length 2. Hence, the vertices are $(\pm 3, 0)$ and the ends of the minor axis are at $(0, \pm 2)$. From this information we can sketch the graph, as in

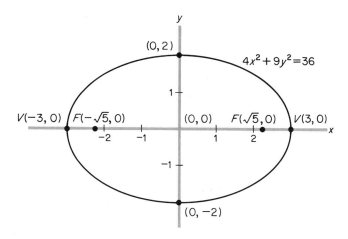

FIGURE 18-14

Fig. 18-14. Also,

$$c^2 = a^2 - b^2 = 9 - 4 = 5$$
$$c = \sqrt{5}.$$

Thus the foci are at $(\pm\sqrt{5}, 0)$.

EXAMPLE
9

Find an equation of the ellipse having a major axis of length eight units, a focus at $(2, 8)$, and center at $(2, 5)$.

An ellipse with a center at $(2, 5)$ and a focus at $(2, 8)$ must have a vertical major axis. The distance from the center to a focus is three units. Thus $c = 3$. Also, since the length of the major axis is $2a\ (= 8)$, $a = 4$. Therefore, by Eq. (18-7), $b^2 = a^2 - c^2 = 16 - 9 = 7$. Thus an equation is

$$\frac{(y - 5)^2}{16} + \frac{(x - 2)^2}{7} = 1.$$

The graph is shown in Fig. 18-15.

Expanding the standard form of an equation of an ellipse, for example Eq. (18-10), will result in an equation in terms of x^2, y^2, x, y, and a constant. Either the x- or y-term may be missing, depending on the location of the center of the ellipse. The **general form** of the ellipse can be written

$$Ax^2 + Cy^2 + Dx + Ey + F = 0. \qquad (18\text{-}12)$$

The signs of A and C *must* be the same, and $A \neq C$. We remark that an ellipse is not the graph of a function. Note that $A = C$ may result in a circle.

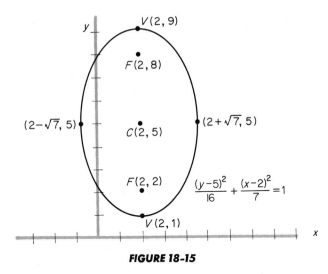

FIGURE 18-15

EXAMPLE
10
Determine the center, foci, vertices, and lengths of the semimajor and semiminor axes of the ellipse

$$x^2 + 2y^2 + 4x - 4y + 2 = 0.$$

Rearranging, we can write

$$(x^2 + 4x) + 2(y^2 - 2y) = -2.$$

Completing the square in x and y, we have

$$(x^2 + 4x + 4) + 2(y^2 - 2y + 1) = -2 + 4 + 2.$$

Note that when 1 was added to $y^2 - 2y$ to complete the square in y, it was equivalent to adding 2 to the left member of the equation. Hence, 2 had to be added to the right member:

$$(x + 2)^2 + 2(y - 1)^2 = 4.$$

Finally, dividing by 4,

$$\frac{(x + 2)^2}{4} + \frac{(y - 1)^2}{2} = 1.$$

Clearly, $a^2 = 4$, $b^2 = 2$, and $c^2 = a^2 - b^2 = 2$. The center is at $(-2, 1)$ and the major axis is horizontal. Since $c = \sqrt{2}$, the foci are at $(-2 \pm \sqrt{2}, 1)$. Also, the semimajor axis $a = 2$ and so the vertices are $(-2 \pm 2, 1)$, that is, at $(0, 1)$ and $(-4, 1)$. Lastly, the semiminor axis $b = \sqrt{2}$.

EXERCISE 18-5

In Problems 1–12, find the center, vertices, foci, length of the semimajor axis, and length of the semiminor axis for the given ellipse. Sketch each curve.

1. $\dfrac{x^2}{25} + \dfrac{y^2}{16} = 1$ **2.** $4y^2 + x^2 = 16$

3. $\dfrac{x^2}{25} + \dfrac{y^2}{100} = 1$

4. $\dfrac{(x-1)^2}{4} + y^2 = 4$

5. $4x^2 + y^2 = 4$

6. $\dfrac{(x+2)^2}{36} + \dfrac{(y-3)^2}{25} = 1$

7. $\dfrac{(x-2)^2}{9} + \dfrac{(y+3)^2}{4} = 1$

8. $9x^2 + 3(y-4)^2 = 4$

9. $4x^2 + y^2 - 16x = 0$

10. $2x^2 + y^2 + 8x + 4y + 6 = 0$

11. $9x^2 + 25y^2 - 54x + 100y = 44$

12. $x^2 + 2y^2 + 4x - 8y - 6 = 0$

In Problems 13–21, determine the standard form of an equation of the ellipse satisfying the given conditions. Assume the center is at the origin unless otherwise stated.

13. Vertex $(6, 0)$, focus $(5, 0)$

14. Major axis 16 units, focus $(6, 0)$

15. Focus $(-8, 1)$, minor axis four units, center $(0, 1)$

16. Focus $(3, 3)$, center on y-axis, vertex $(-5, 3)$

17. Major axis 16 units and horizontal, minor axis 8 units

18. Vertex $(3, 10)$, focus $(3, -6)$, center on x-axis

19. Vertex $(0, 0)$, center $(0, -8)$, minor axis five units

20. Focus $(-4, -5)$, major axis 14 units, center $(-4, 1)$

21. Minor axis horizontal and four units, focus $(-2, -8)$, center $(-2, -1)$

22. The arch of a bridge has the shape of one-half an ellipse. The maximum height of the bridge is 50 ft and the bridge has a span at water level of 120 ft. If the origin of a coordinate system is midway between the ends of the bridge at water level, what is the equation of the ellipse? What is the equation if the origin of the coordinate system is at one end of the bridge at water level?

23. Man-made satellites can be made to orbit the earth in an elliptical path whose center is at the center of the earth. If the altitude of such a satellite— that is, its distance above the surface of the earth—ranges from 1000 to 4000 mi, find the equation of its path. The radius of the earth is approximately 4000 mi.

18-6

THE HYPERBOLA

Definition. A **hyperbola** is the set of all points in a plane the difference of whose distances from two fixed points, called foci, is a positive constant.

Let the distance between foci be $2c$, where $c > 0$, and, as before, locate the foci at points $F_1(c, 0)$ and $F_2(-c, 0)$. Let the constant referred to in the definition be $2a$, where $a > 0$. Selecting a point $P(x, y)$ on the hyperbola (Fig. 18-16), we have by the definition

$$d_1 - d_2 = \pm 2a$$

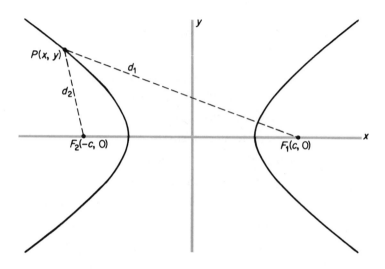

FIGURE 18-16

depending on which of d_1 or d_2 is larger. Therefore, by the distance formula,

$$\sqrt{(x - c)^2 + (y - 0)^2} - \sqrt{(x + c)^2 + (y - 0)^2} = \pm 2a,$$

which by a method similar to that used in the preceding section can be written

$$c^2 x^2 - a^2 x^2 - a^2 y^2 = a^2 c^2 - a^4.$$

Hence

$$x^2(c^2 - a^2) - a^2 y^2 = a^2(c^2 - a^2),$$

and, dividing both members by $a^2(c^2 - a^2)$,

$$\frac{x^2}{a^2} - \frac{y^2}{c^2 - a^2} = 1. \tag{18-13}$$

Following the derivation of an equation of an ellipse, it would seem natural to make the substitution

$$b^2 = c^2 - a^2. \tag{18-14}$$

However, we must be sure that $c^2 - a^2$ is always positive. Recall from geometry that the difference between any two sides of a triangle is less than the third side. Hence, from Fig. 18-16, for the triangle $F_1 F_2 P$ with $d_1 > d_2$ we have

$$2c > d_1 - d_2$$

and by definition it follows that

$$2c > 2a \quad \text{and} \quad c > a.$$

Thus $c^2 > a^2$ and $c^2 - a^2 > 0$. The proof if $d_2 > d_1$ is similar.

Substituting Eq. (18-14) in Eq. (18-13) yields a **standard form** of a hyperbola:

$$\frac{x^2}{a^2} - \frac{y^2}{b^2} = 1, \tag{18-15}$$

where $c^2 = a^2 + b^2$.

When $y = 0$, then $x = \pm a$ and hence the hyperbola crosses the x-axis at the points $V_1(a, 0)$ and $V_2(-a, 0)$ [refer to Fig. 18-17]. These points,

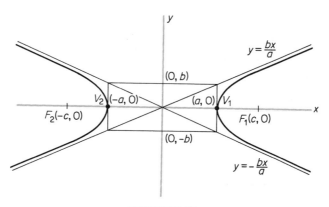

FIGURE 18-17

those for which the hyperbola cuts the line segment joining the foci, are called the **vertices** of the hyperbola. The line segment $V_1 V_2$ joining the vertices is $2a$ units long and is called the **transverse axis**. The midpoint of the transverse axis is called the **center**. In the diagram the center is at the origin.

To determine more precisely the nature of a hyperbola and the geometrical significance of b, we solve Eq. (18-15) for y:

$$b^2 x^2 - a^2 y^2 = a^2 b^2$$

$$y^2 = \frac{b^2 x^2 - a^2 b^2}{a^2} = \frac{b^2 x^2}{a^2}\left(1 - \frac{a^2}{x^2}\right)$$

$$y = \pm \frac{b}{a} x \sqrt{1 - \frac{a^2}{x^2}}.$$

As x increases or decreases without bound, then a^2/x^2 gets close to zero and the expression under the radical sign approaches 1. We can conclude that y will approach the values $(b/a)x$ and $-(b/a)x$. Thus, the straight lines $y = (b/a)x$ and $y = -(b/a)x$ are *asymptotes* for the hyperbola. That is, they are straight lines which the hyperbola approaches as a limiting position. These results are indicated in Fig. 18-17.

From the diagram it should be clear that the asymptotes coincide, so to speak, with diagonals of a rectangle of length $2a$ and width $2b$ and whose vertices are $(a, b), (-a, b), (-a, -b)$, and $(a, -b)$. The line segment between the points $(0, b)$ and $(0, -b)$ has length $2b$ and is called the **conjugate axis** of the hyperbola. Clearly, the asymptotes provide convenient guide lines for sketching a hyperbola. They are most easily sketched by drawing the diagonals of the rectangle whose sides pass through the vertices and the endpoints of the conjugate axis. It can be shown that *the equations of the asymptotes of the hyperbola can be found by setting the left member of Eq. (18-15) equal to zero and solving for y.*

In a similar manner, it can be shown that an equation of a hyperbola with center at the origin, transverse axis along the y-axis, and conjugate axis along the x-axis, is

$$\frac{y^2}{a^2} - \frac{x^2}{b^2} = 1. \tag{18-16}$$

In this case vertices are located at $V_1(0, a)$ and $V_2(0, -a)$, foci at $F_1(0, c)$ and $F_2(0, -c)$, and the lines $y = \pm ax/b$ are asymptotes.

Using the definition, it can be shown that an equation of a hyperbola with center at (h, k) and transverse axis horizontal is given by

$$\frac{(x - h)^2}{a^2} - \frac{(y - k)^2}{b^2} = 1, \tag{18-17}$$

while if the transverse axis is vertical, the equation is

$$\frac{(y - k)^2}{a^2} - \frac{(x - h)^2}{b^2} = 1. \tag{18-18}$$

Note that a^2 is always associated with the positive term.

Finally, a general form of an equation of a hyperbola can be written

$$Ax^2 + Cy^2 + Dx + Ey + F = 0, \tag{18-19}$$

where A and C have opposite signs.

EXAMPLE
11

Discuss and sketch the graph of the hyperbola $\dfrac{y^2}{4} - x^2 = 1$.

From the standard forms of the equation of a hyperbola, a^2 is always associated with the positive term. Hence, in this case, $a^2 = 4$, $b^2 = 1$, or $a = 2$,

$b = 1$. Also $c^2 = a^2 + b^2 = 4 + 1 = 5$ or $c = \sqrt{5}$. Since the equation is of the form of Eq. (18-16), or, equivalently, the form of Eq. (18-18) with $h = k = 0$, we can deduce the hyperbola has vertices $V_1(0, 2)$ and $V_2(0, -2)$, foci $F_1(0, \sqrt{5})$ and $F_2(0, -\sqrt{5})$, a vertical transverse axis of length 4, and a horizontal conjugate axis of length 2. The graph is easily sketched by first locating the vertices and endpoints of the conjugate axis. Then construct the rectangle whose sides pass through these points. Next, sketch the asymptotes by drawing the diagonals of the rectangle. Finally, draw the hyperbola which passes through the vertices and approaches the asymptotes. The results are shown in Fig. 18-18. Note that the equations of the asymptotes are $y = 2x$ and $y = -2x$.

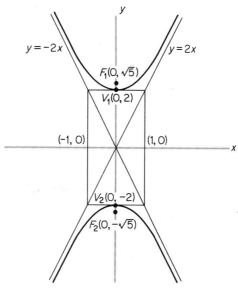

FIGURE 18-18

EXAMPLE 12

Sketch the graph of the equation

$$2x^2 - y^2 - 16x + 4y + 24 = 0.$$

Determine the coordinates of all significant points and the equations of the asymptotes.

The equation can be written

$$(2x^2 - 16x) - (y^2 - 4y) = -24.$$

Completing the square and simplifying, we have

$$2(x^2 - 8x + 16) - (y^2 - 4y + 4) = -24 + 32 - 4$$

$$2(x - 4)^2 - (y - 2)^2 = 4$$

$$\frac{(x - 4)^2}{2} - \frac{(y - 2)^2}{4} = 1. \tag{1}$$

The center of the hyperbola is at $(4, 2)$ and the transverse axis is horizontal. Since $a^2 = 2$ and $b^2 = 4$, then $a = \sqrt{2}$, $b = 2$, and $c^2 = 4 + 2$ or $c = \sqrt{6}$. In Fig. 18-19 the center of the hyperbola, the vertices, and the endpoints of the conjugate axis are located, the rectangle completed, and the asymptotes drawn. The graph is then sketched. As previously mentioned, to find the

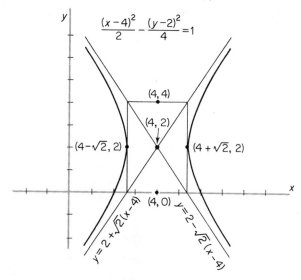

FIGURE 18-19

equations of the asymptotes we set the left member of Eq. (1) equal to zero and solve for y:

$$\frac{(y-2)^2}{4} = \frac{(x-4)^2}{2}$$

$$\frac{y-2}{2} = \pm\frac{x-4}{\sqrt{2}}$$

$$y - 2 = \pm\sqrt{2}(x-4)$$

$$y = 2 \pm \sqrt{2}(x-4).$$

We note that in the standard form of the equation of a hyperbola with center at the origin and transverse axis along the x-axis,

$$\frac{x^2}{a^2} - \frac{y^2}{b^2} = 1,$$

if $a = b$ the result $x^2 - y^2 = a^2$ is a hyperbola with mutually perpendicular asymptotes, $y = \pm x$. The central rectangle is, in this case, a square and the curve is called an **equilateral hyperbola**.

We conclude our discussion here by stating that there is one special form of an equation of an equilateral hyperbola. The equation

$$xy = c, \qquad (18\text{-}20)$$

where c is a nonzero constant, is an equation of an equilateral hyperbola whose asymptotes are the coordinate axes. If c is positive, the foci lie on the line $y = x$, while if c is negative, they lie on the line $y = -x$. The graphs of hyperbolas whose equations are in such forms can be easily sketched by assuming values for one variable and determining the corresponding values of the other variable. The general shapes of the graphs of $xy = c$ for $c > 0$ and $c < 0$ are shown in Figs. 18-20 and 18-21, respectively.

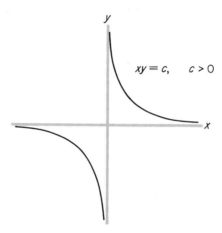

FIGURE 18-20

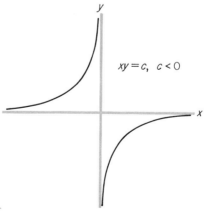

FIGURE 18-21

EXERCISE 18-6

In Problems 1–10, sketch the graph of each hyperbola indicating the center, vertices, foci, and endpoints of the conjugate axis and state the equations of the asymptotes.

1. $\dfrac{x^2}{16} - \dfrac{y^2}{9} = 1$ 　　　　　　　　　**2.** $\dfrac{y^2}{25} - \dfrac{x^2}{144} = 1$

3. $\dfrac{y^2}{36} - x^2 = 1$ 　　　　　　　　　　**4.** $16x^2 - 9y^2 = 1$

5. $25(x - 2)^2 - 4(y + 3)^2 = 100$ 　**6.** $\dfrac{x^2}{4} - \dfrac{(y - 2)^2}{4} = 1$

7. $x^2 - y^2 = 3$ 　　　　　　　　　　**8.** $2y^2 - (x + 3)^2 = 8$

9. $9x^2 - 36x - 16y^2 - 32y - 124 = 0$

10. $y^2 - 4x^2 - 10y + 16x - 7 = 0$

In Problems 11–20, determine the standard form of an equation of the hyperbola satisfying the given conditions. Assume the center is at $(0, 0)$ unless otherwise stated.

11. Focus $(0, 3)$, vertex $(0, 2)$

12. Transverse axis 10 units, focus $(7, 0)$

13. Vertex $(4, 0)$, conjugate axis two units

14. Conjugate axis four units, focus $(0, 6)$

15. Center $(-4, 2)$, focus $(-4, 6)$, vertex on x-axis

16. Center $(-7, 0)$, focus at origin, transverse axis six units

17. Vertex on y-axis, center $(-2, 4)$, conjugate axis three units

18. Vertices at $(4, 0)$ and $(4, 4)$, conjugate axis four units

19. Center on line $y = 4$, vertex $(0, 2)$, focus $(0, 1)$

20. Vertex $(2, 6)$, focus $(4, 6)$, center on y-axis.

In Problems 21–26, sketch the graph of each hyperbola and state the equations of the asymptotes.

21. $xy = 3$ 　　　　　　　　　　　**22.** $xy = -3$

23. $xy = -5$ 　　　　　　　　　　**24.** $5xy = 6$

25. $4xy = 1$ 　　　　　　　　　　**26.** $xy = 120$

27. The speed v of a wave is a function of the frequency f and the wavelength λ, where $v = f\lambda$. For visible light in a vacuum, $v = 3 \times 10^8$ m/sec and values of λ range from about 4×10^{-7} to about 7×10^{-7} m. Sketch a graph of frequency versus wavelength for the visible portion of the spectrum.

28. The relationship between the index of refraction of a material, n, the wavelength of light in the material, λ, and the wavelength of light in air,

λ_a, is $\lambda_a = n\lambda$. If the wavelength of blue-green light in air is 5000 angstroms, sketch a graph of λ versus n for values of n from 1.5 to 2.5, in increments of .1.

29. For an ideal gas at constant temperature, the product of the pressure p and volume v is a constant. A particular sample of such a gas has a pressure of 4 atmospheres and a volume of 5 liters. Sketch a graph of pressure versus volume.

18-7
SUMMARY OF CONIC SECTIONS

It is often important to be able to determine the shape of a curve from a cursory examination of its equation. Summarizing the results of the preceding sections we can say that, assuming no degenerate cases, the graph of

$$Ax^2 + Bxy + Cy^2 + Dx + Ey + F = 0$$

will be a(n)

1. Circle if $A = C$, $B = 0$.

2. Parabola if $A = 0$ or $C = 0$ but not both, and $B = 0$.

3. Ellipse if $A \neq C$ but A and C have same sign, $B = 0$.

4. Hyperbola if
   a. A and C have opposite signs, $B = 0$.
   b. $A = C = 0$, $B \neq 0$.

EXAMPLE
13

Classify each of the following equations as that of a circle, parabola, ellipse, or hyperbola.

a. $x^2 + y^2 + 3x - 6y - 7 = 0$.

Since the coefficients of the x^2- and y^2-terms are equal ($A = C = 1$), the equation is that of a circle.

b. $2y^2 + x^2 - 4y - 4x - 10 = 0$.

Since the coefficients of the x^2- and y^2-terms are unequal but have the same sign ($A = 1 \neq 2 = C$), we identify the equation as that of an ellipse.

c. $2y^2 + 3y - 4x + 9 = 0$.

Since there is a y^2-term but no x^2-term ($A = 0$, $C = 2$), the equation is that of a parabola.

d. $5x^2 - 3y^2 - 5x + 2y + 16 = 0$.

Since the x^2- and y^2-terms are opposite in sign, the equation is that of a hyperbola.

e. $xy = 7$.

The equation is that of an equilateral hyperbola.

EXERCISE 18-7

By inspection, classify each of the following equations as a circle, parabola, ellipse, or hyperbola.

1. $3x^2 + 3y^2 + 2x + 5y - 6 = 0$
2. $3x^2 + 2y^2 + 2x - 7 = 0$
3. $2y^2 + 3x + 2y + 1 = 0$
4. $3xy = 18.21$
5. $x^2 + 3x + 14y - 17 = 0$
6. $-2x^2 - 2y^2 + 3x - 4y + 17 = 0$
7. $2x^2 - 3y^2 + 2x + 6 = 0$
8. $3y^2 - 2x = 4$
9. $4x^2 + 2y^2 + 3y = 8$
10. $2x^2 + 3y = 7$
11. $xy = -3$
12. $4x^2 + 4y^2 - 6x + 3y + 7 = 0$
13. $x^2 + 2y^2 + 3 = 2y - y^2$
14. $4x^2 - 2y^2 - 3x + 4y + 182 = 0$
15. $x^2 + 5y^2 + x + 17y = 1$
16. $1.3x^2 - 7.2 = -8.6y^2$
17. $x^2 + 3x - 4y + 6 = 4$
18. $3x^2 + 3y^2 - 4 = 0$
19. $4x^2 - 7y^2 = 16$
20. $1.2x^2 = 3y + 7.4$

18-8
PARAMETRIC EQUATIONS

Each of the ordered pairs (x, y) associated with the graph of an equation in x and y has been obtained by assigning arbitrary values to one variable from which corresponding values of the other variable were then obtained. It is often convenient, however, to make use of a third variable, say t, for assigning values to **both** x and y. In science and technology, for example, t frequently denotes time, and for each value that t assumes, corresponding values of x and y would then be determined by some given rule.

As an illustration, suppose $x = 2t + 3$ and $y = t + 1$. For $t = 0$ we have, by substitution, $x = 3$ and $y = 1$; that is, with $t = 0$ there is associated the ordered pair $(3, 1)$. Similarly, for $t = 2, 3$, and -1, the corresponding ordered pairs are $(7, 3)$, $(9, 4)$, and $(1, 0)$, respectively. Using these points, and others found in the same manner, the curve can be sketched.

In general, if

$$x = f_1(t) \tag{1}$$

and

$$y = f_2(t), \tag{2}$$

the variable t is called a **parameter**, and Eqs. (1) and (2) are collectively calle'd **parametric equations**.

EXAMPLE
14

Sketch the graph of the curve given by the parametric equations

$$x = t + 1$$
$$y = t^2 - 3.$$

Assigning values to the parameter t we determine the corresponding values of x and y as shown in the accompanying table below.

t	-3	-2	-1	0	1	2
x	-2	-1	0	1	2	3
y	6	1	-2	-3	-2	1

The graph is shown in Fig. 18-22. Observe that the curve appears parabolic and note also that the parameter t does not appear in the graph.

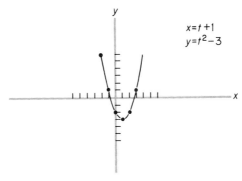

FIGURE 18-22

If a pair of parametric equations is given, the direct relationship between x and y may sometimes be determined by eliminating the parameter by means of previously developed techniques.

EXAMPLE
15

Eliminate the parameter from

$$x = t + 1 \tag{1}$$
$$y = t^2 - 3. \tag{2}$$

From (1), $t = x - 1$. Substituting in Eq. (2),

$$y = (x - 1)^2 - 3 = x^2 - 2x - 2,$$

which is a familiar form of a parabola. Hence, the curve in Fig. 18-22 is indeed a parabola.

EXAMPLE
16

Eliminate the parameter and sketch the graph of

$$x = 2 - \sin^2 \pi t \tag{1}$$
$$y = \cos \pi t. \tag{2}$$

From Eq. (2),

$$y^2 = \cos^2 \pi t = 1 - \sin^2 \pi t$$

and hence

$$\sin^2 \pi t = 1 - y^2.$$

Substituting in Eq. (1) we have

$$x = 2 - (1 - y^2)$$
$$x = 1 + y^2$$
$$y^2 = x - 1, \tag{3}$$

which is a parabola, sketched in Fig. 18-23. Although every point on the curve described by the original parametric equations lies on $y^2 = x - 1$, the converse is not true. For in Eq. (3), $x \geq 1$, but as seen in Eq. (1), $1 \leq x \leq 2$ since $-1 \leq \sin \pi t \leq 1$. The curve given by Eqs. (1) and (2) is *not* the complete parabola but only a portion of it, as shown in Fig. 18-24.

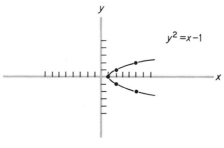

FIGURE 18-23

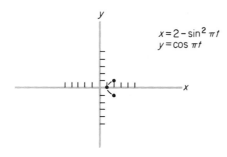

$$x = 2 - \sin^2 \pi t$$
$$y = \cos \pi t$$

t	0	$\frac{1}{2}$	1
x	2	1	2
y	1	0	-1

FIGURE 18-24

EXAMPLE
17

Determine a set of parametric equations for
$$x^3 + y^3 = xy.$$
One technique is to let $y = tx$. Then
$$x^3 + t^3 x^3 = x(tx)$$
or
$$x + t^3 x = t$$
$$x = \frac{t}{1 + t^3}. \tag{1}$$
Hence,
$$y = tx$$
$$y = \frac{t^2}{1 + t^3}. \tag{2}$$

EXERCISE 18-8

*In Problems 1–10, sketch the graph of the curve represented by the given parametric
equations and then eliminate the parameter if convenient.*

1. $x = 2t,\ y = t + 1$ **2.** $x = t - 3,\ y = 3t + 1$

3. $x = t,\ y = \dfrac{1}{t}$ **4.** $x = t - 1,\ y = t^2 - 2t$

5. $x = 3 - t,\ y = t - t^2$ **6.** $x = 2 \cos t,\ y = 2 \sin t$

7. $x = 4 \cos t,\ y = 2 \cos t$ **8.** $x = \cos 2t,\ y = \cos t$

9. $x = 2 \cos t - 1,\ y = 2 \sin t - 2$

10. $x = t^3 - 3t - 2,\ y = t^2 - t - 2$

*In Problems 11–16, in the same manner as Example 17 find parametric representations
for the curves given by the equations.*

11. $x = y + xy$ **12.** $x^2 + y^2 = 9x$

13. $x^3 + y^3 = 2xy$ **14.** $y^2 - 2x^2 = 8y$

15. $x^2 + 2xy + y = x$ **16.** $x^3 - 3xy + y^3 = 0$

17. A particle moves in the xy-plane such that the coordinates of its position as functions of time are given by

$$x = r \cos \omega t$$
$$y = r \sin \omega t,$$

where r and ω are constants. By eliminating the parameter determine the type of path the body follows.

18. If a projectile having initial velocity v_0 is fired at an angle α with the horizontal, it can be shown that at any time t, its position is given by

$$x = (v_0 \cos \alpha)t$$
$$y = (v_0 \sin \alpha)t - \tfrac{1}{2}gt^2,$$

where g is the acceleration due to gravity and can be assumed constant. By eliminating the parameter, show that the path of the projectile, illustrated in Fig. 18-25 is that of a parabola.

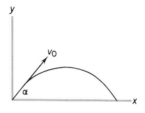

FIGURE 18-25

19. For the projectile motion of Problem 18, for what value of t will the projectile reach its maximum height?

20. If the projectile in Problem 18 has an initial velocity of 64 ft/sec directed at an angle of 45° above the ground, how far from its starting point will the projectile strike the ground assuming level terrain? Take $g = 32$ ft/sec².

21. Show that the parametric equations

$$x = a \cos \theta$$
$$y = b \sin \theta$$

represent an ellipse.

18-9
REVIEW

REVIEW QUESTIONS

The set of all points in a plane that are equidistant from a fixed point in the plane is a(n) _____(1)_____ .

(1) **circle**

The equation $(x - 2)^2 + (y + 4)^2 = 3$ defines a circle of radius
_____(2)_____ whose center is at _____(3)_____ .

(2) $\sqrt{3}$ (3) **(2, −4)**

The graph of $y^2 = x$ is called a _____(4)_____ .

(4) **parabola**

The vertex of the graph of $(y - 2)^2 = 9x$ is at _____(5)_____ .

(5) **(0, 2)**

The parabola $x^2 = -4y$ opens (downward)(to the left).
                                        (6)

(6) **downward**

The graph of $x^2/4 - y^2/9 = 1$ is a(n) _____(7)_____ and that of $x^2/4 +$
$y^2/9 = 1$ is a(n) _____(8)_____ .

(7) **hyperbola**   (8) **ellipse**

The major axis of $(y - 2)^2/9 + (x + 3)^2/4 = 1$ is (horizontal)(vertical).
                                                              (9)

(9) **vertical**

The graph of $xy = -7$ is a(n) _____(10)_____ .

(10) **hyperbola (equilateral)**

In sketching the graph determined by parametric equations, the parameter
 (does)(does not) appear in the graph.
        (11)

(11) **does not**

The graph of a parabola is that of a function of x if the directrix is
(parallel)(perpendicular) to the x-axis.
        (12)

(12) **parallel**

In a plane the set of all points the sum of whose distances from two fixed points is a constant is called a(n) _____(13)_____ .

(13) **ellipse**

In a plane the equation $x^2/5 + y^2/5 = 1$ defines a(n) _____(14)_____ .

(14) **circle**

The center of the circle $x^2 + 4x + y^2 + 8y = 1$ is _____(15)_____ .

(15) **(−2, −4)**

The graph of $x^2 + y^2 = 4$, where $x \geq 0$, $\underset{(16)}{\underline{\text{(is)(is not)}}}$ that of a function of x.

(16) **is not**

If a conic section has at least one asymptote, then it is a(n) _____(17)_____ .

(17) **hyperbola**

REVIEW PROBLEMS

In Problems 1–15, identify and sketch the graph of each of the given equations. Indicate coordinates of any existing center, radius, vertices, foci, or asymptotes.

1. $(x - 2)^2 + (y + 3)^2 = 9$

2. $2x^2 - 2y^2 + 4x + 10 = 0$

3. $\dfrac{(x + 1)^2}{2} - \dfrac{y^2}{8} = 2$

4. $y^2 - 2x - 2y = 4$

5. $xy = -8$ (Sketch only.)

6. $4x^2 + 9y^2 - 16x + 18y = 11$

7. $(y - 2)^2 = 12x$

8. $4x^2 + 2y^2 - 3x + 4y + 16 = 0$

9. $(x + 5)^2 + \frac{1}{2}(y - 2)^2 = 4$

10. $x^2 = 36 - (y + 2)^2$

11. $\frac{1}{2}y - (x - 3)^2 = \frac{3}{2}y$

12. $3x^2 + 4y^2 = 48$

13. $x^2 = \frac{1}{3}y$

14. $(y - 6)^2 - (x + 2)^2 = 1$

15. $2x^2 + 2y^2 - 2x + 6 = 0$

16. Sketch the curve and eliminate the parameter:

$$x = 2(2 - t)$$
$$y = t + t^2$$

17. In the manner of Example 17, find a parametric representation for the curve given by the equation

$$2x^3 - y^3 = 3xy.$$

18. Identify each of the following curves:
   a. $x^2 + 2y^2 + 4x + 2y + 3 = 0$
   b. $y^2 - 6x^2 + 2y + 36x - 59 = 0$
   c. $2y^2 + 4y - x - 4 = 0$
   d. $3x^2 - 12x - y + 12 = 0$
   e. $2x^2 + 2y^2 + 8x + 4y + 6 = 0$

19

sequences and series

19-1
SEQUENCES

When an object starting from rest travels in a straight-line path with a constant acceleration of 4 ft/sec², it is shown in physics that the distance s (in feet) of the object from its starting point at time t (in seconds) is given by the function $s = f(t) = 2t^2$. Let us restrict t to positive integral values; that is, we define the domain of f to be the set **N** of positive integers. We can easily determine that when

$$
\begin{aligned}
t &= 1 \text{ sec}, & s &= 2 \text{ ft} \\
t &= 2 \text{ sec}, & s &= 8 \text{ ft} \\
t &= 3 \text{ sec}, & s &= 18 \text{ ft} \\
t &= 4 \text{ sec}, & s &= 32 \text{ ft} \\
&\ \ \vdots & &\ \ \vdots \\
t &= n \text{ sec} & s &= 2n^2 \text{ ft.} \\
&\ \ \vdots & &\ \ \vdots
\end{aligned}
$$

Note that to each positive integer n there corresponds the number which is the functional value $f(n) = 2n^2$; that is, we have the correspondence

$$
\begin{array}{ccccccc}
1 & 2 & 3 & 4 & \cdots & n & \cdots \\
\downarrow & \downarrow & \downarrow & \downarrow & & \downarrow & \\
2 & 8 & 18 & 32 & \cdots & 2n^2 & \cdots.
\end{array}
$$

Speaking in general terms, a function whose domain is the set of positive integers is referred to as an *infinite sequence*. By considering the natural ordering of the integers in the domain of a sequence f, we can list the corresponding functional values of f in an orderly fashion as follows:

$$ f(1), f(2), f(3), f(4), \ldots, f(n), \ldots. \tag{1} $$

In fact, we can go one step further by dropping the functional notation in (1) and, instead, adopting a subscript notation. That is, for (1) we can equivalently write

$$ a_1, a_2, a_3, a_4, \ldots, a_n, \ldots, \tag{2} $$

where $a_n = f(n)$. Thus, if $f(n) = 2n^2$, then $a_1 = 2$, $a_2 = 8$, $a_3 = 18$, etc., and the functional values can be written

$$ 2, 8, 18, 32, \ldots, 2n^2, \ldots. \tag{3} $$

Since (3) essentially defines a particular sequence, it is commonly referred to as an infinite sequence itself. Similarly, by the infinite sequence

$$ \frac{3}{1}, \frac{4}{2}, \frac{5}{3}, \ldots, \frac{n+2}{n}, \ldots $$

we mean the function, say g, defined by $g(n) = (n+2)/n$, where $n \in \mathbf{N}$. More generally we have the following definition.

Definition.   An **infinite sequence,** denoted

$$ a_1, a_2, a_3, \ldots, a_n, \ldots, $$

is a function f whose domain is the set of positive integers and where $a_n = f(n)$.

Corresponding to the integer 1 is the **first term** a_1, to the integer 2 the **second term** a_2, and so forth. The **nth term**, or **general term**, is denoted a_n and usually denotes a rule of the function. That is, $a_n = f(n)$. A sequence with the general term a_n is often denoted by the symbol $\{a_n\}$. For example, the infinite sequence in (3) can be denoted $\{2n^2\}$.

EXAMPLE
1

Determine the first four terms of the infinite sequence having the general term $a_n = 2n + 3$.

We find the first four terms of the infinite sequence $\{2n + 3\}$ by successively replacing n in the expression $2n + 3$ by the integers 1, 2, 3, and 4:

$$n = 1, \quad a_1 = 2(1) + 3 = 5$$
$$n = 2, \quad a_2 = 2(2) + 3 = 7$$
$$n = 3, \quad a_3 = 2(3) + 3 = 9$$
$$n = 4, \quad a_4 = 2(4) + 3 = 11.$$

Therefore, the first four terms of the sequence are 5, 7, 9, and 11, and we can write

$$\{2n + 3\} = 5, 7, 9, 11, \ldots, 2n + 3, \ldots.$$

EXAMPLE
2

Determine the first four terms of the infinite sequence $\{(-1)^n(n^2 + 1)\}$.

$$n = 1, \quad a_1 = (-1)^1(1^2 + 1) = -2$$
$$n = 2, \quad a_2 = (-1)^2(2^2 + 1) = 5$$
$$n = 3, \quad a_3 = (-1)^3(3^2 + 1) = -10$$
$$n = 4, \quad a_4 = (-1)^4(4^2 + 1) = 17.$$

Hence,

$$\{(-1)^n(n^2 + 1)\} = -2, 5, -10, 17, \ldots, (-1)^n(n^2 + 1), \ldots.$$

EXAMPLE
3

Find a general term for an infinite sequence whose first six terms are

$$1, \sqrt{2}, \sqrt{3}, 2, \sqrt{5}, \sqrt{6}, \ldots.$$

By inspection, a general term is $a_n = \sqrt{n}$.

EXAMPLE
4

Write *all* the terms of the **finite sequence**

$$\{n(n + 1)\}, \quad \text{where } n = 1, 2, 3, 4.$$

In this case, the domain of the sequence is a finite set of consecutive positive integers. Successively substituting 1, 2, 3, and 4 for n in the general term, we have

$$\{n(n + 1)\} = 2, 6, 12, 20.$$

Note that there are exactly four terms. Similarly, the finite sequence $a_1, a_2, \ldots, a_{25}$ has 25 terms. In contrast to an infinite sequence, it is to be noted that a finite sequence has a first term *and* a last term.

EXERCISE 19-1

In Problems 1–16, write the first four terms of the given sequence.

1. $\{3n\}$ **2.** $\{\frac{1}{2}n\}$

3. $\{2n - 1\}$ **4.** $\{n^2 + 4\}$

5. $\left\{\dfrac{n}{n + 1}\right\}$ **6.** $\left\{\dfrac{n - 1}{n}\right\}$

7. $\left\{\dfrac{n-1}{n+1}\right\}$

8. $\left\{\dfrac{n^2+1}{n^2-2}\right\}$

9. $\left\{\dfrac{n}{2^n}\right\}$

10. $\left\{\dfrac{3^n}{n}\right\}$

11. $\left\{\sin\dfrac{n\pi}{2}\right\}$

12. $\left\{\cos\left(\dfrac{n\pi}{2}\right)\right\}$

13. $\left\{\dfrac{e^n}{2}\right\}$

14. $\left\{\dfrac{\sin nx}{n^2}\right\}$

15. $\{(-1)^{n+1}(n^2)\}$

16. $\{(-1)^n(2n)^2\}$

In Problems 17–24, find, by inspection, an nth term for the given infinite sequence.

17. $4, 8, 12, 16, \ldots$

18. $0, -1, -2, -3, \ldots$

19. $4, 6, 8, 10, \ldots$

20. $1, \frac{3}{2}, 2, \frac{5}{2}, \ldots$

21. $1, \frac{1}{3}, \frac{1}{9}, \frac{1}{27}, \ldots$

22. $1, \frac{1}{2}, \frac{1}{4}, \frac{1}{8}, \ldots$

23. $\frac{1}{2}, -\frac{1}{3}, \frac{1}{4}, -\frac{1}{5}, \ldots$

24. $-\frac{1}{2}, \frac{1}{3}, -\frac{1}{4}, \frac{1}{5}, \ldots$

19-2
ARITHMETIC AND GEOMETRIC PROGRESSIONS

A. THE ARITHMETIC PROGRESSION

If the difference between every two consecutive terms of a sequence is a constant d, that is,

$$a_{n+1} - a_n = d \qquad \text{for } n \geq 1,$$

the sequence is called an **arithmetic progression** with **common difference d**. This means that each term of an arithmetic progression can be obtained by adding the common difference d to the preceding term.

EXAMPLE
5

a. The arithmetic progression $1, 3, 5, 7, \ldots$ has a common difference $d = 2$ since the differences $3 - 1, 5 - 3, 7 - 5$, etc., are all 2.

b. The arithmetic progression $6, 11, 16, 21, \ldots$ has $d = 5$.

c. The arithmetic progression $2, -1, -4, \ldots$ has $d = -3$.

We can list the terms of an arithmetic progression in the following manner:

$$a_1, a_1 + d, a_1 + 2d, \ldots, a_1 + (n-1)d, \ldots,$$

where a_1 is the first term and $a_1 + (n-1)d$ is the nth term. Denoting the nth term by a_n, we arrive at the formula

$$a_n = a_1 + (n-1)d. \tag{19-1}$$

EXAMPLE
6

Find the eighteenth term of the arithmetic progression
$$7, 13, 19, \ldots.$$
Here, $a_1 = 7$, $d = 13 - 7 = 6$, and $n = 18$. Hence, by Eq. (19-1),
$$a_{18} = 7 + (18 - 1)6 = 109.$$

EXAMPLE
7

The first term of an arithmetic progression is 3 and the thirteenth term is -45. Find the common difference and the first four terms.

Here $a_1 = 3$, and for $n = 13$ we have $a_n = -45$. Substituting in Eq. (19-1) we obtain
$$a_n = a_1 + (n - 1)d$$
$$-45 = 3 + 12d$$
$$d = -4.$$
The sequence is $3, -1, -5, -9, \ldots.$

EXAMPLE
8

A particle has an initial speed of 5 m/sec and travels in a straight line with an acceleration of 2 m/sec². The values of the speed of the particle in meters per second at positive integral values of time, in seconds, form the arithmetic progression
$$7, 9, 11, 13, \ldots, 5 + 2t, \ldots.$$
Find the speed of the particle at $t = 100$ sec.

Clearly, the first term $a_1 = 7$ corresponds to the time $t = 1$; the second term $a_2 = 9$ corresponds to $t = 2$. Hence, we desire to find a_{100}, which corresponds to $t = 100$. Thus, $a_1 = 7$, $d = 2$, $n = 100$, and
$$a_{100} = 7 + (100 - 1)(2)$$
$$= 205 \text{ m/sec}.$$

B. THE GEOMETRIC PROGRESSION

If the ratio of every two consecutive terms in a sequence is a constant r, that is,
$$\frac{a_{n+1}}{a_n} = r, \qquad n \geq 1,$$
then the sequence is called a **geometric progression** with **common ratio** r. This means that each term of a geometric progression can be obtained by multiplying the preceding term by the common ratio r.

EXAMPLE
9

a. The geometric progression
$$1, \tfrac{1}{2}, \tfrac{1}{4}, \tfrac{1}{8}, \ldots$$

has a common ratio $r = \frac{1}{2}$ since the ratios $(\frac{1}{2})/1$, $(\frac{1}{4})/(\frac{1}{2})$, $(\frac{1}{8})/(\frac{1}{4})$, etc., are all equal to $\frac{1}{2}$.

b. The geometric progression 3, $-\sqrt{3}$, 1, $-\sqrt{3}/3$, ... has common ratio $r = -\sqrt{3}/3$.

c. If the first term of a geometric progression with common ratio -2 is $a_1 = -1$, the progression is

$$-1, 2, -4, 8, -16, \ldots.$$

It should be clear that if a_1 is the first term of a geometric progression with common ratio r, we can write the progression as

$$a_1, a_1 r, a_1 r^2, \ldots, a_1 r^{n-1}, \ldots.$$

Hence, the nth term a_n of a geometric progression is given by

$$a_n = a_1 r^{n-1}. \qquad (19\text{-}2)$$

EXAMPLE
10

a. Find the fifth term of the geometric progression 1, -3, 9, ...

Here $a_1 = 1$ and $r = -3/1 = 9/-3 = -3$. Hence, by Eq. (19-2) we have for $n = 5$,

$$a_5 = 1(-3)^{5-1} = 1(-3)^4 = 81.$$

b. Find the seventh term of the geometric progression $\sqrt{2}$, 2, $2\sqrt{2}$, 4, ...

Here $a_1 = \sqrt{2}$, $r = 2/\sqrt{2} = \sqrt{2}$, and $n = 7$. Thus, by Eq. (19-2) we have

$$a_7 = (\sqrt{2})(\sqrt{2})^{7-1} = (\sqrt{2})^7 = 8\sqrt{2}.$$

EXAMPLE
11

The first term of a geometric progression is 3 and the sixth term is $\frac{3}{32}$. Find the common ratio.

Here $a_1 = 3$, and for $n = 6$ we have $a_n = \frac{3}{32}$. From Eq. (19-2) we obtain

$$a_6 = 3r^{6-1} = \frac{3}{32}$$

$$r^5 = \frac{3}{3(32)} = \frac{1}{32}$$

$$r = \sqrt[5]{\frac{1}{32}} = \frac{1}{2}.$$

EXERCISE 19-2

In Problems 1–16, determine which of the given sequences are arithmetic or geometric progressions. For those that are of these types, find the indicated term.

1. 13, 1, -11, ..., eighth term

2. $13, 0, -13, \ldots$, seventh term

3. $6, -3, \frac{3}{2}, \ldots$, sixth term

4. $4, 1, \frac{1}{4}, \ldots$, seventh term

5. $-1, 3, -9, \ldots$, sixth term

6. $12, 16, 20, \ldots$, tenth term

7. $\frac{1}{3}, \frac{2}{3}, 1, \ldots$, eleventh term

8. $a, -a, -3a, \ldots$, ninth term

9. $-4, 2, -1, \ldots$, tenth term

10. $.3, .03, .003, \ldots$, seventh term

11. $3, 15, 24, \ldots$, tenth term

12. $12, -4, \frac{4}{3}, \ldots$, seventh term

13. $.4, .8, 1.2, \ldots$, tenth term

14. $.9, 1, 1.1, \ldots$, sixth term

15. $\frac{3}{2}, \frac{9}{4}, \frac{27}{8}, \ldots$, sixth term

16. $6, 2, \frac{1}{3}, \ldots$, fifth term

17. The fourth term of an arithmetic progression with $d = 14$ is 86. Find the eighth term.

18. The first term of an arithmetic progression is 6. The tenth term is 10. Find d.

19. The sixth term of a geometric progression is 16. The seventh term is 12. Find the first term.

20. The sixteenth term of an arithmetic progression is 28. The first term is -4. Find d.

21. In an arithmetic progression, $a_1 = 6$, $a_n = 26$, and $d = 4$. Find n.

22. If the fourth term of a geometric progression is 18 and the seventh term is $\frac{2}{81}$, find a_1.

23. The first swing of a pendulum is 10 ft, and because of resistive effects each succeeding swing is $\frac{1}{4}$ ft less. What is the length of the thirteenth pendulum swing?

24. How many swings of the pendulum in Problem 23 are completed before the pendulum comes to rest?

25. Suppose a distant star has a surface temperature of 10,000°C and that observations indicate the temperature decreases by 10 percent every 1000 years. What will be the temperature after 4000 years?

26. For the star of Problem 25, after how many years will the surface temperature be 7290°C?

Definition. An **infinite series** is an expression of the form

$$a_1 + a_2 + a_3 + \cdots + a_n + \cdots, \qquad (1)$$

where a_1, a_2, a_3, etc., are terms of an infinite sequence.

You might at first be alarmed by the notion of an "infinite sum" as indicated in (1). The word *sum* probably has meaning to you only as far as a finite number of quantities is concerned, and the thought of "infinitely many additions" may seem awesome to you. Let us reassure you that we can, in a reasonable way, attach a meaning to such a "sum." In fact, it will be done, in part, in terms of finite sums. We must postpone that development for a while, however, until other topics have been considered. First we introduce to you some mathematical shorthand for the indicated sum in (1).

The infinite series in (1) is often denoted by the **summation** or **sigma notation**,

$$\sum_{n=1}^{\infty} a_n,$$

which means the *sum* of all terms a_n, where n is replaced successively by the integers 1, 2, 3, ... Hence,

$$\sum_{n=1}^{\infty} a_n = a_1 + a_2 + a_3 + \cdots + a_n + \cdots.$$

For the series $\sum_{n=1}^{\infty} a_n$, a_1 is the first term, a_2 the second term, etc., and a_n is the *n*th or *general term*. We point out that the letter *n*, called the **index**, is merely used to indicate the subscripts of the terms in the sum and that any other letter can be used. Hence,

$$\sum_{n=1}^{\infty} a_n = \sum_{k=1}^{\infty} a_k = a_1 + a_2 + a_3 + \cdots + a_k + \cdots.$$

EXAMPLE
12

Write the first four terms of each infinite series.

a. $\sum_{n=1}^{\infty} (n - 1)$.

$$\sum_{n=1}^{\infty} (n - 1) = (1 - 1) + (2 - 1) + (3 - 1) + (4 - 1) + \cdots$$
$$= 0 + 1 + 2 + 3 + \cdots.$$

b. $\sum_{k=2}^{\infty} k^2 = 2^2 + 3^2 + 4^2 + 5^2 + \cdots$
$$= 4 + 9 + 16 + 25 + \cdots.$$

c. $\sum_{m=1}^{\infty} (\tfrac{1}{2})^m = (\tfrac{1}{2}) + (\tfrac{1}{2})^2 + (\tfrac{1}{2})^3 + (\tfrac{1}{2})^4 + \cdots$
$$= \tfrac{1}{2} + \tfrac{1}{4} + \tfrac{1}{8} + \tfrac{1}{16} + \cdots.$$

d. $\displaystyle\sum_{j=1}^{\infty}(-1)^{j}(j+3) = -4+5-6+7-\cdots.$

e. $\displaystyle\sum_{n=1}^{\infty}2 = 2+2+2+2+\cdots.$

f. $\displaystyle\sum_{n=1}^{\infty}x^{n} = x+x^{2}+x^{3}+x^{4}+\cdots.$

If a sequence is finite, the series corresponding to it is called a **finite series**. For example,

$$a_2 + a_3 + a_4 + a_5 + a_6$$

is a finite series comprised of five terms. Using sigma notation we can designate this series by

$$\sum_{k=2}^{6} a_k = a_2 + a_3 + a_4 + a_5 + a_6,$$

where the summation begins with $k = 2$ and ends with $k = 6$.

EXAMPLE
13

Write all the terms of each finite series and find the sum.

a. $\displaystyle\sum_{k=2}^{4}\frac{k+1}{k-1} = \frac{2+1}{2-1} + \frac{3+1}{3-1} + \frac{4+1}{4-1}$

$$= 3 + 2 + \frac{5}{3} = \frac{20}{3}.$$

b. $\displaystyle\sum_{k=1}^{5}(-1)^{k+1}(k) = 1-2+3-4+5 = 3.$

If we are given a series $\displaystyle\sum_{n=1}^{\infty} a_n$, then we can form its so-called **partial sums**:

$$S_1 = a_1$$
$$S_2 = a_1 + a_2 = S_1 + a_2$$
$$S_3 = a_1 + a_2 + a_3 = S_2 + a_3$$
$$\vdots$$
$$S_n = a_1 + a_2 + \cdots + a_n = S_{n-1} + a_n.$$

S_1 is the first partial sum, S_2 the second partial sum, etc. Clearly, the nth partial sum is the sum of the first n terms of the series. Observe that $S_2 = S_1 + a_2,\ S_3 = S_2 + a_3,\ \ldots,\ S_n = S_{n-1} + a_n$, and $S_n = \displaystyle\sum_{k=1}^{n} a_k.$ The sequence

$$\{S_n\} = S_1, S_2, S_3, \ldots, S_n, \ldots,$$

is called the **sequence of partial sums** of the series.

EXAMPLE
14

Determine the first five terms of the sequence of partial sums for the

series $\sum\limits_{n=1}^{\infty} \dfrac{1}{n(n+1)}$.

$$S_1 = \frac{1}{1(1+1)} = \frac{1}{2}$$

$$S_2 = S_1 + \frac{1}{2(2+1)} = \frac{1}{2} + \frac{1}{6} = \frac{2}{3}$$

$$S_3 = S_2 + \frac{1}{3(4)} = \frac{2}{3} + \frac{1}{12} = \frac{3}{4}$$

$$S_4 = S_3 + \frac{1}{4(5)} = \frac{3}{4} + \frac{1}{20} = \frac{4}{5}$$

$$S_5 = S_4 + \frac{1}{5(6)} = \frac{4}{5} + \frac{1}{30} = \frac{5}{6}.$$

Thus, $\{S_n\} = \frac{1}{2}, \frac{2}{3}, \frac{3}{4}, \frac{4}{5}, \frac{5}{6}, \ldots$.

Let us now consider how to find the sum of the first n terms of an arithmetic progression with common difference d. That is, we want a convenient way to evaluate the nth partial sum of a series whose terms form an arithmetic progression. The nth partial sum can be written

$$S_n = a_1 + a_2 + a_3 + \cdots + a_n.$$

However, since $a_2 = a_1 + d$, $a_3 = a_2 + d = a_1 + 2d$, etc., we have,

$$S_n = a_1 + (a_1 + d) + (a_1 + 2d) + \cdots + [a_1 + (n-1)d], \quad (2)$$

which can be written in the alternative form

$$S_n = a_n + (a_n - d) + (a_n - 2d) + \cdots + [a_n - (n-1)d]. \quad (3)$$

Adding corresponding members of Eq. (2) to Eq. (3) yields

$$2S_n = (a_1 + a_n) + (a_1 + a_n) + \cdots + (a_1 + a_n),$$

where the right member has exactly n terms, each equal to $a_1 + a_n$. Hence,

$$2S_n = n(a_1 + a_n).$$

Thus, **the sum of the first n terms of an arithmetic progression is**

$$S_n = \frac{n(a_1 + a_n)}{2}. \qquad (19\text{-}3)$$

EXAMPLE
15

Find the sum of the first eight terms of the series

$$1 + 4 + 7 + \cdots.$$

Since the terms of this series form an arithmetic progression with $d = 3$, by Eq. (19-1) we have

$$a_n = a_1 + (n-1)d$$

$$a_8 = 1 + (8-1)3 = 22.$$

The sum can now be determined by Eq. (19-3):

$$S_8 = \frac{8(a_1 + a_8)}{2} = \frac{8(1 + 22)}{2} = 92.$$

EXAMPLE
16

Find the sum of the odd integers between 20 and 60.

The odd integers between 20 and 60 form an arithmetic progression with $d = 2$. Since $a_1 = 21$ and $a_n = 59$, by Eq. (19-1) we have

$$a_n = a_1 + (n - 1)d$$
$$59 = 21 + (n - 1)2,$$

from which we find $n = 20$. That is, the number 59 is the twentieth term of the arithmetic progression. Hence,

$$S_{20} = \frac{20(21 + 59)}{2} = 800.$$

EXAMPLE
17

Derive a formula for the sum of the first k positive integers.

The terms of the series

$$\sum_{n=1}^{k} n = 1 + 2 + 3 + \cdots + (k - 1) + k$$

form an arithmetic progression with $a_1 = 1$, $d = 1$, and $n = k$. Thus, by Eq. (19-3) we have

$$\sum_{n=1}^{k} n = \frac{k(k + 1)}{2}. \tag{19-4}$$

For example, to find the sum of the first twenty-five positive integers we set $k = 25$:

$$\sum_{n=1}^{25} n = \frac{25(25 + 1)}{2} = 325.$$

Let us now determine the nth partial sum (the sum of the first n terms) of a series whose terms form a geometric progression. If the first term of the progression is a_1 and the common ratio is r, then the nth partial sum of the series is

$$S_n = a_1 + a_1r + a_1r^2 + \cdots + a_1r^{n-1}. \tag{4}$$

Multiplying both members of Eq. (4) by r we have

$$rS_n = a_1r + a_1r^2 + a_1r^3 + \cdots + a_1r^n. \tag{5}$$

Subtracting corresponding members of Eq. (5) from Eq. (4),

$$S_n - rS_n = a_1 - a_1r^n$$
$$S_n(1 - r) = a_1(1 - r^n).$$

Thus, **the sum of the first n terms of a geometric progression with** common ratio r is

$$S_n = \frac{a_1(1 - r^n)}{1 - r}. \qquad (19\text{-}5)$$

Equivalently,

$$S_n = \frac{a_1 - a_1 r^n}{1 - r} = \frac{a_1 - (a_1 r^{n-1})r}{1 - r}.$$

Hence,

$$S_n = \frac{a_1 - r a_n}{1 - r}. \qquad (19\text{-}6)$$

EXAMPLE
18

Find the sum of the first ten terms of the geometric progression 12, 6, 3, ...

Here,

$$a_1 = 12, \qquad r = \tfrac{1}{2}, \qquad n = 10.$$

By Eq. (19-5),

$$S_{10} = \frac{12[1 - (\tfrac{1}{2})^{10}]}{1 - \tfrac{1}{2}} = \frac{12(1 - \tfrac{1}{1024})}{\tfrac{1}{2}}$$

$$= \frac{3069}{128}.$$

EXERCISE 19-3

In Problems 1–10, write the first three terms of the given series.

1. $\displaystyle\sum_{i=1}^{\infty} (i + 1)$ **2.** $\displaystyle\sum_{k=1}^{\infty} (2k + 1)$

3. $\displaystyle\sum_{k=1}^{5} (k^2 - 2k)$ **4.** $\displaystyle\sum_{k=2}^{20} (-k^2 + 3k)$

5. $\displaystyle\sum_{n=2}^{\infty} (-1)^n(n^2 + 1)$ **6.** $\displaystyle\sum_{k=0}^{\infty} (-1)^{k+1}(k^2 + k)$

7. $\displaystyle\sum_{k=1}^{\infty} 2k$ **8.** $2\displaystyle\sum_{k=1}^{\infty} k^2$

9. $\displaystyle\sum_{n=1}^{\infty} (\tfrac{1}{2})^n$ **10.** $\displaystyle\sum_{k=1}^{\infty} (\tfrac{3}{4})^{k+1}$

In Problems 11–26, determine whether the terms of the given series form an arithmetic or geometric progression and find the specified partial sum.

11. $2 + 4 + 6 + \cdots; S_{10}$ **12.** $1 + 5 + 9 + \cdots; S_9$

13. $15 + 10 + 5 + \cdots; S_{16}$ **14.** $-\tfrac{1}{3} + \tfrac{1}{3} + 1 + \cdots; S_{12}$

15. $2 + 0 - 2 - 4 + \cdots; S_{12}$ **16.** $14 + 7 + 0 + \cdots; S_{10}$

17. $-10 + 6 + 22 + \cdots; S_{10}$ **18.** $\tfrac{3}{6} + \tfrac{2}{6} + \tfrac{1}{6} + \cdots; S_9$

19. $3 + 9 + 27 + \cdots; S_6$

20. $1 - \frac{1}{4} + \frac{1}{16} - \cdots; S_6$

21. $6 - 12 + 24 - \cdots; S_6$

22. $.1 + .02 + .004 + \cdots; S_{11}$

23. $.3 + .03 + .003 + \cdots; S_{10}$

24. $24 + 12 + 6 + \cdots; S_{10}$

25. $2 + 2\sqrt{2} + 4 + \cdots; S_{10}$

26. $2\sqrt{3} + 6 + 6\sqrt{3} + \cdots; S_8$

In Problems 27–32, find the values of the indicated quantities if the arithmetic progression has the given properties.

27. $a_n = 12, d = \frac{1}{4}, S_n = 99; n, a_1$

28. $a_9 = 20, d = 1; S_{16}, a_{16}$

29. $a_{13} = 42, d = 2; S_{16}, a_{16}$

30. $a_1 = 12, a_n = 42, d = 2; n, S_n$

31. $a_1 = -30, d = 3, S_n = 69; n, a_n$

32. $a_n = 14, d = \frac{1}{2}, S_n = 98; n$

In Problems 33–38, find the values of the indicated quantities if the geometric progression has the given properties.

33. $r = \frac{1}{2}, S_6 = 126; a_1, a_6$

34. $a_1 = 1, a_6 = 32; r, S_6$

35. $a_1 = \frac{3}{4}, a_n = -96, S_n = -\frac{255}{4}; n$

36. $a_1 = \frac{1}{2}, a_{10} = 256; a_{12}, S_{12}$

37. $a_6 = 1, a_8 = 9; S_5$

38. $r = \frac{1}{3}, S_5 = 121; a_1, a_5$

39. If a person saves 1 cent the first day, 2 cents the next day, 3 cents the next day, etc., how much money will have been saved after 30 days?

40. If a person saves 1 cent the first day, 2 cents the next day, 4 cents the next, 8 cents the next, etc., how much money will have been saved after 30 days?

41. It can be shown that the sum of the squares of the first k positive integers is given by

$$\sum_{n=1}^{k} n^2 = \frac{k(k + 1)(2k + 1)}{6}.$$

Find the sum of the squares of the first ten positive integers.

42. It can be shown that the sum of the cubes of the first k positive integers is given by

$$\sum_{n=1}^{k} n^3 = \frac{k^2(k + 1)^2}{4}.$$

Find the sum of the cubes of the first ten positive integers.

43. Find the sum of the first 100 positive integers.

44. A 12-hr clock strikes 1 at 1 o'clock, 2 at 2 o'clock, etc. How many strikes will it make in 24 *consecutive* hours?

45. In a vacuum, an object falls approximately 16 ft the first second, 48 ft the next, 80 the next, etc. How far does it fall in 12 sec?

46. The tip of a pendulum moves 4 in. the first second, 2 in. the next, 1 in. the next, etc. How far has it moved in 10 sec?

47. A tank full of alcohol is emptied of one-fourth of its contents. The tank is then filled with water. This is repeated three more times. What part of the volume of the tank is now alcohol?

48. By means of a pump, air is being removed from a container in such a way that each second, one-tenth of the remaining air in the container is removed. After 5 sec, what percentage of air is left?

19-4
LIMITS OF SEQUENCES

It is our purpose here to consider one of the very fundamental ideas of mathematics—the concept of a limit. The ideas and techniques we shall develop here are useful in both the main topics of calculus, finding derivatives and finding integrals.

We begin by examining rather closely a sequence which conveniently typifies a situation that arises quite naturally in mechanics. Consider a pendulum which is displaced from its equilibrium (rest) position and released. The pendulum will oscillate back and forth. Let us assume that the length of the first swing was $\frac{1}{2}$ m and, furthermore, that the length of any subsequent swing is exactly one-half the length of the swing which preceded it. If we let a_1 be the length of the first swing, a_2 the length of the second swing, etc., we have

$$a_1 = \frac{1}{2} = \frac{1}{2^1}$$

$$a_2 = \frac{1}{2}\left(\frac{1}{2}\right) = \frac{1}{2^2}$$

$$a_3 = \frac{1}{2}\left(\frac{1}{2^2}\right) = \frac{1}{2^3}$$

$$a_4 = \frac{1}{2}\left(\frac{1}{2^3}\right) = \frac{1}{2^4}.$$

Clearly, for the nth swing of the pendulum we have

$$a_n = \frac{1}{2^n}.$$

Thus, the sequence whose terms correspond to the lengths of the swings of the pendulum is

$$\left\{\frac{1}{2^n}\right\} = \frac{1}{2^1}, \frac{1}{2^2}, \frac{1}{2^3}, \frac{1}{2^4}, \cdots, \frac{1}{2^n}, \cdots$$

$$= \frac{1}{2}, \frac{1}{4}, \frac{1}{8}, \frac{1}{16}, \cdots, \frac{1}{2^n}, \cdots$$

Some of the terms of this sequence are indicated on the real number line in Fig. 19-1.

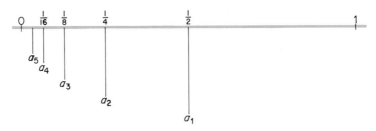

FIGURE 19-1

We observe that as n increases, the terms of the sequence get closer to zero. Moreover, although $\frac{1}{2^n}$ will never equal zero, by going far enough in the sequence we can find a term such that all the remaining terms will be as close to zero as we wish. Suppose, for example, we wish to find how many swings of the pendulum are necessary so that the length of a swing will be less than $\frac{1}{10,000}$ m. Then we must have the inequality

$$\frac{1}{2^n} < \frac{1}{10,000},$$

which when solved yields $n > 13$. That is, beginning with the fourteenth term (the fourteenth swing of the pendulum), every term in the sequence $\left\{\frac{1}{2^n}\right\}$ will satisfy the stated requirement.

The ideas of the preceding discussion are verbally expressed in mathematical terms by saying that as n increases indefinitely through positive integral values, the sequence $\left\{\frac{1}{2^n}\right\}$ has zero for a **limit**. Symbolically we write

$$\lim_{n \to \infty} \frac{1}{2^n} = 0. \tag{1}$$

The notation $n \rightarrow \infty$ means that n is increasing indefinitely through positive integral values; it does not imply that infinity is a number. Equation (1) can be read "the limit of the sequence $\left\{\dfrac{1}{2^n}\right\}$ as n increases without bound is equal to zero." An equivalent statement is that *every* interval containing zero, no matter how small, contains all the terms of the sequence $\left\{\dfrac{1}{2^n}\right\}$ from some term on.

In our discussion we have seen that the interval $\left[0, \dfrac{1}{10,000}\right)$ contains all terms of the sequence $\left\{\dfrac{1}{2^n}\right\}$ from the fourteenth term on. In more general terms we write

$$\lim_{n \to \infty} a_n = a,$$

which means that the limit of the sequence $\{a_n\}$ as $n \rightarrow \infty$ is the number a. That is, *every* interval containing a contains all the terms of the sequence $\{a_n\}$ from some term on. For n sufficiently large, a_n is arbitrarily close to a.

We must note that not every sequence has a limit. For example, the terms in the sequence $\{n\} = 1, 2, 3, \ldots$ increase without bound as $n \rightarrow \infty$. In terms of intervals, no matter how large a chosen number B is, and no matter how large an interval containing B, all the terms of $\{n\}$ from some term on will be outside the interval if n is sufficiently large. We denote this situation by writing

$$\lim_{n \to \infty} n = \infty,$$

which is read "as n increases without bound, the terms of the sequence $\{n\}$ increase without bound." If a sequence has a finite limit, the sequence is said to **converge** or be **convergent**. Otherwise, it **diverges**. Thus, $\left\{\dfrac{1}{2^n}\right\}$ is a convergent sequence, but $\{n\}$ is a divergent sequence. When a sequence converges, its limit must necessarily be unique.

EXAMPLE
19

The sequence

$$7, 6, 3, 2, -5, -5, \ldots, -5, \ldots$$

converges to -5, since *every* interval containing -5 must contain all the terms of the sequence from the fifth term on. For most other sequences, however, you may have to ignore many, perhaps a million, of the terms in the sequence before all the remaining terms lie in a given interval if the interval is quite small.

EXAMPLE
20

The terms of the sequence

$$\left\{\frac{n+1}{n}\right\} = \frac{2}{1}, \frac{3}{2}, \frac{4}{3}, \frac{5}{4}, \frac{6}{5}, \dots$$

are clearly getting close to 1 as n increases. In fact, for $n = 1000$, $a_{1000} = \frac{1001}{1000} = 1 + \frac{1}{1000}$. Later it will be shown that 1 is indeed the limit and, hence, we can write

$$\lim_{n \to \infty} \frac{n+1}{n} = 1.$$

EXAMPLE
21

a. The terms of the arithmetic sequence $\{-5n\} = -5, -10, -15, \dots$ decrease without bound and the sequence is said to diverge; that is, it has no finite limit. Symbolically,

$$\lim_{n \to \infty} (-5n) = -\infty.$$

In fact, owing to the common difference d, where $|d| > 0$, *every arithmetic progression is divergent.*

b. The sequence $\{(-1)^n\} = -1, 1, -1, 1, \dots$ has no limit as $n \to \infty$ and hence is divergent.

Without going into their proofs, we shall state some theorems on limits and illustrate their use in determining the behavior of various sequences.

Theorem I. If $\{a_n\}$ and $\{b_n\}$ are convergent sequences, then

(a) $$\lim_{n \to \infty}(a_n \pm b_n) = \lim_{n \to \infty} a_n \pm \lim_{n \to \infty} b_n.$$

(b) $$\lim_{n \to \infty}(a_n \cdot b_n) = \lim_{n \to \infty} a_n \cdot \lim_{n \to \infty} b_n.$$

(c) $$\lim_{n \to \infty} \frac{a_n}{b_n} = \frac{\lim_{n \to \infty} a_n}{\lim_{n \to \infty} b_n}, \quad \text{if } b_n \neq 0 \text{ and } \lim_{n \to \infty} b_n \neq 0.$$

(d) $$\lim_{n \to \infty} c = c, \quad \text{where } c \text{ is a constant.}$$

(e) $$\lim_{n \to \infty} c a_n = c \lim_{n \to \infty} a_n.$$

Theorem II. If $|r| < 1$, then $\lim_{n \to \infty} r^n = 0$. If $|r| > 1$, then $\lim_{n \to \infty} r^n$ does not exist and $\{r^n\}$ diverges.

Theorem III. If $a_n > 0$, then

$$\lim_{n \to \infty} a_n = 0 \text{ if and only if } \lim_{n \to \infty} \frac{1}{a_n} = \infty.$$

Also, if $a_n > 0$,

$$\lim_{n \to \infty} a_n = \infty \text{ if and only if } \lim_{n \to \infty} \frac{1}{a_n} = 0.$$

EXAMPLE 22

Establish the convergence or divergence of the following sequences. In the case of convergence, find the limit.

a. $\{(-\frac{1}{2})^n\} = -\frac{1}{2}, \frac{1}{4}, -\frac{1}{8}, \dots$

Since $|-\frac{1}{2}| < 1$, by Theorem II we conclude that

$$\lim_{n \to \infty} (-\frac{1}{2})^n = 0.$$

The sequence converges to zero.

b. $\{6\} = 6, 6, 6, \dots$

From Theorem I(d), we obtain

$$\lim_{n \to \infty} 6 = 6.$$

The sequence converges to 6.

c. $\left\{ \frac{3}{4^n} \right\}$.

Since $\frac{3}{4^n} = 3\left(\frac{1}{4}\right)^n$ and $\left|\frac{1}{4}\right| < 1$, by Theorems I(e) and II we have

$$\lim_{n \to \infty} \frac{3}{4^n} = 3\left[\lim_{n \to \infty} \left(\frac{1}{4}\right)^n\right] = 3(0) = 0.$$

d. $\{(\frac{3}{2})^n\}$.

Since $|\frac{3}{2}| > 1$, by Theorem II the sequence diverges. Note that the terms in the sequence increase without bound as $n \to \infty$:

$$\{(\frac{3}{2})^n\} = \frac{3}{2}, \frac{9}{4}, \frac{27}{8}, \frac{81}{16}, \frac{243}{32}, \dots$$

EXAMPLE 23

a. Since $\lim_{n \to \infty} n^3 = \infty$, then by Theorem III, $\lim_{n \to \infty} \frac{1}{n^3} = 0$.

b.

$$\lim_{n \to \infty}\left(2 + \frac{1}{n^3}\right) = \lim_{n \to \infty} 2 + \lim_{n \to \infty} \frac{1}{n^3} \quad \text{[Theorem I(a)]}$$
$$= 2 + 0 = 2 \quad \text{[Theorem I(d) and Example 23a].}$$

EXAMPLE 24

Evaluate $\lim_{n \to \infty} \frac{n+1}{n}$.

As $n \to \infty$, the numerator and denominator get arbitrarily large and we say that the quotient $(n + 1)/n$ is of the form ∞/∞. However, we can find the limit by first performing an algebraic operation.

$$\lim_{n \to \infty} \frac{n+1}{n} = \lim_{n \to \infty} \left(\frac{n}{n} + \frac{1}{n} \right)$$

$$= \lim_{n \to \infty} 1 + \lim_{n \to \infty} \frac{1}{n} \qquad \text{[Theorem I(a)]}$$

$$= 1 + 0 = 1 \qquad \text{[Theorems I(d) and III]}.$$

Our result means that as $n \longrightarrow \infty$, the number $(n+1)/n$ gets arbitrarily close to 1.

EXAMPLE
25

Evaluate

$$\lim_{n \to \infty} \frac{2n^2 + 3n}{3n^2 + 4n}.$$

As $n \longrightarrow \infty$, the quotient takes on the form ∞/∞. A situation such as this can be remedied by dividing the numerator and denominator by the highest power of n that occurs (in our case, n^2) and applying Theorem I(c). We have

$$\lim_{n \to \infty} \frac{2n^2 + 3n}{3n^2 + 4n} = \lim_{n \to \infty} \frac{\dfrac{2n^2 + 3n}{n^2}}{\dfrac{3n^2 + 4n}{n^2}}$$

$$= \lim_{n \to \infty} \frac{2 + \dfrac{3}{n}}{3 + \dfrac{4}{n}}$$

$$= \frac{\lim\limits_{n \to \infty} \left(2 + \dfrac{3}{n} \right)}{\lim\limits_{n \to \infty} \left(3 + \dfrac{4}{n} \right)}$$

$$= \frac{\lim\limits_{n \to \infty} 2 + 3 \lim\limits_{n \to \infty} \dfrac{1}{n}}{\lim\limits_{n \to \infty} 3 + 4 \lim\limits_{n \to \infty} \dfrac{1}{n}} \qquad \text{[Theorem I(a), (e)]}$$

$$= \frac{2 + 0}{3 + 0} = \frac{2}{3} \qquad \text{[Theorems I(d) and III]}.$$

EXAMPLE
26

Evaluate $\lim\limits_{n \to \infty} \dfrac{2n^2}{n^3 + 1}$.

$$\lim_{n \to \infty} \frac{2n^2}{n^3 + 1} = \lim_{n \to \infty} \frac{\dfrac{2n^2}{n^3}}{\dfrac{n^3 + 1}{n^3}} = \lim_{n \to \infty} \frac{\dfrac{2}{n}}{1 + \dfrac{1}{n^3}}$$

$$= \frac{2 \lim\limits_{n \to \infty} \dfrac{1}{n}}{\lim\limits_{n \to \infty} 1 + \lim\limits_{n \to \infty} \dfrac{1}{n^3}}$$

$$= \frac{2 \cdot 0}{1 + 0} = \frac{0}{1} = 0.$$

Our approach to the concept of a limit has been intuitive and is in no way rigorous. However, since the limit concept is a fundamental one in mathematics and lies at the very foundation of higher mathematics, you should have some "feeling" for this notion.

EXERCISE 19-4

In Problems 1–18, determine whether the sequence converges or diverges and in the case of convergence, find the limit.

1. $\lim_{n \to \infty} (2n)$

2. $\lim_{n \to \infty} (3n + 4)$

3. $\lim_{n \to \infty} \left(\dfrac{6}{n} \right)$

4. $\lim_{n \to \infty} (7 + \tfrac{1}{3})$

5. $\lim_{n \to \infty} (3 - \tfrac{1}{4})$

6. $\lim_{n \to \infty} \left(\dfrac{-6}{n} \right)$

7. $\lim_{n \to \infty} (\tfrac{3}{4})^n$

8. $\lim_{n \to \infty} (\tfrac{17}{16})^n$

9. $\lim_{n \to \infty} 3(\tfrac{1}{2})^n$

10. $\lim_{n \to \infty} \left(\dfrac{n+1}{2n} \right)$

11. $\lim_{n \to \infty} \left(\dfrac{3n - 1}{2n} \right)$

12. $\lim_{n \to \infty} \left(1 + \dfrac{n-1}{n} \right)$

13. $\lim_{n \to \infty} \left(2 + \dfrac{n^3 - 1}{n^2} \right)$

14. $\lim_{n \to \infty} \left(\dfrac{n^2 - n + 1}{2n^2} \right)$

15. $\lim_{n \to \infty} \dfrac{(100)^n}{(101)^n}$

16. $\lim_{n \to \infty} \left(\dfrac{3n^2 + 2n + 5}{4n^2} \right)$

17. $\lim_{n \to \infty} \left(\dfrac{2n^4 - 6n^2 + 5}{n^5} \right)$

18. $\lim_{n \to \infty} \dfrac{8n^5 - 6n^4 + 3n^3 + 1}{n^6 - 4n^3 + 1}$

19-5
THE INFINITE GEOMETRIC SERIES

We have found that the nth partial sum of a geometric series is

$$S_n = \frac{a_1(1 - r^n)}{1 - r},$$

where a_1 is the first term and r is the common ratio. We shall now attach a meaning to the *sum* of the infinite geometric series:

$$a_1 + a_1 r + a_1 r^2 + \cdots + a_1 r^{n-1} + \cdots, \qquad a_1 \neq 0.$$

That is, if the number of terms increases indefinitely, is there a limiting value to the sequence $\{S_n\}$ of partial sums? To answer this question, the possible values of r will be considered.

a.　$r = 1$. This gives

$$S_n = a_1 + a_1 + \cdots + a_1$$
$$= na_1,$$

which has no limit as $n \to \infty$.

b.　$r = -1$. This gives

$$S_n = a_1 - a_1 + a_1 - a_1 + \cdots + (\pm a_1).$$

If n is odd, $S_n = a_1$, while if n is even, $S_n = 0$. Hence the series diverges since a limit of a sequence must be unique.

c.　$|r| > 1$. Now,

$$S_n \doteq \frac{a_1(1 - r^n)}{1 - r}$$

$$= \frac{a_1}{1 - r}(1 - r^n).$$

Now, since $a_1/(1 - r)$ is a constant, we have

$$\lim_{n \to \infty} S_n = \lim_{n \to \infty} \frac{a_1}{1 - r}(1 - r^n)$$

$$= \frac{a_1}{1 - r} \lim_{n \to \infty}(1 - r^n)$$

$$= \frac{a_1}{1 - r}(\lim_{n \to \infty} 1 - \lim_{n \to \infty} r^n)$$

$$\lim_{n \to \infty} S_n = \frac{a_1}{1 - r}(1 - \lim_{n \to \infty} r^n). \tag{1}$$

But for $|r| > 1$, $\lim_{n \to \infty} r^n$ does not exist. Thus, $\lim_{n \to \infty} S_n$ does not exist for $|r| > 1$.

d.　$|r| < 1$. In this case $\lim_{n \to \infty} r^n = 0$ and Eq. (1) becomes

$$\lim_{n \to \infty} S_n = \lim_{n \to \infty} \frac{a_1}{1 - r}(1 - 0) = \frac{a_1}{1 - r}.$$

Letting S represent this limit, we have

$$S = \frac{a_1}{1 - r} \qquad \text{for } |r| < 1. \tag{19-7}$$

Hence, we may now state that for $|r| < 1$, the limit S of the sum of n terms of an infinite geometric series as n increases without bound is $a_1/(1 - r)$. We shall consider this to be the **sum** of the series in the sense that it can be approached as closely as one wishes by adding a sufficient number of terms.

EXAMPLE
27

Test the following series for convergence or divergence. In the case of convergence, find the sum.

a. $\sum_{n=1}^{\infty} 8(\frac{1}{2})^n = 4 + 2 + 1 + \frac{1}{2} + \cdots$.

The series is geometric with $a_1 = 4$ and $r = \frac{1}{2}$. Since $|r| < 1$, the series converges to

$$\frac{a_1}{1 - r} = \frac{4}{1 - \frac{1}{2}} = 8.$$

b. $6 - 1 + \frac{1}{6} - \frac{1}{36} + \cdots = \sum_{n=1}^{\infty} 6(-\frac{1}{6})^{n-1}$.

The series is geometric with $a_1 = 6$ and $r = -\frac{1}{6}$. Since $|-\frac{1}{6}| < 1$, the series converges to

$$\frac{6}{1 - (-\frac{1}{6})} = \frac{36}{7}.$$

c. $\sum_{n=1}^{\infty} 4(\frac{5}{3})^n$.

The series is geometric with $|r| = \frac{5}{3} > 1$ and hence diverges.

EXAMPLE
28

Find the sum of the geometric series

$$\sum_{n=1}^{\infty} (\frac{2}{3})^n = \frac{2}{3} + \frac{4}{9} + \frac{8}{27} + \frac{16}{81} + \cdots.$$

Here $a_1 = \frac{2}{3}$ and $r = \frac{2}{3}$. Thus, since $|r| < 1$, the series converges to

$$S = \frac{a_1}{1 - r} = \frac{\frac{2}{3}}{1 - \frac{2}{3}} = 2.$$

EXAMPLE
29

The pendulum discussed in Sec. 19-4 had an initial swing of $\frac{1}{2}$ m; the length of every other swing was one-half the length of the previous swing. How far did the tip of the pendulum move before coming to rest?

We showed that the sequence whose terms were the length of each swing was

$$\left\{ \frac{1}{2^n} \right\} = \frac{1}{2^1}, \frac{1}{2^2}, \frac{1}{2^3}, \frac{1}{2^1}, \ldots .$$

The corresponding geometric series is

$$\frac{1}{2} + \frac{1}{4} + \frac{1}{8} + \cdots \frac{1}{2^n} + \cdots.$$

Here $a_1 = \frac{1}{2}$ and $r = \frac{1}{2}$ and, hence,

$$S = \frac{\frac{1}{2}}{1 - \frac{1}{2}} = 1 \text{m}.$$

EXAMPLE
30

Determine the rational number corresponding to the repeating decimal $.\overline{123}$.

The line above the digits 123 is called a **vinculum** and indicates those digits that repeat. Hence,

$$.\overline{123} = .123123123\cdots$$
$$= .123 + .000123 + .000000123 + \cdots.$$

Thus, the repeating decimal has been expressed as an infinite geometric series with $a_1 = .123$ and $r = .001$. Thus,

$$S = \frac{.123}{1 - .001} = \frac{.123}{.999} = \frac{123}{999} = \frac{41}{333}.$$

That is, $.\overline{123} = 41/333$.

EXERCISE 19-5

In Problems 1–20, find the sum of each series.

1. $\displaystyle\sum_{i=1}^{5} (i^2 + 1)$
    2. $\displaystyle\sum_{i=2}^{6} (2i + 1)$

3. $\displaystyle\sum_{n=1}^{4} \frac{n^2 + n}{n + 1}$
    4. $\displaystyle\sum_{j=6}^{10} (-1)^j (j - 1)$

5. $\displaystyle\sum_{n=1}^{\infty} \left(\tfrac{1}{2}\right)^n$
    6. $\displaystyle\sum_{n=1}^{\infty} \left(\tfrac{1}{3}\right)^n$

7. $\displaystyle\sum_{k=1}^{\infty} \left(\tfrac{2}{3}\right)^k$
    8. $\displaystyle\sum_{n=1}^{\infty} \left(\tfrac{3}{5}\right)^n$

9. $\displaystyle\sum_{k=3}^{\infty} \left(\tfrac{3}{5}\right)^k$
    10. $\displaystyle\sum_{k=5}^{\infty} \left(\tfrac{4}{9}\right)^k$

11. $3 + \tfrac{3}{2} + \tfrac{3}{4} + \cdots$
    12. $4 + 1 + \tfrac{1}{4} + \tfrac{1}{16} + \cdots$

13. $12 + 4 + \tfrac{4}{3} + \cdots$
    14. $\dfrac{1}{1.2} + \dfrac{1}{(1.2)^2} + \dfrac{1}{(1.2)^3} + \cdots$

15. $-4 + 2 - 1 + \cdots$
    16. $100 - 10 + 1 - .1 + \cdots$

17. $\tfrac{5}{3} + \tfrac{1}{6} + \tfrac{1}{60} + \cdots$
    18. $\dfrac{3}{4} + \dfrac{3}{4^2} + \dfrac{3}{4^3} + \cdots$

19. $5 + \dfrac{1}{5} + \dfrac{1}{5^2} + \dfrac{1}{5^3} + \cdots$
    20. $.03 + .003 + .0003 + \cdots$

In Problems 21–26, change the repeating decimals into their equivalent fractions.

21. $.\overline{24}$
    22. $.\overline{42}$

23. $3.\overline{212}$
    24. $2.0\overline{46}$

25. $.021\overline{32}$
    26. $.15\overline{6}$

27. When dropped from a height of 4 ft, a ball after the first bounce reaches a height of 2 ft, after the second bounce a height of 1 ft, etc. What is the total distance traveled by the ball before coming to rest?

28. The tip of a pendulum moves through a distance of 3 in., after which the distance is constantly decreased on each swing by 10 percent. Before coming to rest, through what total distance has the tip moved?

29. The first oscillation of a mass suspended on a vertical spring is 20 in. long. If it is observed that the length of each succeeding oscillation decreases by 20 percent, how far does the mass travel before coming to rest?

30. When the power to a motor is turned off, a flywheel attached to the motor is observed to "coast" to a stop. In the first second it made 190 revolutions, and in each succeeding second it made nine-tenths as many revolutions as the preceding second. How many revolutions did the flywheel make before coming to rest?

31. When a small object is projected up an inclined plane it is observed to move 10 ft in the first second and in any succeeding second it moves four-fifths as far as it did in the preceding second. How far does it travel before coming to rest?

32. The midpoints of the sides of a 1-in. square are joined to form an inscribed square, and this process is continued indefinitely. Find the sum of the areas of all the squares including the original one.

19-6
THE BINOMIAL THEOREM

Expanding some of the positive integral powers of $a + b$, we obtain

$$(a + b)^1 = a + b$$
$$(a + b)^2 = a^2 + 2ab + b^2$$
$$(a + b)^3 = a^3 + 3a^2b + 3ab^2 + b^3$$
$$(a + b)^4 = a^4 + 4a^3b + 6a^2b^2 + 4ab^3 + b^4.$$

From the results above, we make the following observations about $(a + b)^n$ for $n \in \{1, 2, 3, 4\}$:

1. $(a + b)^n$ has $n + 1$ terms.

2. The first term is $a^nb^0 = a^n$.

3. The last term is $a^0b^n = b^n$.

4. As the terms progress from the first to the last, the exponents of a decrease by 1 while those of b increase by 1, the sum of the exponents of a and b being n in each term.

5. The exponent of b is equal to one less than the number of the term in which it appears.

It remains to find a way in which the coefficients can be determined. Notice that if the numerical coefficient of any term is multiplied by the exponent of a in that term and that product divided by the number of the term, the result is the coefficient of the next term. For example, the second term of $(a + b)^4$ is $4a^3b$. Multiplying the coefficient 4 by the exponent of a, 3, and dividing by the number of the term, 2, we get

$$\frac{4 \cdot 3}{2} = 6,$$

which is the coefficient of the third term. Assuming these properties are true for any positive integer n, we have for $n = 6$,

$$(a + b)^6$$
$$= a^6 + \tfrac{6}{1}a^5b + \tfrac{6}{1} \cdot \tfrac{5}{2}a^4b^2 + \tfrac{6}{1} \cdot \tfrac{5}{2} \cdot \tfrac{4}{3}a^3b^3 + \tfrac{6}{1} \cdot \tfrac{5}{2} \cdot \tfrac{4}{3} \cdot \tfrac{3}{4}a^2b^4 +$$
$$\tfrac{6}{1} \cdot \tfrac{5}{2} \cdot \tfrac{4}{3} \cdot \tfrac{3}{4} \cdot \tfrac{2}{5}ab^5 + \tfrac{6}{1} \cdot \tfrac{5}{2} \cdot \tfrac{4}{3} \cdot \tfrac{3}{4} \cdot \tfrac{2}{5} \cdot \tfrac{1}{6}b^6$$
$$= a^6 + 6a^5b + 15a^4b^2 + 20a^3b^3 + 15a^2b^4 + 6ab^5 + b^6.$$

For convenience of notation, we introduce the symbol n **factorial**, denoted $n!$, and defined as follows. If n is a positive integer, then

$$n! = 1 \cdot 2 \cdot 3 \cdots n.$$

If $n = 0$, then

$$n! = 0! = 1.$$

Thus, from the above it appears that we can express the $(r + 1)$st term of $(a + b)^n$ as

$$\overbrace{\frac{n(n-1)(n-2) \cdots (n-r+1)}{r!}}^{r \text{ factors}}a^{n-r}b^r \tag{19-8}$$

and we have the **binomial formula**:

Theorem IV. If n is a positive integer, then

$$(a + b)^n = a^n + \frac{na^{n-1}b}{1!} + \cdots + \frac{n(n-1) \cdots (n-r+1)a^{n-r}b^r}{r!}$$
$$+ \cdots + b^n. \tag{19-9}$$

EXAMPLE
31

Write the first four terms of $(a + b)^{20}$.

$$(a + b)^{20} = a^{20} + \frac{20a^{19}b}{1!} + \frac{20(19)a^{18}b^2}{2!} + \frac{20(19)(18)a^{17}b^3}{3!} + \cdots$$
$$= a^{20} + 20a^{19}b + 190a^{18}b^2 + 1140a^{17}b^3 + \cdots.$$

EXAMPLE
32

Expand $(2x - 3y^2)^5$.

Note that $(2x - 3y^2)^5 = [(2x) + (-3y^2)]^5$:

$$(2x - 3y^2)^5 = (2x)^5 + \frac{5(2x)^4(-3y^2)}{1!} + \frac{5 \cdot 4(2x)^3(-3y^2)^2}{2!} +$$

$$\frac{5 \cdot 4 \cdot 3(2x)^2(-3y^2)^3}{3!} + \frac{5 \cdot 4 \cdot 3 \cdot 2(2x)^1(-3y^2)^4}{4!} + (-3y^2)^5$$

$$= 32x^5 - 240x^4y^2 + 720x^3y^4 - 1080x^2y^6 + 810xy^8 - 243y^{10}.$$

EXAMPLE
33

Find the eighteenth term of $(ab + c)^{21}$.

Substituting in Eq. (19-8) the value $r = 17$ for the $(17 + 1)$st term, we obtain

$$\frac{(21)(20)(19) \cdots (5)(ab)^4(c)^{17}}{17!} = 5985a^4b^4c^{17}.$$

Setting $a = 1$ and $b = x$ in the binomial formula, we get the **binomial series**

$$(1 + x)^n = 1 + nx + \frac{n(n - 1)x^2}{2!} + \cdots$$

$$+ \frac{n(n - 1) \cdots (n - r + 1)x^r}{r!} + \cdots,$$

which for $|x| < 1$ can be shown to be a valid equation for **any real number n**. When n is not a positive integer, the series is unending, but nevertheless we can get a reasonable approximation to $(1 + x)^n$ in most cases by considering a few terms only.

EXAMPLE
34

Approximate $\sqrt{104}$ to three decimal places.

$$\sqrt{104} = \sqrt{100 + 4} = \sqrt{100(1 + \tfrac{4}{100})}$$
$$= 10(1 + \tfrac{1}{25})^{1/2}.$$

Since $|\tfrac{1}{25}| < 1$, we can use the binomial series with $x = \tfrac{1}{25}$ and $n = \tfrac{1}{2}$:

$$\sqrt{104} = 10\left[1 + \frac{\tfrac{1}{2}\left(\tfrac{1}{25}\right)}{1!} + \frac{\tfrac{1}{2}\left(-\tfrac{1}{2}\right)\left(\tfrac{1}{25}\right)^2}{2!} + \cdots \right]$$

$$= 10\left(1 + \frac{1}{50} - \frac{1}{5000} + \cdots \right)$$

$$\approx 10(1.0198).$$

$$\sqrt{104} \approx 10.198.$$

Considering more terms in the series would not have contributed to the accuracy of the desired approximation.

EXAMPLE
35

Write the first four terms of $(1 + x)^{-2}$.

$$(1 + x)^{-2} = 1 + \frac{(-2) \cdot x}{1!} + \frac{(-2)(-3) \cdot x^2}{2!}$$

$$+ \frac{(-2)(-3)(-4) \cdot x^3}{3!} + \cdots$$

$$= 1 - 2x + 3x^2 - 4x^3 + \cdots .$$

EXERCISE 19-6

In Problems 1–16, expand and simplify the given expression.

1. $(x + 3)^5$ 2. $(x - y)^4$

3. $\left(a - \dfrac{1}{b}\right)^5$ 4. $\left(x + \dfrac{1}{b}\right)^5$

5. $(x - 2y)^6$ 6. $(2x - y)^6$

7. $\left(x + \dfrac{y}{2}\right)^6$ 8. $(1 - 2x)^5$

9. $(x + \tfrac{1}{2})^7$ 10. $(3x^2 - 1)^5$

11. $(2x - 5)^4$ 12. $(2a^2b + cd)^7$

13. $\left(1 + \dfrac{x}{y^2}\right)^6$ 14. $(xyz - yz)^5$

15. $\left(\dfrac{x}{y} + \dfrac{y}{x}\right)^7$ 16. $(x^2y - 3x)^6$

In Problems 17–34, find the first four terms in the given expansion.

17. $(x + y)^{15}$ 18. $(x - y)^{13}$

19. $(2x - b^2)^{21}$ 20. $(2a - ab^2)^{25}$

21. $(x - y^{-1})^{16}$ 22. $(x^{-1} + y^{-1})^{11}$

23. $(x - 2w^2y)^{14}$ 24. $(2 - x^2y)^{21}$

25. $(a^2 + b^2)^{20}$ 26. $(2a^2 - ab^3)^{20}$

27. $(1 + x)^{-2}$ 28. $(2x + y)^{-1}$

29. $(x + y)^{-4}$ 30. $(x - y)^{-3}$

31. $(1 + x)^{-2/3}$ 32. $(x - 2y)^{-1/3}$

33. $(2a - 3c)^{-1/2}$ 34. $(1 + x)^{1/x}$

In Problems 35–50, find the indicated terms(s).

35. The sixth term of $(x + y)^{15}$

36. The eighth term of $(x - y)^{18}$

37. The sixteenth term of $(x - y)^{20}$

38. The fourteenth term of $(x + 2y)^{17}$

39. The fifth term of $(2x - 3y)^{12}$

40. The eighth term of $(a + b)^{20}$

41. The ninth term of $(x^{-2} - y)^{15}$

42. The tenth term of $(x - y^{-2})^{14}$

43. The middle term of $(x - 2y)^{12}$

44. The middle terms of $(x^{-1} + y^{-1})^{11}$

45. The middle term of $(a^2 + b^2)^{20}$

46. The middle term of $(2a - ab^2)^{10}$

47. The term involving x^{10} in $(x - y)^{13}$

48. The term involving b^{10} in $(a^2 + b^2)^{20}$

49. The term involving x^6 in $(2y - 3x^2)^{15}$

50. The term involving x^7 in $(2y - x)^{12}$

In Problems 51–60, approximate the given number to three decimal places using the binomial series.

51. $\sqrt{50}$ **52.** $\sqrt{26}$

53. $\sqrt{61}$ **54.** $\sqrt{101}$

55. $\sqrt[3]{29}$ **56.** $\sqrt[3]{66}$

57. $\sqrt[4]{80}$ **58.** $(1.02)^{-4}$

59. $(1.01)^{-8}$ **60.** $(1.01)^{1/5}$

19-7
REVIEW

REVIEW QUESTIONS

The domain of an infinite sequence is the set of _____ (1) _____ .

(1) **positive integers (or natural numbers)**

The sixth term of the sequence $\{(-1)^n(2^n - 1)\}$ is _____ (2) _____ .

(2) **63**

The sequence $\frac{1}{2}, \frac{2}{4}, \frac{3}{8}, \ldots$ has a general term given by _____ (3) _____ .

(3) $\dfrac{n}{2^n}$

The sequence $-6, 2, 10, \ldots$ is a(n) _____(4)_____ progression whose _____(5)_____ is 8.

(4) **arithmetic**   (5) **common difference**

The sequence $-2, 4, -8, \ldots$ is a _____(6)_____ progression whose _____(7)_____ is -2.

(6) **geometric**   (7) **common ratio**

For an arithmetic progression, the nth term is given by the formula _____(8)_____ and the sum of the first n terms is given by the formula _____(9)_____.

(8) $a_n = a_1 + (n-1)d$ (9) $S_n = \dfrac{n}{2}(a_1 + a_n)$

For a geometric progression, the nth term is given by the formula _____(10)_____ and the sum of the first n terms is given by the formula _____(11)_____.

(10) $a_n = a_1 r^{n-1}$ (11) $S_n = \dfrac{a_1(1 - r^n)}{1 - r}$

The sixth term of the series $\sum\limits_{i=1}^{10} (-1)^{i+1} x^{2i}$ is _____(12)_____.

(12) $-x^{12}$

As n increases without bound, the sequence $\{1/n\}$ converges to _____(13)_____.

(13) **0**

The limit of any arithmetic progression with difference $d \neq 0$ $\underline{\text{(does)(does not)}}$ exist.
 (14)

(14) **does not**

$\sum\limits_{n=1}^{\infty} r^n$ converges if _____(15)_____.

(15) $|r| < 1$

The last term of $(a + 2b)^{19}$ is _____(16)_____.

(16) $2^{19}b^{19}$

True or false: For a sequence to have the number 2 as a limit, all the terms from some point on must be equal to 2. _____(17)_____

(17) **false.** $\left(\textbf{To see this, consider the counterexample}\right.$
$\left.\left\{2 - \dfrac{1}{n}\right\}.\right)$

Does the sequence $7, -7, 7, -7, 7, -7, \ldots$ converge or diverge? _____(18)_____

(18) **diverge**

For what value(s) of a does the sequence $a, -a, a, -a, \ldots$ converge? _____(19)_____

(19) $a = 0$

The sum of the geometric series $2 + 1 + \frac{1}{2} + \cdots$ equals _____(20)_____.

(20) **4**

The value of $4!$ is _____(21)_____.

(21) **24**

REVIEW PROBLEMS

1. Find the sum of all integers between 29 and 124 that are divisible by 6.

2. How many terms of $-16, -12, -8, \ldots$ must be added to give a sum of 44?

3. Express $.2\overline{32}$ in fractional form.

4. Find the fifth term of $(a + 2b)^{2/3}$.

5. The twelfth term and twenty-third term of an arithmetic progression are -12 and 20, respectively. Find the sixteenth term.

6. Find the term involving b^8 in $(a - 2b^2)^{-1/2}$.

7. Find the middle term of $(2 - \frac{1}{2}x^2)^{12}$.

8. Find $2 + 1 + \frac{1}{2} + \cdots + \frac{1}{64}$.

9. Evaluate $\lim\limits_{n \to \infty} \left(2 - \dfrac{n^2 - 6n}{2n^2} \right)$.

10. By using the binomial series, approximate $\sqrt[3]{124}$.

11. Find $\sum\limits_{k=1}^{\infty} \left(\frac{4}{9} \right)^k$.

12. In the series $1 + \frac{1}{2} + \frac{1}{4} + \cdots$, what is the numerical difference between the nth partial sum and the sum to infinity?

13. Find the sum of the coefficients in $(x + y)^{15}$.

14. Find $\lim\limits_{n \to \infty} \dfrac{2 - 3n}{3n - 2}$.

15. Which of the following sequences converge?
    a. $-3, -2, -1, 0, 1, 0, 1, 0, 1, \ldots$
    b. $0, \frac{1}{4}, 0, \frac{1}{8}, 0, \frac{1}{16}, \ldots$
    c. $2, \sqrt{2}, 1, \sqrt{2}/2, \ldots$

16. The sum of an infinite geometric progression is 6 and the first term is 2. What is the second term?

17. The value of the fifth term of the expansion for $(a + 2)^{12}$ is 48 for a particular value of a. Find this value.

18. A ball is released from an initial height of 8 ft and it is observed that after each contact with the floor the ball rebounds to a height equal to three-fourths of the height from which it last fell. What height does the ball reach on its fifth bounce?

19. For the ball of Problem 18, what total distance does it travel before coming to rest?

appendices

THE SLIDE RULE

A-1
INTRODUCTION

The slide rule, a mathematical instrument used in performing a variety of calculations, has a long and distinguished ancestry. In 1614 John Napier's early conception of the importance of simplifying calculations resulted in his invention of logarithms. In 1630 William Oughtred placed two straight logarithmic scales, similar to those designed ten years before that date by Edmund Gunter, in a side-by-side arrangement. Kept together by hand, the scales were able to slide along each other and this was, in effect, the birth of the slide rule.

The purpose of this appendix is to provide you with basic instruction in some of the fundamental operations that can be performed with a slide rule. Although many special slide rules have been developed over the years to meet modern requirements, the entire discussion which follows has reference to a typical 10-in. multiple-purpose slide rule.* Photographs

*The diagrams and descriptions in this appendix have been adapted from the instruction manual for the Deci-Lon slide rule by permission of Keuffel and Esser Co.

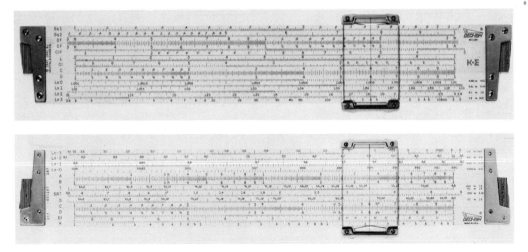

FIGURE A-1 (Photographs of the Deci-Lon slide rule courtesy of Keuffel & Esser Co.)

of the two sides of such a slide rule are shown in Fig. A-1. For your convenience, answers to *all* exercises in this appendix are given at the back of the book.

A-2
READING THE SCALES

Everyone has read a ruler in measuring a length. The number of inches is shown by a number appearing on the ruler; then small divisions are counted to get the number of sixteenths of an inch; and finally in a close measurement, a fraction of a sixteenth of an inch may be estimated. Exactly the same method is used in reading the slide rule. Although the divisions on the slide rule are not uniform in length as with the ruler, the same principle applies.

Figure A-2 represents, in skeleton form, the fundamental scale of the slide rule, namely the D scale. An examination of this actual scale on the slide rule will show that it is divided into nine parts by *primary marks* which are numbered 1, 2, 3, . . . , 9, 1. The space between any two primary marks is divided into ten parts by nine *secondary marks*. These are not

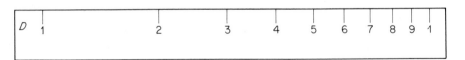

FIGURE A-2

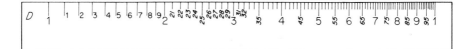

FIGURE A-3

numbered on the actual scale except between the primary marks numbered 1 and 2. Figure A-3 shows the secondary marks lying between the primary marks of the D scale. Each italicized number drawn on the scale in this illustration (they do not appear on the actual scale) gives the reading to be associated with its corresponding secondary mark. Thus, the first secondary mark after 2 is numbered 21, the second 22, the third 23, etc.; the first secondary mark after 3 is numbered 31, the second 32, etc. Between the primary marks numbered 1 and 2, the secondary marks are numbered 1, 2, . . . , 9. Evidently the readings associated with these marks are 11, 12, 13, . . . , 19. Finally between the secondary marks (see Fig. A-4) appear smaller or *tertiary marks* which aid in obtaining the third

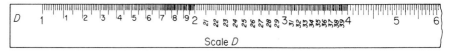

Scale D

FIGURE A-4

digit of a reading. Thus between the secondary marks numbered 22 and 23 there are four tertiary marks. If we think of the end marks as representing 220 and 230, the four tertiary marks divide the interval into five parts each representing *two* units. Hence with these marks we associate the numbers 222, 224, 226, and 228; similarly the tertiary marks between the secondary marks numbered 32 and 33 are read 322, 324, 326, and 328, and the tertiary marks between the primary mark numbered 3 and the first succeeding secondary mark are read 302, 304, 306, and 308. Between any pair of secondary marks to the right of the primary mark numbered 4, there is only one tertiary mark. Hence, each smallest space represents *five* units. Thus the tertiary mark between the secondary marks representing 41 and 42 is read 415, that between the secondary marks representing 55 and 56 is read 555, and the first tertiary mark to the right of the primary mark numbered 4 is read 405.

The reading of any position between a pair of successive tertiary marks must be based on an estimate. Thus a position halfway between the tertiary marks associated with 222 and 224 is read 223, and a position two-fifths of the way from the tertiary mark representing 415 to the next mark is read 417. The principle illustrated by these readings applies in all cases.

Consider the process of finding on the D scale the position representing 246. The first digit on the left, namely 2, tells us that the position lies between the primary marks numbered 2 and 3. This region is indicated by the brace in Fig. A-5. The second digit from the left, namely 4, tells us that the position lies between the secondary marks associated with 24 and 25. This region is indicated by the brace in Fig. A-6. Now there are four marks between the secondary marks associated with 24 and 25. With these are associated the numbers 242, 244, 246, and 248, respectively. Thus, the position representing 246 is indicated by the arrow in Fig. A-7. Figure A-8 gives a condensed summary of the process.

It is important to note that the decimal point has no bearing on the position associated with a number on the C and D scales. Consequently, the arrow in Fig. A-8 may represent 246, 2.46, 0.000246, 24,600, or any

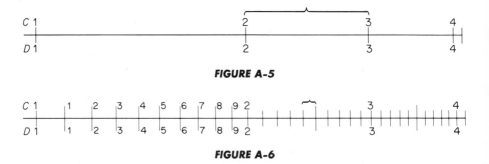

FIGURE A-5

FIGURE A-6

FIGURE A-7

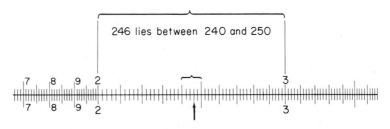

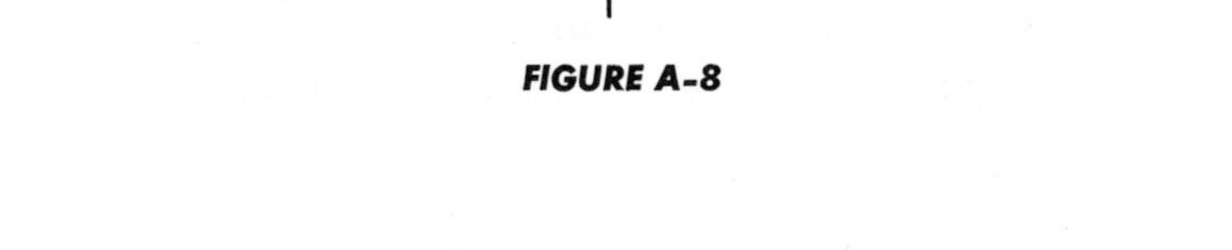

246 lies between 240 and 250

FIGURE A-8

other number whose sequence of principal digits are 2, 4, 6. The placing of the decimal point will be explained later.

For a position between the primary marks numbered 1 and 2, four digits should be read; the first three will be exact and the last one estimated. **No attempt should be made to read more than three digits for positions to the right of the primary mark numbered 4.**

While making a reading, you should have definitely in mind the number associated with the smallest space under consideration. Thus between primary numbers 1 and 2, the smallest division has a value of 10 in the fourth place; between 2 and 4, the smallest division has a value of 2 in the third place; while to the right of 4, the smallest division has a value of 5 in the third place.

You now should read from Fig. A-9 the numbers associated with the marks lettered A, B, C, . . . and compare your readings with the following numbers: A 365, B 327, C 263, D 1750, E 1347, F 305, G 207, H 1075, I 435, J 427.

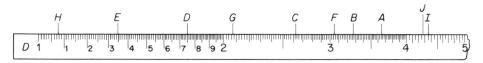

FIGURE A-9

A-3
THE PARTS OF THE SLIDE RULE AND DEFINITIONS

The middle sliding part of the slide rule (Fig. A-10) is called the **slide,** the other part the **body.**

The transparent runner is referred to as the **indicator,** and the line on it is called the **hairline.**

The mark associated with the primary number 1 on the D and C scale

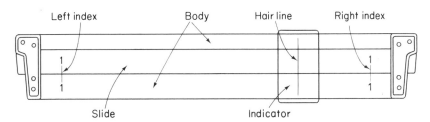

FIGURE A-10

is called the **index** of the scale. The C and D scales have two indexes, one at the left end called the **left index**, the other at the right end called the **right index**.

A number on one scale is said to be **opposite** a number on another scale if the hairline can cover both numbers at the same time. Each number is said to be **opposite** the other.

The slide rule is said to be **closed** when the slide is in such a position that the left index of the C scale is opposite the left index of the D scale.

Mathematical calculations are accomplished on the slide rule by moving the hairline or the slide or both. A description of these movements and of the resulting positions of the hairline and slide will be referred to as the **setting**.

Many settings will be described in what follows. In these descriptions two expressions, **"push the hairline"** and **"draw the number,"** will appear frequently. These two phrases are virtually idiomatic in slide rule language.

The meaning of the first phrase, "push the hairline," is obvious. The phrase "draw the number" is used to describe the operation of moving the slide to bring a number on the slide into a new position relative to the body. Therefore the word *draw*, when used in these settings, should always be associated with movement of the slide.

Such words as *close* and *opposite* will be used repeatedly. Furthermore, the abbreviation C will be used for C scale, D for D scale, etc.

A-4
ACCURACY OF THE SLIDE RULE

From the discussion of Sec. A-2 it appears that we read four digits of a number on one part of the scale and three digits on the remaining part. Assuming that the error of a reading is one-tenth of the smallest interval following the left-hand index of D, we conclude that the error is roughly 1 part in 1000 or one-tenth of 1 percent. The effect of the assumed error in judging a distance is inversely proportional to the length of the rule. Hence, we associate with a 10-in. slide rule an error of one-tenth of 1 percent; with a 20-in. slide rule an error of one-twentieth of 1 percent or 1 part in 2000. We remark that the accuracy obtainable with a 10-in. slide rule is sufficient for many purposes.

A-5
THE LOCATION OF THE DECIMAL POINT

In performing slide rule operations such as multiplication and division, the sequence of digits in the answer is obtained without regard to the position of the decimal point. The location of the decimal point is determined by rounding off the numbers and making a mental calculation. Users of the slide rule soon learn to use common sense for this part of the problem.

For example, if the slide rule is used to multiply 16.75 by 2.83, the three digits of the answer produced by the rule will be 474. To place the decimal point it could be noted that the answer is approximately $16 \times 3 = 48$. Thus, the answer is obviously 47.4.

A-6
MULTIPLICATION

The process of multiplication is performed on the slide rule by using scales C and D. The C scale is on the slide, but in other respects it is just like the D scale and is read in the same manner. We illustrate the process by some examples.

EXAMPLE
1

To multiply 2 by 4 (Fig. A-11),

> to 2 on D set left index of C,
> push hairline to 4 on C,
> at the hairline read **8** on D.

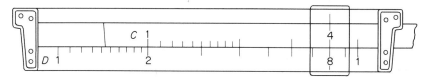

FIGURE A-11

EXAMPLE
2

To multiply 3×3 (Fig. A-12),

> to 3 on D set left index of C,
> push hairline to 3 on C,
> at the hairline read **9** on D.

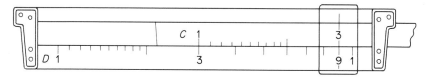

FIGURE A-12

EXAMPLE
3

To multiply 1.5×3.5, disregard the decimal point and

> to 15 on D set left index of C,

push hairline to 35 on C,
at the hairline read 525 on D.

By inspection we know that the answer is approximately 5. Hence, the answer is **5.25**.

EXAMPLE 4

To find the value of 16.75×2.83 (Fig. A-13), disregard the decimal point and

to 1675 on D set left index of C,
push hairline to 283 on C,
at the hairline read 474 on D.

To place the decimal point we note that the answer is approximately $3 \times 16 = 48$. Hence, the answer is **47.4**.

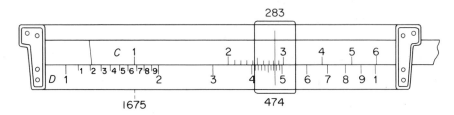

FIGURE A-13

EXAMPLE 5

To find the value of 0.001753×12.17,

to 1753 on D set left index of C,
push hairline to 1217 on C,
at the hairline read 2133 on D.

Here, as in many instances, scientific notation, discussed in Sec. 1-9, provides a convenient means of determining the location of the decimal point. Note:

$$0.001753 \times 12.17 \approx (2 \times 10^{-3})(1.2 \times 10^{1}) \approx 2.4 \times 10^{-2}.$$

Hence, the answer is $2.133 \times 10^{-2} = \mathbf{0.02133}$.

These examples illustrate the use of the following rule:

To find the product of two numbers, disregard the decimal points, opposite either of the numbers on the D scale set the index of the C scale, push the hairline of the indicator to the second number on the C scale, and read the answer under the hairline on the D scale. The decimal point is placed in accordance with the result of a mental approximation or use of scientific notation.

EXERCISE A-6

In Problems 1–14, find the indicated products.

1.	(3)(2)	**2.**	(3.5)(2)
3.	(5)(2)	**4.**	(2)(4.55)
5.	(4.5)(1.5)	**6.**	(1.75)(5.5)
7.	(4.33)(11.5)	**8.**	(2.03)(167.3)
9.	(1.536)(30.6)	**10.**	(0.0756)(1.093)
11.	(1.047)(3080)	**12.**	(0.00205)(408)
13.	$(3.142)^2$	**14.**	$(1.756)^2$

A-7
INTERCHANGING THE INDEXES

It may happen that a product cannot be read when the left index of the C scale is used in the rule for multiplication. It may happen that the second number of the product is on the part of the slide projecting beyond the body. For example, try to multiply 5 times 8 using the left index of the C scale. In such cases we reset the slide using the right index of C in place of the left index, that is, we *interchange the indexes.*

> **When a number is to be read on the D scale opposite a number on the C scale and cannot be read, push the hairline to the index of the C scale inside the body and draw the other index of the C scale under the hairline. Then make the desired reading. This operation is called *interchanging the indexes.***

To illustrate this rule consider the problem of finding the product of 2 and 6. We begin by setting the left index of the C scale opposite 2 on the D scale. We find we cannot read the answer on the D scale opposite 6 on the C scale. Hence, we push the hairline to the left index of C and draw the right index of C under the hairline. Then, opposite 6 on C read the answer, **12**, on D.

Although interchanging the indexes can be performed at any point in a calculation, it can often be avoided by simply choosing an appropriate index. For example, to find 0.0314×564,

> to 314 on D set the *right* index of C,
> push hairline to 564 on C,
> at the hairline read 1771 on D.

An approximation is obtained by finding $0.03 \times 600 = 18$. Hence, the product is **17.71**.

EXERCISE A-7

In Problems 1–16, find the indicated products.

1. (3)(5)
2. (3.05)(5.17)
3. (5.56)(634)
4. (743)(0.0567)
5. (0.0495)(0.0267)
6. (1.876)(926)
7. (1.876)(5.32)
8. (42.3)(31.7)
9. (912)(0.267)
10. (48.7)(1.173)
11. (0.298)(0.544)
12. (0.0456)(4.40)
13. (8640)(0.01973)
14. $(75.0)^2$
15. $(83.0)^2$
16. (4.98)(576)

A-8
DIVISION

The process of division is performed by using the C and D scales.

EXAMPLE
6

To divide 8 by 4 (Fig. A-14),

> push hairline to 8 on D,
> draw 4 of C under the hairline,
> opposite index of C read **2** on D.

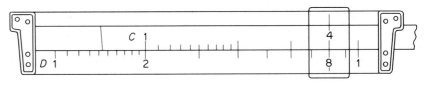

FIGURE A-14

EXAMPLE
7

To divide 876 by 20.4,

> push hairline to 876 on D,
> draw 204 of C under the hairline,
> opposite index of C read 429 on D.

The mental calculation $800 \div 20 = 40$ shows that the decimal point must be placed after the 2. Hence the answer is **42.9**.

These examples illustrate the use of the following rule:

To find the quotient of two numbers, disregard the decimal points, opposite the numerator on the *D* scale set the denominator on the *C* scale, opposite the index of the *C* scale read the quotient on the *D* scale. The position of the decimal point is determined from information gained by making a mental calculation or use of scientific notation.

EXERCISE A-8

In Problems 1–16, find the indicated quotients.

1. $87.5 \div 37.7$
2. $3.75 \div .0227$
3. $.685 \div 8.93$
4. $1029 \div 9.70$
5. $.00377 \div 5.29$
6. $2875 \div 37.1$
7. $871 \div .468$
8. $.0385 \div .001462$
9. $3.14 \div 2.72$
10. $3.42 \div 81.7$
11. $529 \div 565$
12. $.0456 \div .0297$
13. $396 \div .643$
14. $.0592 \div 1.983$
15. $.378 \div .0762$
16. $10.05 \div 30.3$

A-9
THE FOLDED SCALES—*DF* AND *CF*

The *DF* and the *CF* scales are the same as the *D* and the *C* scales, respectively, except in the position of their indexes. The fundamental fact concerning the folded scales may be stated as follows: *If for any setting of the slide, a number M of the C scale is opposite a number N on the D scale, then the number M of the CF scale is opposite the number N on the DF scale.* Thus, if you will draw 1 of the *CF* scale opposite 1.5 on the *DF* scale, you will find the following opposites on the *CF* and *DF* scales and the same oppo-

DF	1.5	3	6	7.5	9	1
CF	1	2	4	5	6	6.67

sites will appear on the *C* and *D* scales.

The following statement relating to the folded scales is basic.

The process of setting the hairline to a number *N* on scale *C*

to find its opposite M on scale D may be replaced by setting the hairline to N on scale CF to find its opposite M on scale DF.

The statement holds true if letters C and D are interchanged.

In accordance with the principle stated above, if you wish to read a number on the D scale opposite a number N on the C scale but cannot do so, you can generally read the required number on the DF scale opposite N on the CF scale. For example, to find 2×6,

> to 2 on D set left index of C,
> push hairline to 6 on CF,
> at the hairline read **12** on DF.

By using the CF and DF scales we saved the trouble of moving the slide as well as the attendant source of error. This saving, entering as it does in many ways, is the main reason for using the folded scales.

EXAMPLE
8

The folded scales may be used to perform multiplications and divisions just as the C and D scales are used. Thus to find 6.17×7.34,

> to 617 on DF set index of CF,
> push hairline to 734 on CF,
> at the hairline read **45.3** on DF;

or, alternatively,

> to 617 on DF set index of CF,
> push hairline to 734 on C,
> at the hairline read **45.3** on D.

EXAMPLE
9

To find the quotient $7.68/8.43$,

> push hairline to 768 on DF,
> draw 843 of CF under the hairline,
> opposite the index of CF read **0.911** on DF;

or, alternatively,

> push hairline to 768 on DF,
> draw 843 of CF under the hairline,
> opposite the index of C read **0.911** on D.

It now appears that we may perform a multiplication or a division in several ways by using two or more of the scales C, D, CF, and DF. As an additional comment, we point out that the DF scale may be conveniently used to multiply or divide a number by π. For a number a under the hairline on D, the product πa is read under the hairline on DF. Similarly, for a number b under the hairline on DF, the quotient b/π is read under the hairline on D. Using this technique, you should verify that $3\pi = 9.42$ and $3/\pi = .955$.

EXERCISE A-9

Perform the indicated operations, reading the answers, wherever possible without resetting, on the D scale and the DF scale.

1. 5.78×6.35	**2.** 7.84×1.065
3. $.00465 \div 73.6$	**4.** $.0634 \times 53,600$
5. $1.769 \div 496$	**6.** $946 \div .0677$
7. 813×1.951	**8.** $.00755 \div .338$
9. $.0948 \div 7.23$	**10.** $149.0 \div 63.3$
11. $2.718 \div 65.7$	**12.** $1.072 \div 10.97$

A-10
COMBINED OPERATIONS—MULTIPLICATION
AND DIVISION

Combined operations of multiplication and division can be performed by making use of the techniques already developed. The following examples will illustrate.

EXAMPLE 10

Find the value of $\dfrac{7.36 \times 8.44}{92}$.

Reason as follows: First divide 736 by 92 and then multiply the result by 844. This would suggest that we

> push hairline to 736 on D,
> draw 92 of C under the hairline,
> opposite 844 on C, read **0.675** on D.

EXAMPLE 11

Find the value of $\dfrac{18 \times 45 \times 37}{23 \times 29}$.

Reason as follows: (*a*) Divide 18 by 23, (*b*) multiply the result by 45, (*c*) divide this second result by 29, and (*d*) multiply this third result by 37. This argument suggests that we

> push hairline to 18 on D,
> draw 23 of C under the hairline,
> push hairline to 45 on C,
> draw 29 of C under the hairline,
> push hairline to 37 on C,
> at the hairline read 449 on D.

To determine the position of the decimal point write $\dfrac{20 \times 40 \times 40}{20 \times 30}$ which is approximately 50. Hence the answer is **44.9**.

A little reflection on the procedure of Example 11 will enable you to evaluate, by the shortest method, expressions similar to the one just considered. You should observe that the D scale was used only twice, once at the beginning of the process and once at its end; *the process for each number of the denominator consisted in drawing that number, located on the C scale, under the hairline; and the process for each number of the numerator consisted in pushing the hairline to that number located on the C scale.*

If at any time the indicator cannot be placed because of the projection of the slide, interchange the indexes or carry on the operations using the folded scales.

EXERCISE A-10

In Problems 1–8, perform the indicated operations.

1. $\dfrac{9 \times 14}{5}$

2. $\dfrac{37.4 \times 5.96\pi}{75.6}$

3. $\dfrac{146.2 \times 8.50}{3290\pi}$

4. $\dfrac{11 \times 12 \times 27\pi}{7 \times 13}$

5. $\dfrac{65.6 \times .842}{4.63}$

6. $\dfrac{76.6 \times 63.4 \times 96}{3.23\pi}$

7. $\dfrac{47.2 \times 18.3\pi}{32.6 \times 16.4}$

8. $\dfrac{3.82 \times 6.94 \times 7.82 \times 426}{77.8 \times .0322 \times 642}$

9. Multiply 312 successively by 1.44, 2.62, 3.18, 4.6, 5.12, 6.72, 7.46, 8.12, 9.62. *Hint*: Draw the left index of C to 312 on D, push hairline in succession to the given numbers on C or CF, and read the answers under the hairline on D or DF, respectively.

A-11
A VISUAL SUMMARY OF SOME FUNDAMENTAL OPERATIONS

In the visual summaries which follow, the letters a, b, and c, represent known quantities, x represents unknown quantities to be found, and the capital letters A, B, C, etc. designate scales.

To multiply a by b: $x = a \times b$.

1. To a on D set either index of C,

2. at b on C (CF) read ab on D (DF).

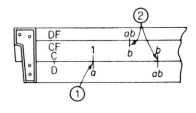

To divide a by b: $x = \dfrac{a}{b}$.

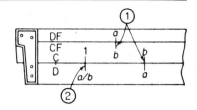

1. To a on D (DF) set b on C (CF),

2. at index of C read a/b on D.

To multiply or divide by π: $x = \pi a$; $x = \dfrac{b}{\pi}$.

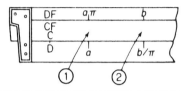

1. At a on D read πa on DF;

2. at b on DF read $\dfrac{b}{\pi}$ on D.

Combined multiplication and division: $x = \dfrac{ab}{c}$.

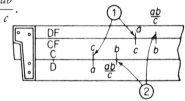

1. To a on D (DF) set c on C (CF),

2. at b on C (CF) read $\dfrac{ab}{c}$ on D (DF).

A-12
THE RECIPROCAL SCALES—*DI, CI,* AND *CIF*

The reciprocal scales *CI* and *CIF*, on the front face of the slide rule, and *DI*, on the reverse face, are marked and numbered like the *C, CF,* and *D* scales, respectively, but in the reverse (or inverted) order; that is, the numbers represented by the marks on these scales increase from right to left. The red numbers associated with the reciprocal scales enable you to recognize these scales easily.

When the hairline is set to a number on the *C* scale, the reciprocal (or inverse) of the number is at the hairline on the *CI* scale, conversely, when the hairline is set to a number on the *CI* scale, its reciprocal is at the hairline on the *C* scale.

The same relation exists between the *D* and *DI* scales and between the *CF* and *CIF* scales.

To fix this relation in mind push the hairline in succession to the numbers on *DI* in the second row of the diagram and read on *D* the respective reciprocals written in the first row. Also opposite the numbers

D	1	.5	.25	.2	.125	.1111
		$(=1/2)$	$(=1/4)$	$(=1/5)$	$(=1/8)$	$(=1/9)$
DI	1	2	4	5	8	9

1, 2, 4, 5, 8, and 9 on CI read their respective reciprocals on C. Again, find the same opposites on CIF and CF.

By using the facts just mentioned, we can multiply a number or divide it by the reciprocal of another number.

EXAMPLE
12

We may think of $\frac{28}{7}$ as $28 \times \frac{1}{7}$ and thus

to 28 on D set index of C,
opposite 7 on CI, read 4 on D.

EXAMPLE
13

To find 12×3, we may think of it as $12 \div \frac{1}{3}$ and

push hairline to 12 on D,
draw 3 of CI under the hairline,
opposite index of C, read **36** on D.

EXAMPLE
14

The DF and CIF scales may be used to perform multiplications and divisions in the same manner as the D and CI scales; thus to multiply 40.3 by 1/9.04,

to 403 on DF set index of CF,
opposite 904 on CIF, read **4.46** on DF.

Again, to multiply 40.3 by 1/0.207,

to 403 on D set left index of C,
opposite 207 on CIF, read **194.7** on DF.

It should be noted that when the hairline is set to any number on a scale on one face of the slide rule, the rule may be turned over, without changing the position of the indicator, to read the opposite number on a scale on the other face of the slide rule.

EXERCISE A-12

1. Use the DI scale to find the reciprocals of 16, 260, .72, .065, 17.4, 18.5, 67.1.

2. Find 18.2×21.7 in the usual way and then read $1/(18.2 \times 21.7)$ on DI opposite the first answer on D. Similarly find the values of $1/(2.87 \times 623)$, and $1/(.324 \times .497)$.

3. Using the D scale and the CI scale, multiply 18 by $1/9$ and divide 18 by $1/9$.

4. Using the D scale and the CI scale, multiply 28.5 by $1/.385$ and divide 28.5 by $1/.385$. Also find $28.5/.385$ and $28.5 \times .385$ by using the C scale and the D scale.

5. Using the D scale and the CI scale, multiply 41.3 by $1/.207$ and divide 41.3 by $1/.207$.

6. Perform the operations of problems 2, 3, and 4 by using the CIF scale and the DF scale.

7. Set the hairline to 8.62 on DF, and read at the hairline $8.62/\pi$ on D and $\pi/8.62$ on DI. Also find the values of $1.23/\pi$, $\pi/1.23$, $39.4/\pi$, and $\pi/39.4$.

A-13
SQUARE ROOTS—SCALES *A* AND *B*

Scale A consists of two parts which differ only in slight details. We shall refer to the left-hand part as A *left* and to the right-hand part as A *right*. Similar reference will be made to the B scale. Scale A is on the body. Scale B is on the slide. In all other respects the two scales are identical.

The A scale is so designed that when the hairline is set to a number on the A scale the square root of the number is on scale D under the hairline. Hence:

To find the square root of a number between 1 and 10 push the hairline to the number on A left and read its square root under the hairline on scale D. To find the square root of a number between 10 and 100 push the hairline to the number on A right and read its square root under the hairline on scale D. In either case place the decimal point in the square root after the first digit.

This statement also applies if A and D are replaced by B and C, respectively; that is, a square root may be determined by using the A and D scales on the body, or the B and C scales on the slide.

EXAMPLE
15

a. To find the square root of 9.00,

> push hairline to 9 on A *left*,
> under hairline read **3.00** on D.

b. To find the square root of 16.00,

> push hairline to 16 on A *right*,
> under hairline read **4.00** on D.

When a number is outside the range from 1 to 100, its square root can still be found on the slide rule by using the following fact: Moving the decimal point **two** places in a **number** results in moving the decimal point **one** place in the **square root** of the number. Hence:

> To obtain the square root of any number outside the range of 1 to 100, move the decimal point an *even number* of places to obtain a number between 1 and 100, find the square root of this latter number, then move the decimal point in this square root *one half* as many places as it was moved in the original number but in the *opposite* direction.

EXAMPLE
16

Find $\sqrt{432}$.

Move the decimal point two places to the **left** to obtain $\sqrt{4.32}$, and

> push hairline to 432 on A *left*,
> under hairline read 208 on D.

Therefore $\sqrt{4.32}$ is 2.08. Finally, since the decimal point was moved **two** places to the **left** in the original number, move the decimal point in this last result **one** place to the **right** to obtain the answer **20.8**.

EXAMPLE
17

Find $\sqrt{0.432}$.

Move the decimal point **two** places to the **right** to obtain $\sqrt{43.2}$, and

> push hairline to 432 on A *right*,
> under hairline read 658 on D.

Therefore $\sqrt{4.32} = 6.58$; finally move the decimal point in this last result **one** place to the **left** to obtain the answer **.658**.

EXERCISE A-13

Determine the square root of each of the following numbers.

1.	8	**2.**	12
3.	17	**4.**	89
5.	8.90	**6.**	890
7.	.89	**8.**	7280
9.	.0635	**10.**	.0000635
11.	63,500	**12.**	100,000
13.	53,500	**14.**	.0776
15.	32,700,000		

COMBINED OPERATIONS INVOLVING
SQUARE ROOTS

The principles previously explained may be applied to evaluate a fraction containing indicated square roots as well as numbers and reciprocals of numbers. If you will recall that when the hairline is set to a number on the *CI* scale it is automatically set to the reciprocal of the number on the *C* scale and when set to a number on the *B* scale it is automatically set to the square root of the number on the *C* scale, you will easily understand that the method used in this section is essentially the same as that used previously. The principle in determining whether *B left* or *B right* should be used is the same whether we are merely extracting the square root of a number or whether the square root is involved with other numbers.

EXAMPLE 18

Evaluate $\dfrac{915 \times \sqrt{36.5}}{804}$.

Remembering that the hairline is automatically set to $\sqrt{36.5}$ on the *C* scale when it is set to 36.5 on *B right*,

        push hairline to 915 on *D*,
        draw 804 of the *C* scale under the hairline,
        push hairline to 365 on *B right*,
        under hairline read 6.88 on *D*.

EXAMPLE 19

Evaluate $\dfrac{\sqrt{832} \times \sqrt{365} \times 1863}{(1/736) \times 89{,}400}$.

        Push hairline to 832 on *A left*,
        draw 736 of *CI* under the hairline,
        push hairline to 365 on *B left*,
        draw 894 of *C* under the hairline,
        push hairline to 1863 on *CF*,
        under hairline read **8450** on *DF*.

To get an approximate value write $\dfrac{(30)(18)(2000)(700)}{90{,}000} = 8400$.

EXAMPLE 20

Evaluate $\dfrac{0.286 \times 652 \times \sqrt{2350} \times \sqrt{5.53}}{785\sqrt{1288}}$.

Write the expression in the form

$$\frac{0.286 \times \sqrt{2350} \times \sqrt{5.53} \times 1}{(1/652) \times 785 \times \sqrt{1288}}.$$

        Push hairline to .286 on *D*,
        draw 652 of *CI* under the hairline,

push hairline to 235 on B *right*,
draw 785 of C under the hairline,
push hairline to 553 on B *left*,
draw 1288 of B *right* under hairline,
opposite the index of C, read **.755** on D.

As an approximate value use $\dfrac{.3(700)(50)(2)}{800(30)} \approx .9$.

EXERCISE A-14

In Problems 1–16, evaluate the given expression.

1. $42.2\sqrt{.328}$

2. $1.83\sqrt{.0517}$

3. $\sqrt{3.28} \div .212$

4. $\sqrt{51.7} \div 103$

5. $.763 \div \sqrt{.0296}$

6. $\dfrac{\sqrt{277}}{5.34 \times \sqrt{7.02}}$

7. $\dfrac{645}{5.34\sqrt{13.6}}$

8. $14.3 \times 47.5\sqrt{.344}$

9. $20.6 \times \sqrt{7.89} \times \sqrt{.571}$

10. $\dfrac{7.92\sqrt{7.89}}{\sqrt{.571}}$

11. $\dfrac{7.87 \times \sqrt{377}}{2.38}$

12. $\dfrac{86 \times \sqrt{734} \times \pi}{775 \times .685}$

13. $\dfrac{4.25 \times \sqrt{63.5} \times \sqrt{7.75}}{.275 \times \pi}$

14. $\dfrac{189.7 \times \sqrt{.00296} \times \sqrt{347} \times .274}{\sqrt{2.85} \times 165 \times \pi}$

15. $\sqrt{285} \times 667 \times \sqrt{6.65} \times 78.4 \times \sqrt{.00449}$

16. $\dfrac{239 \times \sqrt{.677} \times 374 \times 9.45 \times \pi}{84.3 \times \sqrt{9350} \times \sqrt{28,400}}$

A-15
CUBE ROOTS, THE *K* SCALE

The K scale is on the body of the slide rule. It is divided into three equal and identical sections. We shall refer to the left-hand section, the middle section, and the right-hand section as K *left*, K *middle*, and K *right*, respectively.

The K scale is so constructed that when the hairline is set to a number on the K scale the cube root of the number is on the D scale at the hairline.

To find the cube root of a number between 1 and 10, set the hairline to the number on K left and read its cube root at the hairline on D. To find the cube root of a number between 10 and

100, set the hairline to the number on K middle and read its cube root at the hairline on D. The cube root of a number between 100 and 1000 is found on the D scale opposite the number on K right.

In each of these three cases the decimal point is placed after the first digit.

EXAMPLE
21

a. To find the cube root of 27,

> push hairline to 27 on K *middle*,
> under hairline read **3** on D.

b. To find the cube root of 343,

> push hairline to 343 on K *right*,
> under hairline read **7** on D.

When a number is outside the range from 1 to 1000, its cube root can be found on the slide rule by making use of the following fact: Moving the decimal point **three** places in a **number** results in moving the decimal point **one** place in the **cube root** of the number. Hence:

To obtain the cube root of a number outside the range from 1 to 1000, move the decimal point *three places* at a time until a number between 1 and 1000 is obtained. Find the cube root of this latter number, then move the decimal point in the cube root *one third* as many places as it was moved in the original number but in the *opposite* direction.

EXAMPLE
22

Find $\sqrt[3]{23,400,000}$.

Move the decimal point 6 places to the **left**, thus obtaining 23.4. Since this is between 1 and 1000,

> push hairline to 23.4 on K *middle*,
> under hairline read on D, 2.86 $= \sqrt[3]{23.4}$,
> move the decimal point $\frac{1}{3}(6) = 2$ places to
> the *right* to obtain the answer, **286**.

The decimal point could have been placed by noting that $\sqrt[3]{27,000,000} = 300$.

EXAMPLE
23

Find $\sqrt[3]{.000585}$.

Move the decimal point 6 places to the **right** to obtain $\sqrt[3]{585}$, a number between 1 and 1000, and

> push hairline to 585 on K *right*,
> under hairline read on D, 8.36 $= \sqrt[3]{585}$,
> move the decimal point $\frac{1}{3}(6) = 2$ places to
> *left* to obtain the answer **.0836**.

EXERCISE A-15

Find the cube root of each given number.

1.	8.72	**2.**	30
3.	729	**4.**	850
5.	7630	**6.**	.00763
7.	.0763	**8.**	.763
9.	89,600	**10.**	.625
11.	75×10^7	**12.**	10

A-16
SQUARES AND CUBES

Squaring a number is the inverse operation to extracting its square root. It is not surprising therefore to find that squares can be obtained using the square root scales A and B. In this connection the following rule will be found useful.

> **To find the square of a number using scales A and D, set the hairline to the number on scale D, and under the hairline read on scale A the square of the number. Similarly, to find the square of a number using scales B and C, set the hairline to the number on scale C and under the hairline read on scale B the square of the number.**

To gain familiarity with this use of scales A and B make the following settings:

To find 3^2,

> push hairline to 3 on D,
> under hairline read **9** on A.

To find 4^2,

> push hairline to 4 on D,
> under hairline read **16** on A, or
> push hairline to 4 on C,
> under hairline read **16** on B.

By interchanging the roles of the K and D scales in the operations performed for finding cube roots, we may find the cubes of numbers using scales K and D. In this connection the following rule may be found helpful.

To find the cube of a number, set the hairline to the number on the D scale, and read its cube on the K scale at the hairline.

To convince yourself of this, you should set the hairline to 2 on D and read $2^3 = 8$ at the hairline on K; set the hairline to 3 on D and read $3^3 = 27$ at the hairline on K, etc. To find 21.7^3, set the hairline to 217 on D and read 102 on K. Since $20^3 = 8000$, the answer is near 8000. Hence, the answer is **10,200**. To obtain this answer otherwise, write

$$21.7^3 = \frac{21.7 \times 21.7}{(1/21.7)} = \mathbf{10{,}220}$$

and use the general method of combined operations. *The latter method is more accurate as it is carried out on the full-length scales.*

A-17
THE *S*(SINE) AND *SRT*(SINE, RADIAN, TANGENT) SCALES

The graduations on the sine scales S and SRT represent *angles*. Accordingly, for convenience, we shall speak of *pushing the hairline to an angle* or *drawing an angle under the hairline.*

The S scale serves a double function. When read from left to right (on many slide rules these numbers are in *black*) it covers the angles from about 5.5° to 90° and is used for finding *sines*. When read from right to left (on many slide rules these numbers are in *red*) it covers angles from about 0° to 84.5° and is used for finding *cosines*. In what follows, any reference to an angle on a trigonometric scale will be the angle in black unless otherwise indicated.

The SRT scale covers the angles from about .55° to 6° and is used for finding sines, radian equivalents, and tangents of these small angles. In this section we are primarily concerned with determining the sines and cosines of angles.

Note that the S and SRT scales are essentially one continuous scale (with a slight overlap), read against two continuous cycles of the C scale. Figure A-15 represents this relationship.

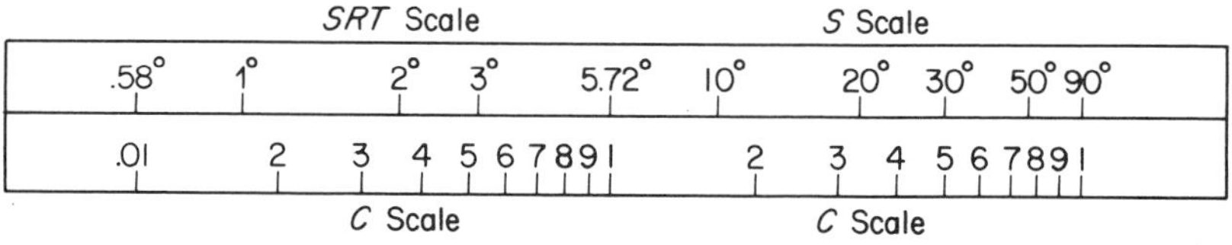

FIGURE A-15

In order to set the hairline to an angle on the S scale, it is necessary to determine the values of the angles represented by the subdivisions. Since there are ten primary intervals between 8° and 9°, each represents .1°; since each of the primary intervals is subdivided into two secondary intervals, each of the latter represents .05°. Again since there are five primary intervals between 20° and 25°, each represents 1°; since each primary interval here is subdivided into five secondary intervals, each of the latter represents .2°. The last mark at the right end represents 90°, the next mark to the left 85°, and the third 80°.

When the hairline is set to an angle on the sine scale (S *black*) or the SRT scale, the sine of the angle is on scale C at the hairline, and hence on scale D when the rule is closed. Also when the hairline is set to an angle on the cosine scale (S *red*) the cosine of the angle is on scale C at the hairline.

Each small inscription at the right end of a scale is called the **legend** of the scale. A legend of a scale specifies a range of values associated with the function represented by the scale. Thus *the legend* **.1 to 1.0** *of scale S specifies that the sines of the angles on S and the cosines of angles on S red range from .1 to 1, and the legend* **.01 to .1** *of the SRT scale indicates that sines (or radian equivalents and tangents) of angles on SRT range from .01 to .1.*

EXAMPLE

24

Evaluate (a) sin 36.4° and (b) sin 3.40°.

a.

> Opposite 36.4° on S,
> read 593 on C (or D when rule is closed).

To locate the decimal point, we note that the legend on the S scale indicates that resulting values must lie between .1 and 1.0. Hence the answer is **.593**.

b.

> Opposite 3.40° on SRT,
> read 593 on C.

To locate the decimal point, we observe that the legend on the SRT scale establishes values as lying between .01 and .1. Therefore the final answer is **.0593**.

Figure A-16 shows scales SRT, S, and D on which certain angles and their sines are indicated. As an exercise, close your slide rule and read the sines of the angles shown in the figure and compare your results with those given. Note that the values of sines appearing in Fig. A-16 conform to the corresponding legends.

Each angle on S *red* is 90° minus the corresponding angle on S *black*.

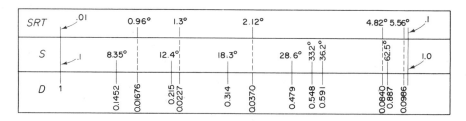

FIGURE A-16

Also, from trigonometry

$$\sin A = \cos (90° - A); \quad \cos A = \sin (90° - A).$$

Hence, when the hairline is set to an angle A on S, it is set to sin A and to cos $(90° - A)$ on scale C. For example,

> set the hairline to 25° on S,
> at the hairline read on C **.423** $= \sin 25° = \cos 65°$.

To find the cosine of an angle greater than 84.5°, use the formula cos A = sin $(90° - A)$.

Thus to find cos 86.9°, write cos 86.9° = sin 3.1° and opposite 3.1° on SRT read on C, .0541 = sin 3.1° = cos 86.9°.

The following is a tabular summary of the rules for finding the values of sines and cosines:

SINES

Angles from .573° to 5.73°:	Use SRT scale; values will lie between .01 and .10.
Angles from 5.73° to 90°:	Use S scale, black numbers, reading left to right; values will lie between .1 and 1.0.

COSINES

Angles from 0° to 84.25°:	Use S scale, red numbers, reading right to left; values will lie between .1 and 1.0.
Angles from 84.25° to 89.427°:	Use SRT scale to find the sine of 90° minus the angle; values will lie between .01 and .10.

EXERCISE A-17

1. By examination of the slide rule, verify that on the S scale from the left index to 10° the smallest subdivision represents .05°; from 10° to 20° it represents .1°; from 20° to 30° it represents .2°; from 30° to 60° it repre-

sents .5°; from 60° to 80° it represents 1°; and from 80° to 90° it represents 5°.

2. Find the sine of each of the following angles.

a. 30°
b. 38°
c. 3.33°
d. 90°
e. 88°
f. 1.583°
g. 14.63°
h. 22.4°
i. 11.80°
j. 51.5°

3. Find the cosine of each of the angles in Problem 2.

4. Find a value of x in each equation.

a. $\sin x = .5$
b. $\sin x = .875$
c. $\sin x = .375$
d. $\sin x = .1$
e. $\sin x = .015$
f. $\sin x = .62$
g. $\sin x = .062$
h. $\sin x = .031$
i. $\sin x = .92$

5. Find a value of x in each equation.

a. $\cos x = .5$
b. $\cos x = .875$
c. $\cos x = .375$
d. $\cos x = .1$
e. $\cos x = .015$
f. $\cos x = .62$
g. $\cos x = .062$
h. $\cos x = .031$
i. $\cos x = .92$

A-18
THE *T*(TANGENT) SCALE

On the T scale of most slide rules the *black* numbers represent angles from about 5.5° to 45°; the *red* numbers represent angles from 45° to about 84.5°.

When the hairline is set to an angle A on T *black*, tan A is at the hairline on scale C, and hence on scale D when the rule is closed, when the hairline is set to an angle A on T *red*, tan A is at the hairline on CI (or on DI when the rule is closed).

The (*black*) legend .1 to 1.0 at the right end of the T scale indicates that tangents read on C (*black*) are between .1 and 1; the (*red*) legend 10.0 to 1.0 indicates that tangents on CI (*red*) are between 10 and 1.

For example,

opposite 26° on T *black*, read **.488** = tan 26° on C,
opposite 64° on T *red*, read **2.05** = tan 64° on CI.

The cotangent of an angle may be found by first using the identity

$$\cot A = \frac{1}{\tan A}$$

to express the cotangent in terms of the tangent of an angle and then using the method outlined above. Thus to find cot 26°, write

$$\cot 26° = 1/\tan 26°$$

and opposite 26° on T, read **2.05** = cot 26° on CI. Note here that reading on CI gives the required reciprocal immediately.

It is shown in trigonometry that the sine and the tangent of an angle less than 5.71° are so nearly equal that they may be considered identical for slide rule purposes. Thus to find tan 2.25° and cot 2.25°,

opposite 2.25° on SRT read on C, .0393 = tan 2.25°,
opposite 2.25° on SRT read on CI, 25.5 = 1/tan 2.25° = cot 2.25°.

EXERCISE A-18

1. Complete the following table:

ψ	8.1°	27.25°	62.32°	1.017°	74.25°	87°	47.47°
tan ψ							
cot ψ							

2. The following numbers are tangents of angles. Find the angles.

a.	.24	b.	.785	c.	.92	d.	.54	e.	.059
f.	.082	g.	.432	h.	.043	i.	.0149	j.	.374
k.	3.72	l.	4.67	m.	17.01	n.	1.03	o.	1.232

3. The numbers in Problem 2 are cotangents of angles. Find the angles.

4. Find an angle x from each equation.

   a. $\tan x = \dfrac{3.7}{6.8}$ b. $\tan x = \dfrac{287}{642}$ c. $\tan x = \dfrac{5.72}{2.86}$

   d. $\tan x = \dfrac{8.52}{6.73}$ e. $\cot x = \dfrac{5}{6}$ f. $\cot x = \dfrac{17.2}{143}$

A-19
RADIANS—SMALL ANGLES

The SRT scale is a scale whose marks represent numbers of degrees ranging from .573° to 5.73° approximately. It is so designed that the following rule holds:

When the hairline is set to an angle in degrees on the SRT scale, it is also set to the same angle in radians on the C scale, provided

the number on the C scale is prefixed by ".0" as indicated by the legend .01 to .1 at the end of the SRT scale.

For example, in accordance with the rule,

push hairline to 3.56° on SRT,
at hairline read 621 on C.

Therefore 3.56° = **.0621 radian.**

Observe that if we multiply both members of the equation

$$3.56° = .0621 \text{ radian}$$

by 10, 10^2, $\frac{1}{10}$, and $\frac{1}{10^2}$ in succession, we get

$$(10)(3.56°) = (10)(.0621), \text{ or } 35.6° = .621 \text{ radian}$$

$$(100)(3.56°) = (100)(.0621), \text{ or } 356° = 6.21 \text{ radians}$$

$$(\tfrac{1}{10})(3.56°) = (\tfrac{1}{10})(.0621), \text{ or } .356° = .00621 \text{ radian}$$

$$(\tfrac{1}{100})(3.56°) = (\tfrac{1}{100})(.0621), \text{ or } .0356° = .000621 \text{ radian.}$$

In other words, the range of the scale can be extended by noting that for any integer k, positive or negative,

$$10^k(3.56°) = 10^k(.0621) \text{ radian.}$$

Now using the rule in reverse,

push the hairline to 1176 on C,
at hairline read .674° on SRT,

and conclude that

$$.01176 \text{ radian} = \textbf{.674°.}$$

Multiplying this through by 10^2, $\frac{1}{10}$, and 10^k in succession, we get

$$1.176 \text{ radians} = 67.4°$$

$$.001176 \text{ radian} = .0674°$$

and, in general,

$$10^k(.01176) \text{ radians} = 10^k(.674°).$$

For angles θ in radians, where θ is less than .1 radian (or 5.73°), the following relation holds,

$$\boldsymbol{\theta \text{ (radians)} \approx \sin \theta \approx \tan \theta,} \tag{A-1}$$

where the symbol $\approx$ means "*approximately equals.*" In other words, *the value of an angle in radians found by means of the previous rule is also its sine and its tangent to slide rule accuracy.* Equation (A-1) is often called the *small-angle approximation.*

For example,

> push hairline to 3.84° on *SRT*,
> at hairline read 670 on *C*.

Therefore, in accordance with the rule,

$$\sin 3.84° \approx \tan 3.84° \approx .0670,$$

and, in agreement with Eq. (A-1),

$$\sin 0.384° \approx \ \tan .384° \approx .00670,$$
$$\sin 0.0384° \approx \tan .0384° \approx .000670, \text{ etc.}$$

EXERCISE A-19

1. Express in radians.
    a. 1.416°   b. .833°   c. 2.5°   d. 2.67°

2. Express in degrees.
    a. .01823 radian   b. .0462 radian   c. .0865 radian

3. Express in radians.
    a. 3.59°   b. .0359°   c. 35.9°   d. 359°

4. Express in degrees.
    a. .0296 radian   b. .296 radian   c. .000296 radian

5. Express in radians.
    a. 912°   b. 435°   c. .000314°   d. 2900°

6. Find sin 3.42°, tan 3.42°, and cot 3.42°.

7. Find sin .056°, tan .056°, and cot .056°.

8. Find cos 89.75°, tan 89.75°, and cot 89.75°.

A-20
THE LON SCALES

The eight scales labeled *Ln*0, *Ln*1, *Ln*2, *Ln*3, *Ln*-0, *Ln*-1, *Ln*-2, and *Ln*-3 will be referred to collectively as the *Lon scales*, and any individual scale of the eight as a *Lon scale*. The Lon scales are divided into two groups. The scales labeled *Ln*0, *Ln*1, *Ln*2, and *Ln*3 comprise one of these groups. They are all on the front face of the rule and are usually numbered in black. The group collectively exhibits a continuous sequence of numbers increasing from left to right, ranging from 1.001 to 30,000. Each scale of this group functions with the *C* and *D* scales on the front face of the rule.

The second group consists of the scales labeled *Ln*-0, *Ln*-1, *Ln*-2, and *Ln*-3. This group of scales is on the reverse face of the rule and is usually numbered in red. The group exhibits a continuum of decimal fractions increasing from right to left, ranging from .00003 at the right end of the Lon-minus-three scale to .999 at the left end of the Lon-minus-zero scale. These scales operate with the *C* and *D* scales on the reverse face of the slide rule.

Each calibration mark on the Lon scales represents a single unique number complete with decimal point. This is in contrast to the calibration marks on the *C*, *D* and other numerical scales, on which each mark may represent many numbers. For example, the calibration mark numbered 4 on the *C* scale may represent 4, 40, 400, .04, 004, etc., whereas the calibration mark numbered 4 on the *Ln3* scale represents 4 and no other number.

The left and right indexes of any Lon scale are defined to be the positions on that scale opposite the left and right indexes respectively, of the *D* scale. For convenience of reading, the calibration marks on some Lon scales extend beyond the scale's indexes, that is, to the left of the left index and to the right of the right index. When referring to marks on any Lon scale, unless otherwise indicated, we shall mean the marks between the indexes of the scale.

By using the Lon scales in conjunction with the *D* scale, you can readily find the natural logarithms of numbers. *The Lon scales are so designed that when the hairline is set to a number N on a Lon scale, Ln N is under the hairline on Scale D.*

To use this fact in evaluating Ln 7.39 (see Fig. A-17), push hairline to 7.39 on *Ln3*, and under hairline read **2.00** on *D*.

The position of the decimal point in the logarithm is determined in accord with the following facts:

The left index of the *Ln3* scale represents e^1, and Ln $e = 1$; the right index represents e^{10}, and Ln $e^{10} = 10$; hence the natural logarithms of the numbers represented on *Ln3* range from **1 to 10**. Note that the legend

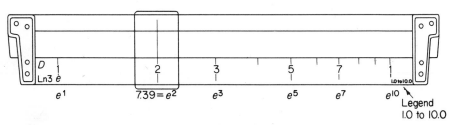

FIGURE A-17

at the right end of the *Ln3* scale is **1.0 to 10.0**. If Ln $N = x$, then the legend numbers **1.0 to 10.0** of *Ln3* are the limits of x when N is on *Ln3*. In like manner the legend of the *Ln-3* scale, **−1.0 to −10.0**, is based on the Lns of e^{-1} and e^{-10}, respectively, represented by the indices of the scale. Hence, the natural logarithms of the numbers represented on *Ln-3* range from **−1.0 to −10.0**. The legend of each Lon scale has a similar relation to the numbers represented on the scale. In short:

> **The legend at the right end of each Lon scale gives the limits of the values of the natural logarithms (read on D) of the numbers represented on that scale.**

In the previous example, 7.39 is found on *Ln3*, so the value of Ln 7.39, according to the legend of the *Ln3* scale, must lie between **1.0 and 10.0**. Hence, Ln 7.39 = 2. This, of course, can be verified by a quick mental estimate.

The rule for finding the logarithm to the base e of any number (Fig. A-18) is

> **To find the value of Ln N, push the hairline to N on a Lon scale, read Ln N on the D scale under the hairline, and place the decimal point in this number so that the result lies between the legend numbers of the Lon scale used.**

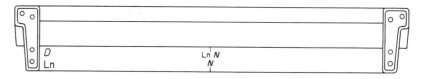

FIGURE A-18

To gain familiarity with the process of finding logarithms to the base e, you should make the suggested setting for finding the Ln of each of the following numbers:

$$.002, \quad .2, \quad 2.0, \quad 20, \quad 200,$$

and place the decimal point in each Ln by means of the appropriate legend.

> Opposite .002 on *Ln-3* read 621 on D, and in accord with the legend of *Ln-3* obtain Ln .002 = **−6.21**;
>
> opposite .2 on *Ln-3* read 161 on D, and in accord with the legend of *Ln-3* obtain Ln .2 = **−1.61**;
>
> opposite 2.0 on *Ln2* read 639 on D, and in accord with the legend of *Ln2* obtain Ln 2.0 = **.693**;

opposite 20 on *Ln3* read 2995 on *D*, and in accord
with the legend of *Ln3* obtain Ln 20 = **2.995**;

opposite 200 on *Ln3* read 530 on *D*, and in accord
with the legend of *Ln3* obtain Ln 200 = **5.30**.

EXAMPLE
25

Find Ln 1.345 and Ln 0.9946.

Push hairline to 1.345 on *Ln2*,
under hairline read on *D*, **.296** = Ln 1.345;
push hairline to .9946 on *Ln-0*,
under hairline read on *D*, −**.0054** = Ln .9946.

Here the decimal point was placed in each answer by observing that .296
lies within the legend range **.1 to 1.0** of scale *Ln2*, the scale on which 1.345
is found, and that −.0054 lies within the legend range −**.001 to** −**.01** of
scale *Ln-0*, the scale on which .9946 is found.

EXERCISE A-20

1. Find the Ln of 500, 50, 2, 1.4, and 1.043.

2. Find the Ln of .002, .02, .5, .714, .9091, and .9804.

3. Find the values of

a. Ln 76. b. Ln 7.6. c. Ln 9.2.
d. Ln 0.84. e. Ln 0.145. f. Ln 0.893.
g. Ln 0.909. h. Ln 1.43. i. Ln 1.043.

4. Show that the first three significant digits in the natural logarithm of each of
the following numbers is 693: 1.0718, .9330, 2, .5, 1024, .000977, .99307.

5. Find Ln 4. Then find seven numbers other than 4, each having as its
natural logarithms the same first three digits as Ln 4.

A-21
POWERS OF e

We have seen that we can locate a number on a Lon scale and opposite it, on *D*, read its natural logarithm. By the inverse process, we can
locate a number *x* on the *D* scale and opposite it read e^x on a Lon scale.

The reason for this relationship is illustrated in Fig. A-19. Since
Ln $N = x$, then by definition $N = e^x$. But, Ln N on *D* is opposite N on
a Lon scale. Therefore, *x* on *D* is opposite e^x on a Lon scale. The following
rule embodies this relation:

To find the value of e^x, push hairline to *x* on *D* and under the

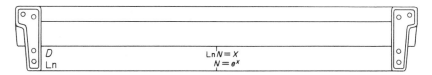

FIGURE A-19

hairline read the value of e^x on the Lon scale which contains x between its legend numbers.

EXAMPLE
26

Evaluate $e^{3.5}$ and $e^{-3.5}$.

In accord with the above rule (see Fig. A-20) make the following setting:

> push hairline to 35 on D,
> under hairline read on $Ln3$, **33.1** $= e^{3.5}$.

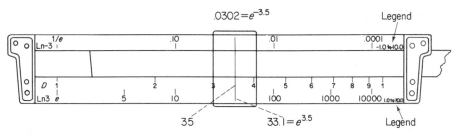

FIGURE A-20

In a similar manner,

> push hairline to 35 on D,
> under the hairline read on Ln-3, **.0302** $= e^{-3.5}$.

Scale $Ln3$ was selected on which to read the value of $e^{3.5}$ because the exponent 3.5 lies within the range specified by the legend of the $Ln3$ scale. Similarly, scale Ln-3 was chosen for $e^{-3.5}$ because -3.5 lies within the range specified by the legend on the Ln-3 scale.

EXAMPLE
27

Evaluate (a) e^2, e^{-2}; (b) $e^{0.2}$, $e^{-0.2}$; (c) $e^{0.02}$, $e^{-0.02}$; and (d) $e^{0.002}$, $e^{-0.002}$.

Push hairline to 2 on D,

a. Under hairline read $\begin{cases} \text{on } Ln3, \textbf{7.39} = e^2, \\ \text{on } Ln\text{-}3, \textbf{.135} = e^{-2}. \end{cases}$

b. Under hairline read $\begin{cases} \text{on } Ln2, \textbf{1.221} = e^{.2}, \\ \text{on } Ln\text{-}2, \textbf{.8187} = e^{-.2}. \end{cases}$

c. Under hairline read $\begin{cases} \text{on } Ln1, \textbf{1.0202} = e^{.02}, \\ \text{on } Ln\text{-}1, \textbf{.9802} = e^{-.02}. \end{cases}$

d. Under hairline read $\begin{cases} \text{on } Ln0, \textbf{1.002} = e^{.002}, \\ \text{on } Ln\text{-}0, \textbf{.998} = e^{-.002}. \end{cases}$

EXERCISE A-21

1. Evaluate.

   a. e^3 b. e^{-3} c. $e^{.4}$

   d. $e^{-.4}$ e. $e^{.035}$ f. $e^{-.035}$

   g. $e^{1.342}$ h. $e^{-1.342}$ i. $e^{-2.46}$

   j. $e^{3.55}$ k. $e^{-.0264}$ l. $e^{.0853}$

2. Find e^x when (a) $x = 2.12$, (b) $x = -2.12$, (c) $x = .212$, (d) $x = -.212$, (e) $x = .0212$, and (f) $x = -.0212$.

3. Evaluate.

   a. e^4 b. $e^{8.2}$ c. $e^{.43}$

   d. $e^{.0214}$ e. $e^{-3.4}$ f. $e^{-.163}$

   g. $e^{-.0185}$ h. $e^{-6.2}$

Appendix B

TABLE OF EXPONENTIAL FUNCTIONS

x	e^x	e^{-x}	x	e^x	e^{-x}
0.00	1.0000	1.0000	2.5	12.182	0.0821
0.05	1.0513	0.9512	2.6	13.464	0.0743
0.10	1.1052	0.9048	2.7	14.880	0.0672
0.15	1.1618	0.8607	2.8	16.445	0.0608
0.20	1.2214	0.8187	2.9	18.174	0.0550
0.25	1.2840	0.7788	3.0	20.086	0.0498
0.30	1.3499	0.7408	3.1	22.198	0.0450
0.35	1.4191	0.7047	3.2	24.533	0.0408
0.40	1.4918	0.6703	3.3	27.113	0.0369
0.45	1.5683	0.6376	3.4	29.964	0.0334
0.50	1.6487	0.6065	3.5	33.115	0.0302
0.55	1.7333	0.5769	3.6	36.598	0.0273
0.60	1.8221	0.5488	3.7	40.447	0.0247
0.65	1.9155	0.5220	3.8	44.701	0.0224
0.70	2.0138	0.4966	3.9	49.402	0.0202
0.75	2.1170	0.4724	4.0	54.598	0.0183
0.80	2.2255	0.4493	4.1	60.340	0.0166
0.85	2.3396	0.4274	4.2	66.686	0.0150
0.90	2.4596	0.4066	4.3	73.700	0.0136
0.95	2.5857	0.3867	4.4	81.451	0.0123
1.0	2.7183	0.3679	4.5	90.017	0.0111
1.1	3.0042	0.3329	4.6	99.484	0.0101
1.2	3.3201	0.3012	4.7	109.55	0.0091
1.3	3.6693	0.2725	4.8	121.51	0.0082
1.4	4.0552	0.2466	4.9	134.29	0.0074
1.5	4.4817	0.2231	5	148.41	0.0067
1.6	4.9530	0.2019	6	403.43	0.0025
1.7	5.4739	0.1827	7	1096.6	0.0009
1.8	6.0496	0.1653	8	2981.0	0.0003
1.9	6.6859	0.1496	9	8103.1	0.0001
2.0	7.3891	0.1353	10	22026	0.00005
2.1	8.1662	0.1225			
2.2	9.0250	0.1108			
2.3	9.9742	0.1003			
2.4	11.023	0.0907			

Appendix C

NATURAL TRIGONOMETRIC FUNCTIONS

Degrees	Radians	Sin	Cos	Tan	Cot	Sec	Csc		
0° 00′	.0000	.0000	1.0000	.0000	——	1.000	——	1.5708	90° 00′
10	029	029	000	029	343.8	000	343.8	679	50
20	058	058	000	058	171.9	000	171.9	650	40
30	.0087	.0087	1.0000	.0087	114.6	1.000	114.6	1.5621	30
40	116	116	.9999	116	85.94	000	85.95	592	20
50	145	145	999	145	68.75	000	68.76	563	10
1° 00′	.0175	.0175	.9998	.0175	57.29	1.000	57.30	1.5533	89° 00′
10	204	204	998	204	49.10	000	49.11	504	50
20	233	233	997	233	42.96	000	42.98	475	40
30	.0262	.0262	.9997	.0262	38.19	1.000	38.20	1.5446	30
40	291	291	996	291	34.37	000	34.38	417	20
50	320	320	995	320	31.24	001	31.26	388	10
2° 00′	.0349	.0349	.9994	.0349	28.64	1.001	28.65	1.5359	88° 00′
10	378	378	993	378	26.43	001	26.45	330	50
20	407	407	992	407	24.54	001	24.56	301	40
30	.0436	.0436	.9990	.0437	22.90	1.001	22.93	1.5272	30
40	465	465	989	466	21.47	001	21.49	243	20
50	495	494	988	495	20.21	001	20.23	213	10
3° 00′	.0524	.0523	.9986	.0524	19.08	1.001	19.11	1.5184	87° 00′
10	553	552	985	553	18.07	002	18.10	155	50
20	582	581	983	582	17.17	002	17.20	126	40
30	.0611	.0610	.9981	.0612	16.35	1.002	16.38	1.5097	30
40	640	640	980	641	15.60	002	15.64	068	20
50	669	669	978	670	14.92	002	14.96	039	10
4° 00′	.0698	.0698	.9976	.0699	14.30	1.002	14.34	1.5010	86° 00′
10	727	727	974	729	13.73	003	13.76	981	50
20	756	756	971	758	13.20	003	13.23	952	40
30	.0785	.0785	.9969	.0787	12.71	1.003	12.75	1.4923	30
40	814	814	967	816	12.25	003	12.29	893	20
50	844	843	964	846	11.83	004	11.87	864	10
5° 00′	.0873	.0872	.9962	.0875	11.43	1.004	11.47	1.4835	85° 00′
10	902	901	959	904	11.06	004	11.10	806	50
20	931	929	957	934	10.71	004	10.76	777	40
30	.0960	.0958	.9954	.0963	10.39	1.005	10.43	1.4748	30
40	989	987	951	992	10.08	005	10.13	719	20
50	.1018	.1016	948	.1022	9.788	005	9.839	690	10
6° 00′	.1047	.1045	.9945	.1051	9.514	1.006	9.567	1.4661	84° 00′
10	076	074	942	080	9.255	006	9.309	632	50
20	105	103	939	110	9.010	006	9.065	603	40
30	.1134	.1132	.9936	.1139	8.777	1.006	8.834	1.4573	30
40	164	161	932	169	8.556	007	8.614	544	20
50	193	190	929	198	8.345	007	8.405	515	10
7° 00′	.1222	.1219	.9925	.1228	8.144	1.008	8.206	1.4486	83° 00′
10	251	248	922	257	7.953	008	8.016	457	50
20	280	276	918	287	7.770	008	7.834	428	40
30	.1309	.1305	.9914	.1317	7.596	1.009	7.661	1.4399	30
40	338	334	911	346	7.429	009	7.496	370	20
50	367	363	907	376	7.269	009	7.337	341	10
8° 00′	.1396	.1392	.9903	.1405	7.115	1.010	7.185	1.4312	82° 00′
10	425	421	899	435	6.968	010	7.040	283	50
20	454	449	894	465	6.827	011	6.900	254	40
30	.1484	.1478	.9890	.1495	6.691	1.011	6.765	1.4224	30
40	513	507	886	524	6.561	012	6.636	195	20
50	542	536	881	554	6.435	012	6.512	166	10
9° 00′	.1571	.1564	.9877	.1584	6.314	1.012	6.392	1.4137	81°00′
		Cos	Sin	Cot	Tan	Csc	Sec	Radians	Degrees

NATURAL TRIGONOMETRIC FUNCTIONS

Degrees	Radians	Sin	Cos	Tan	Cot	Sec	Csc		
9° 00'	.1571	.1564	.9877	.1584	6.314	1.012	6.392	1.4137	81° 00'
10	600	593	872	614	197	013	277	108	50
20	629	622	868	644	084	013	166	079	40
30	.1658	.1650	.9863	.1673	5.976	1.014	6.059	1.4050	30
40	687	679	858	703	871	014	5.955	1.4021	20
50	716	708	853	733	769	015	855	992	10
10° 00'	.1745	.1736	.9848	.1763	5.671	1.015	5.759	1.3963	80° 00'
10	774	765	843	793	576	016	665	934	50
20	804	794	838	823	485	016	575	904	40
30	.1833	.1822	.9833	.1853	5.396	1.017	5.487	1.3875	30
40	862	851	827	883	309	018	403	846	20
50	891	880	822	914	226	018	320	817	10
11° 00'	.1920	.1908	.9816	.1944	5.145	1.019	5.241	1.3788	79° 00'
10	949	937	811	974	066	019	164	759	50
20	978	965	805	.2004	4.989	020	089	730	40
30	.2007	.1994	.9799	.2035	4.915	1.020	5.016	1.3701	30
40	036	.2022	793	065	843	021	4.945	672	20
50	065	051	787	095	773	022	876	643	10
12° 00'	.2094	.2079	.9781	.2126	4.705	1.022	4.810	1.3614	78° 00'
10	123	108	775	156	638	023	745	584	50
20	153	136	769	186	574	024	682	555	40
30	.2182	.2164	.9763	.2217	4.511	1.024	4.620	1.3526	30
40	211	193	757	247	449	025	560	497	20
50	240	221	750	278	390	026	502	468	10
13° 00'	.2269	.2250	.9744	.2309	4.331	1.026	4.445	1.3439	77° 00'
10	298	278	737	339	275	027	390	410	50
20	327	306	730	370	219	028	336	381	40
30	.2356	.2334	.9724	.2401	4.165	1.028	4.284	1.3352	30
40	385	363	717	432	113	029	232	323	20
50	414	391	710	462	061	030	182	294	10
14° 00'	.2443	.2419	.9703	.2493	4.011	1.031	4.134	1.3265	76° 00'
10	473	447	696	524	3.962	031	086	235	50
20	502	476	689	555	914	032	039	206	40
30	.2531	.2504	.9681	.2586	3.867	1.033	3.994	1.3177	30
40	560	532	674	617	821	034	950	148	20
50	589	560	667	648	776	034	906	119	10
15° 00'	.2618	.2588	.9659	.2679	3.732	1.035	3.864	1.3090	75° 00'
10	647	616	652	711	689	036	822	061	50
20	676	644	644	742	647	037	782	032	40
30	.2705	.2672	.9636	.2773	3.606	1.038	3.742	1.3003	30
40	734	700	628	805	566	039	703	974	20
50	763	728	621	836	526	039	665	945	10
16° 00'	.2793	.2756	.9613	.2867	3.487	1.040	3.628	1.2915	74° 00'
10	822	784	605	899	450	041	592	886	50
20	851	812	596	931	412	042	556	857	40
30	.2880	.2840	.9588	.2962	3.376	1.043	3.521	1.2828	30
40	909	868	580	994	340	044	487	799	20
50	938	896	572	.3026	305	045	453	770	10
17° 00'	.2967	.2924	.9563	.3057	3.271	1.046	3.420	1.2741	73° 00'
10	996	952	555	089	237	047	388	712	50
20	.3025	979	546	121	204	048	356	683	40
30	.3054	.3007	.9537	.3153	3.172	1.049	3.326	1.2654	30
40	083	035	528	185	140	049	295	625	20
50	113	062	520	217	108	050	265	595	10
18° 00'	.3142	.3090	.9511	.3249	3.078	1.051	3.236	1.2566	72° 00'
		Cos	Sin	Cot	Tan	Csc	Sec	Radians	Degrees

NATURAL TRIGONOMETRIC FUNCTIONS

Degrees	Radians	Sin	Cos	Tan	Cot	Sec	Csc		
18° 00'	.3142	.3090	.9511	.3249	3.078	1.051	3.236	1.2566	72° 00'
10	171	118	502	281	047	052	207	537	50
20	200	145	492	314	018	053	179	508	40
30	.3229	.3173	.9483	.3346	2.989	1.054	3.152	1.2479	30
40	258	201	474	378	960	056	124	450	20
50	287	228	465	411	932	057	098	421	10
19° 00'	.3316	.3256	.9455	.3443	2.904	1.058	3.072	1.2392	71° 00'
10	345	283	446	476	877	059	046	363	50
20	374	311	436	508	850	060	021	334	40
30	.3403	.3338	.9426	.3541	2.824	1.061	2.996	1.2305	30
40	432	365	417	574	798	062	971	275	20
50	462	393	407	607	773	063	947	246	10
20° 00'	.3491	.3420	.9397	.3640	2.747	1.064	2.924	1.2217	70° 00'
10	520	448	387	673	723	065	901	188	50
20	549	475	377	706	699	066	878	159	40
30	.3578	.3502	.9367	.3739	2.675	1.068	2.855	1.2130	30
40	607	529	356	772	651	069	833	101	20
50	636	557	346	805	628	070	812	072	10
21° 00'	.3665	.3584	.9336	.3839	2.605	1.071	2.790	1.2043	69° 00'
10	694	611	325	872	583	072	769	1.2014	50
20	723	638	315	906	560	074	749	985	40
30	.3752	.3665	.9304	.3939	2.539	1.075	2.729	1.1956	30
40	782	692	293	973	517	076	709	926	20
50	811	719	283	.4006	496	077	689	897	10
22° 00'	.3840	.3746	.9272	.4040	2.475	1.079	2.669	1.1868	68° 00'
10	869	773	261	074	455	080	650	839	50
20	898	800	250	108	434	081	632	810	40
30	.3927	.3827	.9239	.4142	2.414	1.082	2.613	1.1781	30
40	956	854	228	176	394	084	595	752	20
50	985	881	216	210	375	085	577	723	10
23° 00'	.4014	.3907	.9205	.4245	2.356	1.086	2.559	1.1694	67° 00'
10	043	934	194	279	337	088	542	665	50
20	072	961	182	314	318	089	525	636	40
30	.4102	.3987	.9171	.4348	2.300	1.090	2.508	1.1606	30
40	131	.4014	159	383	282	092	491	577	20
50	160	041	147	417	264	093	475	548	10
24° 00'	.4189	.4067	.9135	.4452	2.246	1.095	2.459	1.1519	66° 00'
10	218	094	124	487	229	096	443	490	50
20	247	120	112	522	211	097	427	461	40
30	.4276	.4147	.9100	.4557	2.194	1.099	2.411	1.1432	30
40	305	173	088	592	177	100	396	403	20
50	334	200	075	628	161	102	381	374	10
25° 00'	.4363	.4226	.9063	.4663	2.145	1.103	2.366	1.1345	65° 00'
10	392	253	051	699	128	105	352	316	50
20	422	279	038	734	112	106	337	286	40
30	.4451	.4305	.9026	.4770	2.097	1.108	2.323	1.1257	30
40	480	331	013	806	081	109	309	228	20
50	509	358	001	841	066	111	295	199	10
26° 00'	.4538	.4384	.8988	.4877	2.050	1.113	2.281	1.1170	64° 00'
10	567	410	975	913	035	114	268	141	50
20	596	436	962	950	020	116	254	112	40
30	.4625	.4462	.8949	.4986	2.006	1.117	2.241	1.1083	30
40	654	488	936	.5022	1.991	119	228	054	20
50	683	514	923	059	977	121	215	1.1025	10
27° 00'	.4712	.4540	.8910	.5095	1.963	1.122	2.203	1.0996	63° 00'
		Cos	Sin	Cot	Tan	Csc	Sec	Radians	Degrees

NATURAL TRIGONOMETRIC FUNCTIONS

Degrees	Radians	Sin	Cos	Tan	Cot	Sec	Csc		
27° 00'	.4712	.4540	.8910	.5095	1.963	1.122	2.203	1.0996	63° 00'
10	741	566	897	132	949	124	190	966	50
20	771	592	884	169	935	126	178	937	40
30	.4800	.4617	.8870	.5206	1.921	1.127	2.166	1.0908	30
40	829	643	857	243	907	129	154	879	20
50	858	669	843	280	894	131	142	850	10
28° 00'	.4887	.4695	.8829	.5317	1.881	1.133	2.130	1.0821	62° 00'
10	916	720	816	354	868	134	118	792	50
20	945	746	802	392	855	136	107	763	40
30	.4974	.4772	.8788	.5430	1.842	1.138	2.096	1.0734	30
40	.5003	797	774	467	829	140	085	705	20
50	032	823	760	505	816	142	074	676	10
29° 00'	.5061	.4848	.8746	.5543	1.804	1.143	2.063	1.0647	61° 00'
10	091	874	732	581	792	145	052	617	50
20	120	899	718	619	780	147	041	588	40
30	.5149	.4924	.8704	.5658	1.767	1.149	2.031	1.0559	30
40	178	950	689	696	756	151	020	530	20
50	207	975	675	735	744	153	010	501	10
30° 00'	.5236	.5000	.8660	.5774	1.732	1.155	2.000	1.0472	60° 00'
10	265	025	646	812	720	157	1.990	443	50
20	294	050	631	851	709	159	980	414	40
30	.5323	.5075	.8616	.5890	1.698	161	1.970	1.0385	30
40	352	100	601	930	686	163	961	356	20
50	381	125	587	969	675	165	951	327	10
31° 00'	.5411	.5150	.8572	.6009	1.664	1.167	1.942	1.0297	59° 00'
10	440	175	557	048	653	169	932	268	50
20	469	200	542	088	643	171	923	239	40
30	.5498	.5225	.8526	.6128	1.632	1.173	1.914	1.0210	30
40	527	250	511	168	621	175	905	181	20
50	556	275	496	208	611	177	896	152	10
32° 00'	.5585	.5299	.8480	.6249	1.600	1.179	1.887	1.0123	58° 00'
10	614	324	465	289	590	181	878	094	50
20	643	348	450	330	580	184	870	065	40
30	.5672	.5373	.8434	.6371	1.570	1.186	1.861	1.0036	30
40	701	398	418	412	560	188	853	1.0007	20
50	730	422	403	453	550	190	844	977	10
33° 00'	.5760	.5446	.8387	.6494	1.540	1.192	1.836	.9948	57° 00'
10	789	471	371	536	530	195	828	919	50
20	818	495	355	577	520	197	820	890	40
30	.5847	.5519	.8339	.6619	1.511	1.199	1.812	.9861	30
40	876	544	323	661	501	202	804	832	20
50	905	568	307	703	1.492	204	796	803	10
34° 00'	.5934	.5592	.8290	.6745	1.483	1.206	1.788	.9774	56° 00'
10	963	616	274	787	473	209	781	745	50
20	992	640	258	830	464	211	773	716	40
30	.6021	.5664	.8241	.6873	1.455	1.213	1.766	.9687	30
40	050	688	225	916	446	216	758	657	20
50	080	712	208	959	437	218	751	628	10
35° 00'	.6109	.5736	.8192	.7002	1.428	1.221	1.743	.9599	55° 00'
10	138	760	175	046	419	223	736	570	50
20	167	783	158	089	411	226	729	541	40
30	.6196	.5807	.8141	.7133	1.402	1.228	1.722	.9512	30
40	225	831	124	177	393	231	715	483	20
50	254	854	107	221	385	233	708	454	10
36° 00'	.6283	.5878	.8090	.7265	1.376	1.236	1.701	.9425	54° 00'
		Cos	Sin	Cot	Tan	Csc	Sec	Radians	Degrees

NATURAL TRIGONOMETRIC FUNCTIONS

Degrees	Radians	Sin	Cos	Tan	Cot	Sec	Csc		
36° 00′	.6283	.5878	.8090	.7265	1.376	1.236	1.701	.9425	54° 00′
10	312	901	073	310	368	239	695	396	50
20	341	925	056	355	360	241	688	367	40
30	.6370	.5948	.8039	.7400	1.351	1.244	1.681	.9338	30
40	400	972	021	445	343	247	675	308	20
50	429	995	004	490	335	249	668	279	10
37° 00′	.6458	.6018	.7986	.7536	1.327	1.252	1.662	.9250	53° 00′
10	487	041	969	581	319	255	655	221	50
20	516	065	951	627	311	258	649	192	40
30	.6545	.6088	.7934	.7673	1.303	1.260	1.643	.9163	30
40	574	111	916	720	295	263	636	134	20
50	603	134	898	766	288	266	630	105	10
38° 00′	.6632	.6157	.7880	.7813	1.280	1.269	1.624	.9076	52° 00′
10	661	180	862	860	272	272	618	047	50
20	690	202	844	907	265	275	612	.9018	40
30	.6720	.6225	.7826	.7954	1.257	1.278	1.606	.8988	30
40	749	248	808	.8002	250	281	601	959	20
50	778	271	790	050	242	284	595	930	10
39° 00′	.6807	.6293	.7771	.8098	1.235	1.287	1.589	.8901	51° 00′
10	836	316	753	146	228	290	583	872	50
20	865	338	735	195	220	293	578	843	40
30	.6894	.6361	.7716	.8243	1.213	1.296	1.572	.8814	30
40	923	383	698	292	206	299	567	785	20
50	952	406	679	342	199	302	561	756	10
40° 00′	.6981	.6428	.7660	.8391	1.192	1.305	1.556	.8727	50° 00′
10	.7010	450	642	441	185	309	550	698	50
20	039	472	623	491	178	312	545	668	40
30	.7069	.6494	.7604	.8541	1.171	1.315	1.540	.8639	30
40	098	517	585	591	164	318	535	610	20
50	127	539	566	642	157	322	529	581	10
41° 00′	.7156	.6561	.7547	.8693	1.150	1.325	1.524	.8552	49° 00′
10	185	583	528	744	144	328	519	523	50
20	214	604	509	796	137	332	514	494	40
30	.7243	.6626	.7490	.8847	1.130	1.335	1.509	.8465	30
40	272	648	470	899	124	339	504	436	20
50	301	670	451	952	117	342	499	407	10
42° 00′	.7330	.6691	.7431	.9004	1.111	1.346	1.494	.8378	48° 00′
10	359	713	412	057	104	349	490	348	50
20	389	734	392	110	098	353	485	319	40
30	.7418	.6756	.7373	.9163	1.091	1.356	1.480	.8290	30
40	447	777	353	217	085	360	476	261	20
50	476	799	333	271	079	364	471	232	10
43° 00′	.7505	.6820	.7314	.9325	1.072	1.367	1.466	.8203	47° 00′
10	534	841	294	380	066	371	462	174	50
20	563	862	274	435	060	375	457	145	40
30	.7592	.6884	.7254	.9490	1.054	1.379	1.453	.8116	30
40	621	905	234	545	048	382	448	087	20
50	650	926	214	601	042	386	444	058	10
44° 00′	.7679	.6947	.7193	.9657	1.036	1.390	1.440	.8029	46° 00′
10	709	967	173	713	030	394	435	999	50
20	738	988	153	770	024	398	431	970	40
30	.7767	.7009	.7133	.9827	1.018	1.402	1.427	.7941	30
40	796	030	112	884	012	406	423	912	20
50	825	050	092	942	006	410	418	883	10
45° 00′	.7854	.7071	.7071	1.000	1.000	1.414	1.414	.7854	45° 00′
		Cos	Sin	Cot	Tan	Csc	Sec	Radians	Degrees

Appendix D

LOGARITHMS OF THE TRIGONOMETRIC FUNCTIONS

Attach — 10 to each entry in this table

Angle θ	log sin θ	log csc θ	log tan θ	log cot θ	log sec θ	log cos θ	
0° 00′	No value	No value	No value	No value	10.0000	10.0000	90° 00′
10′	7.4637	12.5363	7.4637	12.5363	.0000	.0000	50′
20′	.7648	.2352	.7648	.2352	.0000	.0000	40′
30′	7.9408	12.0592	7.9409	12.0591	.0000	.0000	30′
40′	8.0658	11.9342	8.0658	11.9342	.0000	.0000	20′
50′	.1627	.8373	.1627	.8373	.0000	10.0000	10′
1° 00′	8.2419	11.7581	8.2419	11.7581	10.0001	9.9999	89° 00′
10′	.3088	.6912	.3089	.6911	.0001	.9999	50′
20′	.3668	.6332	.3669	.6331	.0001	.9999	40′
30′	.4179	.5821	.4181	.5819	.0001	.9999	30′
40′	.4637	.5363	.4638	.5362	.0002	.9998	20′
50′	.5050	.4950	.5053	.4947	.0002	.9998	10′
2° 00′	8.5428	11.4572	8.5431	11.4569	10.0003	9.9997	88° 00′
10′	.5776	.4224	.5779	.4221	.0003	.9997	50′
20′	.6097	.3903	.6101	.3899	.0004	.9996	40′
30′	.6397	.3603	.6401	.3599	.0004	.9996	30′
40′	.6677	.3323	.6682	.3318	.0005	.9995	20′
50′	.6940	.3060	.6945	.3055	.0005	.9995	10′
3° 00′	8.7188	11.2812	8.7194	11.2806	10.0006	9.9994	87° 00′
10′	.7423	.2577	.7429	.2571	.0007	.9993	50′
20′	.7645	.2355	.7652	.2348	.0007	.9993	40′
30′	.7857	.2143	.7865	.2135	.0008	.9992	30′
40′	.8059	.1941	.8067	.1933	.0009	.9991	20′
50′	.8251	.1749	.8261	.1739	.0010	.9990	10′
4° 00′	8.8436	11.1564	8.8446	11.1554	10.0011	9.9989	86° 00′
10′	.8613	.1387	.8624	.1376	.0011	.9989	50′
20′	.8783	.1217	.8795	.1205	.0012	.9988	40′
30′	.8946	.1054	.8960	.1040	.0013	.9987	30′
40′	.9104	.0896	.9118	.0882	.0014	.9986	20′
50′	.9256	.0744	.9272	.0728	.0015	.9985	10′
5° 00′	8.9403	11.0597	8.9420	11.0580	10.0017	9.9983	85° 00′
10′	.9545	.0455	.9563	.0437	.0018	.9982	50′
20′	.9682	.0318	.9701	.0299	.0019	.9981	40′
30′	.9816	.0184	.9836	.0164	.0020	.9980	30′
40′	8.9945	11.0055	8.9966	11.0034	.0021	.9979	20′
50′	9.0070	10.9930	9.0093	10.9907	.0023	.9977	10′
6° 00′	9.0192	10.9808	9.0216	10.9784	10.0024	9.9976	84° 00′
	log cos θ	log sec θ	log cot θ	log tan θ	log csc θ	log sin θ	Angle θ

LOGARITHMS OF THE TRIGONOMETRIC FUNCTIONS

Angle θ	log sin θ	log csc θ	log tan θ	log cot θ	log sec θ	log cos θ	
6° 00'	9.0192	10.9808	9.0216	10.9784	10.0024	9.9976	84° 00'
10'	.0311	.9689	.0336	.9664	.0025	.9975	50'
20'	.0426	.9574	.0453	.9547	.0027	.9973	40'
30'	.0539	.9461	.0567	.9433	.0028	.9972	30'
40'	.0648	.9352	.0678	.9322	.0029	.9971	20'
50'	.0755	.9245	.0786	.9214	.0031	.9969	10'
7° 00'	9.0859	10.9141	9.0891	10.9109	10.0032	9.9968	83° 00'
10'	.0961	.9039	.0995	.9005	.0034	.9966	50'
20'	.1060	.8940	.1096	.8904	.0036	.9964	40'
30'	.1157	.8843	.1194	.8806	.0037	.9963	30'
40'	.1252	.8748	.1291	.8709	.0039	.9961	20'
50'	.1345	.8655	.1385	.8615	.0041	.9959	10'
8° 00'	9.1436	10.8564	9.1478	10.8522	10.0042	9.9958	82° 00'
10'	.1525	.8475	.1569	.8431	.0044	.9956	50'
20'	.1612	.8388	.1658	.8342	.0046	.9954	40'
30'	.1697	.8303	.1745	.8255	.0048	.9952	30'
40'	.1781	.8219	.1831	.8169	.0050	.9950	20'
50'	.1863	.8137	.1915	.8085	.0052	.9948	10'
9° 00'	9.1943	10.8057	9.1997	10.8003	10.0054	9.9946	81° 00'
10'	.2022	.7978	.2078	.7922	.0056	.9944	50'
20'	.2100	.7900	.2158	.7842	.0058	.9942	40'
30'	.2176	.7824	.2236	.7764	.0060	.9940	30'
40'	.2251	.7749	.2313	.7687	.0062	.9938	20'
50'	.2324	.7676	.2389	.7611	.0064	.9936	10'
10° 00'	9.2397	10.7603	9.2463	10.7537	10.0066	9.9934	80° 00'
10'	.2468	.7532	.2536	.7464	.0069	.9931	50'
20'	.2538	.7462	.2609	.7391	.0071	.9929	40'
30'	.2606	.7394	.2680	.7320	.0073	.9927	30'
40'	.2674	.7326	.2750	.7250	.0076	.9924	20'
50'	.2740	.7260	.2819	.7181	.0078	.9922	10'
11° 00'	9.2806	10.7194	9.2887	10.7113	10.0081	9.9919	79° 00'
10'	.2870	.7130	.2953	.7047	.0083	.9917	50'
20'	.2934	.7066	.3020	.6980	.0086	.9914	40'
30'	.2997	.7003	.3085	.6915	.0088	.9912	30'
40'	.3058	.6942	.3149	.6851	.0091	.9909	20'
50'	.3119	.6881	.3212	.6788	.0093	.9907	10'
12° 00'	9.3179	10.6821	9.3275	10.6725	10.0096	9.9904	78° 00'
10'	.3238	.6762	.3336	.6664	.0099	.9901	50'
20'	.3296	.6704	.3397	.6603	.0101	.9899	40'
30'	.3353	.6647	.3458	.6542	.0104	.9896	30'
40'	.3410	.6590	.3517	.6483	.0107	.9893	20'
50'	.3466	.6534	.3576	.6424	.0110	.9890	10'
13° 00'	9.3521	10.6479	9.3634	10.6366	10.0113	9.9887	77° 00'
	log cos θ	log sec θ	log cot θ	log tan θ	log csc θ	log sin θ	Angle θ

Appendix D (Cont'd)

LOGARITHMS OF THE TRIGONOMETRIC FUNCTIONS

Angle θ	log sin θ	log csc θ	log tan θ	log cot θ	log sec θ	log cos θ	
13° 00'	9.3521	10.6479	9.3634	10.6366	10.0113	9.9887	77° 00'
10'	.3575	.6425	.3691	.6309	.0116	.9884	50'
20'	.3629	.6371	.3748	.6252	.0119	.9881	40'
30'	.3682	.6318	.3804	.6196	.0122	.9878	30'
40'	.3734	.6266	.3859	.6141	.0125	.9875	20'
50'	.3786	.6214	.3914	.6086	.0128	.9872	10'
14° 00'	9.3837	10.6163	9.3968	10.6032	10.0131	9.9869	76° 00'
10'	.3887	.6113	.4021	.5979	.0134	.9866	50'
20'	.3937	.6063	.4074	.5926	.0137	.9863	40'
30'	.3986	.6014	.4127	.5873	.0141	.9859	30'
40'	.4035	.5965	.4178	.5822	.0144	.9856	20'
50'	.4083	.5917	.4230	.5770	.0147	.9853	10'
15° 00'	9.4130	10.5870	9.4281	10.5719	10.0151	9.9849	75° 00'
10'	.4177	.5823	.4331	.5669	.0154	.9846	50'
20'	.4223	.5777	.4381	.5619	.0157	.9843	40'
30'	.4269	.5731	.4430	.5570	.0161	.9839	30'
40'	.4314	.5686	.4479	.5521	.0164	.9836	20'
50'	.4359	.5641	.4527	.5473	.0168	.9832	10'
16° 00'	9.4403	10.5597	9.4575	10.5425	10.0172	9.9828	74° 00'
10'	.4447	.5553	.4622	.5378	.0175	.9825	50'
20'	.4491	.5509	.4669	.5331	.0179	.9821	40'
30'	.4533	.5467	.4716	.5284	.0183	.9817	30'
40'	.4576	.5424	.4762	.5238	.0186	.9814	20'
50'	.4618	.5382	.4808	.5192	.0190	.9810	10'
17° 00'	9.4659	10.5341	9.4853	10.5147	10.0194	9.9806	73° 00'
10'	.4700	.5300	.4898	.5102	.0198	.9802	50'
20'	.4741	.5259	.4943	.5057	.0202	.9798	40'
30'	.4781	.5219	.4987	.5013	.0206	.9794	30'
40'	.4821	.5179	.5031	.4969	.0210	.9790	20'
50'	.4861	.5139	.5075	.4925	.0214	.9786	10'
18° 00'	9.4900	10.5100	9.5118	10.4882	10.0218	9.9782	72° 00'
10'	.4939	.5061	.5161	.4839	.0222	.9778	50'
20'	.4977	.5023	.5203	.4797	.0226	.9774	40'
30'	.5015	.4985	.5245	.4755	.0230	.9770	30'
40'	.5052	.4948	.5287	.4713	.0235	.9765	20'
50'	.5090	.4910	.5329	.4671	.0239	.9761	10'
19° 00'	9.5126	10.4874	9.5370	10.4630	10.0243	9.9757	71° 00'
10'	.5163	.4837	.5411	.4589	.0248	.9752	50'
20'	.5199	.4801	.5451	.4549	.0252	.9748	40'
30'	.5235	.4765	.5491	.4509	.0257	.9743	30'
40'	.5270	.4730	.5531	.4469	.0261	.9739	20'
50'	.5306	.4694	.5571	.4429	.0266	.9734	10'
20° 00'	9.5341	10.4659	9.5611	10.4389	10.0270	9.9730	70° 00'
	log cos θ	log sec θ	log cot θ	log tan θ	log csc θ	log sin θ	Angle θ

LOGARITHMS OF THE TRIGONOMETRIC FUNCTIONS

Angle θ	log sin θ	log csc θ	log tan θ	log cot θ	log sec θ	log cos θ	
20° 00'	9.5341	10.4659	9.5611	10.4389	10.0270	9.9730	70° 00'
10'	.5375	.4625	.5650	.4350	.0275	.9725	50'
20'	.5409	.4591	.5689	.4311	.0279	.9721	40'
30'	.5443	.4557	.5727	.4273	.0284	.9716	30'
40'	.5477	.4523	.5766	.4234	.0289	.9711	20'
50'	.5510	.4490	.5804	.4196	.0294	.9706	10'
21° 00'	9.5543	10.4457	9.5842	10.4158	10.298	9.9702	69° 00'
10'	.5576	.4424	.5879	.4121	.0303	.9697	50'
20'	.5609	.4391	.5917	.4083	.0308	.9692	40'
30'	.5641	.4359	.5954	.4046	.0313	.9687	30'
40'	.5673	.4327	.5991	.4009	.0318	.9682	20'
50'	.5704	.4296	.6028	.3972	.0323	.9677	10'
22° 00'	9.5736	10.4264	9.6064	10.3936	10.0328	9.9672	68° 00'
10'	.5767	.4233	.6100	.3900	.0333	.9667	50'
20'	.5798	.4202	.6136	.3864	.0339	.9661	40'
30'	.5828	.4172	.6172	.3828	.0344	.9656	30'
40'	.5859	.4141	.6208	.3792	.0349	.9651	20'
50'	.5889	.4111	.6243	.3757	.0354	.9646	10'
23° 00'	9.5919	10.4081	9.6279	10.3721	10.0360	9.9640	67° 00'
10'	.5948	.4052	.6314	.3686	.0365	.9635	50'
20'	.5978	.4022	.6348	.3652	.0371	.9629	40'
30'	.6007	.3993	.6383	.3617	.0376	.9624	30'
40'	.6036	.3964	.6417	.3583	.0382	.9618	20'
50'	.6065	.3935	.6452	.3548	.0387	.9613	10'
24° 00'	9.6093	10.3907	9.6486	10.3514	10.0393	9.9607	66° 00'
10'	.6121	.3879	.6520	.3480	.0398	.9602	50'
20'	.6149	.3851	.6553	.3447	.0404	.9596	40'
30'	.6177	.3823	.6587	.3413	.0410	.9590	30'
40'	.6205	.3795	.6620	.3380	.0416	.9584	20'
50'	.6232	.3768	.6654	.3346	.0421	.9579	10'
25° 00'	9.6259	10.3741	9.6687	10.3313	10.0427	9.9573	65° 00'
10'	.6286	.3714	.6720	.3280	.0433	.9567	50'
20'	.6313	.3687	.6752	.3248	.0439	.9561	40'
30'	.6340	.3660	.6785	.3215	.0445	.9555	30'
40'	.6366	.3634	.6817	.3483	.0451	.9549	20'
50'	.6392	.3608	.6850	.3150	.0457	.9543	10'
26° 00'	9.6418	10.3582	9.6882	10.3118	10.0463	9.9537	64° 00'
10'	.6444	.3556	.6914	.3086	.0470	.9530	50'
20'	.6470	.3530	.6946	.3054	.0476	.9524	40'
30'	.6495	.3505	.6977	.3023	.0482	.9518	30'
40'	.6521	.3479	.7009	.2991	.0488	.9512	20'
50'	.6546	.3454	.7040	.2960	.0495	.9505	10'
27° 00'	9.6570	10.3430	9.7072	10.2928	10.0501	9.9499	63° 00'
	log cos θ	log sec θ	log cot θ	log tan θ	log csc θ	log sin θ	Angle θ

Appendix D (Cont'd)

LOGARITHMS OF THE TRIGONOMETRIC FUNCTIONS

Angle θ	log sin θ	log csc θ	log tan θ	log cot θ	log sec θ	log cos θ	
27° 00'	9.6570	10.3430	9.7072	10.2928	10.0501	9.9499	63° 00'
10'	.6595	.3405	.7103	.2897	.0508	.9492	50'
20'	.6620	.3380	.7134	.2866	.0514	.9486	40'
30'	.6644	.3356	.7165	.2835	.0521	.9479	30'
40'	.6668	.3332	.7196	.2804	0.527	.9473	20'
50'	.6692	.3308	.7226	.2774	0.534	.9466	10'
28° 00'	9.6716	10.3284	9.7257	10.2743	10.0541	9.9459	62° 00'
10'	.6740	.3260	.7287	.2713	.0547	.9453	50'
20'	.6763	.3237	.7317	.2683	.0554	.9446	40'
30'	.6787	.3213	.7348	.2652	.0561	.9439	30'
40'	.6810	.3190	.7378	.2622	.0568	.9432	20'
50'	.6833	.3167	.7408	.2592	.0575	.9425	10'
29° 00'	9.6856	10.3144	9.7438	10.2562	10.0582	9.9418	61° 00'
10'	.6878	.3122	.7467	.2533	.0589	.9411	50'
20'	.6901	.3099	.7497	.2503	.0596	.9404	40'
30'	.6923	.3077	.7526	.2474	.0603	.9397	30'
40'	.6946	.3054	.7556	.2444	.0610	.9390	20'
50'	.6968	.3032	.7585	.2415	.0617	.9383	10'
30° 00'	9.6990	10.3010	9.7614	10.2386	10.0625	9.9375	60° 00'
10'	.7012	.2988	.7644	.2356	.0632	.9368	50'
20'	.7033	.2967	.7673	.2327	.0639	.9361	40'
30'	.7055	.2945	.7701	.2299	.0647	.9353	30'
40'	.7076	.2924	.7730	.2270	.0654	.9346	20'
50'	.7097	.2903	.7759	.2241	.0662	.9338	10'
31° 00'	9.7118	10.2882	9.7788	10.2212	10.0669	9.9331	59° 00'
10'	.7139	.2861	.7816	.2184	.0677	.9323	50'
20'	.7160	.2840	.7845	.2155	.0685	.9315	40'
30'	.7181	.2819	.7873	.2127	.0692	.9308	30'
40'	.7201	.2799	.7902	.2098	.0700	.9300	20'
50'	.7222	.2778	.7930	.2070	.0708	.9292	10'
32° 00'	9.7242	10.2758	9.7958	10.2042	10.0716	9.9284	58° 00'
10'	.7262	.2738	.7986	.2014	.0724	.9276	50'
20'	.7282	.2718	.8014	.1986	.0732	.9268	40'
30'	.7302	.2698	.8042	.1958	.0740	.9260	30'
40'	.7322	.2678	.8070	.1930	.0748	.9252	20'
50'	.7342	.2658	.8097	.1903	.0756	.9244	10'
33° 00'	9.7361	10.2639	9.8125	10.1875	10.0764	9.9236	57° 00'
10'	.7380	.2620	.8153	.1847	.0772	.9228	50'
20'	.7400	.2600	.8180	.1820	.0781	.9219	40'
30'	.7419	.2581	.8208	.1792	.0789	.9211	30'
40'	.7438	.2562	.8235	.1765	.0797	.9203	20'
50'	.7457	.2543	.8263	.1737	.0806	.9194	10'
34° 00'	9.7476	10.2524	9.8290	10.1710	10.0814	9.9186	56° 00'
	log cos θ	log sec θ	log cot θ	log tan θ	log csc θ	log sin θ	Angle θ

LOGARITHMS OF THE TRIGONOMETRIC FUNCTIONS

Angle θ	log sin θ	log csc θ	log tan θ	log cot θ	log sec θ	log cos θ	
34° 00′	9.7476	10.2524	9.8290	10.1710	10.0814	9.9186	56° 00′
10′	.7494	.2506	.8317	.1683	.0823	.9177	50′
20′	.7513	.2487	.8344	.1656	.0831	.9169	40′
30′	.7531	.2469	.8371	.1629	.0840	.9160	30′
40′	.7550	.2450	.8398	.1602	.0849	.9151	20′
50′	.7568	.2432	.8425	.1575	.0858	.9142	10′
35° 00′	9.7586	10.2414	9.8452	10.1548	10.0866	9.9134	55° 00′
10′	.7604	.2396	.8479	.1521	.0875	.9125	50′
20′	.7622	.2378	.8506	.1494	.0884	.9116	40′
30′	.7640	.2360	.8533	.1467	.0893	.9107	30′
40′	.7657	.2343	.8559	.1441	.0902	.9098	20′
50′	.7675	.2325	.8586	.1414	.0911	.9089	10′
36° 00′	9.7692	10.2308	9.8613	10.1387	10.0920	9.9080	54° 00′
10′	.7710	.2290	.8639	.1361	.0930	.9070	50′
20′	.7727	.2273	.8666	.1334	.0939	.9061	40′
30′	.7744	.2256	.8692	.1308	.0948	.9052	30′
40′	.7761	.2239	.8718	.1282	.0958	.9042	20′
50′	.7778	.2222	.8745	.1255	.0967	.9033	10′
37° 00′	9.7795	10.2205	9.8771	10.1229	10.0977	9.9023	53° 00′
10′	.7811	.2189	.8797	.1203	.0986	.9014	50′
20′	.7828	.2172	.8824	.1176	.0996	.9004	40′
30′	.7844	.2156	.8850	.1150	.1005	.8995	30′
40′	.7861	.2139	.8876	.1124	.1015	.8985	20′
50′	.7877	.2123	.8902	.1098	.1025	.8975	10′
38° 00′	9.7893	10.2107	9.8928	10.1072	10.1035	9.8965	52° 00′
10′	.7910	.2090	.8954	.1046	.1045	.8955	50′
20′	.7926	.2074	.8980	.1020	.1055	.8945	40′
30′	.7941	.2059	.9006	.0994	.1065	.8935	30′
40′	.7957	.2043	.9032	.0968	.1075	.8925	20′
50′	.7973	.2027	.9058	.0942	.1085	.8915	10′
39° 00′	9.7989	10.2011	9.9084	10.0916	10.1095	9.8905	51° 00′
10′	.8004	.1996	.9110	.0890	.1105	.8895	50′
20′	.8020	.1980	.9135	.0865	.1116	.8884	40′
30′	.8035	.1965	.9161	.0839	.1126	.8874	30′
40′	.8050	.1950	.9187	.0813	.1136	.8864	20′
50′	.8066	.1934	.9212	.0788	.1147	.8853	10′
40° 00′	9.8081	10.1919	9.9238	10.0762	10.1157	9.8843	50° 00′
10′	.8096	.1904	.9264	.0736	.1168	.8832	50′
20′	.8111	.1889	.9289	.0711	.1179	.8821	40′
30′	.8125	.1875	.9315	.0685	.1190	.8810	30′
40′	.8140	.1860	.9341	.0659	.1200	.8800	20′
50′	.8155	.1845	.9366	.0634	.1211	.8789	10′
41° 00′	9.8169	10.1831	9.9392	10.0608	10.1222	9.8778	49° 00′
	log cos θ	log sec θ	log cot θ	log tan θ	log csc θ	log sin θ	Angle θ

Appendix D (Cont'd)

LOGARITHMS OF THE TRIGONOMETRIC FUNCTIONS

Angle θ	log sin θ	log csc θ	log tan θ	log cot θ	log sec θ	log cos θ	
41° 00′	9.8169	10.1831	9.9392	10.0608	10.1222	9.8778	49° 00′
10′	.8184	.1816	.9417	.0583	.1233	.8767	50′
20′	.8198	.1802	.9443	.0557	.1244	.8756	40′
30′	.8213	.1787	.9468	.0532	.1255	.8745	30′
40′	.8227	.1773	.9494	.0506	.1267	.8733	20′
50′	.8241	.1759	.9519	.0481	.1278	.8722	10′
42° 00′	9.8255	10.1745	9.9544	10.0456	10.1289	9.8711	48° 00′
10′	.8269	.1731	.9570	.0430	.1301	.8699	50′
20′	.8283	.1717	.9595	.0405	.1312	.8688	40′
30′	.8297	.1703	.9621	.0379	.1324	.8676	30′
40′	.8311	.1689	.9646	.0354	.1335	.8665	20′
50′	.8324	.1676	.9671	.0329	.1347	.8655	10′
43° 00′	9.8338	10.1662	9.9697	10.0303	10.1359	9.8641	47° 00′
10′	.8351	.1649	.9722	.0278	.1371	.8629	50′
20′	.8365	.1635	.9747	.0253	.1382	.8618	40′
30	.8378	.1622	.9772	.0228	.1394	.8606	30
40	.8391	.1609	.9798	.0202	.1406	.8594	20
50′	.8405	.1595	.9823	.0177	.1418	.8582	10′
44° 00′	9.8418	10.1582	9.9848	10.0152	10.1431	9.8569	46° 00′
10′	.8431	.1569	.9874	.0126	.1443	.8557	50′
20′	.8444	.1556	.9899	.0101	.1455	.8545	40′
30′	.8457	.1543	.9924	.0076	.1468	.8532	30′
40′	.8469	.1531	.9949	.0051	.1480	.8520	20′
50′	.8482	.1518	9.9975	.0025	.1493	.8507	10′
45° 00′	9.8495	10.1505	10.0000	10.0000	10.1505	9.8495	45° 00′
	log cos θ	log sec θ	log cot θ	log tan θ	log csc θ	log sin θ	Angle θ

$x = b^w \quad y = b^z$
$xy = b^{w+z}$
$\log_b xy = w+z = \log_b xy = \log_b x + \log_b y$

COMMON LOGARITHMS

100 — 150

N.		0	1	2	3	4	5	6	7	8	9
100	00	000	043	087	130	173	217	260	303	346	389
101		432	475	518	561	604	647	689	732	775	817
102		860	903	945	988	*030	*072	*115	*157	*199	*242
103	01	284	326	368	410	452	494	536	578	620	662
104		703	745	787	828	870	912	953	995	*036	*078
105	02	119	160	202	243	284	325	366	407	449	490
106		531	572	612	653	694	735	776	816	857	898
107		938	979	*019	*060	*100	*141	*181	*222	*262	*302
108	03	342	383	423	463	503	543	583	623	663	703
109		743	782	822	862	902	941	981	*021	*060	*100
110	04	139	179	218	258	297	336	376	415	454	493
111		532	571	610	650	689	727	766	805	844	883
112		922	961	999	*038	*077	*115	*154	*192	*231	*269
113	05	308	346	385	423	461	500	538	576	614	652
114		690	729	767	805	843	881	918	956	994	*032
115	06	070	108	145	183	221	258	296	333	371	408
116		446	483	521	558	595	633	670	707	744	781
117		819	856	893	930	967	*004	*041	*078	*115	*151
118	07	188	225	262	298	335	372	408	445	482	518
119		555	591	628	664	700	737	773	809	846	882
120		918	954	990	*027	*063	*099	*135	*171	*207	*243
121	08	279	314	350	386	422	458	493	529	565	600
122		636	672	707	743	778	814	849	884	920	955
123		991	*026	*061	*096	*132	*167	*202	*237	*272	*307
124	09	342	377	412	447	482	517	552	587	621	656
125		691	726	760	795	830	864	899	934	968	*003
126	10	037	072	106	140	175	209	243	278	312	346
127		380	415	449	483	517	551	585	619	653	687
128		721	755	789	823	857	890	924	958	992	*025
129	11	059	093	126	160	193	227	261	294	327	361
130		394	428	461	494	528	561	594	628	661	694
131		727	760	793	826	860	893	926	959	992	*024
132	12	057	090	123	156	189	222	254	287	320	352
133		385	418	450	483	516	548	581	613	646	678
134		710	743	775	808	840	872	905	937	969	*001
135	13	033	066	098	130	162	194	226	258	290	322
136		354	386	418	450	481	513	545	577	609	640
137		672	704	735	767	799	830	862	893	925	956
138		988	*019	*051	*082	*114	*145	*176	*208	*239	*270
139	14	301	333	364	395	426	457	489	520	551	582
140		613	644	675	706	737	768	799	829	860	891
141		922	953	983	*014	*045	*076	*106	*137	*168	*198
142	15	229	259	290	320	351	381	412	442	473	503
143		534	564	594	625	655	685	715	746	776	806
144		836	866	897	927	957	987	*017	*047	*077	*107
145	16	137	167	197	227	256	286	316	346	376	406
146		435	465	495	524	554	584	613	643	673	702
147		732	761	791	820	850	879	909	938	967	997
148	17	026	056	085	114	143	173	202	231	260	289
149		319	348	377	406	435	464	493	522	551	580
150		609	638	667	696	725	754	782	811	840	869
N.		0	1	2	3	4	5	6	7	8	9

Proportional parts

	44	43	42
1	4.4	4.3	4.2
2	8.8	8.6	8.4
3	13.2	12.9	12.6
4	17.6	17.2	16.8
5	22.0	21.5	21.0
6	26.4	25.8	25.2
7	30.8	30.1	29.4
8	35.2	34.4	33.6
9	39.6	38.7	37.8

	41	40	39
1	4.1	4.0	3.9
2	8.2	8.0	7.8
3	12.3	12.0	11.7
4	16.4	16.0	15.6
5	20.5	20.0	19.5
6	24.6	24.0	23.4
7	28.7	28.0	27.3
8	32.8	32.0	31.2
9	36.9	36.0	35.1

	38	37	36
1	3.8	3.7	3.6
2	7.6	7.4	7.2
3	11.4	11.1	10.8
4	15.2	14.8	14.4
5	19.0	18.5	18.0
6	22.8	22.2	21.6
7	26.6	25.9	25.2
8	30.4	29.6	28.8
9	34.2	33.3	32.4

	35	34	33
1	3.5	3.4	3.3
2	7.0	6.8	6.6
3	10.5	10.2	9.9
4	14.0	13.6	13.2
5	17.5	17.0	16.5
6	21.0	20.4	19.8
7	24.5	23.8	23.1
8	28.0	27.2	26.4
9	31.5	30.6	29.7

	32	31	30
1	3.2	3.1	3.0
2	6.4	6.2	6.0
3	9.6	9.3	9.0
4	12.8	12.4	12.0
5	16.0	15.5	15.0
6	19.2	18.6	18.0
7	22.4	21.7	21.0
8	25.6	24.8	24.0
9	28.8	27.9	27.0

Proportional parts

.00 000 — .17 869

$\frac{x}{y} = \frac{b^w}{b^z} \quad b^{w-z}$
$\frac{x}{y} = b^{w-z}$
$\log_b\left(\frac{x}{y}\right) = w-z \quad \log_b\left(\frac{x}{y}\right) = \log_b x - \log_b y$

COMMON LOGARITHMS

150 — 200

Handwritten margin notes:
$X = b^w$
$X^N = (b^w)^N$
$X^N = b^{wN}$
$\log x^n = wN$
$\log_b x^n = \log_b x (N)$

N.	0	1	2	3	4	5	6	7	8	9
150	17 609	638	667	696	725	754	782	811	840	869
151	898	926	955	984	*013	*041	*070	*099	*127	*156
152	18 184	213	241	270	298	327	355	384	412	441
153	469	498	526	554	583	611	639	667	696	724
154	752	780	808	837	865	893	921	949	977	*005
155	19 033	061	089	117	145	173	201	229	257	285
156	312	340	368	396	424	451	479	507	535	562
157	590	618	645	673	700	728	756	783	811	838
158	866	893	921	948	976	*003	*030	*058	*085	*112
159	20 140	167	194	222	249	276	303	330	358	385
160	412	439	466	493	520	548	575	602	629	656
161	683	710	737	763	790	817	844	871	898	925
162	952	978	*005	*032	*059	*085	*112	*139	*165	*192
163	21 219	245	272	299	325	352	378	405	431	458
164	484	511	537	564	590	617	643	669	696	722
165	748	775	801	827	854	880	906	932	958	985
166	22 011	037	063	089	115	141	167	194	220	246
167	272	298	324	350	376	401	427	453	479	505
168	531	557	583	608	634	660	686	712	737	763
169	789	814	840	866	891	917	943	968	994	*019
170	23 045	070	096	121	147	172	198	223	249	274
171	300	325	350	376	401	426	452	477	502	528
172	553	578	603	629	654	679	704	729	754	779
173	805	830	855	880	905	930	955	980	*005	*030
174	24 055	080	105	130	155	180	204	229	254	279
175	304	329	353	378	403	428	452	477	502	527
176	551	576	601	625	650	674	699	724	748	773
177	797	822	846	871	895	920	944	969	993	*018
178	25 042	066	091	115	139	164	188	212	237	261
179	285	310	334	358	382	406	431	455	479	503
180	527	551	575	600	624	648	672	696	720	744
181	768	792	816	840	864	888	912	935	959	983
182	26 007	031	055	079	102	126	150	174	198	221
183	245	269	293	316	340	364	387	411	435	458
184	482	505	529	553	576	600	623	647	670	694
185	717	741	764	788	811	834	858	881	905	928
186	951	975	998	*021	*045	*068	*091	*114	*138	*161
187	27 184	207	231	254	277	300	323	346	370	393
188	416	439	462	485	508	531	554	577	600	623
189	646	669	692	715	738	761	784	807	830	852
190	875	898	921	944	967	989	*012	*035	*058	*081
191	28 103	126	149	171	194	217	240	262	285	307
192	330	353	375	398	421	443	466	488	511	533
193	556	578	601	623	646	668	691	713	735	758
194	780	803	825	847	870	892	914	937	959	981
195	29 003	026	048	070	092	115	137	159	181	203
196	226	248	270	292	314	336	358	380	403	425
197	447	469	491	513	535	557	579	601	623	645
198	667	688	710	732	754	776	798	820	842	863
199	885	907	929	951	973	994	*016	*038	*060	*081
200	30 103	125	146	168	190	211	233	255	276	298
N.	0	1	2	3	4	5	6	7	8	9

Proportional parts

	29	28
1	2.9	2.8
2	5.8	5.6
3	8.7	8.4
4	11.6	11.2
5	14.5	14.0
6	17.4	16.8
7	20.3	19.6
8	23.2	22.4
9	26.1	25.2

	27	26
1	2.7	2.6
2	5.4	5.2
3	8.1	7.8
4	10.8	10.4
5	13.5	13.0
6	16.2	15.6
7	18.9	18.2
8	21.6	20.8
9	24.3	23.4

	25
1	2.5
2	5.0
3	7.5
4	10.0
5	12.5
6	15.0
7	17.5
8	20.0
9	22.5

	24	23
1	2.4	2.3
2	4.8	4.6
3	7.2	6.9
4	9.6	9.2
5	12.0	11.5
6	14.4	13.8
7	16.8	16.1
8	19.2	18.4
9	21.6	20.7

	22	21
1	2.2	2.1
2	4.4	4.2
3	6.6	6.3
4	8.8	8.4
5	11.0	10.5
6	13.2	12.6
7	15.4	14.7
8	17.6	16.8
9	19.8	18.9

.17 609 — .30 298

COMMON LOGARITHMS

200 — 250

N.		0	1	2	3	4	5	6	7	8	9
200	30	103	125	146	168	190	211	233	255	276	298
201		320	341	363	384	406	428	449	471	492	514
202		535	557	578	600	621	643	664	685	707	728
203		750	771	792	814	835	856	878	899	920	942
204		963	984	*006	*027	*048	*069	*091	*112	*133	*154
205	31	175	197	218	239	260	281	302	323	345	366
206		387	408	429	450	471	492	513	534	555	576
207		597	618	639	660	681	702	723	744	765	785
208		806	827	848	869	890	911	931	952	973	994
209	32	015	035	056	077	098	118	139	160	181	201
210		222	243	263	284	305	325	346	366	387	408
211		428	449	469	490	510	531	552	572	593	613
212		634	654	675	695	715	736	756	777	797	818
213		838	858	879	899	919	940	960	980	*001	*021
214	33	041	062	082	102	122	143	163	183	203	224
215		244	264	284	304	325	345	365	385	405	425
216		445	465	486	506	526	546	566	586	606	626
217		646	666	686	706	726	746	766	786	806	826
218		846	866	885	905	925	945	965	985	*005	*025
219	34	044	064	084	104	124	143	163	183	203	223
220		242	262	282	301	321	341	361	380	400	420
221		439	459	479	498	518	537	557	577	596	616
222		635	655	674	694	713	733	753	772	792	811
223		830	850	869	889	908	928	947	967	986	*005
224	35	025	044	064	083	102	122	141	160	180	199
225		218	238	257	276	295	315	334	353	372	392
226		411	430	449	468	488	507	526	545	564	583
227		603	622	641	660	679	698	717	736	755	774
228		793	813	832	851	870	889	908	927	946	965
229		984	*003	*021	*040	*059	*078	*097	*116	*135	*154
230	36	173	192	211	229	248	267	286	305	324	342
231		361	380	399	418	436	455	474	493	511	530
232		549	568	586	605	624	642	661	680	698	717
233		736	754	773	791	810	829	847	866	884	903
234		922	940	959	977	996	*014	*033	*051	*070	*088
235	37	107	125	144	162	181	199	218	236	254	273
236		291	310	328	346	365	383	401	420	438	457
237		475	493	511	530	548	566	585	603	621	639
238		658	676	694	712	731	749	767	785	803	822
239		840	858	876	894	912	931	949	967	985	*003
240	38	021	039	057	075	093	112	130	148	166	184
241		202	220	238	256	274	292	310	328	346	364
242		382	399	417	435	453	471	489	507	525	543
243		561	578	596	614	632	650	668	686	703	721
244		739	757	775	792	810	828	846	863	881	899
245		917	934	952	970	987	*005	*023	*041	*058	*076
246	39	094	111	129	146	164	182	199	217	235	252
247		270	287	305	322	340	358	375	393	410	428
248		445	463	480	498	515	533	550	568	585	602
249		620	637	655	672	690	707	724	742	759	777
250		794	811	829	846	863	881	898	915	933	950
N.		0	1	2	3	4	5	6	7	8	9

Proportional parts

	22	21
1	2.2	2.1
2	4.4	4.2
3	6.6	6.3
4	8.8	8.4
5	11.0	10.5
6	13.2	12.6
7	15.4	14.7
8	17.6	16.8
9	19.8	18.9

	20
1	2.0
2	4.0
3	6.0
4	8.0
5	10.0
6	12.0
7	14.0
8	16.0
9	18.0

	19
1	1.9
2	3.8
3	5.7
4	7.6
5	9.5
6	11.4
7	13.3
8	15.2
9	17.1

	18
1	1.8
2	3.6
3	5.4
4	7.2
5	9.0
6	10.8
7	12.6
8	14.4
9	16.2

	17
1	1.7
2	3.4
3	5.1
4	6.8
5	8.5
6	10.2
7	11.9
8	13.6
9	15.3

Proportional parts

.30 103 — .39 950

COMMON LOGARITHMS

250 — 300

N.	0	1	2	3	4	5	6	7	8	9	Proportional parts
250	39 794	811	829	846	863	881	898	915	933	950	18
251	967	985	*002	*019	*037	*054	*071	*088	*106	*123	
252	40 140	157	175	192	209	226	243	261	278	295	1 1.8
253	312	329	346	364	381	398	415	432	449	466	2 3.6
254	483	500	518	535	552	569	586	603	620	637	3 5.4
											4 7.2
255	654	671	688	705	722	739	756	773	790	807	5 9.0
256	824	841	858	875	892	909	926	943	960	976	6 10.8
257	993	*010	*027	*044	*061	*078	*095	*111	*128	*145	7 12.6
258	41 162	179	196	212	229	246	263	280	296	313	8 14.4
259	330	347	363	380	397	414	430	447	464	481	9 16.2
260	497	514	531	547	564	581	597	614	631	647	17
261	664	681	697	714	731	747	764	780	797	814	
262	830	847	863	880	896	913	929	946	963	979	1 1.7
263	996	*012	*029	*045	*062	*078	*095	*111	*127	*144	2 3.4
264	42 160	177	193	210	226	243	259	275	292	308	3 5.1
											4 6.8
265	325	341	357	374	390	406	423	439	455	472	5 8.5
266	488	504	521	537	553	570	586	602	619	635	6 10.2
267	651	667	684	700	716	732	749	765	781	797	7 11.9
268	813	830	846	862	878	894	911	927	943	959	8 13.6
269	975	991	*008	*024	*040	*056	*072	*088	*104	*120	9 15.3
270	43 136	152	169	185	201	217	233	249	265	281	16
271	297	313	329	345	361	377	393	409	425	441	
272	457	473	489	505	521	537	553	569	584	600	1 1.6
273	616	632	648	664	680	696	712	727	743	759	2 3.2
274	775	791	807	823	838	854	870	886	902	917	3 4.8
											4 6.4
275	933	949	965	981	996	*012	*028	*044	*059	*075	5 8.0
276	44 091	107	122	138	154	170	185	201	217	232	6 9.6
277	248	264	279	295	311	326	342	358	373	389	7 11.2
278	404	420	436	451	467	483	498	514	529	545	8 12.8
279	560	576	592	607	623	638	654	669	685	700	9 14.4
280	716	731	747	762	778	793	809	824	840	855	15
281	871	886	902	917	932	948	963	979	994	*010	
282	45 025	040	056	071	086	102	117	133	148	163	1 1.5
283	179	194	209	225	240	255	271	286	301	317	2 3.0
284	332	347	362	378	393	408	423	439	454	469	3 4.5
											4 6.0
285	484	500	515	530	545	561	576	591	606	621	5 7.5
286	637	652	667	682	697	712	728	743	758	773	6 9.0
287	788	803	818	834	849	864	879	894	909	924	7 10.5
288	939	954	969	984	*000	*015	*030	*045	*060	*075	8 12.0
289	46 090	105	120	135	150	165	180	195	210	225	9 13.5
290	240	255	270	285	300	315	330	345	359	374	14
291	389	404	419	434	449	464	479	494	509	523	
292	538	553	568	583	598	613	627	642	657	672	1 1.4
293	687	702	716	731	746	761	776	790	805	820	2 2.8
294	835	850	864	879	894	909	923	938	953	967	3 4.2
											4 5.6
295	982	997	*012	*026	*041	*056	*070	*085	*100	*114	5 7.0
296	47 129	144	159	173	188	202	217	232	246	261	6 8.4
297	276	290	305	319	334	349	363	378	392	407	7 9.8
298	422	436	451	465	480	494	509	524	538	553	8 11.2
299	567	582	596	611	625	640	654	669	683	698	9 12.6
300	712	727	741	756	770	784	799	813	828	842	$\log e = 0.43429$
N.	0	1	2	3	4	5	6	7	8	9	Proportional parts

.39 794 — .47 842

COMMON LOGARITHMS

300 — 350

N.		0	1	2	3	4	5	6	7	8	9
300	47	712	727	741	756	770	784	799	813	828	842
301		857	871	885	900	914	929	943	958	972	986
302	48	001	015	029	044	058	073	087	101	116	130
303		144	159	173	187	202	216	230	244	259	273
304		287	302	316	330	344	359	373	387	401	416
305		430	444	458	473	487	501	515	530	544	558
306		572	586	601	615	629	643	657	671	686	700
307		714	728	742	756	770	785	799	813	827	841
308		855	869	883	897	911	926	940	954	968	982
309		996	*010	*024	*038	*052	*066	*080	*094	*108	*122
310	49	136	150	164	178	192	206	220	234	248	262
311		276	290	304	318	332	346	360	374	388	402
312		415	429	443	457	471	485	499	513	527	541
313		554	568	582	596	610	624	638	651	665	679
314		693	707	721	734	748	762	776	790	803	817
315		831	845	859	872	886	900	914	927	941	955
316		969	982	996	*010	*024	*037	*051	*065	*079	*092
317	50	106	120	133	147	161	174	188	202	215	229
318		243	256	270	284	297	311	325	338	352	365
319		379	393	406	420	433	447	461	474	488	501
320		515	529	542	556	569	583	596	610	623	637
321		651	664	678	691	705	718	732	745	759	772
322		786	799	813	826	840	853	866	880	893	907
323		920	934	947	961	974	987	*001	*014	*028	*041
324	51	055	068	081	095	108	121	135	148	162	175
325		188	202	215	228	242	255	268	282	295	308
326		322	335	348	362	375	388	402	415	428	441
327		455	468	481	495	508	521	534	548	561	574
328		587	601	614	627	640	654	667	680	693	706
329		720	733	746	759	772	786	799	812	825	838
330		851	865	878	891	904	917	930	943	957	970
331		983	996	*009	*022	*035	*048	*061	*075	*088	*101
332	52	114	127	140	153	166	179	192	205	218	231
333		244	257	270	284	297	310	323	336	349	362
334		375	388	401	414	427	440	453	466	479	492
335		504	517	530	543	556	569	582	595	608	621
336		634	647	660	673	686	699	711	724	737	750
337		763	776	789	802	815	827	840	853	866	879
338		892	905	917	930	943	956	969	982	994	*007
339	53	020	033	046	058	071	084	097	110	122	135
340		148	161	173	186	199	212	224	237	250	263
341		275	288	301	314	326	339	352	364	377	390
342		403	415	428	441	453	466	479	491	504	517
343		529	542	555	567	580	593	605	618	631	643
344		656	668	681	694	706	719	732	744	757	769
345		782	794	807	820	832	845	857	870	882	895
346		908	920	933	945	958	970	983	995	*008	*020
347	54	033	045	058	070	083	095	108	120	133	145
348		158	170	183	195	208	220	233	245	258	270
349		283	295	307	320	332	345	357	370	382	394
350		407	419	432	444	456	469	481	494	506	518
N.		0	1	2	3	4	5	6	7	8	9

Proportional parts

	15		14		13		12
1	1.5	1	1.4	1	1.3	1	1.2
2	3.0	2	2.8	2	2.6	2	2.4
3	4.5	3	4.2	3	3.9	3	3.6
4	6.0	4	5.6	4	5.2	4	4.8
5	7.5	5	7.0	5	6.5	5	6.0
6	9.0	6	8.4	6	7.8	6	7.2
7	10.5	7	9.8	7	9.1	7	8.4
8	12.0	8	11.2	8	10.4	8	9.6
9	13.5	9	12.6	9	11.7	9	10.8

$\log \pi = 0.49715$

.47 712 — .54 518

COMMON LOGARITHMS

350 — 400

N.	0	1	2	3	4	5	6	7	8	9	Proportional parts
350	54 407	419	432	444	456	469	481	494	506	518	
351	531	543	555	568	580	593	605	617	630	642	
352	654	667	679	691	704	716	728	741	753	765	
353	777	790	802	814	827	839	851	864	876	888	
354	900	913	925	937	949	962	974	986	998	*011	
355	55 023	035	047	060	072	084	096	108	121	133	
356	145	157	169	182	194	206	218	230	242	255	
357	267	279	291	303	315	328	340	352	364	376	
358	388	400	413	425	437	449	461	473	485	497	
359	509	522	534	546	558	570	582	594	606	618	
360	630	642	654	666	678	691	703	715	727	739	
361	751	763	775	787	799	811	823	835	847	859	
362	871	883	895	907	919	931	943	955	967	979	
363	991	*003	*015	*027	*038	*050	*062	*074	*086	*098	
364	56 110	122	134	146	158	170	182	194	205	217	
365	229	241	253	265	277	289	301	312	324	336	
366	348	360	372	384	396	407	419	431	443	455	
367	467	478	490	502	514	526	538	549	561	573	
368	585	597	608	620	632	644	656	667	679	691	
369	703	714	726	738	750	761	773	785	797	808	
370	820	832	844	855	867	879	891	902	914	926	
371	937	949	961	972	984	996	*008	*019	*031	*043	
372	57 054	066	078	089	101	113	124	136	148	159	
373	171	183	194	206	217	229	241	252	264	276	
374	287	299	310	322	334	345	357	368	380	392	
375	403	415	426	438	449	461	473	484	496	507	
376	519	530	542	553	565	576	588	600	611	623	
377	634	646	657	669	680	692	703	715	726	738	
378	749	761	772	784	795	807	818	830	841	852	
379	864	875	887	898	910	921	933	944	955	967	
380	978	990	*001	*013	*024	*035	*047	*058	*070	*081	
381	58 092	104	115	127	138	149	161	172	184	195	
382	206	218	229	240	252	263	274	286	297	309	
383	320	331	343	354	365	377	388	399	410	422	
384	433	444	456	467	478	490	501	512	524	535	
385	546	557	569	580	591	602	614	625	636	647	
386	659	670	681	692	704	715	726	737	749	760	
387	771	782	794	805	816	827	838	850	861	872	
388	883	894	906	917	928	939	950	961	973	984	
389	995	*006	*017	*028	*040	*051	*062	*073	*084	*095	
390	59 106	118	129	140	151	162	173	184	195	207	
391	218	229	240	251	262	273	284	295	306	318	
392	329	340	351	362	373	384	395	406	417	428	
393	439	450	461	472	483	494	506	517	528	539	
394	550	561	572	583	594	605	616	627	638	649	
395	660	671	682	693	704	715	726	737	748	759	
396	770	780	791	802	813	824	835	846	857	868	
397	879	890	901	912	923	934	945	956	966	977	
398	988	999	*010	*021	*032	*043	*054	*065	*076	*086	
399	60 097	108	119	130	141	152	163	173	184	195	
400	206	217	228	239	249	260	271	282	293	304	
N.	0	1	2	3	4	5	6	7	8	9	Proportional parts

Proportional parts

	13
1	1.3
2	2.6
3	3.9
4	5.2
5	6.5
6	7.8
7	9.1
8	10.4
9	11.7

	12
1	1.2
2	2.4
3	3.6
4	4.8
5	6.0
6	7.2
7	8.4
8	9.6
9	10.8

	11
1	1.1
2	2.2
3	3.3
4	4.4
5	5.5
6	6.6
7	7.7
8	8.8
9	9.9

	10
1	1.0
2	2.0
3	3.0
4	4.0
5	5.0
6	6.0
7	7.0
8	8.0
9	9.0

.54 407 — .60 304

COMMON LOGARITHMS

400 — 450

N.	0	1	2	3	4	5	6	7	8	9
400	60 206	217	228	239	249	260	271	282	293	304
401	314	325	336	347	358	369	379	390	401	412
402	423	433	444	455	466	477	487	498	509	520
403	531	541	552	563	574	584	595	606	617	627
404	638	649	660	670	681	692	703	713	724	735
405	746	756	767	778	788	799	810	821	831	842
406	853	863	874	885	895	906	917	927	938	949
407	959	970	981	991	*002	*013	*023	*034	*045	*055
408	61 066	077	087	098	109	119	130	140	151	162
409	172	183	194	204	215	225	236	247	257	268
410	278	289	300	310	321	331	342	352	363	374
411	384	395	405	416	426	437	448	458	469	479
412	490	500	511	521	532	542	553	563	574	584
413	595	606	616	627	637	648	658	669	679	690
414	700	711	721	731	742	752	763	773	784	794
415	805	815	826	836	847	857	868	878	888	899
416	909	920	930	941	951	962	972	982	993	*003
417	62 014	024	034	045	055	066	076	086	097	107
418	118	128	138	149	159	170	180	190	201	211
419	221	232	242	252	263	273	284	294	304	315
420	325	335	346	356	366	377	387	397	408	418
421	428	439	449	459	469	480	490	500	511	521
422	531	542	552	562	572	583	593	603	613	624
423	634	644	655	665	675	685	696	706	716	726
424	737	747	757	767	778	788	798	808	818	829
425	839	849	859	870	880	890	900	910	921	931
426	941	951	961	972	982	992	*002	*012	*022	*033
427	63 043	053	063	073	083	094	104	114	124	134
428	144	155	165	175	185	195	205	215	225	236
429	246	256	266	276	286	296	306	317	327	337
430	347	357	367	377	387	397	407	417	428	438
431	448	458	468	478	488	498	508	518	528	538
432	548	558	568	579	589	599	609	619	629	639
433	649	659	669	679	689	699	709	719	729	739
434	749	759	769	779	789	799	809	819	829	839
435	849	859	869	879	889	899	909	919	929	939
436	949	959	969	979	988	998	*008	*018	*028	*038
437	64 048	058	068	078	088	098	108	118	128	137
438	147	157	167	177	187	197	207	217	227	237
439	246	256	266	276	286	296	306	316	326	335
440	345	355	365	375	385	395	404	414	424	434
441	444	454	464	473	483	493	503	513	523	532
442	542	552	562	572	582	591	601	611	621	631
443	640	650	660	670	680	689	699	709	719	729
444	738	748	758	768	777	787	797	807	816	826
445	836	846	856	865	875	885	895	904	914	924
446	933	943	953	963	972	982	992	*002	*011	*021
447	65 031	040	050	060	070	079	089	099	108	118
448	128	137	147	157	167	176	186	196	205	215
449	225	234	244	254	263	273	283	292	302	312
450	321	331	341	350	360	369	379	389	398	408
N.	0	1	2	3	4	5	6	7	8	9

Proportional parts

	11		10		9
1	1.1	1	1.0	1	0.9
2	2.2	2	2.0	2	1.8
3	3.3	3	3.0	3	2.7
4	4.4	4	4.0	4	3.6
5	5.5	5	5.0	5	4.5
6	6.6	6	6.0	6	5.4
7	7.7	7	7.0	7	6.3
8	8.8	8	8.0	8	7.2
9	9.9	9	9.0	9	8.1

.60 206 — .65 408

COMMON LOGARITHMS

450 — 500

N.		0	1	2	3	4	5	6	7	8	9
450	65	321	331	341	350	360	369	379	389	398	408
451		418	427	437	447	456	466	475	485	495	504
452		514	523	533	543	552	562	571	581	591	600
453		610	619	629	639	648	658	667	677	686	696
454		706	715	725	734	744	753	763	772	782	792
455		801	811	820	830	839	849	858	868	877	887
456		896	906	916	925	935	944	954	963	973	982
457		992	*001	*011	*020	*030	*039	*049	*058	*068	*077
458	66	087	096	106	115	124	134	143	153	162	172
459		181	191	200	210	219	229	238	247	257	266
460		276	285	295	304	314	323	332	342	351	361
461		370	380	389	398	408	417	427	436	445	455
462		464	474	483	492	502	511	521	530	539	549
463		558	567	577	586	596	605	614	624	633	642
464		652	661	671	680	689	699	708	717	727	736
465		745	755	764	773	783	792	801	811	820	829
466		839	848	857	867	876	885	894	904	913	922
467		932	941	950	960	969	978	987	997	*006	*015
468	67	025	034	043	052	062	071	080	089	099	108
469		117	127	136	145	154	164	173	182	191	201
470		210	219	228	237	247	256	265	274	284	293
471		302	311	321	330	339	348	357	367	376	385
472		394	403	413	422	431	440	449	459	468	477
473		486	495	504	514	523	532	541	550	560	569
474		578	587	596	605	614	624	633	642	651	660
475		669	679	688	697	706	715	724	733	742	752
476		761	770	779	788	797	806	815	825	834	843
477		852	861	870	879	888	897	906	916	925	934
478		943	952	961	970	979	988	997	*006	*015	*024
479	68	034	043	052	061	070	079	088	097	106	115
480		124	133	142	151	160	169	178	187	196	205
481		215	224	233	242	251	260	269	278	287	296
482		305	314	323	332	341	350	359	368	377	386
483		395	404	413	422	431	440	449	458	467	476
484		485	494	502	511	520	529	538	547	556	565
485		574	583	592	601	610	619	728	637	646	655
486		664	673	681	690	699	708	717	726	735	744
487		753	762	771	780	789	797	806	815	824	833
488		842	851	860	869	878	886	895	904	913	922
489		931	940	949	958	966	975	984	993	*002	*011
490	69	020	028	037	046	055	064	073	082	090	099
491		108	117	126	135	144	152	161	170	179	188
492		197	205	214	223	232	241	249	258	267	276
493		285	294	302	311	320	329	338	346	355	364
494		373	381	390	399	408	417	425	434	*43	452
495		461	469	478	487	496	504	513	522	531	539
496		548	557	566	574	583	592	601	609	618	627
497		636	644	653	662	671	679	688	697	705	714
498		723	732	740	749	758	767	775	784	793	801
499		810	819	827	836	845	854	862	871	880	888
500		897	906	914	923	932	940	949	958	966	975
N.		0	1	2	3	4	5	6	7	8	9

Proportional parts

	10
1	1.0
2	2.0
3	3.0
4	4.0
5	5.0
6	6.0
7	7.0
8	8.0
9	9.0

	9
1	0.9
2	1.8
3	2.7
4	3.6
5	4.5
6	5.4
7	6.3
8	7.2
9	8.1

	8
1	0.8
2	1.6
3	2.4
4	3.2
5	4.0
6	4.8
7	5.6
8	6.4
9	7.2

.65 321 — .69 975

COMMON LOGARITHMS

500 — 550

N.	0	1	2	3	4	5	6	7	8	9
500	69 897	906	914	923	932	940	949	958	966	975
501	984	992	*001	*010	*018	*027	*036	*044	*053	*062
502	70 070	079	088	096	105	114	122	131	140	148
503	157	165	174	183	191	200	209	217	226	234
504	243	252	260	269	278	286	295	303	312	321
505	329	338	346	355	364	372	381	389	398	406
506	415	424	432	441	449	458	467	475	484	492
507	501	509	518	526	535	544	552	561	569	578
508	586	595	603	612	621	629	638	646	655	663
509	672	680	689	697	706	714	723	731	740	749
510	757	766	774	783	791	800	808	817	825	834
511	842	851	859	868	876	885	893	902	910	919
512	927	935	944	952	961	969	978	986	995	*003
513	71 012	020	029	037	046	054	063	071	079	088
514	096	105	113	122	130	139	147	155	164	172
515	181	189	198	206	214	223	231	240	248	257
516	265	273	282	290	299	307	315	324	332	341
517	349	357	366	374	383	391	399	408	416	425
518	433	441	450	458	466	475	483	492	500	508
519	517	525	533	542	550	559	567	575	584	592
520	600	609	617	625	634	642	650	659	667	675
521	684	692	700	709	717	725	734	742	750	759
522	767	775	784	792	800	809	817	825	834	842
523	850	858	867	875	883	892	900	908	917	925
524	933	941	950	958	966	975	983	991	999	*008
525	72 016	024	032	041	049	057	066	074	082	090
526	099	107	115	123	132	140	148	156	165	173
527	181	189	198	206	214	222	230	239	247	255
528	263	272	280	288	296	304	313	321	329	337
529	346	354	362	370	378	387	395	403	411	419
530	428	436	444	452	460	469	477	485	493	501
531	509	518	526	534	542	550	558	567	575	583
532	591	599	607	616	624	632	640	648	656	665
533	673	681	689	697	705	713	722	730	738	746
534	754	762	770	779	787	795	803	811	819	827
535	835	843	852	860	868	876	884	892	900	908
536	916	925	933	941	949	957	965	973	981	989
537	997	*006	*014	*022	*030	*038	*046	*054	*062	*070
538	73 078	086	094	102	111	119	127	135	143	151
539	159	167	175	183	191	199	207	215	223	231
540	239	247	255	263	272	280	288	296	304	312
541	320	328	336	344	352	360	368	376	384	392
542	400	408	416	424	432	440	448	456	464	472
543	480	488	496	504	512	520	528	536	544	552
544	560	568	576	584	592	600	608	616	624	632
545	640	648	656	664	672	679	687	695	703	711
546	719	727	735	743	751	759	767	775	783	791
547	799	807	815	823	830	838	846	854	862	870
548	878	886	894	902	910	918	926	933	941	949
549	957	965	973	981	989	997	*005	*013	*020	*028
550	74 036	044	052	060	068	076	084	092	099	107
N.	0	1	2	3	4	5	6	7	8	9

Proportional parts

	9		8		7
1	0.9	1	0.8	1	0.7
2	1.8	2	1.6	2	1.4
3	2.7	3	2.4	3	2.1
4	3.6	4	3.2	4	2.8
5	4.5	5	4.0	5	3.5
6	5.4	6	4.8	6	4.2
7	6.3	7	5.6	7	4.9
8	7.2	8	6.4	8	5.6
9	8.1	9	7.2	9	6.3

.69 897 — .74 107

COMMON LOGARITHMS

550 — 600

N.	0	1	2	3	4	5	6	7	8	9
550	74 036	044	052	060	068	076	084	092	099	107
551	115	123	131	139	147	155	162	170	178	186
552	194	202	210	218	225	233	241	249	257	265
553	273	280	288	296	304	312	320	327	335	343
554	351	359	367	374	382	390	398	406	414	421
555	429	437	445	453	461	468	476	484	492	500
556	507	515	523	531	539	547	554	562	570	578
557	586	593	601	609	617	624	632	640	648	656
558	663	671	679	687	695	702	710	718	726	733
559	741	749	757	764	772	780	788	796	803	811
560	819	827	834	842	850	858	865	873	881	889
561	896	904	912	920	927	935	943	950	958	966
562	974	981	989	997	*005	*012	*020	*028	*035	*043
563	75 051	059	066	074	082	089	097	105	113	120
564	128	136	143	151	159	166	174	182	189	197
565	205	213	220	228	236	243	251	259	266	274
566	282	289	297	305	312	320	328	335	343	351
567	358	366	374	381	389	397	404	412	420	427
568	435	442	450	458	465	473	481	488	496	504
569	511	519	526	534	542	549	557	565	572	580
570	587	595	603	610	618	626	633	641	648	656
571	664	671	679	686	694	702	709	717	724	732
572	740	747	755	762	770	778	785	793	800	808
573	815	823	831	838	846	853	861	868	876	884
574	891	899	906	914	921	929	937	944	952	959
575	967	974	982	989	997	*005	*012	*020	*027	*035
576	76 042	050	057	065	072	080	087	095	103	110
577	118	125	133	140	148	155	163	170	178	185
578	193	200	208	215	223	230	238	245	253	260
579	268	275	283	290	298	305	313	320	328	335
580	343	350	358	365	373	380	388	395	403	410
581	418	425	433	440	448	455	462	470	477	485
582	492	500	507	515	522	530	537	545	552	559
583	567	574	582	589	597	604	612	619	626	634
584	641	649	656	664	671	678	686	693	701	708
585	716	723	730	738	745	753	760	768	775	782
586	790	797	805	812	819	827	834	842	849	856
587	864	871	879	886	893	901	908	916	923	930
588	938	945	953	960	967	975	982	989	997	*004
589	77 012	019	026	034	041	048	056	063	070	078
590	085	093	100	107	115	122	129	137	144	151
591	159	166	173	181	188	195	203	210	217	225
592	232	240	247	254	262	269	276	283	291	298
593	305	313	320	327	335	342	349	357	364	371
594	379	386	393	401	408	415	422	430	437	444
595	452	459	466	474	481	488	495	503	510	517
596	525	532	539	546	554	561	568	576	583	590
597	597	605	612	619	627	634	641	648	656	663
598	670	677	685	692	699	706	714	721	728	735
599	743	750	757	764	772	779	786	793	801	808
600	815	822	830	837	844	851	859	866	873	880
N.	0	1	2	3	4	5	6	7	8	9

Proportional parts

	8
1	0.8
2	1.6
3	2.4
4	3.2
5	4.0
6	4.8
7	5.6
8	6.4
9	7.2

	7
1	0.7
2	1.4
3	2.1
4	2.8
5	3.5
6	4.2
7	4.9
8	5.6
9	6.3

.74 036 — .77 880

COMMON LOGARITHMS

600 — 650

N.		0	1	2	3	4	5	6	7	8	9
600	77	815	822	830	837	844	851	859	866	873	880
601		887	895	902	909	916	924	931	938	945	952
602		960	967	974	981	988	996	*003	*010	*017	*025
603	78	032	039	046	053	061	068	075	082	089	097
604		104	111	118	125	132	140	147	154	161	168
605		176	183	190	197	204	211	219	226	233	240
606		247	254	262	269	276	283	290	297	305	312
607		319	326	333	340	347	355	362	369	376	383
608		390	398	405	412	419	426	433	440	447	455
609		462	469	476	483	490	497	504	512	519	526
610		533	540	547	554	561	569	576	583	590	597
611		604	611	618	625	633	640	647	654	661	668
612		675	682	689	696	704	711	718	725	732	739
613		746	753	760	767	774	781	789	796	803	810
614		817	824	831	838	845	852	859	866	873	880
615		888	895	902	909	916	923	930	937	944	951
616		958	965	972	979	986	993	*000	*007	*014	*021
617	79	029	036	043	050	057	064	071	078	085	092
618		099	106	113	120	127	134	141	148	155	162
619		169	176	183	190	197	204	211	218	225	232
620		239	246	253	260	267	274	281	288	295	302
621		309	316	323	330	337	344	351	358	365	372
622		379	386	393	400	407	414	421	428	435	442
623		449	456	463	470	477	484	491	498	505	511
624		518	525	532	539	546	553	560	567	574	581
625		588	595	602	609	616	623	630	637	644	650
626		657	664	671	678	685	692	699	706	713	720
627		727	734	741	748	754	761	768	775	782	789
628		796	803	810	817	824	831	837	844	851	858
629		865	872	879	886	893	900	906	913	920	927
630		934	941	948	955	962	969	975	982	989	996
631	80	003	010	017	024	030	037	044	051	058	065
632		072	079	085	092	099	106	113	120	127	134
633		140	147	154	161	168	175	182	188	195	202
634		209	216	223	229	236	243	250	257	264	271
635		277	284	291	298	305	312	318	325	332	339
636		346	353	359	366	373	380	387	393	400	407
637		414	421	428	434	441	448	455	462	468	475
638		482	489	496	502	509	516	523	530	536	543
639		550	557	564	570	577	584	591	598	604	611
640		618	625	632	638	645	652	659	665	672	679
641		686	693	699	706	713	720	726	733	740	747
642		754	760	767	774	781	787	794	801	808	814
643		821	828	835	841	848	855	862	868	875	882
644		889	895	902	909	916	922	929	936	943	949
645		956	963	969	976	983	990	996	*003	*010	*017
646	81	023	030	037	043	050	057	064	070	077	084
647		090	097	104	111	117	124	131	137	144	151
648		158	164	171	178	184	191	198	204	211	218
649		224	231	238	245	251	258	265	271	278	285
650		291	298	305	311	318	325	331	338	345	351
N.		0	1	2	3	4	5	6	7	8	9

Proportional parts

	8
1	0.8
2	1.6
3	2.4
4	3.2
5	4.0
6	4.8
7	5.6
8	6.4
9	7.2

	7
1	0.7
2	1.4
3	2.1
4	2.8
5	3.5
6	4.2
7	4.9
8	5.6
9	6.3

	6
1	0.6
2	1.2
3	1.8
4	2.4
5	3.0
6	3.6
7	4.2
8	4.8
9	5.4

.77 815 — .81 351

COMMON LOGARITHMS

650 — 700

N.	0	1	2	3	4	5	6	7	8	9
650	81 291	298	305	311	318	325	331	338	345	351
651	358	365	371	378	385	391	398	405	411	418
652	425	431	438	445	451	458	465	471	478	485
653	491	498	505	511	518	525	531	538	544	551
654	558	564	571	578	584	591	598	604	611	617
655	624	631	637	644	651	657	664	671	677	684
656	690	697	704	710	717	723	730	737	743	750
657	757	763	770	776	783	790	796	803	809	816
658	823	829	836	842	849	856	862	869	875	882
659	889	895	902	908	915	921	928	935	941	948
660	954	961	968	974	981	987	994	*000	*007	*014
661	82 020	027	033	040	046	053	060	066	073	079
662	086	092	099	105	112	119	125	132	138	145
663	151	158	164	171	178	184	191	197	204	210
664	217	223	230	236	243	249	256	263	269	276
665	282	289	295	302	308	315	321	328	334	341
666	347	354	360	367	373	380	387	393	400	406
667	413	419	426	432	439	445	452	458	465	471
668	478	484	491	497	504	510	517	523	530	536
669	543	549	556	562	569	575	582	588	595	601
670	607	614	620	627	633	640	646	653	659	666
671	672	679	685	692	698	705	711	718	724	730
672	737	743	750	756	763	769	776	782	789	795
673	802	808	814	821	827	834	840	847	853	860
674	866	872	879	885	892	898	905	911	918	924
675	930	937	943	950	956	963	969	975	982	988
676	995	*001	*008	*014	*020	*027	*033	*040	*046	*052
677	83 059	065	072	078	085	091	097	104	110	117
678	123	129	136	142	149	155	161	168	174	181
679	187	193	200	206	213	219	225	232	238	245
680	251	257	264	270	276	283	289	296	302	308
681	315	321	327	334	340	347	353	359	366	372
682	378	385	391	398	404	410	417	423	429	436
683	442	448	455	461	467	474	480	487	493	499
684	506	512	518	525	531	537	544	550	556	563
685	569	575	582	588	594	601	607	613	620	626
686	632	639	645	651	658	664	670	677	683	689
687	696	702	708	715	721	727	734	740	746	753
688	759	765	771	778	784	790	797	803	809	816
689	822	828	835	841	847	853	860	866	872	879
690	885	891	897	904	910	916	923	929	935	942
691	948	954	960	967	973	979	985	992	998	*004
692	84 011	017	023	029	036	042	048	055	061	067
693	073	080	086	092	098	105	111	117	123	130
694	136	142	148	155	161	167	173	180	186	192
695	198	205	211	217	223	230	236	242	248	255
696	261	267	273	280	286	292	298	305	311	317
697	323	330	336	342	348	354	361	367	373	379
698	386	392	398	404	410	417	423	429	435	442
699	448	454	460	466	473	479	485	491	497	504
700	510	516	522	528	535	541	547	553	559	566
N.	0	1	2	3	4	5	6	7	8	9

Proportional parts

	7
1	0.7
2	1.4
3	2.1
4	2.8
5	3.5
6	4.2
7	4.9
8	5.6
9	6.3

	6
1	0.6
2	1.2
3	1.8
4	2.4
5	3.0
6	3.6
7	4.2
8	4.8
9	5.4

.81 291 — .84 566

COMMON LOGARITHMS

700 — 750

N.	0	1	2	3	4	5	6	7	8	9
700	84 510	516	522	528	535	541	547	553	559	566
701	572	578	584	590	597	603	609	615	621	628
702	634	640	646	652	658	665	671	677	683	689
703	696	702	708	714	720	726	733	739	745	751
704	757	763	770	776	782	788	794	800	807	813
705	819	825	831	837	844	850	856	862	868	874
706	880	887	893	899	905	911	917	924	930	936
707	942	948	954	960	967	973	979	985	991	997
708	85 003	009	016	022	028	034	040	046	052	058
709	065	071	077	083	089	095	101	107	114	120
710	126	132	138	144	150	156	163	169	175	181
711	187	193	199	205	211	217	224	230	236	242
712	248	254	260	266	272	278	285	291	297	303
713	309	315	321	327	333	339	345	352	358	364
714	370	376	382	388	394	400	406	412	418	425
715	431	437	443	449	455	461	467	473	479	485
716	491	497	503	509	516	522	528	534	540	546
717	552	558	564	570	576	582	588	594	600	606
718	612	618	625	631	637	643	649	655	661	667
719	673	679	685	691	697	703	709	715	721	727
720	733	739	745	751	757	763	769	775	781	788
721	794	800	806	812	818	824	830	836	842	848
722	854	860	866	872	878	884	890	896	902	908
723	914	920	926	932	938	944	950	956	962	968
724	974	980	986	992	998	*004	*010	*016	*022	*028
725	86 034	040	046	052	058	064	070	076	082	088
726	094	100	106	112	118	124	130	136	141	147
727	153	159	165	171	177	183	189	195	201	207
728	213	219	225	231	237	243	249	255	261	267
729	273	279	285	291	297	303	308	314	320	326
730	332	338	344	350	356	362	368	374	380	386
731	392	398	404	410	415	421	427	433	439	445
732	451	457	463	469	475	481	487	493	499	504
733	510	516	522	528	534	540	546	552	558	564
734	570	576	581	587	593	599	605	611	617	623
735	629	635	641	646	652	658	664	670	676	682
736	688	694	700	705	711	717	723	729	735	741
737	747	753	759	764	770	776	782	788	794	800
738	806	812	817	823	829	835	841	847	853	859
739	864	870	876	882	888	894	900	906	911	917
740	923	929	935	941	947	953	958	964	970	976
741	982	988	994	999	*005	*011	*017	*023	*029	*035
742	87 040	046	052	058	064	070	075	081	087	093
743	099	105	111	116	122	128	134	140	146	151
744	157	163	169	175	181	186	192	198	204	210
745	216	221	227	233	239	245	251	256	262	268
746	274	280	286	291	297	303	309	315	320	326
747	332	338	344	349	355	361	367	373	379	384
748	390	396	402	408	413	419	425	431	437	442
749	448	454	460	466	471	477	483	489	495	500
750	506	512	518	523	529	535	541	547	552	558
N.	0	1	2	3	4	5	6	7	8	9

Proportional parts

	7		6		5
1	0.7	1	0.6	1	0.5
2	1.4	2	1.2	2	1.0
3	2.1	3	1.8	3	1.5
4	2.8	4	2.4	4	2.0
5	3.5	5	3.0	5	2.5
6	4.2	6	3.6	6	3.0
7	4.9	7	4.2	7	3.5
8	5.6	8	4.8	8	4.0
9	6.3	9	5.4	9	4.5

.84 510 — .87 558

COMMON LOGARITHMS

750 — 800

N.	0	1	2	3	4	5	6	7	8	9
750	87 506	512	518	523	529	535	541	547	552	558
751	564	570	576	581	587	593	599	604	610	616
752	622	628	633	639	645	651	656	662	668	674
753	679	685	691	697	703	708	714	720	726	731
754	737	743	749	754	760	766	772	777	783	789
755	795	800	806	812	818	823	829	835	841	846
756	852	858	864	869	875	881	887	892	898	904
757	910	915	921	927	933	938	944	950	955	961
758	967	973	978	984	990	996	*001	*007	*013	*018
759	88 024	030	036	041	047	053	058	064	070	076
760	081	087	093	098	104	110	116	121	127	133
761	138	144	150	156	161	167	173	178	184	190
762	195	201	207	213	218	224	230	235	241	247
763	252	258	264	270	275	281	287	292	298	304
764	309	315	321	326	332	338	343	349	355	360
765	366	372	377	383	389	395	400	406	412	417
766	423	429	434	440	446	451	457	463	468	474
767	480	485	491	497	502	508	513	519	525	530
768	536	542	547	553	559	564	570	576	581	587
769	593	598	604	610	615	621	627	632	638	643
770	649	655	660	666	672	677	683	689	694	700
771	705	711	717	722	728	734	739	745	750	756
772	762	767	773	779	784	790	795	801	807	812
773	818	824	829	835	840	846	852	857	863	868
774	874	880	885	891	897	902	908	913	919	925
775	930	936	941	947	953	958	964	969	975	981
776	986	992	997	*003	*009	*014	*020	*025	*031	*037
777	89 042	048	053	059	064	070	076	081	087	092
778	098	104	109	115	120	126	131	137	143	148
779	154	159	165	170	176	182	187	193	198	204
780	209	215	221	226	232	237	243	248	254	260
781	265	271	276	282	287	293	298	304	310	315
782	321	326	332	337	343	348	354	360	365	371
783	376	382	387	393	398	404	409	415	421	426
784	432	437	443	448	454	459	465	470	476	481
785	487	492	498	504	509	515	520	526	531	537
786	542	548	553	559	564	570	575	581	586	592
787	597	603	609	614	620	625	631	636	642	647
788	653	658	664	669	675	680	686	691	697	702
789	708	713	719	724	730	735	741	746	752	757
790	763	768	774	779	785	790	796	801	807	812
791	818	823	829	834	840	845	851	856	862	867
792	873	878	883	889	894	900	905	911	916	922
793	927	933	938	944	949	955	960	966	971	977
794	982	988	993	998	*004	*009	*015	*020	*026	*031
795	90 037	042	048	053	059	064	069	075	080	086
796	091	097	102	108	113	119	124	129	135	140
797	146	151	157	162	168	173	179	184	189	195
798	200	206	211	217	222	227	233	238	244	249
799	255	260	266	271	276	282	287	293	298	304
800	309	314	320	325	331	336	342	347	352	358
N.	0	1	2	3	4	5	6	7	8	9

Proportional parts

	6
1	0.6
2	1.2
3	1.8
4	2.4
5	3.0
6	3.6
7	4.2
8	4.8
9	5.4

	5
1	0.5
2	1.0
3	1.5
4	2.0
5	2.5
6	3.0
7	3.5
8	4.0
9	4.5

.87 506 — .90 358

COMMON LOGARITHMS

800 — 850

N.	0	1	2	3	4	5	6	7	8	9
800	90 309	314	320	325	331	336	342	347	352	358
801	363	369	374	380	385	390	396	401	407	412
802	417	423	428	434	439	445	450	455	461	466
803	472	477	482	488	493	499	504	509	515	520
804	526	531	536	542	547	553	558	563	569	574
805	580	585	590	596	601	607	612	617	623	628
806	634	639	644	650	655	660	666	671	677	682
807	687	693	698	703	709	714	720	725	730	736
808	741	747	752	757	763	768	773	779	784	789
809	795	800	806	811	816	822	827	832	838	843
810	849	854	859	865	870	875	881	886	891	897
811	902	907	913	918	924	929	934	940	945	950
812	956	961	966	972	977	982	988	993	998	*004
813	91 009	014	020	025	030	036	041	046	052	057
814	062	068	073	078	084	089	094	100	105	110
815	116	121	126	132	137	142	148	153	158	164
816	169	174	180	185	190	196	201	206	212	217
817	222	228	233	238	243	249	254	259	265	270
818	275	281	286	291	297	302	307	312	318	323
819	328	334	339	344	350	355	360	365	371	376
820	381	387	392	397	403	408	413	418	424	429
821	434	440	445	450	455	461	466	471	477	482
822	487	492	498	503	508	514	519	524	529	535
823	540	545	551	556	561	566	572	577	582	587
824	593	598	603	609	614	619	624	630	635	640
825	645	651	656	661	666	672	677	682	687	693
826	698	703	709	714	719	724	730	735	740	745
827	751	756	761	766	772	777	782	787	793	798
828	803	808	814	819	824	829	834	840	845	850
829	855	861	866	871	876	882	887	892	897	903
830	908	913	918	924	929	934	939	944	950	955
831	960	965	971	976	981	986	991	997	*002	*007
832	92 012	018	023	028	033	038	044	049	054	059
833	065	070	075	080	085	091	096	101	106	111
834	117	122	127	132	137	143	148	153	158	163
835	169	174	179	184	189	195	200	205	210	215
836	221	226	231	236	241	247	252	257	262	267
837	273	278	283	288	293	298	304	309	314	319
838	324	330	335	340	345	350	355	361	366	371
839	376	381	387	392	397	402	407	412	418	423
840	428	433	438	443	449	454	459	464	469	474
841	480	485	490	495	500	505	511	516	521	526
842	531	536	542	547	552	557	562	567	572	578
843	583	588	593	598	603	609	614	619	624	629
844	634	639	645	650	655	660	665	670	675	681
845	686	691	696	701	706	711	716	722	727	732
846	737	742	747	752	758	763	768	773	778	783
847	788	793	799	804	809	814	819	824	829	834
848	840	845	850	855	860	865	870	875	881	886
849	891	896	901	906	911	916	921	927	932	937
850	942	947	952	957	962	967	973	978	983	988
N.	0	1	2	3	4	5	6	7	8	9

Proportional parts

	6
1	0.6
2	1.2
3	1.8
4	2.4
5	3.0
6	3.6
7	4.2
8	4.8
9	5.4

	5
1	0.5
2	1.0
3	1.5
4	2.0
5	2.5
6	3.0
7	3.5
8	4.0
9	4.5

.90 309 — .92 988

COMMON LOGARITHMS

850 — 900

N.	0	1	2	3	4	5	6	7	8	9
850	92 942	947	952	957	962	967	973	978	983	988
851	993	998	*003	*008	*013	*018	*024	*029	*034	*039
852	93 044	049	054	059	064	069	075	080	085	090
853	095	100	105	110	115	120	125	131	136	141
854	146	151	156	161	166	171	176	181	186	192
855	197	202	207	212	217	222	227	232	237	242
856	247	252	258	263	268	273	278	283	288	293
857	298	303	308	313	318	323	328	334	339	344
858	349	354	359	364	369	374	379	384	389	394
859	399	404	409	414	420	425	430	435	440	445
860	450	455	460	465	470	475	480	485	490	495
861	500	505	510	515	520	526	531	536	541	546
862	551	556	561	566	571	576	581	586	591	596
863	601	606	611	616	621	626	631	636	641	646
864	651	656	661	666	671	676	682	687	692	697
865	702	707	712	717	722	727	732	737	742	747
866	752	757	762	767	772	777	782	787	792	797
867	802	807	812	817	822	827	832	837	842	847
868	852	857	862	867	872	877	882	887	892	897
869	902	907	912	917	922	927	932	937	942	947
870	952	957	962	967	972	977	982	987	992	997
871	94 002	007	012	017	022	027	032	037	042	047
872	052	057	062	067	072	077	082	086	091	096
873	101	106	111	116	121	126	131	136	141	146
874	151	156	161	166	171	176	181	186	191	196
875	201	206	211	216	221	226	231	236	240	245
876	250	255	260	265	270	275	280	285	290	295
877	300	305	310	315	320	325	330	335	340	345
878	349	354	359	364	369	374	379	384	389	394
879	399	404	409	414	419	424	429	433	438	443
880	448	453	458	463	468	473	478	483	488	493
881	498	503	507	512	517	522	527	532	537	542
882	547	552	557	562	567	571	576	581	586	591
883	596	601	606	611	616	621	626	630	635	640
884	645	650	655	660	665	670	675	680	685	689
885	694	699	704	709	714	719	724	729	734	738
886	743	748	753	758	763	768	773	778	783	787
887	792	797	802	807	812	817	822	827	832	836
888	841	846	851	856	861	866	871	876	880	885
889	890	895	900	905	910	915	919	924	929	934
890	939	944	949	954	959	963	968	973	978	983
891	988	993	998	*002	*007	*012	*017	*022	*027	*032
892	95 036	041	046	051	056	061	066	071	075	080
893	085	090	095	100	105	109	114	119	124	129
894	134	139	143	148	153	158	163	168	173	177
895	182	187	192	197	202	207	211	216	221	226
896	231	236	240	245	250	255	260	265	270	274
897	279	284	289	294	299	303	308	313	318	323
898	328	332	337	342	347	352	357	361	366	371
899	376	381	386	390	395	400	405	410	415	419
900	424	429	434	439	444	448	453	458	463	468
N.	0	1	2	3	4	5	6	7	8	9

Proportional parts

6	
1	0.6
2	1.2
3	1.8
4	2.4
5	3.0
6	3.6
7	4.2
8	4.8
9	5.4

5	
1	0.5
2	1.0
3	1.5
4	2.0
5	2.5
6	3.0
7	3.5
8	4.0
9	4.5

4	
1	0.4
2	0.8
3	1.2
4	1.6
5	2.0
6	2.4
7	2.8
8	3.2
9	3.6

.92 942 — .95 468

COMMON LOGARITHMS

900 — 950

N.		0	1	2	3	4	5	6	7	8	9
900	95	424	429	434	439	444	448	453	458	463	468
901		472	477	482	487	492	497	501	506	511	516
902		521	525	530	535	540	545	550	554	559	564
903		569	574	578	583	588	593	598	602	607	612
904		617	622	626	631	636	641	646	650	655	660
905		665	670	674	679	684	689	694	698	703	708
906		713	718	722	727	732	737	742	746	751	756
907		761	766	770	775	780	785	789	794	799	804
908		809	813	818	823	828	832	837	842	847	852
909		856	861	866	871	875	880	885	890	895	899
910		904	909	914	918	923	928	933	938	942	947
911		952	957	961	966	971	976	980	985	990	995
912		999	*004	*009	*014	*019	*023	*028	*033	*038	*042
913	96	047	052	057	061	066	071	076	080	085	090
914		095	099	104	109	114	118	123	128	133	137
915		142	147	152	156	161	166	171	175	180	185
916		190	194	199	204	209	213	218	223	227	232
917		237	242	246	251	256	261	265	270	275	280
918		284	289	294	298	303	308	313	317	322	327
919		332	336	341	346	350	355	360	365	369	374
920		379	384	388	393	398	402	407	412	417	421
921		426	431	435	440	445	450	454	459	464	468
922		473	478	483	487	492	497	501	506	511	515
923		520	525	530	534	539	544	548	553	558	562
924		567	572	577	581	586	591	595	600	605	609
925		614	619	624	628	633	638	642	647	652	656
926		661	666	670	675	680	685	689	694	699	703
927		708	713	717	722	727	731	736	741	745	750
928		755	759	764	769	774	778	783	788	792	797
929		802	806	811	816	820	825	830	834	839	844
930		848	853	858	862	867	872	876	881	886	890
931		895	900	904	909	914	918	923	928	932	937
932		942	946	951	956	960	965	970	974	979	984
933		988	993	997	*002	*007	*011	*016	*021	*025	*030
934	97	035	039	044	049	053	058	063	067	072	077
935		081	086	090	095	100	104	109	114	118	123
936		128	132	137	142	146	151	155	160	165	169
937		174	179	183	188	192	197	202	206	211	216
938		220	225	230	234	239	243	248	253	257	262
939		267	271	276	280	285	290	294	299	304	308
940		313	317	322	327	331	336	340	345	350	354
941		359	364	368	373	377	382	387	391	396	400
942		405	410	414	419	424	428	433	437	442	447
943		451	456	460	465	470	474	479	483	488	493
944		497	502	506	511	516	520	525	529	534	539
945		543	548	552	557	562	566	571	575	580	585
946		589	594	598	603	607	612	617	621	626	630
947		635	640	644	649	653	658	663	667	672	676
948		681	685	690	695	699	704	708	713	717	722
949		727	731	736	740	745	749	754	759	763	768
950		772	777	782	786	791	795	800	804	809	813
N.		0	1	2	3	4	5	6	7	8	9

Proportional parts

	5
1	0.5
2	1.0
3	1.5
4	2.0
5	2.5
6	3.0
7	3.5
8	4.0
9	4.5

	4
1	0.4
2	0.8
3	1.2
4	1.6
5	2.0
6	2.4
7	2.8
8	3.2
9	3.6

.95 424 — .97 813

COMMON LOGARITHMS

950 — 1000

N.	0	1	2	3	4	5	6	7	8	9	Proportional parts
950	97 772	777	782	786	791	795	800	804	809	813	
951	818	823	827	832	836	841	845	850	855	859	
952	864	868	873	877	882	886	891	896	900	905	
953	909	914	918	923	928	932	937	941	946	950	
954	955	959	964	968	973	978	982	987	991	996	
955	98 000	005	009	014	019	023	028	032	037	041	
956	046	050	055	059	064	068	073	078	082	087	
957	091	096	100	105	109	114	118	123	127	132	
958	137	141	146	150	155	159	164	168	173	177	
959	182	186	191	195	200	204	209	214	218	223	
960	227	232	236	241	245	250	254	259	263	268	
961	272	277	281	286	290	295	299	304	308	313	
962	318	322	327	331	336	340	345	349	354	358	
963	363	367	372	376	381	385	390	394	399	403	
964	408	412	417	421	426	430	435	439	444	448	
965	453	457	462	466	471	475	480	484	489	493	
966	498	502	507	511	516	520	525	529	534	538	
967	543	547	552	556	561	565	570	574	579	583	
968	588	592	597	601	605	610	614	619	623	628	
969	632	637	641	646	650	655	659	664	668	673	
970	677	682	686	691	695	700	704	709	713	717	
971	722	726	731	735	740	744	749	753	758	762	
972	767	771	776	780	784	789	793	798	802	807	
973	811	816	820	825	829	834	838	843	847	851	
974	856	860	865	869	874	878	883	887	892	896	
975	900	905	909	914	918	923	927	932	936	941	
976	945	949	954	958	963	967	972	976	981	985	
977	989	994	998	*003	*007	*012	*016	*021	*025	*029	
978	99 034	038	043	047	052	056	061	065	069	074	
979	078	083	087	092	096	100	105	109	114	118	
980	123	127	131	136	140	145	149	154	158	162	
981	167	171	176	180	185	189	193	198	202	207	
982	211	216	220	224	229	233	238	242	247	251	
983	255	260	264	269	273	277	282	286	291	295	
984	300	304	308	313	317	322	326	330	335	339	
985	344	348	352	357	361	366	370	374	379	383	
986	388	392	396	401	405	410	414	419	423	427	
987	432	436	441	445	449	454	458	463	467	471	
988	476	480	484	489	493	498	502	506	511	515	
989	520	524	528	533	537	542	546	550	555	559	
990	564	568	572	577	581	585	590	594	599	603	
991	607	612	616	621	625	629	634	638	642	647	
992	651	656	660	664	669	673	677	682	686	691	
993	695	699	704	708	712	717	721	726	730	734	
994	739	743	747	752	756	760	765	769	774	778	
995	782	787	791	795	800	804	808	813	817	822	
996	826	830	835	839	843	848	852	856	861	865	
997	870	874	878	883	887	891	896	900	904	909	
998	913	917	922	926	930	935	939	944	948	952	
999	957	961	965	970	974	978	983	987	991	996	
1000	00 000	004	009	013	017	022	026	030	035	039	
N.	0	1	2	3	4	5	6	7	8	9	Proportional parts

Proportional parts:

	5
1	0.5
2	1.0
3	1.5
4	2.0
5	2.5
6	3.0
7	3.5
8	4.0
9	4.5

	4
1	0.4
2	0.8
3	1.2
4	1.6
5	2.0
6	2.4
7	2.8
8	3.2
9	3.6

.97 772 — .99 996

Appendix F

NATURAL LOGARITHMS

To extend this table for a number less than 1.0 or greater than 10.9 write the number in the form $x = y \cdot 10^n$ where $1.0 \leq y < 10$ and use the fact that $\ln x = \ln y + n \ln 10$

$1 \ln 10 = 2.30259$	$6 \ln 10 = 13.81551$
$2 \ln 10 = 4.60517$	$7 \ln 10 = 16.11810$
$3 \ln 10 = 6.90776$	$8 \ln 10 = 18.42068$
$4 \ln 10 = 9.21034$	$9 \ln 10 = 20.72327$
$5 \ln 10 = 11.51293$	$10 \ln 10 = 23.02585$

N	0	1	2	3	4	5	6	7	8	9
1.0	0.0 0000	0995	1980	2956	3922	4879	5827	6766	7696	8618
1.1	9531	*0436	*1333	*2222	*3103	*3976	*4842	*5700	*6551	*7395
1.2	0.1 8232	9062	9885	*0701	*1511	*2314	*3111	*3902	*4686	*5464
1.3	0.2 6236	7003	7763	8518	9267	*0010	*0748	*1481	*2208	*2930
1.4	0.3 3647	4359	5066	5767	6464	7156	7844	8526	9204	9878
1.5	0.4 0547	1211	1871	2527	3178	3825	4469	5108	5742	6373
1.6	7000	7623	8243	8858	9470	*0078	*0672	*1282	*1879	*2473
1.7	0.5 3063	3649	4232	4812	5389	5962	6531	7098	7661	8222
1.8	8779	9333	9884	*0432	*0977	*1519	*2058	*2594	*3127	*3658
1.9	0.6 4185	4710	5233	5752	6269	6783	7294	7803	8310	8813
2.0	9315	9813	*0310	*0804	*1295	*1784	*2271	*2755	*3237	*3716
2.1	0.7 4194	4669	5142	5612	6081	6547	7011	7473	7932	8390
2.2	8846	9299	9751	*0200	*0648	*1093	*1536	*1978	*2418	*2855
2.3	0.8 3291	3725	4157	4587	5015	5442	5866	6289	6710	7129
2.4	7547	7963	8377	8789	9200	9609	*0016	*0422	*0826	*1228
2.5	0.9 1629	2028	2426	2822	3216	3609	4001	4391	4779	5166
2.6	5551	5935	6317	6698	7078	7456	7833	8208	8582	8954
2.7	9325	9695	*0063	*0430	*0796	*1160	*1523	*1885	*2245	*2604
2.8	1.0 2962	3318	3674	4028	4380	4732	5082	5431	5779	6126
2.9	6471	6815	7158	7500	7841	8181	8519	8856	9192	9527
3.0	9861	*0194	*0526	*0856	*1186	*1514	*1841	*2168	*2493	*2817
3.1	1.1 3140	3462	3783	4103	4422	4740	5057	5373	5688	6002
3.2	6315	6627	6938	7248	7557	7865	8173	8479	8784	9089
3.3	9392	9695	9996	*0297	*0597	*0896	*1194	*1491	*1788	*2083
3.4	1.2 2378	2671	2964	3256	3547	3837	4127	4415	4703	4990
3.5	5276	5562	5846	6130	6413	6695	6976	7257	7536	7815
3.6	8093	8371	8647	8923	9198	9473	9746	*0019	*0291	*0563
3.7	1.3 0833	1103	1372	1641	1909	2176	2442	2708	2972	3237
3.8	3500	3763	4025	4286	4547	4807	5067	5325	5584	5841
3.9	6098	6354	6609	6864	7118	7372	7624	7877	8128	8379
4.0	8629	8879	9128	9377	9624	9872	*0118	*0364	*0610	*0854
4.1	1.4 1099	1342	1585	1828	2070	2311	2552	2792	3031	3270
4.2	3508	3746	3984	4220	4456	4692	4927	5161	5395	5629
4.3	5862	6094	6326	6557	6787	7018	7247	7476	7705	7933
4.4	8160	8387	8614	8840	9065	9290	9515	9739	9962	*0185
4.5	1.5 0408	0630	0851	1072	1293	1513	1732	1951	2170	2388
4.6	2606	2823	3039	3256	3471	3687	3902	4116	4330	4543
4.7	4756	4969	5181	5393	5604	5814	6025	6235	6444	6653
4.8	6862	7070	7277	7485	7691	7898	8104	8309	8515	8719
4.9	8924	9127	9331	9534	9737	9939	*0141	*0342	*0543	*0744
N	0	1	2	3	4	5	6	7	8	9

NATURAL LOGARITHMS

N	0	1	2	3	4	5	6	7	8	9
5.0	1.6 0944	1144	1343	1542	1741	1939	2137	2334	2531	2728
5.1	2924	3120	3315	3511	3705	3900	4094	4287	4481	4673
5.2	4866	5058	5250	5441	5632	5823	6013	6203	6393	6582
5.3	6771	6959	7147	7335	7523	7710	7896	8083	8269	8455
5.4	8640	8825	9010	9194	9378	9562	9745	9928	*0111	*0293
5.5	1.7 0475	0656	0838	1019	1199	1380	1560	1740	1919	2098
5.6	2277	2455	2633	2811	2988	3166	3342	3519	3695	3871
5.7	4047	4222	4397	4572	4746	4920	5094	5267	5440	5613
5.8	5786	5958	6130	6302	6473	6644	6815	6985	7156	7326
5.9	7495	7665	7834	8002	8171	8339	8507	8675	8842	9009
6.0	1.7 9176	9342	9509	9675	9840	*0006	*0171	*0336	*0500	*0665
6.1	1.8 0829	0993	1156	1319	1482	1645	1808	1970	2132	2294
6.2	2455	2616	2777	2938	3098	3258	3418	3578	3737	3896
6.3	4055	4214	4372	4530	4688	4845	5003	5160	5317	5473
6.4	5630	5786	5942	6097	6253	6408	6563	6718	6872	7026
6.5	7180	7334	7487	7641	7794	7947	8099	8251	8403	8555
6.6	8707	8858	9010	9160	9311	9462	9612	9762	9912	*0061
6.7	1.9 0211	0360	0509	0658	0806	0954	1102	1250	1398	1545
6.8	1692	1839	1986	2132	2279	2425	2571	2716	2862	3007
6.9	3152	3297	3442	3586	3730	3874	4018	4162	4305	4448
7.0	4591	4734	4876	5019	5161	5303	5445	5586	5727	5869
7.1	6009	6150	6291	6431	6571	6711	6851	6991	7130	7269
7.2	7408	7547	7685	7824	7962	8100	8238	8376	8513	8650
7.3	8787	8924	9061	9198	9334	9470	9606	9742	9877	*0013
7.4	2.0 0148	0283	0418	0553	0687	0821	0956	1089	1223	1357
7.5	1490	1624	1757	1890	2022	2155	2287	2419	2551	2683
7.6	2815	2946	3078	3209	3340	3471	3601	3732	3862	3992
7.7	4122	4252	4381	4511	4640	4769	4898	5027	5156	5284
7.8	5412	5540	5668	5796	5924	6051	6179	6306	6433	6560
7.9	6686	6813	6939	7065	7191	7317	7443	7568	7694	7819
8.0	7944	8069	8194	8318	8443	8567	8691	8815	8939	9063
8.1	9186	9310	9433	9556	9679	9802	9924	*0047	*0169	*0291
8.2	2.1 0413	0535	0657	0779	0900	1021	1142	1263	1384	1505
8.3	1626	1746	1866	1986	2106	2226	2346	2465	2585	2704
8.4	2823	2942	3061	3180	3298	3417	3535	3653	3771	3889
8.5	4007	4124	4242	4359	4476	4593	4710	4827	4943	5060
8.6	5176	5292	5409	5524	5640	5756	5871	5987	6102	6217
8.7	6332	6447	6562	6677	6791	6905	7020	7134	7248	7361
8.8	7475	7589	7702	7816	7929	8042	8155	8267	8380	8493
8.9	8605	8717	8830	8942	9054	9165	9277	9389	9500	9611
9.0	9722	9834	9944	*0055	*0166	*0276	*0387	*0497	*0607	*0717
9.1	2.2 0827	0937	1047	1157	1266	1375	1485	1594	1703	1812
9.2	1920	2029	2138	2246	2354	2462	2570	2678	2786	2894
9.3	3001	3109	3216	3324	3431	3538	3645	3751	3858	3965
9.4	4071	4177	4284	4390	4496	4601	4707	4813	4918	5024
9.5	5129	5234	5339	5444	5549	5654	5759	5863	5968	6072
9.6	6176	6280	6384	6488	6592	6696	6799	6903	7006	7109
9.7	7213	7316	7419	7521	7624	7727	7829	7932	8034	8136
9.8	8238	8340	8442	8544	8646	8747	8849	8950	9051	9152
9.9	9253	9354	9455	9556	9657	9757	9858	9958	*0058	*0158
10.0	2.3 0259	0358	0458	0558	0658	0757	0857	0956	1055	1154
N	0	1	2	3	4	5	6	7	8	9

Appendix G

POWER—ROOTS—RECIPROCALS

n	n^2	$\sqrt{n}$	$\sqrt{10n}$	n^3	$\sqrt[3]{n}$	$\sqrt[3]{10n}$	$\sqrt[3]{100n}$	$1/n$
1.0	1.0000	1.0000	3.1623	1.0000	1.0000	2.1544	4.6416	1.0000
1.1	1.2100	1.0488	3.3166	1.3310	1.0323	2.2240	4.7914	.9091
1.2	1.4400	1.0954	3.4641	1.7280	1.0627	2.2894	4.9324	.8333
1.3	1.6900	1.1402	3.6056	2.1970	1.0914	2.3513	5.0658	.7692
1.4	1.9600	1.1832	3.7417	2.7440	1.1187	2.4101	5.1925	.7143
1.5	2.2500	1.2247	3.8730	3.3750	1.1447	2.4662	5.3133	.6667
1.6	2.5600	1.2649	4.0000	4.0960	1.1696	2.5198	5.4288	.6250
1.7	2.8900	1.3038	4.1231	4.9130	1.1935	2.5713	5.5397	.5882
1.8	3.2400	1.3416	4.2426	5.8320	1.2164	2.6207	5.6462	.5556
1.9	3.6100	1.3784	4.3589	6.8590	1.2386	2.6684	5.7489	.5263
2.0	4.0000	1.4142	4.4721	8.0000	1.2599	2.7144	5.8480	.5000
2.1	4.4100	1.4491	4.5826	9.2610	1.2806	2.7589	5.9439	.4762
2.2	4.8400	1.4832	4.6904	10.6480	1.3006	2.8020	6.0368	.4545
2.3	5.2900	1.5166	4.7958	12.1670	1.3200	2.8439	6.1269	.4348
2.4	5.7600	1.5492	4.8990	13.8240	1.3389	2.8845	6.2145	.4167
2.5	6.2500	1.5811	5.0000	15.6250	1.3572	2.9240	6.2996	.4000
2.6	6.7600	1.6125	5.0990	17.5760	1.3751	2.9625	6.3825	.3846
2.7	7.2900	1.6432	5.1962	19.6830	1.3925	3.0000	6.4633	.3704
2.8	7.8400	1.6733	5.2915	21.9520	1.4095	3.0366	6.5421	.3571
2.9	8.4100	1.7029	5.3852	24.3890	1.4260	3.0723	6.6191	.3448
3.0	9.0000	1.7321	5.4772	27.0000	1.4422	3.1072	6.6943	.3333
3.1	9.6100	1.7607	5.5678	29.7910	1.4581	3.1414	6.7679	.3226
3.2	10.2400	1.7889	5.6569	32.7680	1.4736	3.1748	6.8399	.3125
3.3	10.8900	1.8166	5.7446	35.9370	1.4888	3.2075	6.9104	.3030
3.4	11.5600	1.8439	5.8310	39.3040	1.5037	3.2396	6.9795	.2941
3.5	12.2500	1.8708	5.9161	42.8750	1.5183	3.2711	7.0473	.2857
3.6	12.9600	1.8974	6.0000	46.6560	1.5326	3.3019	7.1138	.2778
3.7	13.6900	1.9235	6.0828	50.6530	1.5467	3.3322	7.1791	.2703
3.8	14.4400	1.9494	6.1644	54.8720	1.5605	3.3620	7.2432	.2632
3.9	15.2100	1.9748	6.2450	59.3190	1.5741	3.3912	7.3061	.2564
4.0	16.0000	2.0000	6.3246	64.0000	1.5874	3.4200	7.3681	.2500
4.1	16.8100	2.0248	6.4031	68.9210	1.6005	3.4482	7.4290	.2439
4.2	17.6400	2.0494	6.4807	74.0880	1.6134	3.4760	7.4889	.2381
4.3	18.4900	2.0736	6.5574	79.5070	1.6261	3.5034	7.5478	.2326
4.4	19.3600	2.0976	6.6333	85.1840	1.6386	3.5303	7.6059	.2273
4.5	20.2500	2.1213	6.7082	91.1250	1.6510	3.5569	7.6631	.2222
4.6	21.1600	2.1448	6.7823	97.3360	1.6631	3.5830	7.7194	.2174
4.7	22.0900	2.1679	6.8557	103.823	1.6751	3.6088	7.7750	.2128
4.8	23.0400	2.1909	6.9282	110.592	1.6869	3.6342	7.8297	.2083
4.9	24.0100	2.2136	7.0000	117.649	1.6985	3.6593	7.8837	.2041
5.0	25.0000	2.2361	7.0711	125.000	1.7100	3.6840	7.9370	.2000
5.1	26.0100	2.2583	7.1414	132.651	1.7213	3.7084	7.9896	.1961
5.2	27.0400	2.2804	7.2111	140.608	1.7325	3.7325	8.0415	.1923
5.3	28.0900	2.3022	7.2801	148.877	1.7435	3.7563	8.0927	.1887
5.4	29.1600	2.3238	7.3485	157.464	1.7544	3.7798	8.1433	.1852

Appendix G (Cont'd)

POWER—ROOTS—RECIPROCALS

n	n^2	$\sqrt{n}$	$\sqrt{10n}$	n^3	$\sqrt[3]{n}$	$\sqrt[3]{10n}$	$\sqrt[3]{100n}$	$1/n$
5.5	30.2500	2.3452	7.4162	166.375	1.7652	3.8030	8.1932	.1818
5.6	31.3600	2.3664	7.4833	175.616	1.7758	3.8259	8.2426	.1786
5.7	32.4900	2.3875	7.5498	185.193	1.7863	3.8485	8.2913	.1754
5.8	33.6400	2.4083	7.6158	195.112	1.7967	3.8709	8.3396	.1724
5.9	34.8100	2.4290	7.6811	205.379	1.8070	3.8930	8.3872	.1695
6.0	36.0000	2.4495	7.7460	216.000	1.8171	3.9149	8.4343	.1667
6.1	37.2100	2.4698	7.8102	226.981	1.8272	3.9365	8.4809	.1639
6.2	38.4400	2.4900	7.8740	238.328	1.8371	3.9579	8.5270	.1613
6.3	39.6900	2.5100	7.9372	250.047	1.8469	3.9791	8.5726	.1587
6.4	40.9600	2.5298	8.0000	262.144	1.8566	4.0000	8.6177	.1563
6.5	42.2500	2.5495	8.0623	274.625	1.8663	4.0207	8.6624	.1538
6.6	43.5600	2.5690	8.1240	287.496	1.8758	4.0412	8.7066	.1515
6.7	44.8900	2.5884	8.1854	300.763	1.8852	4.0615	8.7503	.1493
6.8	46.2400	2.6077	8.2462	314.432	1.8945	4.0817	8.7937	.1471
6.9	47.6100	2.6268	8.3066	328.509	1.9038	4.1016	8.8366	.1449
7.0	49.0000	2.6458	8.3666	343.000	1.9129	4.1213	8.8790	.1429
7.1	50.4100	2.6646	8.4261	357.911	1.9220	4.1408	8.9211	.1408
7.2	51.8400	2.6833	8.4853	373.248	1.9310	4.1602	8.9628	.1389
7.3	53.2900	2.7019	8.5440	389.017	1.9399	4.1793	9.0041	.1370
7.4	54.7600	2.7203	8.6023	405.224	1.9487	4.1983	9.0450	.1351
7.5	56.2500	2.7386	8.6603	421.875	1.9574	4.2172	9.0856	.1333
7.6	57.7600	2.7568	8.7178	438.976	1.9661	4.2358	9.1258	.1316
7.7	59.2900	2.7749	8.7750	456.533	1.9747	4.2543	9.1657	.1299
7.8	60.8400	2.7928	8.8318	474.552	1.9832	4.2727	9.2052	.1282
7.9	62.4100	2.8107	8.8882	493.039	1.9916	4.2908	9.2443	.1266
8.0	64.0000	2.8284	8.9443	512.000	2.0000	4.3089	9.2832	.1250
8.1	65.6100	2.8460	9.0000	531.441	2.0083	4.3267	9.3217	.1235
8.2	67.2400	2.8636	9.0554	551.368	2.0165	4.3445	9.3599	.1220
8.3	68.8900	2.8810	9.1104	571.787	2.0247	4.3621	9.3978	.1205
8.4	70.5600	2.8983	9.1652	592.704	2.0328	4.3795	9.4354	.1190
8.5	72.2500	2.9155	9.2195	614.125	2.0408	4.3968	9.4727	.1176
8.6	73.9600	2.9326	9.2736	636.056	2.0488	4.4140	9.5097	.1163
8.7	75.6900	2.9496	9.3274	658.503	2.0567	4.4310	9.5464	.1149
8.8	77.4400	2.9665	9.3808	681.472	2.0646	4.4480	9.5828	.1136
8.9	79.2100	2.9833	9.4340	704.969	2.0723	4.4647	9.6190	.1124
9.0	81.0000	3.0000	9.4868	729.000	2.0801	4.4814	9.6549	.1111
9.1	82.8100	3.0166	9.5394	753.571	2.0878	4.4979	9.6905	.1099
9.2	84.6400	3.0332	9.5917	778.688	2.0954	4.5144	9.7259	.1087
9.3	86.4900	3.0496	9.6436	804.357	2.1029	4.5307	9.7610	.1075
9.4	88.3600	3.0659	9.6954	830.584	2.1105	4.5468	9.7959	.1064
9.5	90.2500	3.0822	9.7468	857.375	2.1179	4.5629	9.8305	.1053
9.6	92.1600	3.0984	9.7980	884.736	2.1253	4.5789	9.8648	.1042
9.7	94.0900	3.1145	9.8489	912.673	2.1327	4.5947	9.8990	.1031
9.8	96.0400	3.1305	9.8995	941.192	2.1400	4.6104	9.9329	.1020
9.9	98.0100	3.1464	9.9499	970.299	2.1472	4.6261	9.9666	.1010
10.0	100.000	3.1623	10.000	1000.00	2.1544	4.6416	10.0000	.1000

answers to odd-numbered problems

EXERCISE 0-2

1. 14 **3.** 550 **5.** 61 **7.** 36 **9.** 393, 372 **11.** 25
13. 4968 **15.** 32 **17.** 10, 902, 460 **19.** 104 **21.** 36
23. 1 **25.** ten-thousand's place

EXERCISE 0-3

1. $\dfrac{49}{24}$ **3.** $\dfrac{437}{264}$ **5.** $\dfrac{35}{12}$ **7.** $\dfrac{8}{15}$ **9.** $\dfrac{161}{300}$ **11.** $\dfrac{32}{15}$

13. $\dfrac{336}{145}$ **15.** $\dfrac{4150}{3}$ **17.** $\dfrac{125}{168}$ **19.** $\dfrac{22}{10}, \dfrac{3}{4}, \dfrac{2}{3}, \dfrac{3}{5}, \dfrac{15}{28}$

EXERCISE 0-4

1. a. 135.7416 b. 3.035 c. 75.7306 d. 155.6534 **3.** a. 1.7 b. 537.6
c. 0.0002 d. 50 **5.** a. $\dfrac{6241}{10,000}$ b. $\dfrac{10,241}{10,000}$ c. $\dfrac{3}{500}$ d. $\dfrac{2201}{100}$ e. $\dfrac{12,483}{20,000}$

f. $\dfrac{1161}{100}$ **7.** a. $\dfrac{35}{4}$ b. $\dfrac{288}{385}$

EXERCISE 1-1

1. true **3.** true **5.** true **7.** false **9.** false **11.** finite
13. finite **15.** infinite **17.** finite **19.** $\{14, 16, 18\}$

EXERCISE 1-2

1. false **3.** false **5.** true **7.** false, $\dfrac{22}{7}$ is an *approximation* to π

9. true **11.** true **13.** true **15.** finite **17.** infinite
19. infinite **21.** true **23.** true **25.** false **27.** true

EXERCISE 1-3

1. $\{x \mid 3 \leq x \leq 6\}$

3. $\{x \mid -2 < x < 3\}$

5. $\{x \mid x < 6\}$

7. $\{x \mid x \geq -7\}$

9. $[4, \infty)$

11. $[2, 5]$

13. $\{x \mid 3 \leq x < 4\}$

15. $(5, 12]$

EXERCISE 1-4

1. a. symmetric axiom   b. commutative axiom of multiplication   c. distributive axiom   d. commutative and associative axioms of multiplication   e. distributive axiom   f. commutative axiom of addition   g. axiom of closure for multiplication   h. associative and commutative axioms of multiplication   i. commutative axiom of multiplication   j. additive identity axiom   k. multiplicative identity axiom   l. commutative and associative axioms of addition   m. definition of subtraction   n. definition of division   **3.** a. 0 b. undefined   c. undefined

EXERCISE 1-5

1. 6 **3.** 270 **5.** 0 **7.** -6 **9.** 2 **11.** 11 **13.** -2
15. -63 **17.** -6 **19.** $6 - x$ **21.** $-12x + 12y$ (or $12y - 12x$)
23. $-\dfrac{1}{3}$ **25.** -2 **27.** 18 **29.** 25 **31.** -16 **33.** $-\dfrac{2}{3}$

35. $-8; -2; 2; 15; -\dfrac{5}{3}$ **37.** $10; -4; 4; 21; \dfrac{3}{7}$

39. $12; -12; 12; 0; 0$ **41.** $-\dfrac{7}{2}; \dfrac{9}{2}; -\dfrac{9}{2}; -2; \dfrac{1}{8}$

43. $-\dfrac{3}{4}; -\dfrac{1}{4}; \dfrac{1}{4}; \dfrac{1}{8}; -2$ **45.** $-3a$ **47.** $-\dfrac{67}{7}$ **49.** -12

51. $\dfrac{128}{3}$ **53.** 148 **55.** $\dfrac{19}{9}$

EXERCISE 1-6

1. $2^5 (=32)$ **3.** x^{15} **5.** w^{12} **7.** x^{13} **9.** $\dfrac{x^8}{y^{17}}$ **11.** x^{48}

13. $\dfrac{x^{10}}{y^{50}}$ **15.** $\dfrac{x^{10}}{y^{15}}$ **17.** $8x^6y^9$ **19.** $\dfrac{w^4s^6}{y^4}$ **21.** x^6 **23.** $\dfrac{1}{8x^6}$

25. x^{14} **27.** 1 **29.** $x^{ac}y^{bc}$ **31.** -729 **33.** $-x^4$ **35.** x^{10}

37. $\dfrac{1}{x^8y^{12}}$

EXERCISE 1-8

1. $\dfrac{1}{x^5}$ **3.** $\dfrac{x}{y^4}$ **5.** x^7 **7.** x **9.** $\dfrac{x^3}{y^2z^2}$ **11.** $\dfrac{2x^2}{b^4}$ **13.** $\dfrac{2}{x}$

15. $\dfrac{1}{9x^2}$ **17.** $\dfrac{x^6}{y^{33}}$ **19.** $\dfrac{z^2}{x^3y^5}$ **21.** $\dfrac{wxz}{y^4}$ **23.** $\dfrac{2x^4z^2}{3y^{11}}$

25. $\dfrac{512x^{12}}{y^{18}}$

EXERCISE 1-9

1. 6.0214×10^{-6} **3.** 1.04×10 **5.** 2.62451001×10^4
7. 1.42×10^2 **9.** 7.6×10^{-1} **11.** 262,000,000 **13.** .000000000624
15. .2020 **17.** 761100 **19.** 5.983×10^{24} kilograms
21. 6.67×10^{-11} **23.** 9.999×10^9 years **25.** 2.40×10^{12} volts

EXERCISE 1-10

1. 9 **3.** -4 **5.** 1 **7.** 0 **9.** $\dfrac{1}{5}$ **11.** x^2 **13.** 12

15. 0 **17.** -1 **19.** .15

EXERCISE 1-11

1. 5 **3.** $\sqrt[8]{61}$ **5.** $\sqrt[5]{x^3}$ **7.** $x^{1/3}$ **9.** $\dfrac{1}{4}$ **11.** $\dfrac{1}{x^2}$

13. $x^{3/5}$ **15.** x^4 **17.** $x^{2/5}$ **19.** $\dfrac{y^{1/3}}{x^2}$ **21.** 7 **23.** x^8y^9

25. $x^{1/5}y^{2/5}$ **27.** $5(3)^{1/2}$ **29.** xy^2z^3 **31.** $x^{3/2}$ **33.** $x^{8/3}y^{4/3}$

35. $\dfrac{x^2}{y^4}$ **37.** 2 **39.** x^2 **41.** $2^{1/10}$ **43.** $a^{3/4}b^{3/2}c^{9/4}$

45. $5x^2(3)^{1/2}$ **47.** $\dfrac{3^{1/2}}{3}$ **49.** $\dfrac{6^{2/3}}{6}$ **51.** $\dfrac{2x^{1/3}}{x}$ **53.** $\dfrac{x^{2/7}}{x^2}$

55. $\dfrac{1}{27}$ **57.** $\dfrac{x^{4/3}y}{zw^2}$ **59.** $9xy^2w^{1/2}$ **61.** $\dfrac{x^{1/2}}{12x}$ **63.** $x^{m/2}$

65. $\dfrac{x^{1/4}y^{1/2}z^{3/4}}{y}$ **67.** $\dfrac{3^{2/3}n^{2/3}h^2\pi^{1/3}}{8m\pi}$ **69.** It is halved.

REVIEW PROBLEMS—CHAPTER 1

1. $\dfrac{33}{2}$ 2. $x^{45}y^{24}z^{20}$ 3. 2   4. -1 5. $\dfrac{\sqrt[5]{256}}{4}$ 6. 2

7. $-4x^5$ 8. $x^{a+b+r+t}$ 9. x^2 10. $x^{10}y^2z^4$ 11. 0   12. 2

13. $\dfrac{x^6}{y^4}$ 14. undefined   15. .2   16. 16   17. $\dfrac{x^8z^{48}}{y^{12}}$

18. $\dfrac{1}{x^5}$ 19. $\{x \mid 3 < x < 5\}$ 20. $\{x \mid 2 \le x < 6\}$ 21. $\{x \mid x \ge 2\}$

22. $\{x \mid -6 < x \le 6\}$ 23. 8.76×10^{-4}

EXERCISE 2-1

1. (a), (e); 2   3. (b), (d), (e); 2   5. (b), (c), (e); 4   7. (a), (e); 1
9. (b), (d)   11. (b), (c), (e); 4   13. (b)   15. (b), (e); 5

17. (b), (d)   19. 3; 4   21. $2y; -2$ 23. $4a; 1$ 25. $0; \dfrac{7}{3}$

EXERCISE 2-2

1. $7 + x - y$ 3. $a + 2b + 2c$ 5. $x - y + z$ 7. $a - b - c + d$
9. $2x + 3 + 2y - x^2$ 11. $3a - 3b - 2x - 6y + 4$
13. $-x - 6 + 3x^2$ 15. $2ab + 6 - 2x^2 + 2y^2$
17. $3x - 6 + 6x^2 - 12y - x^3$
19. $-2x + 2y + 3 - 6x^2 + 6y^2 - 6xy + 6y^3 - 3x^3$
21. $-(-x - 2x^2 + 3x^3) + 2b + 1$
23. $-(2x^2 - 3x^2y + 5x + 3xy) - 6y - 1$
25. $Kt_1 - Kt_2 - \dfrac{v}{c^2}Kx_1 + \dfrac{v}{c^2}Kx_2$

EXERCISE 2-3

1. $11x - 2y - 3$ 3. $-5x + 6y - 7$ 5. $5 - b + a$ 7. $-y^2 - 1$
9. $2b + c$ 11. $x^2 + 2a$ 13. $-2x - 4y - 3xy + 1$
15. $-7x - 2y - 19$ 17. (a) $5a^2x + 2a^2bz$ (b) $-a^2x + 6a^2y - 4a^2bz$
(c) $a^2x - 6a^2y + 4a^2bz$ 19. (a) $16z^2xy + x - 3y$
(b) $-2z^2xy + 16z^2 - 5x - 11y$ (c) $2z^2xy - 16z^2 + 5x + 11y$
21. (a) $\dfrac{3}{2}a^2x + \dfrac{7}{2}bc - f + g$ (b) $-\dfrac{1}{2}a^2x + \dfrac{5}{2}bc + 5f - g$

(c) $\dfrac{1}{2}a^2x - \dfrac{5}{2}bc - 5f + g$ 23. (a) $-5x^4 + 10x^2 - 10x - 4$

(b) $-x^4 + 4x^2 + 6x + 6$ (c) $-7x^4 + 8x^2 - 8x - 6$
25. (a) $-21xy + 18x - 7y - 8z$ (b) $47xy + 7y - 6z + 2$
(c) $11xy + 18x - 7y - 8z + 2$ 27. $2a + 4b + 4c$

EXERCISE 2-4

1. $8x^3y^2z^2$ 3. $3x^3y^4z^2$ 5. $-x^3yz^4$ 7. $6a^3b^4c^3$ 9. $-144x^{10}y^5z$
11. $36x^8y^8$ 13. $ab^2c^5d^3$ 15. $x^3 - 2x^2 + 4x$

17. $-2a^3b^3 + a^3b^2 - 2a^2b$ **19.** $-2x^2y^4 + 2x^2y^3 - 4x^2y^2 - 2xy^2$

21. $a^5bx^2y^2 - a^5x^3y^2 - a^5x^2y^4$ **23.** $x^2 + 5x + 6$ **25.** $2y^2 - 5y - 12$

27. $10x^2 + 19x + 6$ **29.** $x^3 + 4x^2 - 3x - 12$

31. $6x^4 + 11x^3 - 7x^2 - 10x + 5$ **33.** $2x^3 + 12x^2 + 8x + 48$

35. $3x^2 + 2y^2 + 5xy + 2x - 8$ **37.** $9a^2b^6 - 12ab^3rt + 4r^2t^2$

39. $12x^4 + 12x^3 - 17x^2 - 20x - 5$ **41.** $x^2 + x - 6$

43. $6x^3y + 9x^2y^2$ **45.** $-3xyw^2$ **47.** $5x^3 + 5x^2 + 6x$ **49.** 0

51. $b + c + 5ab + 5ac$

EXERCISE 2-5

1. $2x + y$ **3.** $\dfrac{xw}{2}$ **5.** $-3y^3$ **7.** $\dfrac{10yz^2}{x}$ **9.** $\dfrac{1}{x}$ **11.** $\dfrac{4x}{z^3}$

13. $4a^8b^7$ **15.** $-xy^4$ **17.** $1 + \dfrac{x}{y}$ **19.** $\dfrac{3x}{y} - 1 + \dfrac{1}{xy^2}$

21. $-2x^6 - \dfrac{3x}{y} + \dfrac{1}{y^4}$ **23.** $4y^2 + 2x^3y^2z - x^4y$

25. $-2xy + \dfrac{2y^2}{3x} - \dfrac{7}{3} + \dfrac{4}{3x}$ **27.** $\dfrac{1}{625s^8r^6z^{22}}$ **29.** $x + 3$

31. $x^3 + x^2 + 3x + 3 + \dfrac{4}{x - 1}$ **33.** $x + 2 + \dfrac{2x + 1}{x^2 - x + 1}$

35. $5x^2 - 13x - 36 + \dfrac{39x + 220}{x^2 - x + 6}$ **37.** $x^3 - x^2y + xy^2 - y^3$

39. $1 - \dfrac{T_2}{T_1}; \ T_1 \neq 0$

REVIEW PROBLEMS—CHAPTER 2

1. $4a - 4c$ **2.** $7a + 2b - 2$ **3.** $x^5 + 8x^4 - 17x^3 - x^2 + 12x + 8$

4. $-25x^3 - 2x^2 + 3xy + 1$ **5.** $-15 - 4a^2b^3c^2$ **6.** $30u^8v^{10}w^6$

7. $\dfrac{49b^6w^6y^2}{x^2}$ **8.** $3x^3y^2 - x^4 + 18x^2y^2$ **9.** $9p^6q^2r^2 + 9p^4q^2r^5 - 9p^5q^3r^3$

10. $x^3 + 2x^2 - 2x - 3$ **11.** $x^2 - y^2 + 6y - 9$ **12.** $-16x^{13}y^{20}$

13. $-\dfrac{3x^5y^2}{z} + \dfrac{9x^3y^3}{z} - \dfrac{6y^{12}}{xz}$ **14.** $x^2 - 8x + 17 + \dfrac{-38}{x + 2}$

15. $x^4 - 2x^3 - 2x^2 - 2x - 1$

EXERCISE 3-1

1. $3x - 3y$ **3.** $x^2 + 10x + 25$ **5.** $x^2 - 8x + 16$

7. $x^2 + x + \dfrac{1}{4}$ **9.** $x^2 + 4xy + 4y^2$ **11.** $3x^3 - 12bx^2 + 12b^2x$

13. $4x^2 + 12x + 9$ **15.** $9x^2 - 24xy + 16y^2$ **17.** $64x^2 + 32xy + 4y^2$

19. $9a^2b^2c^2 - 24abcef + 16e^2f^2$ **21.** $36x^2 - 60xy + 25y^2$

23. $x^2 + 11x + 24$ **25.** $x^2 - x - 2$ **27.** $x^2 + 2x - 35$

29. $x^2 - 5x + 6$ **31.** $x^2 - 9$ **33.** $9x^2 - 16$ **35.** $\dfrac{1}{4}x^2 - 4$

37. $10x^2 + 19x + 6$ **39.** $12x^2 - 25x + 12$ **41.** $x^4 - 9$

43. $\frac{2}{9}a^2 - a - 2$ **45.** $2t^2 + 5t - 3$ **47.** $x^2y^2z^2 - a^2$

49. $49 - x^2$ **51.** $6x^2 - 6x - 120$ **53.** $x^3 + 15x^2 + 75x + 125$

55. $6y^2 + by - b^2$ **57.** $2x^4 - 18x^3 + 54x^2 - 54x$

59. $12x^2y^2z^2 - 2axyz - 2a^2$ **61.** $x^2 - 2xy + y^2 - 4x + 4y + 4$

63. $x^2 - 3$ **65.** $\frac{3}{4}Rcz^2 - \frac{3}{2}Rcz + \frac{3}{4}Rc$

EXERCISE 3-2

1. $5x(2y + z)$ **3.** $3xy(1 + 2xy + 3x^2y^2)$ **5.** $7zy(3z + w^2 + 2z^2y^2w^3)$

7. $(x + 4)(x - 4)$ **9.** $(x + 2)(x + 4)$ **11.** $(x + 5)(x - 3)$

13. $(x - 7)(x - 5)$ **15.** $(x + 9)^2$ **17.** $(x - 3)^2$

19. $4(x + 2)(x - 2)$ **21.** prime **23.** $(2x + 1)(x + 3)$

25. $(x - 3)(3x + 2)$ **27.** $(2x + 3)^2$ **29.** $(3x - 4)^2$ **31.** prime

33. $(3 + y)(3 - y)$ **35.** $(7 - x)(2 + x)$

37. $32x^2y^2mn^2(mn^2 - 2xymn - 3)$ **39.** $(7 + 2x)(7 - 2x)$

41. $4(x - 5)(x + 3)$ **43.** $(2x + 1)^2$ **45.** $(x + 3)(2 - x)$

47. $4x(3x + 1)(x + 2)$ **49.** $(x + 2)(x^2 - 2x + 4)$

51. $(2a^4bc^8 + 1)(2a^4bc^8 - 1)$ **53.** $(2xy + w)(2xy - w)$

55. $y(y - 3)(y - 1)$ **57.** $2(8x + 3)(2x - 1)$ **59.** prime

61. $(x + y - b)(x^2 + 2xy + y^2 + bx + by + b^2)$

63. $3a^{4/3}x^{2/3}y^{5/2}(3xy - 1 - 2a)$ **65.** $2(3ab - 2c)(3ab + c)$

67. $(x - y)(x + y + 1)$

69. $(x - y)(x^2 + xy + y^2)(x + y)(x^2 - xy + y^2)$

71. $(a + b - c)(a - b + c)$ **73.** $(x^{a+b} + y^{a-b})(x^{a+b} - y^{a-b})$

75. $(x + y)(x - y + 1)$ **77.** $(3z + 2x - 1)(3z - 2x + 1)$

79. $(x + 1)(x^2 + 1)$ **81.** $2(a + b)(x + y)(x - y)$

83. $(2x + 3 + a)(2x + 3 - a)$ **85.** $4(2x + 1)(2x - 1)(x + 2)(x - 2)$

87. $(x - y)(4x^2 + 1)(x + y)$ **89.** $(x + y)(n - m)(y - x - m - n)$

91. $\frac{2\pi^2me^4}{h^3}\left(\frac{1}{n_2} + \frac{1}{n_1}\right)\left(\frac{1}{n_2} - \frac{1}{n_1}\right)$ **93.** $\frac{1}{5}\pi\sigma hr^2\left(\frac{1}{4}r^2 + h^2\right)$

95. $\frac{\mu W}{c + d}\left(d + \frac{ha}{g}\right)$

REVIEW PROBLEMS—CHAPTER 3

1. $x^2 - 49$ **2.** $36 - 4x^2$ **3.** $4x^2 + 12x + 9$

4. $x^2 + 4xy + 4y^2$ **5.** $16x^2 - 56x + 49$ **6.** $4y^2 - 12xy + 9x^2$

7. $x^2 + 2x - 3$ **8.** $x^2 - 9x + 14$ **9.** $6x^2 + 8x - 8$

10. $10x^2 - 17x + 6$ **11.** $8x^3 + 12x^2 + 6x + 1$

12. $x^3 - 6x^2y + 12xy^2 - 8y^3$ **13.** $5x(2y + z)$ **14.** $3x^2y(1 - 3xy^2)$

15. $4bc(2a^3 - 3ab^2d + b^3cd^2)$ **16.** $(x + 5)(x - 5)$

17. $(2x - 3)(x + 5)$ **18.** $(x + 3)^2$ **19.** $(2x - 3)(4x^2 + 6x + 9)$

20. $2(x + 4)(x + 2)$ **21.** $x(xy - 5)^2$ **22.** $(x + 2)(x - 2)^2$
23. $(2x + 3z)(2x - 3z)$ **24.** $(xy + z^2)(x + 1)(x - 1)$

EXERCISE 4-1

1. true   **3.** true   **5.** false   **7.** true   **9.** false   **11.** $\dfrac{x}{2}$

13. $\dfrac{x + 2}{3x + 5}$ **15.** 3 **17.** $\dfrac{1}{3a}$ **19.** $\dfrac{x - 9}{x}$ **21.** $\dfrac{x - 12}{x - 3}$

23. $\dfrac{x + 3}{x - 4}$ **25.** -1 **27.** $\dfrac{x + 7}{7 - x}$ **29.** $3 - x$

31. $\dfrac{x^2 - 2x + 4}{x + 1}$ **33.** $\dfrac{1}{2y - x}$ **35.** $(a^2 + b^2)(a - b)$ **37.** $\dfrac{2x - 1}{3x - 2}$

39. $\dfrac{y}{x - y}$ **41.** $3w - 2x + 1$ **43.** $2(a - b)$

45. $\bar{x} = \dfrac{x + 6}{8};\ \bar{y} = y$

EXERCISE 4-2

1. $x + 3$ **3.** $\dfrac{2x^2 + 3x + 12}{(2x - 1)(x + 3)}$ **5.** $\dfrac{x + 2}{2}$ **7.** $\dfrac{x^2 + 2xy - y^2}{(x - y)(x + y)}$

9. 0 **11.** $\dfrac{-1}{(x + 1)(x - 1)}$ **13.** $\dfrac{1}{x^2 + x + 1}$

15. $\dfrac{-x^2 - 2x + 1}{(x + 2)(x + 5)(x + 1)}$ **17.** $\dfrac{(2x - 1)(x + 1)}{x - 1}$

19. $\dfrac{3y^2 - 7y - 2}{(3y + 1)(y - 2)(3y - 1)}$ **21.** $\dfrac{x^2 + 4xy + y^2}{(x + y)^2(x - y)}$

23. $\dfrac{x^4 - 3y^4 - 2x^2y^2 + 2x^3y - 2xy^3}{(x + y)(x - y)(x^2 + y^2)}$ **25.** $\dfrac{9x - 16a - 17}{6}$ **27.** 0

29. $\dfrac{x}{(x - 2a)^2}$ **31.** $\dfrac{48a^3}{(x - a)(x + a)(x + 3a)(x - 3a)}$

33. $\dfrac{kq_1(x + 2)(x + 3) + kq_2x(x + 3) + kq_3x(x + 2)}{x(x + 2)(x + 3)}$

35. $\dfrac{1}{4\pi\epsilon_0}\left(\dfrac{q_1r_2 - q_2r_1}{r_1r_2}\right)$

EXERCISE 4-3

1. $\dfrac{7}{xy}$ **3.** $\dfrac{4}{x^3}$ **5.** $\dfrac{ab^2}{xy^2}$ **7.** $-\dfrac{y^2}{(y - 3)(y + 2)}$ **9.** $\dfrac{x + 3}{x - 2}$

11. $-\dfrac{2x - 3}{2x + 3}$ **13.** $-(x + y)^2$ **15.** $\dfrac{x + 2}{x - 4}$ **17.** $\dfrac{x - 3}{x}$

19. $\dfrac{2(x + 4)}{(x - 4)(x + 2)}$ **21.** $\dfrac{(x - 3)(x - 1)}{(x - 2)(x + 2)}$ **23.** $\dfrac{x}{6y}$ **25.** $\dfrac{8}{27xy}$

27. $4z^2$ **29.** $-\dfrac{1}{x(x - y)}$ **31.** $\dfrac{2}{x^2}$ **33.** $\dfrac{FL}{eA}$

EXERCISE 4-4

1. $\dfrac{y(x+1)}{x(2y-1)}$ **3.** $-\dfrac{b}{a}$ **5.** $\dfrac{(6x-1)(x+2)}{2x^2(x+3)}$ **7.** $\dfrac{x^4+x^2+1}{x^2(x^2+x+1)}$

9. $x-y$ **11.** $\dfrac{2y+x}{y+2x}$ **13.** $\dfrac{x-y+1}{y-1}$ **15.** $\dfrac{a(2d+1)}{2d+2c+1}$

EXERCISE 4-5

1. 1.60 m   **3.** 2.78 yd   **5.** 1.86×10^4 cm²   **7.** 26.8 m/sec

9. 1.60×10^4 lb/m³   **11.** 8.93 atm   **13.** 5.87×10^{12} mi

REVIEW PROBLEMS—CHAPTER 4

1. 0   **2.** $\dfrac{3}{2}$ **3.** -1 **4.** $\dfrac{(x+3)(x-1)}{x}$ **5.** $\dfrac{4y}{x-1}$

6. $-\dfrac{2x}{x-3}$ **7.** $\dfrac{1}{x}$ **8.** $\dfrac{(3x+1)(x-2)(3x^2-2)}{x^2}$

9. $\dfrac{x+4}{(x-2)(x+5)}$ **10.** $\dfrac{(x-3)(x^2+81)}{x+9}$ **11.** $\dfrac{1}{x+6}$

12. $\dfrac{7}{(x+1)(x-1)}$ **13.** $\dfrac{(x-3)^2}{(x-1)^2}$ **14.** -1 **15.** 1.86×10^5 mm²

16. 26.8 m/sec

EXERCISE 5-1

1. a. $0, 2$ b. $5, 4$ c. $\dfrac{10}{3}$ d. $2, -4$ e. none   f. $\sqrt{2}, -\sqrt{2}$

g. $-1, -2, 3$ **3.** a. $x - 10 = 0$; no   b. $5x - 3 = 14$; no

c. $x(a+b) = bx + ax$; yes   d. $3 - 15x = 3 + 4x$; no

e. $2(x+3) = 5(x-9)$; no   f. $(x+2)(x-2) = -4 + x^2$; yes

g. $\dfrac{1}{t_1} + \dfrac{1}{t_2} = \dfrac{1}{T}$ h. $x + (x+1) + (x+2) = 36$ i. $s = vt$

j. $x(.05) + (8000 - x)(.06) = 420$ k. $z = y - 4z + 5\left[z - \dfrac{y}{5}\right]$

l. $F = \dfrac{9}{5}c + 32$ m. $E = mc^2$ n. $W = mg$ o. $pv = k$

EXERCISE 5-2

1. Adding 5 to both members; equivalence guaranteed

3. Squaring both members; equivalence not guaranteed

5. Dividing both members by x; equivalence not guaranteed

7. Multiplying both members by $x - 1$; equivalence not guaranteed

9. Multiplying both members by $\dfrac{x-5}{x}$; equivalence not guaranteed

EXERCISE 5-3

1. $\left\{\dfrac{9}{2}\right\}$ 3. $\{0\}$ 5. $\left\{-\dfrac{6}{5}\right\}$ 7. $\{4\}$ 9. $\{2\}$ 11. $\left\{\dfrac{10}{3}\right\}$

13. $\{90\}$ 15. $\{8\}$ 17. $\left\{-\dfrac{26}{9}\right\}$ 19. $\left\{\dfrac{60}{17}\right\}$ 21. $\left\{\dfrac{14}{3}\right\}$

23. $\{3\}$ 25. $\left\{\dfrac{29}{12}\right\}$ 27. $\left\{\dfrac{20}{3}\right\}$ 29. $\left\{\dfrac{1}{2}\right\}$ 31. $V_2 = \dfrac{P_1 V_1}{P_2}$

33. $a = \dfrac{v_0 - v}{t}$ 35. $m = \dfrac{2K}{v^2}$ 37. $h = \dfrac{V}{\pi r^2}$ 39. $R = \dfrac{P}{i^2}$

41. $i = \dfrac{E_2 - E_1}{2R_1 + 3R_2 - 4R_4}$ 43. $Q' = \dfrac{Fr^2}{kQ}$ 45. $T_2 = \dfrac{VP_2 T_1}{V_0 P_1}$

47. $E = \dfrac{IR}{1 - e^{-Rt/L}}$

49. $t_2 = \dfrac{Q + mct_1 - mL}{mc}$, $t_1 = \dfrac{mct_2 + mL - Q}{mc}$, $c = \dfrac{Q - mL}{m(t_2 - t_1)}$

51. $m = \dfrac{I\omega^2}{2gh - v^2}$, $I = \dfrac{m(2gh - v^2)}{\omega^2}$ 53. $h = \dfrac{V - 2\pi r^2}{2\pi r}$

55. $f = \dfrac{f_0(v_0 U + \epsilon)}{\epsilon}$ 57. $n_0 = \dfrac{\sigma\lambda + n_e L}{L}$, $n_e = \dfrac{n_0 L - \sigma\lambda}{L}$ 59. $53°C$

EXERCISE 5-4

1. $\left\{\dfrac{1}{4}\right\}$ 3. $\left\{\dfrac{3}{2}\right\}$ 5. $\{0\}$ 7. $\left\{\dfrac{5}{3}\right\}$ 9. $\left\{\dfrac{1}{8}\right\}$ 11. $\{3\}$

13. $\left\{\dfrac{5}{13}\right\}$ 15. $\varnothing$ 17. $\left\{\dfrac{7}{5}\right\}$ 19. $x = \dfrac{b(4a - 1)}{a(a + 2b)}$

21. $t_1 = \dfrac{p_1 v_1 t_2}{p_2 v_2}$ 23. $P_2 = \dfrac{V_0 P_1 T_2}{V T_1}$ 25. $q = \dfrac{pf}{p - f}$, $f = \dfrac{pq}{p + q}$

27. $R_1 = \dfrac{R_t R_2}{R_2 - R_t}$ 29. $m^2 = \dfrac{p^2}{v^2}$

31. $R_1 = \dfrac{f(n - 1)R_2}{R_2 + f(n - 1)}$, $R_2 = \dfrac{f(n - 1)R_1}{f(n - 1) - R_1}$ 33. 12 ohms

35. $\dfrac{40}{3}$ cm ≈ 13.3 cm

EXERCISE 5-5

1. $\left\{\dfrac{8}{7}\right\}$ 3. $\left\{\dfrac{8}{3}\right\}$ 5. $\left\{\dfrac{21}{19}\right\}$ 7. $\left\{-\dfrac{4}{3}\right\}$ 9. $\left\{\dfrac{3}{4}\right\}$ 11. $\dfrac{17}{9}$

13. 1   15. $7, \dfrac{35}{3}$ units   17. $\dfrac{32}{3}$ 19. 24 cm per sec

21. $77°F = 25°C$; $-40°F = -40°C$ 23. $\dfrac{5}{9}$ atm.   25. 259 cycles/sec

27. 543 cycles/sec   29. $\dfrac{\alpha}{1 + \alpha\beta}$

EXERCISE 5-6

1. 18 gallons
3. 420 gallons of the 20% solution and 280 gallons of the 30% solution
5. $4\frac{1}{2}$ miles   7. $2\frac{2}{9}$ hours   9. $37\frac{1}{2}$ gm of A, $137\frac{1}{2}$ gm of B
11. $3\frac{3}{4}$ lb   13. $5\frac{3}{5}$ ft   15. 2100 ft; $\frac{12}{11}$ sec   17. 45 degrees
19. 4 days; 56 inches   21. 80 ft   23. 60 inches

REVIEW PROBLEMS—CHAPTER 5

1. $\left\{\frac{1}{4}\right\}$ 2. $\left\{-\frac{2}{15}\right\}$ 3. $\{0\}$ 4. $\left\{-\frac{41}{30}\right\}$ 5. $\left\{-\frac{1}{2}\right\}$
6. $\{0\}$ 7. $\left\{-\frac{9}{7}\right\}$ 8. $\left\{\frac{47}{9}\right\}$ 9. $\varnothing$ 10. $\{0\}$ 11. $\left\{\frac{5}{2}\right\}$
12. $\varnothing$ 13. -15 14. 60 gallons   15. 8 gallons   16. 3.76 lb

EXERCISE 6-1

1. x; $\mathbf{R}$; $11, \frac{13}{2}, 19, 7$ 3. x; $\mathbf{R}$; $-\frac{2}{3}, -\frac{2}{3}, -\frac{2}{3}, -\frac{2}{3}$
5. x; $\mathbf{R}$; $1, 2, 2, 5, x^2 + 2hx + h^2 + 1$ 7. s; $\mathbf{R}$; $0, 3, 8, 15, 48$
9. t; $\mathbf{R}$; $8, \frac{16}{3}, 8 - t, 8 - \frac{2}{t}$ 11. t; $\mathbf{R}$; $3, 0, 6$
13. x; $\{x \mid x > 0\}$; $1, \frac{1}{2}$ 15. r; $\mathbf{R}$; $-2, 2, 8, 2$ 17. yes; no
19. $0, 16, 64$; $\{t \mid t \geq 0\}$ 21. $A = f(a) = \frac{73a^2}{4}$
23. $V = f(x) = 4x(6 - x)(8 - x)$

EXERCISE 6-2

3.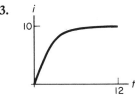

5.

7.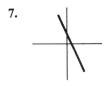

9.

11.

13.

15.

17.

19.

21.

23.

25.

27.

29.

31.

33.

35.

37.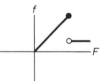

39.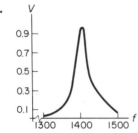

EXERCISE 6-3

1. $\dfrac{3}{2}$ **3.** $-\dfrac{4}{5}$ **5.** undefined   **7.** $\dfrac{1}{2}$ ohm   **9.** 1.2×10^{-5}

EXERCISE 6-4

1. $9x - y - 23 = 0$ **3.** $y - 5 = 0$ **5.** $x - 2y + 6 = 0$

7. $6x - y - 4 = 0$

9. $5x + y = 0$

11. $2x - y - 3 = 0$

13. $x - 7 = 0$

15. $(5, -4)$

17. $s = -\dfrac{5}{2}t + 70$

19. $2; \left(\dfrac{1}{2}, 0\right); (0, -1)$

21. $\dfrac{3}{8}; \left(\dfrac{8}{3}, 0\right);$ $(0, -1)$

23. slope not defined; $(-5, 0)$; no y-intercept

25. 0; no x-intercept; $(0, 1)$

27. $-\dfrac{1}{2}; (3, 0);$ $\left(0, \dfrac{3}{2}\right)$

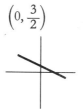

29. $x + 2y - 4 = 0;$ $y = -\dfrac{1}{2}x + 2;$ $\dfrac{x}{4} + \dfrac{y}{2} = 1$

31. $9x - 28y - 3 = 0; y = \dfrac{9}{28}x - \dfrac{3}{28}; \dfrac{x}{\frac{1}{3}} + \dfrac{y}{\frac{-3}{28}} = 1$

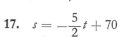
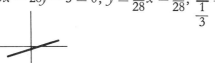

33. $x = 2$

35. a.

b.

c.

37. $P = \dfrac{T}{4} + 80$　　**39.** a. acc $= \dfrac{90}{5}$; acc $= 0$　b. $V = \dfrac{90}{5}t$; $V = 90$

REVIEW PROBLEMS—CHAPTER 6

1. $7, 46, 62, 3x^4 - 4x^2 + 7$　　**2.** $0, -\dfrac{4}{3}, -\dfrac{1}{6}, \dfrac{x+h-3}{x+h+4}$

3. 　　**4.** 　　**5.**

6. 　　**7.** 　　**8.**

9. 　　**10.** 　　**11.**

12. 　　**13.** $y = \dfrac{3}{2}x - 2$; $\dfrac{x}{\frac{4}{3}} + \dfrac{4}{-2} = 1$

14. $y = -\dfrac{1}{3}x + \dfrac{4}{3}$; $\dfrac{x}{4} + \dfrac{y}{\frac{4}{3}} = 1$　　**15.** $y = -x - \dfrac{3}{2}$; $\dfrac{x}{-3} + \dfrac{y}{-3} = 1$

16. $3y - x = 11$　　**17.** $5y + 3x = 13$　　**18.** $y - 2x + 1 = 0$
19. $y - 3x = 20$　　**20.** a. $\{x \mid x \neq 2\}$, b. $(-\infty, \infty)$　　**21.** no

EXERCISE 7-2

1. $\dfrac{\pi}{3}$　　**3.** $90°$　　**5.** $135°$　　**7.** $\dfrac{11\pi}{6}$　　**9.** $\dfrac{5\pi}{4}$　　**11.** $210°$

13. $-22\dfrac{1}{2}°$　　**15.** $-\dfrac{\pi}{3}$　　**17.** $\left(\dfrac{8100}{\pi}\right)°$　　**19.** $1080°$　　**21.** $208°$

23. $50°$　　**25.** $\dfrac{4\pi}{3}, 80\pi$　　**27.** $395°, 755°, -325°, -685°$

29. $581° \, 51' \, 5'', 941° \, 51' \, 5'', -138° \, 8' \, 55'', -498° \, 8' \, 55''$
31. $289° \, 39' \, 40'', 649° \, 39' \, 40'', -430° \, 20' \, 20'', -790° \, 20' \, 20''$
33. $\dfrac{7\pi}{3}, \dfrac{13\pi}{3}, -\dfrac{5\pi}{3}, -\dfrac{11\pi}{3}$　　**35.** $40°$　　**37.** $\dfrac{\pi}{3600}$ radians

EXERCISE 7-3

Answers to Problems 1–15 are given in the order $\sin \theta$, $\cos \theta$, $\tan \theta$, $\cot \theta$, $\sec \theta$, and $\csc \theta$.

1. $\dfrac{3}{5}, \dfrac{4}{5}, \dfrac{3}{4}, \dfrac{4}{3}, \dfrac{5}{4}, \dfrac{5}{3}$ **3.** $\dfrac{\sqrt{26}}{26}, \dfrac{5\sqrt{26}}{26}, \dfrac{1}{5}, 5, \dfrac{\sqrt{26}}{5}, \sqrt{26}$

5. $\dfrac{\sqrt{3}}{2}, \dfrac{1}{2}, \sqrt{3}, \dfrac{\sqrt{3}}{3}, 2, \dfrac{2\sqrt{3}}{3}$ **7.** $\dfrac{1}{2}, \dfrac{\sqrt{3}}{2}, \dfrac{\sqrt{3}}{3}, \sqrt{3}, \dfrac{2\sqrt{3}}{3}, 2$

9. $\dfrac{1}{2}, \dfrac{\sqrt{3}}{2}, \dfrac{\sqrt{3}}{3}, \sqrt{3}, \dfrac{2\sqrt{3}}{3}, 2$ **11.** $\dfrac{\sqrt{2}}{2}, \dfrac{\sqrt{2}}{2}, 1, 1, \sqrt{2}, \sqrt{2}$

13. $\dfrac{\sqrt{3}}{3}, \dfrac{\sqrt{6}}{3}, \dfrac{\sqrt{2}}{2}, \sqrt{2}, \dfrac{\sqrt{6}}{2}, \sqrt{3}$

15. $\dfrac{a\sqrt{a^2+b^2}}{a^2+b^2}, \dfrac{b\sqrt{a^2+b^2}}{a^2+b^2}, \dfrac{a}{b}, \dfrac{b}{a}, \dfrac{\sqrt{a^2+b^2}}{b}, \dfrac{\sqrt{a^2+b^2}}{a}$

17. $\cos \theta = \dfrac{\sqrt{21}}{5}$, $\tan \theta = \dfrac{2\sqrt{21}}{21}$, $\cot \theta = \dfrac{\sqrt{21}}{2}$, $\sec \theta = \dfrac{5\sqrt{21}}{21}$, $\csc \theta = \dfrac{5}{2}$

19. $\sin \theta = \dfrac{4\sqrt{17}}{17}$, $\cos \theta = \dfrac{\sqrt{17}}{17}$, $\cot \theta = \dfrac{1}{4}$, $\sec \theta = \sqrt{17}$, $\csc \theta = \dfrac{\sqrt{17}}{4}$

21. $\sin \theta = .8$, $\cos \theta = .6$, $\tan \theta = 1.33$, $\cot \theta = .75$, $\sec \theta = 1.67$

EXERCISE 7-4

1. 0, 1, 0, undefined, 1, undefined **3.** 1, 0, undefined, 0, undefined, 1

5. $\dfrac{\sqrt{2}}{2}$ **7.** $\dfrac{\sqrt{3}}{3}$ **9.** 0 **11.** $\dfrac{\sqrt{3}}{2}$ **13.** $\sqrt{2}$ **15.** $\dfrac{1}{2}$

17. 0 **19.** false **21.** false **23.** true **25.** $\sqrt{2}$

EXERCISE 7-5

1. .5299 **3.** 5.145 **5.** .5430 **7.** .1736 **9.** .8007 **11.** 11.03
13. 1.516 **15.** 4.123 **17.** .8766 **19.** 1.000 **21.** .5934
23. .4763 **25.** 41° 40′ **27.** 74° 20′ **29.** 45° **31.** 52° 57′
33. 37° 14′ **35.** 40° 17′ **37.** 76° 2′ **39.** 41° 49′ **41.** 17° 45′

EXERCISE 7-6

1. $\dfrac{1}{2}$ **3.** $-\dfrac{1}{2}$ **5.** -1 **7.** -1 **9.** $\dfrac{-2\sqrt{3}}{3}$ **11.** $\sqrt{2}$

13. $-.4245$ **15.** 9.567 **17.** .9838 **19.** 2.747 **21.** 9.567

23. $-.9744$ **25.** $\dfrac{\sqrt{3}}{3}$ **27.** $-.3330$ **29.** $-.3420$

31. 277° 10′ **33.** 203° **35.** 330° **37.** 135°, 225°
39. 27°, 333° **41.** 140° **43.** 240°, 300° **45.** 138° **47.** 189°
49. 52°, 232° **51.** impossible **53.** 30° or 150°

EXERCISE 7-7

1. $A = 60°, B = 30°, a = 4\sqrt{3}$ 3. $A = 30°, b = 3\sqrt{3}, c = 6$
5. $B = 45°, a = 3\sqrt{2}, b = 3\sqrt{2}$ 7. $A = 53° 8', B = 36° 52', c = 10$
9. $B = 67°, a = 9.38, b = 22.09$ 11. $A = 62° 35', a = 47.4, c = 53.4$
13. $A = 45° 34', B = 44° 26', a = 7.1$
15. $B = 56° 27', a = 6.85, b = 10.33$
17. $A = 63° 26', B = 26° 34', c = 15.65$
19. $A = 3° 31', a = 4.93, b = 80.2$ 21. $A = 52° 17', B = 37° 43', c = 29.6$
23. $41° 49'$ 25. 370.5 ft   27. $20° 33'$

EXERCISE 7-8

1. $V_x = -75\sqrt{2}, V_y = 75\sqrt{2}$ 3. $V_x = .5282, V_y = -10.09$
5. $V_x = -31.92, V_y = -2.234$ 7. $V_x = -38.16, V_y = 196.3$
9. $V_x = 88.17, V_y = -121.4$

EXERCISE 7-9

1. $R = 141.1, \theta = 245° 06'$ 3. $R = 486.0, \theta = 109° 22'$
5. $R = 362.9, \theta = 67° 32'$ 7. $R = 507.3, \theta = 117° 11'$
9. $R = 210.2, \theta = 49° 29'$ 11. $135°$ 13. 10.61 lbs

REVIEW PROBLEMS—CHAPTER 7

1. a. $18° 31'$ b. $-341° 29'$ c. $.3176, .9482, .3349, 2.986, 1.054, 3.149$
2. a. $125° 42'$ b. $-234° 18'$ c. $.8121, -.5836, -1.391, -.7186, -1.714,$
1.231 3. a. $142° 12'$ b. $-217° 48'$ c. $.6129, -.7902, -.7757, -1.289,$
$-1.265, 1.631$ 4. a. $30°$ b. $-330°$ c. $\frac{1}{2}, \frac{\sqrt{3}}{2}, \frac{\sqrt{3}}{3}, \sqrt{3}, \frac{2\sqrt{3}}{3}, 2$
5. a. $343° 50'$ b. $-16° 10'$ c. $-.2784, .9605, -.2899, -3.450, 1.041,$
-3.592 6. a. $22° 22'$ b. $-337° 38'$ c. $.3805, .9248, .4115, 2.430, 1.081,$
2.628 7. a. $60°$ b. $-300°$ c. $\frac{\sqrt{3}}{2}, \frac{1}{2}, \sqrt{3}, \frac{\sqrt{3}}{3}, 2, \frac{2\sqrt{3}}{3}$
8. a. $67° 06'$ b. $-292° 54'$ c. $.9212, .3891, 2.367, .4224, 2.570, 1.085$
9. a. $117° 45'$ b. $-242° 15'$ c. $.8850, -.4656, -1.900, -.5262, -2.148,$
1.130 10. a. $302° 50'$ b. $-57° 10'$ c. $-.8403, .5422, -1.550, -.6453,$
$1.190, -1.844$ 11. a. $262° 40'$ b. $-97° 20'$ c. $-.9918, -.1276, 7.770,$
$.1287, -7.834, -1.008$ 12. a. $267° 45'$ b. $-92° 15'$ c. $-.9992,$
$-.0393, 25.48, .0393, -25.50, -1.001$ 13. a. $261° 30'$ b. $-98° 30'$
c. $-.9890, -.1478, 6.691, .1495, -6.765, -1.011$ 14. a. $120°$ b. $-240°$
c. $\frac{\sqrt{3}}{2}, -\frac{1}{2}, -\sqrt{3}, -\frac{\sqrt{3}}{3}, -2, \frac{2\sqrt{3}}{3}$ 15. $A = 21° 48', B = 68° 12',$
$c = 10.77$ 16. $A = 13° 46', B = 76° 14', a = 20.40$ 17. $A = 75°,$
$b = 1.607, c = 6.210$ 18. $B = 66°, a = 4.007, c = 9.855$ 19. $B = 44°,$
$a = 7.252, c = 10.08$ 20. $A = 54° 50', a = 9.810, b = 6.912$

21. $B = 57° 20'$, $a = 10.80$, $b = 16.84$ **22.** $B = 30°$, $b = 15\sqrt{3}$,
$c = 30\sqrt{3}$ **23.** 6.7 ft   **24.** $R_x = 50\sqrt{3} + 60\sqrt{2} + 25$;
$R_y = 25\sqrt{3} + 60\sqrt{2} - 50$ **25.** $R_x = 5\sqrt{2} + 5$;
$R_y = 5\sqrt{3} - 25\sqrt{2}$

EXERCISE 8-1

1. consistent, independent; $\{(-1, 1)\}$

3. inconsistent; $\varnothing$

5. consistent, independent; $\{(1, 2)\}$

7. consistent, dependent

9. consistent, independent; $\{(2.8, .6)\}$

EXERCISE 8-2

1. $\{(-1, 1)\}$ **3.** $\{(3, -1)\}$ **5.** $\varnothing$, inconsistent

7. $\{(x, y) \mid y = 6x - 3\}$ **9.** $\{(12, -12)\}$ **11.** $\left\{\left(\dfrac{7}{8}, \dfrac{5}{8}\right)\right\}$

13. $\{(4, -1)\}$ **15.** $\{(0, 0)\}$ **17.** $\varnothing$ **19.** $\left\{\left(\dfrac{a^2 b}{a^2 + b^2}, \dfrac{-ab^2}{a^2 + b^2}\right)\right\}$

EXERCISE 8-3

1. $\{(-1, 1)\}$ **3.** $\{(3, 5)\}$ **5.** $\varnothing$ **7.** $\left\{\left(\dfrac{14}{5}, \dfrac{3}{5}\right)\right\}$ **9.** $\{(0, 18)\}$

11. $\left\{\left(\dfrac{7}{8}, \dfrac{5}{8}\right)\right\}$ **13.** $\{(4, -1)\}$ **15.** $\{(0, 0)\}$ **17.** $\varnothing$

19. $N = 16, F = 4.8$

EXERCISE 8-4

1. $\{(4, 2, 0)\}$ **3.** $\{(1, 3, 5)\}$ **5.** $\left\{\left(\dfrac{1}{2}, \dfrac{1}{2}, \dfrac{1}{4}\right)\right\}$ **7.** $\{(1, -7, 4)\}$

9. $\left\{\left(\dfrac{26}{9}, \dfrac{22}{9}, \dfrac{16}{3}\right)\right\}$ **11.** $\{(-10, 56, -8, -98)\} = \{(x, y, z, w)\}$

EXERCISE 8-5

1. a. 1 b. -2 c. 0 d. $-3b - a$ e. $az - eh$ f. $-4 - a^2$ g. $-\dfrac{2}{7}$

h. $\dfrac{7}{9}$ 3. $\left\{\left(\dfrac{9}{5}, -\dfrac{2}{5}\right)\right\}$ 5. $\left\{\left(\dfrac{7}{16}, \dfrac{13}{8}\right)\right\}$ 7. $\varnothing$ 9. $\{(0, -4)\}$

11. $\left\{\left(-\dfrac{2}{25}, -\dfrac{3}{35}\right)\right\}$ 13. $\{(2, 1)\}$ 15. $\left\{\left(\dfrac{6}{5}, \dfrac{16}{5}\right)\right\}$

EXERCISE 8-6

1. a. -16 b. -48 c. 98 d. 2 e. -89 f. -8 3. $\{(4, 2, 0)\}$

5. $\left\{\left(\dfrac{2}{3}, -\dfrac{28}{15}, -\dfrac{26}{15}\right)\right\}$ 7. $\{(1, 3, 5)\}$ 9. $\left\{\left(\dfrac{1}{2}, \dfrac{1}{2}, \dfrac{1}{4}\right)\right\}$

11. $\{(1, 1, 1)\}$

EXERCISE 8-7

1. \$4000 at 5 percent, \$8000 at $5\dfrac{1}{2}$ percent 3. 8 ft by 5 ft

5. $\dfrac{300}{7}$ cc of 25 percent solution, $\dfrac{225}{7}$ cc of 32 percent solution

7. Len takes 18 hours; Dick takes $22\dfrac{1}{2}$ hours 9. $\left(\dfrac{225}{2}\right)^\circ$ and $\left(\dfrac{135}{2}\right)^\circ$

11. $y = \dfrac{5}{3}x^2 + \dfrac{2}{3}x$ 13. $H = 50$, $T = 50\sqrt{2}$, $V = 50$ (all in lbs)

15. $i_1 = 5$, $i_2 = -8$, $i_3 = -3$ (all in amps)

REVIEW PROBLEMS—CHAPTER 8

1. $\left\{\left(\dfrac{17}{7}, -\dfrac{8}{7}\right)\right\}$ 2. $\varnothing$ 3. $\left\{\left(0, -\dfrac{6}{5}\right)\right\}$ 4. $\varnothing$

5. $\left\{\left(\dfrac{1}{2}, -\dfrac{9}{2}\right)\right\}$ 6. $\{(3, 5)\}$ 7. $\{(0, 18)\}$ 8. $\{(13, -10)\}$

9. $\{(12, -12)\}$ 10. $\left\{\left(\dfrac{1}{2}, \dfrac{1}{3}\right)\right\}$ 11. $\left\{\left(\dfrac{37}{22}, -\dfrac{3}{11}\right)\right\}$

12. $\left\{\left(\dfrac{13}{7}, \dfrac{3}{7}\right)\right\}$ 13. $\left\{\left(\dfrac{17}{20}, \dfrac{31}{20}\right)\right\}$ 14. $\left\{\left(\dfrac{260}{29}, \dfrac{308}{29}\right)\right\}$

15. $\left\{\left(-\dfrac{16}{11}, \dfrac{32}{11}\right)\right\}$ 16. $\left\{\left(\dfrac{106}{3}, \dfrac{111}{4}\right)\right\}$ 17. $\left\{\left(\dfrac{14}{55}, -\dfrac{7}{19}\right)\right\}$

18. $\left\{\left(\dfrac{84}{29}, \dfrac{29}{5}\right)\right\}$ 19. $\left\{\left(\dfrac{19}{13}, \dfrac{93}{13}, \dfrac{73}{13}\right)\right\}$ 20. $\left\{\left(-\dfrac{10}{3}, \dfrac{91}{15}, \dfrac{143}{15}\right)\right\}$

21. $\left\{\left(\dfrac{722}{305}, \dfrac{3}{61}, -\dfrac{432}{305}\right)\right\}$ 22. $\left\{\left(-\dfrac{810}{31}, -\dfrac{344}{31}, \dfrac{280}{31}\right)\right\}$ 23. 18

24. -24 25. 3 26. -14 27. $\{(1, 2)\}$ 28. $\{(0, 0, 0)\}$

EXERCISE 9-1

1. x^2 3. x 5. $\dfrac{x}{y}$ 7. $x^{5/6}$ 9. x^4 11. $x^{3/2}$

13. $-x^{3/2}$ 15. $\dfrac{8y}{x^6}$ 17. $\dfrac{x^{1/2}}{x^2}$ 19. $\dfrac{x^{2/3}}{x^2}$ 21. $\dfrac{y^2}{x^4}$ 23. $\dfrac{2x^2}{y}$

25. $\dfrac{1}{3x^5}$ 27. $3x^5$ 29. $-\dfrac{1}{3x^5}$ 31. $-\dfrac{x^2}{8}$ 33. 1

35. 6×10^{-13} cm

EXERCISE 9-2

1. $\dfrac{x+y}{xy}$ 3. $-3x^{1/2}$ 5. $\dfrac{2(x+1)}{x}$ 7. $\dfrac{2x^{2/3}}{x}$ 9. $\dfrac{3+xy}{x}$

11. $x^3 + 4x^{3/2} + 4$ 13. $x^2 - 2x^{1/2} + x^{3/2} - 2$ 15. $x + 2x^2 + x^3$

17. $x + 2x^{1/2}y^{1/2} + y$ 19. $\dfrac{3(1+xy)}{x^3}$ 21. $\dfrac{x+y}{x}$

23. $x + x^{5/3} - x^{11/3}$ 25. $x^{1/4} - 3x^{1/12}$ 27. $\dfrac{9y}{x^5}$ 29. $\dfrac{x^{1/2}y^{2/3}}{2xy}$

31. $\dfrac{18y^4w^2}{x^2}$ 33. $\dfrac{22}{3}$ 35. $\dfrac{4-x}{4+x}$ 37. $16x^3 - 1$

EXERCISE 9-3

1. $5\sqrt{2}$ 3. $.1x^2$ 5. $\dfrac{2x^2y^3}{3}$ 7. $2x^3y^2\sqrt{2x}$ 9. $2xy\sqrt[3]{y}$

11. $x + 2$ 13. $x^2y\sqrt[3]{2}$ 15. $(x^2-1)\sqrt{2}$ 17. $x^2w^4\sqrt[6]{y^5w}$

19. $\dfrac{2x}{y^2}\sqrt[4]{x}$ 21. $\sqrt{9xy}$ 23. $\sqrt{75a^2x}$ 25. $\sqrt[3]{27x^4y}$

27. $\sqrt[4]{(x+2)^4(x-2)}$ 29. $\sqrt{x^{2a}y}$ 31. $\sqrt{3}$ 33. $2x\sqrt{y}$

35. $x\sqrt{2x}$ 37. $5\sqrt{\dfrac{x}{y}}$ 39. $\sqrt[6]{xy^2z^5}$ 41. $\sqrt{x}$ 43. $h\sqrt{\dfrac{2\pi}{MkT}}$

EXERCISE 9-4

1. $3\sqrt[3]{3}$ 3. $-2\sqrt{6}$ 5. $(x^2+4)\sqrt[3]{y}$ 7. $(x^2+3x+3)\sqrt{6}$

9. $11\sqrt{3}$ 11. $14\sqrt{2} - 1$ 13. $-y\sqrt{x}$ 15. $-8\sqrt[3]{2}$

17. 11 19. $(y - z + x)\sqrt{xyz}$ 21. $5x\sqrt{2}$

23. $\sqrt{(x+y)(x-y)}$ 25. $15x^2\sqrt{3y}$

EXERCISE 9-5

1. 4 3. $2\sqrt{3}$ 5. $18\sqrt{2}$ 7. $3\sqrt[3]{2}$ 9. 3 11. $x\sqrt{6x}$

13. $x\sqrt[6]{27x}$ 15. $3\sqrt[6]{3x^5}$ 17. $\dfrac{1}{16}x\sqrt[3]{x}$ 19. $7 + 4\sqrt{3}$

21. $31 - 10\sqrt{6}$ 23. $2\sqrt{3} - \sqrt{6} + 2\sqrt{2} - 2$

25. $6 - 3\sqrt{3} - 2\sqrt{5} + \sqrt{15}$ 27. $3xy^3\sqrt{2x}$

29. $6\sqrt[3]{xy^2} - 4\sqrt[3]{x^2y^2}$ **31.** $6x\sqrt[6]{8xy^2}$ **33.** $46\sqrt{6}$ **35.** $y^2 - 8$
37. $2y + 6\sqrt{2y} + 9$ **39.** $4x^5y^2\sqrt{2y}$

EXERCISE 9-6

1. 4 **3.** $a\sqrt{2}$ **5.** $2x^4$ **7.** $\dfrac{\sqrt{21}}{7}$ **9.** $\sqrt{6}$ **11.** $\dfrac{\sqrt[6]{72}}{2}$

13. $\dfrac{\sqrt[3]{12}}{2}$ **15.** $\sqrt[6]{8x}$ **17.** $\dfrac{\sqrt[6]{8x^3y^4}}{y}$ **19.** $2\sqrt[4]{xy}$ **21.** $2 - \sqrt{3}$

23. $2(\sqrt{3} + \sqrt{2})$ **25.** $-2(2 + \sqrt{6})$ **27.** $\dfrac{\sqrt{3}}{3}$

29. $10 + 3\sqrt{6}$ **31.** $\dfrac{x - \sqrt{5}}{x^2 - 5}$ **33.** $\dfrac{5\sqrt{3} - 4\sqrt{2} - 13}{2}$

35. $\dfrac{\sqrt{2mgh(mb^2 + I)}}{mb^2 + I}$ **37.** $\dfrac{\sqrt{35}}{5}$

EXERCISE 9-7

1. $\{27\}$ **3.** $\left\{\dfrac{41}{2}\right\}$ **5.** $\left\{\dfrac{85}{19}\right\}$ **7.** $\{4\}$ **9.** $\{43\}$ **11.** $\left\{\dfrac{8}{3}\right\}$

13. $\varnothing$ **15.** $\{6^6 + 5\}$ **17.** $\dfrac{4\pi^2 L}{T^2}$

REVIEW PROBLEMS—CHAPTER 9

1. $-27xy^8\sqrt{x}$ **2.** $81x^6y^5\sqrt{2xy}$ **3.** $\dfrac{1}{2y^3}$ **4.** $27xy\sqrt{y}$

5. $\dfrac{2x^4y^3 + 1}{x^3y^3}$ **6.** $x - y$ **7.** $2\sqrt{35}$ **8.** 3 **9.** $\sqrt{2} - 2\sqrt{3}$

10. $\sqrt{6} - 1$ **11.** $\dfrac{2\sqrt{3} - \sqrt{2}}{10}$ **12.** $12 - 4\sqrt{6}$ **13.** $\sqrt[3]{x^2}$

14. $\sqrt[9]{x}$ **15.** $\dfrac{\sqrt{(x+y)(x-y)}}{x+y}$ **16.** $\dfrac{x\sqrt{y}}{y^2}$ **17.** 0

18. $6 + 3\sqrt{3}$ **19.** $\dfrac{\sqrt[6]{72x}}{2}$ **20.** $\dfrac{\sqrt[15]{xy^{13}}}{y}$ **21.** $x = 10$

22. $x = 9$ **23.** $x = 2$ **24.** $x = 5$ **25.** $x = 3$

26. $z = -\dfrac{9}{4}$ **27.** $E_{1y} = \dfrac{kq_1\sqrt{35}}{49}$, $E_{2y} = \dfrac{kq_2\sqrt{85}}{289}$

28. $E_{1x} = \dfrac{kq_1\sqrt{14}}{49}$

EXERCISE 10-1

1. 24 **3.** $\dfrac{9}{2}$ **5.** 96 **7.** 10.6 lb **9.** 1.42 mi/sec

11. 30 calories **13.** It must be six times as great. **15.** The resulting potential V is related to the original potential V_0 by $V = -\dfrac{2}{3}V_0$.

EXERCISE 10-2

1. 9   3. $\dfrac{256}{\pi}$ 5. $\dfrac{81}{10}$ 7. 30 dynes

9. The result is half the original illumination.   11. 36,000 mi

13. $(6.67)(10^{-7})$ N   15. $x = \dfrac{12\sqrt{15}}{\sqrt{y}}$

REVIEW PROBLEMS—CHAPTER 10

1. $\dfrac{21}{8}$ 2. $\dfrac{4}{243}$ 3. 27   4. $\sqrt{2}$ 5. $(1.5)(10^4)$ lb/in.^{2}

6. 16 ft/sec   7. 32 newtons   8. 3 m   9. 48,000 cal/sec

10. (a) 4 (b) 1

EXERCISE 11-2

1. j 3. $-j$ 5. -1 7. 1   9. $9j$ 11. $-5j$

13. $4j\sqrt{2}$ 15. $\dfrac{1}{2}j$ 17. $.3j$ 19. $\pi^3 j$ 21. (a) -6 (b) 5

(c) $-6 - 5j$ 23. (a) -2 (b) -31 (c) $-2 + 31j$ 25. (a) 0

(b) -3 (c) $3j$ 27. (a) 7 (b) $\dfrac{3}{5}$ (c) $7 - \dfrac{3}{5}j$ 29. (a) 0 (b) 1

(c) $-j$ 31. (a) 7 (b) 0 (c) 7   33. (a) -2 (b) 0 (c) -2

35. (a) 0 (b) -7 (c) $7j$ 37. $x = 2, y = -6$ 39. $x = 17, y = 4$

41. $x = 2, y = 16$ 43. $x = -\dfrac{11}{2}, y = -4$ 45. (a) true (b) true

(c) false (d) true   47. $R_x = \dfrac{R_2 R_3}{R_1}, L_x = R_2 R_3 C_1$

EXERCISE 11-3

1. $4\sqrt{2}(\cos 45° + j \sin 45°)$ 3. $4(\cos 300° + j \sin 300°)$

5. $5(\cos 90° + j \sin 90°)$ 7. $3\sqrt{2}(\cos 225° + j \sin 225°)$

9. $\cos 0° + j \sin 0°$ 11. $2(\cos 120° + j \sin 120°)$

13. $2(\cos 210° + j \sin 210°)$ 15. $5(\cos 270° + j \sin 270°)$

17. $2\sqrt{5}(\cos 63° 26' + j \sin 63° 26')$ 19. $\sqrt{3} + j$

21. $-1 + j\sqrt{3}$ 23. $\sqrt{2} - j\sqrt{2}$ 25. $-3j$ 27. 4

29. $\dfrac{3\sqrt{3}}{2} - \dfrac{3j}{2}$ 31. $1.932 + .5176j$ 33. $-.8452 - 1.813j$

35. a circle of radius 1 with its center at the origin

37. c is any nonzero constant   39. $\dfrac{-3\sqrt{3}}{2} + \dfrac{3}{2}j$

41. $\dfrac{3\sqrt{2}}{8} - \dfrac{3j\sqrt{2}}{8}$

EXERCISE 11-4

1. $5j - 3$ 3. $-3 - j$ 5. $-2 + 3j$ 7. $6\sqrt{3} + (2 + 4\sqrt{2})j$
9. $8 - 9j$ 11. $-10j\sqrt{2}$ 13. -216 15. 64 17. $4 + 7j$
19. $26 + 7j$ 21. $16 + 2j$ 23. $-38 - 26j$ 25. $5 + 12j$
27. $-16 - 30j$ 29. $\dfrac{99}{280} - \dfrac{43}{210}j$ 31. $\dfrac{11}{17} - \dfrac{7}{17}j$

33. $\dfrac{7}{10} - \dfrac{1}{10}j$ 35. $1 + j$ 37. $-\dfrac{3}{2}j$ 39. $2j$

41. $\dfrac{12}{29} + \dfrac{30}{29}j$ 43. $\dfrac{1}{6} - \dfrac{\sqrt{5}}{6}j$ 45. $-27 + 11j$

47. $\dfrac{14}{25} - \dfrac{48}{25}j$ 49. -1 51. $-\dfrac{3}{25} - \dfrac{4}{25}j$ 53. $-68 - 576j$

55. -1 57. $\dfrac{7}{5} - \dfrac{1}{5}j$ 59. $3 + 2j, \dfrac{9}{13} + \dfrac{7}{13}j$ 61. $2 - 7j$

63. Yes

EXERCISE 11-5

1. $4 + 4j\sqrt{3}$ 3. $-3\sqrt{3} - 3j$ 5. $2.294 + 3.277j$ 7. $21j$
9. $5\sqrt{2} + 5j\sqrt{2}$ 11. $-3\sqrt{2} - 3j\sqrt{2}$ 13. $-.3728 + .5527j$
15. $-\sqrt{3} - j$ 17. $-\dfrac{8}{3}$ 19. $\dfrac{27\sqrt{2}}{2} + \dfrac{27j\sqrt{2}}{2}$ 21. -2
23. $1 - 2j$ 25. $(50\sqrt{3} + 60\sqrt{2} + 25) + (25\sqrt{3} + 60\sqrt{2} - 50)j$
27. $18\sqrt{2} + (120 - 67\sqrt{2})j$

EXERCISE 11-6

1. $8(\cos 30° + j \sin 30°), 4\sqrt{3} + 4j$
3. $16(\cos 120° + j \sin 120°), -8 + 8j\sqrt{3}$
5. $\cos(-240°) + j \sin(-240°), -\dfrac{1}{2} + \dfrac{j\sqrt{3}}{2}$
7. $32(\cos 225° + j \sin 225°), -16\sqrt{2} - 16j\sqrt{2}$
9. $16(\cos 240° + j \sin 240°), -8 - 8j\sqrt{3}$
11. $16(\cos 120° + j \sin 120°), -8 + 8j\sqrt{3}$
13. $64(\cos 90° + j \sin 90°), 64j$
15. $32(\cos 270° + j \sin 270°), -32j$
17. $1(\cos 150° + j \sin 150°), -\dfrac{\sqrt{3}}{2} + \dfrac{j}{2}$
19. $\left(\dfrac{1}{2^{12}}\right)(\cos 0° + j \sin 0°), \dfrac{1}{4096}$
21. $1, -\dfrac{1}{2} + \dfrac{j\sqrt{3}}{2}, -\dfrac{1}{2} - \dfrac{j\sqrt{3}}{2}$ 23. $-2 + 2j, 2 - 2j$
25. $\dfrac{\sqrt{2}}{2} + \dfrac{j\sqrt{2}}{2}, -\dfrac{\sqrt{2}}{2} - \dfrac{j\sqrt{2}}{2}$

27. $\dfrac{\sqrt{6}}{2} + \dfrac{j\sqrt{2}}{2}$, $-\dfrac{\sqrt{2}}{2} + \dfrac{j\sqrt{6}}{2}$, $-\dfrac{\sqrt{6}}{2} - \dfrac{j\sqrt{2}}{2}$, $\dfrac{\sqrt{2}}{2} - \dfrac{j\sqrt{6}}{2}$

29. $\sqrt[3]{2}\,(\cos 50° + j \sin 50°)$, $\sqrt[3]{2}\,(\cos 110° + j \sin 110°)$,
  $\sqrt[3]{2}\,(\cos 170° + j \sin 170°)$, $\sqrt[3]{2}\,(\cos 230° + j \sin 230°)$,
  $\sqrt[3]{2}\,(\cos 290° + j \sin 290°)$, $\sqrt[3]{2}\,(\cos 350° + j \sin 350°)$

31. $1, -\dfrac{1}{2} + \dfrac{j\sqrt{3}}{2}, -\dfrac{1}{2} - \dfrac{j\sqrt{3}}{2}$

33. $(\cos 0° + j \sin 0°) = 1$, $(\cos 72° + j \sin 72°)$,
  $(\cos 144° + j \sin 144°)$, $(\cos 216° + j \sin 216°)$,
  $(\cos 288° + j \sin 288°)$

35. $\sqrt[6]{2}\,(\cos 15° + j \sin 15°)$, $\sqrt[6]{2}\,(\cos 135° + j \sin 135°)$,
  $\sqrt[6]{2}\,(\cos 255° + j \sin 255°)$

REVIEW PROBLEMS—CHAPTER 11

1. $(10 + \sqrt{5}) + (8 - 2\sqrt{2})j$ 2. $-9j$

3. $(5 - 2\sqrt{10}) + (12\sqrt{2} - 47)j$ 4. $4 - j$ 5. $\dfrac{7}{25} - \dfrac{24j}{25}$

6. $x = \dfrac{137}{42}, y = \dfrac{31}{21}$ 7. $8(\cos 45° + j \sin 45°)$

8. $\dfrac{6}{5}(\cos 300° + j \sin 300°)$ 9. $1(\cos 150° + j \sin 150°)$

10. $27(\cos 270° + j \sin 270°)$ 11. $\dfrac{\sqrt{6}}{2} - \dfrac{j\sqrt{6}}{2}$ 12. $-1 + j$

13. $-\dfrac{\sqrt{3}}{4} - \dfrac{j}{4}$ 14. $\dfrac{1 + j\sqrt{3}}{3}$ 15. $6j$ 16. $-\sqrt{2} - j\sqrt{2}$

17. $-164.6 - 98.88j$ 18. $1, -1, j, -j, \dfrac{\sqrt{2}}{2} + \dfrac{j\sqrt{2}}{2},$
  $\dfrac{\sqrt{2}}{2} - \dfrac{j\sqrt{2}}{2}, -\dfrac{\sqrt{2}}{2} + \dfrac{j\sqrt{2}}{2}, -\dfrac{\sqrt{2}}{2} - \dfrac{j\sqrt{2}}{2}$

19. $1 + j\sqrt{3}, -1 - j\sqrt{3}$

20. $\sqrt[6]{2}\left[\cos\left(\dfrac{225° + k\cdot360°}{6}\right) + j \sin\left(\dfrac{225° + k\cdot360°}{6}\right)\right]$ where $k \in \{0, 1,$
  $2, 3, 4, 5\}$

EXERCISE 12-2

1. $\{-2, -1\}$ 3. $\{-7, -2\}$ 5. $\{4, 3\}$ 7. $\{3, -1\}$ 9. $\{6\}$

11. $\{\pm 2\}$ 13. $\{0, 8\}$ 15. $\{\pm 4j\}$ 17. $\{-1\}$ 19. $\left\{-\dfrac{3}{2}, \dfrac{1}{3}\right\}$

21. $\left\{\dfrac{1}{2}, -4\right\}$ 23. $\{5, -2\}$ 25. $\left\{\dfrac{1}{2}\right\}$ 27. $\{1, -5\}$

29. $\{2a, 3a\}$ 31. $\{-6\}$ 33. $\{-8, 8, 0\}$ 35. $\{-3, -1, 2\}$

37. $\pm\sqrt{\dfrac{2(c - mgh)}{m}}$ 39. $\pm 2\pi\sqrt{\dfrac{L}{g}}$ 41. $\pm\sqrt{\dfrac{2mgh - mv^2}{I}}$

43. $r = 4$ in.   45. 289 sq ft

EXERCISE 12-3

1. $\{-2 \pm \sqrt{7}\}$ 3. $\left\{\dfrac{-1 \pm \sqrt{5}}{2}\right\}$ 5. $\left\{\dfrac{-2 \pm \sqrt{14}}{2}\right\}$

7. real, distinct; $\{-5, 3\}$ 9. real, equal; $\{4\}$

11. imaginary, distinct; $\{2 \pm j\}$ 13. real, distinct; $\left\{\dfrac{5 \pm \sqrt{13}}{2}\right\}$

15. imaginary, distinct; $\{1 \pm j\sqrt{3}\}$ 17. $\left\{\dfrac{1}{2}, -\dfrac{5}{3}\right\}$

19. $\left\{4, -\dfrac{5}{2}\right\}$ 21. $\{\pm 2j\}$ 23. $\{0, -2\}$ 25. $\{4 \pm \sqrt{11}\}$

27. $\left\{\dfrac{-4 \pm j\sqrt{2}}{6}\right\}$ 29. $\left\{0, -\dfrac{4}{11}\right\}$ 31. $\left\{\dfrac{5 \pm j\sqrt{23}}{6}\right\}$

33. $\left\{\dfrac{1 \pm j\sqrt{431}}{12}\right\}$ 35. $\{-10 \pm 4\sqrt{10}\}$ 37. $\left\{\dfrac{\sqrt{3} \pm j\sqrt{29}}{4}\right\}$

39. $\left\{1, -\dfrac{5}{6}\right\}$ 41. $3j, -2j$ 43. $\dfrac{5y \pm y\sqrt{73}}{6}$

45. 1.28 or 2.40 sec   47. 9 cc   49. $m = \dfrac{-D \pm \sqrt{D^2 - 4JS}}{2J}$

EXERCISE 12-4

1. minimum when $x = 1; f(1) = -3$

3. maximum when $x = \dfrac{1}{4}; f\left(\dfrac{1}{4}\right) = -\dfrac{15}{4}$

5. minimum when $x = \dfrac{1}{2}; f\left(\dfrac{1}{2}\right) = \dfrac{47}{4}$

7. $x = 3$ 9. 500 ft by 1000 ft

EXERCISE 12-5

1. $\left\{1, -\dfrac{5}{6}\right\}$ 3. $\{6, -2\}$ 5. $\left\{-1, \dfrac{13}{10}\right\}$ 7. $\{5, -2\}$

9. $\left\{4, -\dfrac{1}{2}\right\}$ 11. $\left\{\dfrac{-5 \pm \sqrt{13}}{6}\right\}$ 13. $\{21\}$ 15. $\left\{\dfrac{64}{9}\right\}$

17. $\{2\}$ 19. $\{0, 4\}$ 21. $\{4\}$ 23. $\{\pm\sqrt{2399}\}$

25. $x = \pm\sqrt{3}, \pm j\sqrt{2}$ 27. $x = -\dfrac{1}{4}, -\dfrac{1}{2}$

29. $x = \dfrac{\pm j\sqrt{7}}{7}, \dfrac{\pm j\sqrt{2}}{2}$ 31. $x = 9$ 33. $x = -4, 1$

35. $x = \dfrac{15}{7}, \dfrac{11}{5}$ 37. $x = \pm j\sqrt{13}, \pm j\sqrt{3}$ 39. $y = -8, 1$

41. $y = 0, -2, 1 \pm j\sqrt{3}$ 43. 6 ohms   45. 86.8 in. or 33.2 in.
47. 11 : 09   49. 6 hrs for first, 2 hrs for second, 3 hrs for third

EXERCISE 12-6

1. $\{(4, 3), (-4, 3), (4, -3), (-4, -3)\}$ 3. $\{(2, 0), (-2, 0)\}$
5. $\{(5, 2\sqrt{2}), (5, -2\sqrt{2}), (-5, 2\sqrt{2}), (-5, -2\sqrt{2})\}$ 7. $\{(6, -8)\}$
9. $x = \dfrac{1}{2}, y = -4$, or $x = 6, y = 7$

11. $x = \dfrac{1}{2}, y = 8$, or $x = 4, y = 1$

13. $x = 3, y = 4$, or $x = 2, y = 1$
15. 7 by 7 and 3 by 3, or 5 by 5 and 9 by 9

17. $V_A = \dfrac{9}{7}$ m/sec, $V_B = \dfrac{30}{7}$ m/sec

REVIEW PROBLEMS—CHAPTER 12

1. $\left\{0, \dfrac{1}{2}\right\}$ 2. $\{5\}$ 3. $\{-5 \pm 5\sqrt{2}\}$ 4. $\left\{\dfrac{3 \pm j\sqrt{23}}{4}\right\}$

5. $\left\{\dfrac{1 \pm j\sqrt{31}}{4}\right\}$ 6. $\left\{\dfrac{5 \pm \sqrt{13}}{6}\right\}$ 7. $\left\{\dfrac{5}{8}, -3\right\}$ 8. $\{0, 9\}$

9. $\{1 \pm \sqrt{3}\}$ 10. $\{-4 \pm \sqrt{19}\}$ 11. $\left\{0, \dfrac{7}{5}\right\}$ 12. $\left\{-\dfrac{5}{3}, 1\right\}$

13. $\{64, -27\}$ 14. $\{\pm j\sqrt{3}, \pm j\sqrt{2}\}$ 15. $\left\{-\dfrac{1}{5}\right\}$ 16. $\varnothing$

17. $\{576\}$ 18. $\varnothing$ 19. $\{2, -1 \pm j\sqrt{3}\}$
20. $\{(\sqrt{14}, 1), (\sqrt{14}, -1), (-\sqrt{14}, 1), (-\sqrt{14}, -1)\}$
21. $\{(-4 + 2\sqrt{6}, -4 + 3\sqrt{6}), (-4 - 2\sqrt{6}, -4 - 3\sqrt{6})\}$
22. $\{(1, 7), (-1, -7), (3, -3), (-3, 3)\}$
23. $\{(4, 4), (-4, -4), (4j\sqrt{2}, -2j\sqrt{2}), (-4j\sqrt{2}, 2j\sqrt{2})\}$
24. $\{(-3 \pm \sqrt{10})j\}$ 25. -1 26. -22

EXERCISE 13-1

1. $\log_4 256 = 4$ 3. $\log_2 32 = 5$ 5. $\log_3\left(\dfrac{1}{9}\right) = -2$

7. $3^3 = 27$ 9. $4^{-2} = \dfrac{1}{16}$ 11. $2^8 = 256$ 13. $2^{-7} = \dfrac{1}{128}$

15. $(.5)^4 = \dfrac{1}{16}$ 17. $\log_{10} 151.4 = 2.1801$ 19. $(.5)^{-4} = 16$

21. $\log_{10} 10,000 = 4$ 23. $\left(\dfrac{1}{6}\right)^4 = \dfrac{1}{1296}$

25.   27.

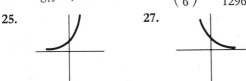

29.   **31.**

33. 81 **35.** 3 **37.** 4 **39.** 3 **41.** 8, −12 **43.** 7

45. $\dfrac{3}{4}$ **47.** −1

EXERCISE 13-2

1. $\log_{10} x + \log_{10} y$ **3.** $\log_{10} x - 2 \log_{10} z$ **5.** $\dfrac{1}{2} \log_{10} x$

7. $6(\log_{10} x + 2 \log_{10} y)$ **9.** $\dfrac{1}{6}(2 \log_{10} x + 3 \log_{10} y - 5 \log_{10} z)$

11. $\log_2 21$ **13.** $\log_{10}\left(\dfrac{7^3}{23^5}\right)$ **15.** $\log_3 32$ **17.** $\log_4\left(\dfrac{1}{50}\right)$

19. $\log_4 3^{10}$ **21.** $\log_2\left(\dfrac{2x}{x+1}\right)$

23. (a) $\dfrac{1}{2}[2 \log_{10} x + 3 \log_{10} (x - 3) - \log_{10} (x + 2)]$

(b) $\log_{10} (x - 3) + 2 \log_{10} (x + 2) + \dfrac{1}{2}[\log_{10} (x) + \log_{10} (x + 2)]$

(c) $-\log_{10} x - 2 \log_{10} (x - 3) - 3 \log_{10} (x + 2)$

EXERCISE 13-3

1. 1.10106 **3.** 6.93556 − 10 **5.** 1.45332 **7.** 4.96379
9. 2.10278 **11.** 9.89020 − 10 **13.** 5.91455 − 10
15. 7.79556 − 10 **17.** 2.13239 **19.** 8.80838 − 10 **21.** 0.30177
23. 4.42073 **25.** 9.626 **27.** 66230 **29.** .0002819 **31.** 128.6
33. .036453 **35.** 55.714 **37.** .0042869 **39.** .00046415
41. .00033335 **43.** 6365.7

EXERCISE 13-4

1. .0078246 **3.** 26.79 **5.** 1.6674 **7.** .000030971
9. 2.5749 **11.** .18525 **13.** .30122 **15.** 124860 **17.** 11.793
19. .012018 **21.** $-(1.0607 \times 10^{-3})$ **23.** 5.2151×10^4
25. 7.9318×10^{-1} **27.** 8.4878 **29.** 6.7352×10^{-1} **31.** 24.8 db
33. 3.0238×10^{-13} cm **35.** 651 Hz.

EXERCISE 13-6

1. 4.80402 **3.** 15.97191 **5.** −5.74149 **7.** 3.48585
9. 15.64006 **11.** 3.6812 **13.** 0.48244 **15.** −6.6301
17. 3.9069 **19.** −4.4912 **21.** 68,600 joules **23.** 6.31 μf/m

EXERCISE 13-7

1. $9.7282 - 10$ **3.** $9.9724 - 10$ **5.** $9.7653 - 10$ **7.** $9.8706 - 10$
9. $9.3882 - 10$ **11.** $36°\,24'$ **13.** $\theta = 45°$ **15.** $.147$

EXERCISE 13-8

1. $\dfrac{1}{2}$ **3.** $\dfrac{81}{2}$ **5.** 1.2789 **7.** 55.479 **9.** 7.1286

11. 3.6229 **13.** 416.7 **15.** 3.071 **17.** $.6483$ **19.** 1.3533

EXERCISE 13-9

1. $y = .4e^{1.1x}$

EXERCISE 13-10

1. $\sqrt{2}\,e^{\pi j/4}$ **3.** $4e^{5\pi j/6}$ **5.** $7e^{3\pi j/2}$ **7.** j

9. $\dfrac{-3\sqrt{2}}{2} + \dfrac{3j\sqrt{2}}{2}$ **11.** $-\sqrt{3} + j$

REVIEW PROBLEMS—CHAPTER 13

1. 2.80265 **2.** $6.80836 - 10$ **3.** 16.93551 **4.** $8.92499 - 10$
5. 3.78155 **6.** $7.98767 - 10$ **7.** $-.45887$ **8.** 6.14847
9. 1.55181 **10.** -11.17646 **11.** 3.48032 **12.** 8.08549
13. 1.3635×10 **14.** 1.581×10^{-2} **15.** 4.7949 **16.** 4.9017×10^{3}
17. 3.0202×10^{-1} **18.** 8.147×10^{-3} **19.** 3.4727×10^{2}
20. 2.6392 **21.** $.39579$ **22.** 1.1447 **23.** $.37539$
24. $9.7509 - 10$ **25.** $9.9169 - 10$ **26.** 1.4371 **27.** 2.6175
28. 2.8813×10^{-43} **29.** $2\sqrt{2}\,e^{5\pi j/4}$ **30.** $-2 + 2j\sqrt{3}$

EXERCISE 14-2

1. $x_b = 0, \dfrac{\pi}{2}, \pi, \dfrac{3\pi}{2}, 2\pi = x_e;\ A = 3;\ p = 2\pi$ **3.** $x_b = 0, \dfrac{\pi}{4}, \dfrac{\pi}{2},$

$\dfrac{3\pi}{4}, \pi = x_e;\ A = 1;\ p = \pi$ **5.** $x_b = 0, \dfrac{3\pi}{2}, 3\pi, \dfrac{9\pi}{2}, 6\pi = x_e;\ A = 2;$

$p = 6\pi$ **7.** $x_b = 0, \dfrac{\pi}{4}, \dfrac{\pi}{2}, \dfrac{3\pi}{4}, \pi = x_e;\ A = 4;\ p = \pi$ **9.** $x_b = 0;$

$\dfrac{\pi}{8}, \dfrac{\pi}{4}, \dfrac{3\pi}{8}, \dfrac{\pi}{2} = x_e;\ A = 2;\ p = \dfrac{\pi}{2};$ inverted   **11.** $x_b = -\dfrac{\pi}{4}, \dfrac{\pi}{4},$

$\dfrac{3\pi}{4}, \dfrac{5\pi}{4}, \dfrac{7\pi}{4} = x_e;\ A = 1;\ p = 2\pi$ **13.** $x_b = \dfrac{\pi}{4}, \dfrac{\pi}{2}, \dfrac{3\pi}{4}, \pi, \dfrac{5\pi}{4} = x_e;$

$A = 4;\ p = \pi$ **15.** $x_b = -\pi, 0, \pi, 2\pi, 3\pi = x_e;\ A = 3;\ p = 4\pi;$

inverted $\qquad$ **17.** $x_b = -\dfrac{1}{\pi}, -\dfrac{1}{\pi}+\dfrac{1}{2}, -\dfrac{1}{\pi}+1, -\dfrac{1}{\pi}+\dfrac{3}{2},$

$-\dfrac{1}{\pi}+2 = x_e;\ A = 1;\ p = 2$

19. **21.** (a) (b) 45°

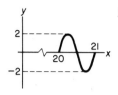

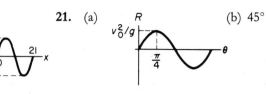

EXERCISE 14-3

1. $x_b = 0, \dfrac{\pi}{4}, \dfrac{\pi}{2}, \dfrac{3\pi}{4}, \pi = x_e;\ A = 1;\ p = \pi$ $\qquad$ **3.** $x_b = 0, \dfrac{\pi}{8}, \dfrac{\pi}{4},$

$\dfrac{3\pi}{8}, \dfrac{\pi}{2} = x_e;\ A = 4;\ p = \dfrac{\pi}{2}$ $\qquad$ **5.** $x_b = 0, \dfrac{\pi}{6}, \dfrac{\pi}{3}, \dfrac{\pi}{2}, \dfrac{2\pi}{3} = x_e;\ A = 4;$

$p = \dfrac{2\pi}{3};$ inverted $\qquad$ **7.** $x_b = -\dfrac{\pi}{3}, \dfrac{\pi}{6}, \dfrac{2\pi}{3}, \dfrac{7\pi}{6}, \dfrac{5\pi}{3} = x_e;\ A = 2;\ p = 2\pi$

9. $x_b = \dfrac{4\pi}{3}, \dfrac{7\pi}{3}, \dfrac{10\pi}{3}, \dfrac{13\pi}{3}, \dfrac{16\pi}{3} = x_e;\ A = 2;\ p = 4\pi$

11. $x_b = \dfrac{1}{2}, 1, \dfrac{3}{2}, 2, \dfrac{5}{2} = x_e;\ A = 1;\ p = 2;$ inverted

13. y_1 leads y_2 by $\dfrac{1}{4}$ cycle or 90° $\qquad$ **15.** 90°

EXERCISE 14-6

1. **3.** **5.**

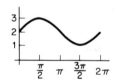 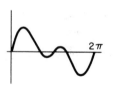 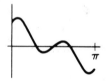

7. **9.**

EXERCISE 14-7

1. $\dfrac{\pi}{3}$ $\qquad$ **3.** π $\qquad$ **5.** $-\dfrac{\pi}{6}$ $\qquad$ **7.** $\dfrac{\pi}{4}$ $\qquad$ **9.** $\dfrac{1}{3}$ $\qquad$ **11.** $\dfrac{\sqrt{3}}{2}$

13. $\dfrac{\pi}{3}$

EXERCISE 14-8

1.   3.

REVIEW PROBLEMS—CHAPTER 14

1. $x_b = -\frac{\pi}{6}$, $x_e = \frac{\pi}{2}$; $A = 2$, $p = \frac{2\pi}{3}$ 2. $x_b = 0$, $x_e = \frac{2\pi}{3}$; $A = 1$,

$p = \frac{2\pi}{3}$, inverted   3. $p = \pi$ 4. $x_b = \frac{\pi}{7}$, $x_e = \frac{3\pi}{7}$; $A = 1$, $p = \frac{2\pi}{7}$

5. $x_b = \frac{\pi}{2}$, $x_e = \frac{5\pi}{2}$; $A = 4$, $p = 2\pi$ 6. $x_b = -\frac{\pi}{12}$, $x_e = \frac{11\pi}{12}$;

$A = 3$, $p = \pi$ 7. $x_b = \frac{2\pi}{3}$, $x_e = \frac{14\pi}{3}$; $A = 1$, $p = 4\pi$ 8. $x_b = 0$,

$x_e = \frac{\pi}{4}$; $A = 3$, $p = \frac{\pi}{4}$ 9. $p = 2\pi$

10. The graph is the straight line $y = \frac{-\sqrt{3}}{2}$.

11.   12.

13. (a) $\frac{\pi}{2}$ (b) π (c) $\frac{\pi}{2}$

EXERCISE 15-2

1. $\frac{\sqrt{6} - \sqrt{2}}{4}$ 3. $\frac{\sqrt{2} - \sqrt{6}}{4}$ 5. $2 - \sqrt{3}$ 7. $\frac{\sqrt{6} - \sqrt{2}}{4}$

9. $\frac{-\sqrt{6} - \sqrt{2}}{4}$ 11. $-\frac{1}{2}$ 13. $\frac{\sqrt{2} + \sqrt{6}}{4}$ 15. $\sqrt{3} - 2$

17. true   19. false   21. true   23. (a) $-\frac{7}{4}$ (b) $\frac{4\sqrt{65}}{65}$

25. $y = -2 \sin x$ 27. $y = 4 \sin x$ 29. $y = \tan x$ 31. $y = \cos x$

33. (a) $x = 2 \cos 2t + 2 \sin 2t$ (b) $2 \cos \frac{\pi}{2} = 0$, $2 \sin \frac{\pi}{2} = 2$

35. $\sqrt{3} \sin \frac{\alpha}{2} + \cos \frac{\alpha}{2}$

EXERCISE 15-3

1. $\frac{\sqrt{3}}{2}$ 3. $-\frac{1}{2}$ 5. $-\sqrt{3}$ 7. $\frac{\sqrt{2} - \sqrt{3}}{2}$

9. $\dfrac{\sqrt{2+\sqrt{2}}}{2}$ 11. $-\sqrt{2}-1$ 13. $\dfrac{4}{5}, \dfrac{3}{5}, \dfrac{4}{3}, \dfrac{24}{25}, -\dfrac{7}{25}, -\dfrac{24}{7}$

15. $-\dfrac{1}{3}, \dfrac{-2\sqrt{2}}{3}, \dfrac{\sqrt{2}}{4}, \dfrac{4\sqrt{2}}{9}, \dfrac{7}{9}, \dfrac{4\sqrt{2}}{7}$ 17. $\dfrac{5}{13}, \dfrac{12}{13}, \dfrac{5}{12}, \dfrac{\sqrt{26}}{26},$

$\dfrac{5\sqrt{26}}{26}, \dfrac{1}{5}$ 19. $-\dfrac{3}{5}, -\dfrac{4}{5}, \dfrac{3}{4}, \dfrac{3\sqrt{10}}{10}, \dfrac{-\sqrt{10}}{10}, -3$

21. $R = \dfrac{2V_0^2 \sin\theta \cos\theta}{g}$ 23. $3\sin x - 4\sin^3 x$

EXERCISE 15-4

1. $\left\{\dfrac{\pi}{4}, \dfrac{3\pi}{4}\right\}$ 3. $\left\{\dfrac{3\pi}{4}, \dfrac{7\pi}{4}\right\}$ 5. $\left\{\dfrac{5\pi}{8}, \dfrac{7\pi}{8}, \dfrac{13\pi}{8}, \dfrac{15\pi}{8}\right\}$

7. $\left\{0, \pi, \dfrac{\pi}{3}, \dfrac{5\pi}{3}\right\}$ 9. $\left\{\dfrac{\pi}{2}\right\}$ 11. $\left\{\dfrac{\pi}{3}, \dfrac{4\pi}{3}\right\}$

13. $\left\{199° 28', 340° 32', \dfrac{\pi}{6}, \dfrac{5\pi}{6}\right\}$ 15. $\left\{\dfrac{\pi}{6}, \dfrac{5\pi}{6}, \dfrac{3\pi}{2}\right\}$

17. $\left\{\dfrac{\pi}{2}, \dfrac{3\pi}{2}, \dfrac{7\pi}{6}, \dfrac{11\pi}{6}\right\}$ 19. $\{0, \pi\}$ 21. $\left\{\dfrac{\pi}{2}\right\}$

REVIEW PROBLEMS—CHAPTER 15

6. $\dfrac{\sqrt{2}+\sqrt{6}}{4}$ 7. $\dfrac{\sqrt{2}+\sqrt{6}}{4}$ 8. $\sqrt{2}-1$

9. $\dfrac{\sqrt{2}-\sqrt{6}}{4}$ 10. $\dfrac{\sqrt{6}-\sqrt{2}}{4}$ 11. undefined

12. $\sqrt{6}+\sqrt{2}$ 13. $2-\sqrt{3}$ 14. $-(\sqrt{2}+\sqrt{6})$

15. $\dfrac{\sqrt{2+\sqrt{2}}}{2}$ 16. $\dfrac{\sqrt{2+\sqrt{2}}}{2}$

17. $-\dfrac{15}{17}, -\dfrac{8}{17}, \dfrac{15}{8}, \dfrac{240}{289}, -\dfrac{161}{289}, -\dfrac{240}{161}, \dfrac{5\sqrt{34}}{34}, \dfrac{-3\sqrt{34}}{34}, -\dfrac{5}{3}$

18. $-\dfrac{3}{5}, \dfrac{4}{5}, -\dfrac{3}{4}, \dfrac{24}{25}, \dfrac{7}{25}, -\dfrac{24}{7}, \dfrac{\sqrt{10}}{10}, \dfrac{-3\sqrt{10}}{10}, -\dfrac{1}{3}$

19. $y = 2\sqrt{2}\sin x + 2\sqrt{2}\cos x$ 20. $\left\{\dfrac{\pi}{2}, \dfrac{3\pi}{2}\right\}$

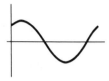

21. $\left\{0, \dfrac{\pi}{4}, \pi, \dfrac{7\pi}{4}\right\}$ 22. $\left\{\dfrac{4\pi}{3}, \dfrac{5\pi}{3}\right\}$

EXERCISE 16-1

1. $C = 30°$, $b = 25.71$, $c = 13.05$ 3. No triangle   5. $B = 90°$, $C = 60°$, $c = 95.26$ 7. $B = 39° 03'$, $C = 59° 57'$, $c = 101.7$

9. (a) $B = 26° 5'$, $C = 133° 55'$, $c = 14.75$ (b) $B = 153° 55'$, $C = 6° 5'$, $c = 2.170$ **11.** $B = 27°$, $C = 126°$, $c = 81.97$ **13.** No triangle **15.** $B = 6° 35'$, $C = 75° 25'$, $c = 591.0$ **17.** $C = 48°$, $a = 93.23$, $b = 102.4$ **19.** No triangle **21.** $B = 13°$, $C = 27°$, $c = 14.13$ **23.** $F = 986.4$, $\theta = 70° 33'$

EXERCISE 16-2

1. $A = 22° 46'$, $B = 129° 14'$, $c = 24.23$ **3.** $A = 54° 19'$, $B = 59° 39'$, $C = 66° 2'$ **5.** $A = 36° 33'$, $C = 28° 27'$, $b = 22.83$ **7.** $A = 77° 50'$, $B = 49° 3'$, $C = 53° 7'$ **9.** $A = 30° 45'$, $B = 24° 09'$, $C = 125° 06'$ **11.** 24.24 lb 67° 11' west of north **13.** 52° 15', 82° 6' **15.** 3.8 ft

EXERCISE 16-3

1. $B = 99° 45'$, $b = 120.1$, $c = 32.61$ **3.** $C = 29° 12'$, $b = 28.93$, $c = 14.36$ **5.** $B = 26° 33'$, $C = 10° 56'$, $c = 5.216$

EXERCISE 16-4

1. 8.208 sq units **3.** 32.17 sq units **5.** 514.2 sq units **7.** 2π in.; 12π sq in. **9.** 4π in.; 36π sq in. **11.** $\left(\dfrac{108}{\pi}\right)°$ **13.** 30π rad/sec **15.** 350.1 ft **17.** 35.2 rad/sec **19.** (a) 40π rad/sec; (b) 1120π radians; (c) $\bar{\omega} = 40\pi$ rad/sec, $\bar{v} = 120\pi$ in./sec; (d) $\bar{\omega} = 40\pi$ rad/sec, $\bar{v} = 200\pi$ in./sec

REVIEW PROBLEMS—CHAPTER 16

1. (a) $C = 80°$, $b = 5.64$, $c = 5.91$, Area = 8.33 sq units; (b) $A = 30°$, $a = 4.79$, $c = 6.16$, Area = 13.86 sq units; (c) $B = 25° 40'$, $C = 94° 20'$, $c = 23.03$, Area = 99.72 sq units (d) Case 1. $B = 13° 23'$, $C = 156° 37'$, $c = 13.72$, Area = 9.53 sq units; Case 2. $B = 166° 37'$, $C = 3° 23'$, $c = 2.04$, Area = 1.42 sq units; (e) No triangle **2.** 31.8 ft, 94.7 ft; 26.6 rev **3.** $\dfrac{10\pi}{3}$ in./sec **4.** $75\sqrt{3}$ sq units

EXERCISE 17-1

1. $(-\infty, 2)$ **3.** $(-\infty, 2]$ **5.** $(-2, \infty)$ **7.** $\left[-\dfrac{2}{7}, \infty\right)$ **9.** $\left[-\dfrac{12}{5}, \infty\right)$ **11.** $\left(-\infty, \dfrac{25}{14}\right]$ **13.** $\varnothing$ **15.** $(-\infty, \infty)$ **17.** $(-\infty, 0)$ **19.** $(-\infty, -2]$ **21.** $\left(-\infty, \dfrac{\sqrt{3}-2}{4}\right)$ **23.** $(0, \infty)$ **25.** $84 < I < 114$ **27.**

36

EXERCISE 17-2

1. $\{-1, 1, 2, 3, 6\}$ **3.** $\{0\}$ **5.** $(-\infty, \infty)$ **7.** $\{x \mid x > 1\}$
9. $(-2, 1)$ **11.** $(-\infty, 0) \cup (2, \infty)$ **13.** $(-\infty, -5] \cup [-2, \infty)$
15. $\left[-\dfrac{3}{2}, \dfrac{1}{3}\right]$ **17.** $(-\infty, 1) \cup (1, \infty)$ **19.** $\varnothing$
21. $(-\infty, -4) \cup (4, \infty)$ **23.** $(-\infty, \infty)$ **25.** $\left(-\dfrac{3}{2}, \dfrac{1}{2}\right)$
27. $(-1, 2) \cup (4, \infty)$ **29.** $(-1, 0) \cup (0, 1)$
31. $(-\infty, -1) \cup (3, \infty)$ **33.** $(-3, 2)$ **35.** $(-\infty, -1) \cup (0, 3)$
37. $(-2, 1) \cup (6, \infty)$ **39.** $(-\infty, -2) \cup (-1, 0)$

EXERCISE 17-3

1. 6 **3.** 4 **5.** $\dfrac{28}{3}$ **7.** $-4 < x < 4$ **9.** (a) $|x - 7| < 3$
(b) $|x - 2| < 3$ (c) $|x - 7| \le 5$ (d) $|x - 7| = 4$ (e) $|x + 4| < 2$
(f) $|x| < 3$ (g) $|x| > 6$ (h) $|x - 6| > 4$ (i) $|x - 105| < 3$
11. $\{\pm 6\}$ **13.** $\{\pm 6\}$ **15.** $\{-3, 13\}$ **17.** $\left\{\dfrac{2}{5}\right\}$ **19.** $\left\{\dfrac{1}{2}, 3\right\}$
21. $(-4, 4)$ **23.** $(-12, 20)$ **25.** $(-\infty, \infty)$
27. $(-\infty, -7) \cup (5, \infty)$ **29.** $(-\infty, -16] \cup [-8, \infty)$ **31.** $[2, 3]$
33. $(-\infty, 0] \cup \left[\dfrac{16}{3}, \infty\right)$ **35.** $|x - .01| \le .005$

EXERCISE 17-4

1. **3.** 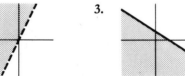 **5.**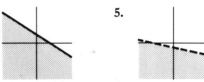

7. 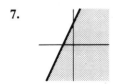 **9.** 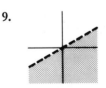 **11.**

13. 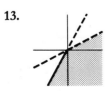 **15.** 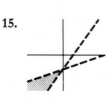 **17.**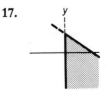

19.   **21.** No solution   **23.**

25.

REVIEW PROBLEMS—CHAPTER 17

1. $\left(-\dfrac{1}{2}, \infty\right)$ **2.** $\left[\dfrac{5}{2}, \infty\right)$ **3.** $(-\infty, \infty)$ **4.** $(-\infty, 0]$

5. $(-4, 5)$ **6.** $(-\infty, -6] \cup [2, \infty)$ **7.** $(-3, 4)$ **8.** $(-\infty, \infty)$

9. $[-4, -2] \cup [0, \infty)$ **10.** $(-4, -3] \cup [0, 2)$

11. $(-\infty, 5) \cup (5, \infty)$ **12.** $\varnothing$ **13.** $\left(-\infty, -\dfrac{9}{8}\right) \cup \left(\dfrac{3}{8}, \infty\right)$

14. $\left[\dfrac{9}{2}, \dfrac{21}{2}\right]$ **15.** $\{-2, 5\}$ **16.** $\left\{\dfrac{8}{5}\right\}$

17.   **18.**   **19.**

20.   **21.** No solution   **22.**

EXERCISE 18-2

1. (a) $\sqrt{53}$ (b) 1   (c) $\sqrt{65}$ (d) 10

EXERCISE 18-3

1. $x^2 + y^2 = 4$; $x^2 + y^2 - 4 = 0$

3. $x^2 + (y - 2)^2 = 2$; $x^2 + y^2 - 4y + 2 = 0$

5. $(x + 2)^2 + (y - 6)^2 = 16$; $x^2 + y^2 + 4x - 12y + 24 = 0$

7. $(x + 4)^2 + (y + 7)^2 = k^2$; $x^2 + y^2 + 8x + 14y + (65 - k^2) = 0$

9. $(x - \sqrt{2})^2 + (y - \sqrt{3})^2 = 4$; $x^2 + y^2 - 2x\sqrt{2} - 2y\sqrt{3} + 1 = 0$

11. $C(0, 0), r = 4$ **13.** $C(2, 0), r = 3$ **15.** $C(-3, 3), r = \sqrt{3}$

17. $C(1, 2), r = 1$ **19.** $C(0, -3), r = 2$ **21.** No real locus

23. $C(-2, 1), r = \sqrt{7}$ **25.** two   **27.** none   **29.** b., c., d., f., g.
31. $x^2 + y^2 = 10^6$ **33.** $x^2 + y^2 = 4.49 \times 10^{-4}$

EXERCISE 18-4

1. $V(0, 0), F(1, 0), D(x = -1)$ **3.** $V(0, 0), F(0, -2), D(y = 2)$

5. $V(0, 0), F\left(0, \dfrac{1}{2}\right), D\left(y = -\dfrac{1}{2}\right)$ **7.** $V(0, -2), F\left(\dfrac{1}{8}, -2\right),$

$D\left(x = -\dfrac{1}{8}\right)$ **9.** $V(4, -3), F(4, -1), D(y = -5)$ **11.** $V(1, -2),$

$F\left(\dfrac{15}{16}, -2\right), D\left(x = \dfrac{17}{16}\right)$ **13.** $(y - 2)^2 = -12(x + 2)$

15. $x^2 = 16(y + 1)$ **17.** $(y - 4)^2 = -8(x - 3)$
19. $(x + 3)^2 = -8(y - 2)$ **21.** $x^2 = 6y$ **23.** $(y + 2)^2 = 4(x + 4)$
25. $(y - 3)^2 = -4(x - 2), V(2, 3), F(1, 3), D(x = 3)$

27. $(x - 2)^2 = \dfrac{1}{3}y, V(2, 0), F\left(2, \dfrac{1}{12}\right), D\left(y = -\dfrac{1}{12}\right)$

29. $(y + 2)^2 = x - 1, V(1, -2), F\left(\dfrac{5}{4}, -2\right), D\left(x = \dfrac{3}{4}\right)$

31. none   **33.** two   **35.** $\dfrac{5}{8}$ sec; 6.25 m   **37.** less than 5 cm

EXERCISE 18-5

1. $C(0, 0), V(\pm 5, 0), F(\pm 3, 0), a = 5, b = 4$
3. $C(0, 0), V(0, \pm 10), F(0, \pm 5\sqrt{3}), a = 10, b = 5$
5. $C(0, 0), V(0, \pm 2), F(0, \pm \sqrt{3}), a = 2, b = 1$
7. $C(2, -3), V(2 \pm 3, -3), F(2 \pm \sqrt{5}, -3), a = 3, b = 2$
9. $C(2, 0), V(2, \pm 4), F(2, \pm 2\sqrt{3}), a = 4, b = 2$
11. $C(3, -2), V(3 \pm 5, -2), F(3 \pm 4, -2), a = 5, b = 3$

13. $\dfrac{x^2}{36} + \dfrac{y^2}{11} = 1$ **15.** $\dfrac{x^2}{68} + \dfrac{(y - 1)^2}{4} = 1$ **17.** $\dfrac{x^2}{64} + \dfrac{y^2}{16} = 1$

19. $\dfrac{x^2}{\frac{25}{4}} + \dfrac{(y + 8)^2}{64} = 1$ **21.** $\dfrac{(x + 2)^2}{4} + \dfrac{(y + 1)^2}{53} = 1$

23. $\dfrac{x^2}{(8000)^2} + \dfrac{y^2}{(5000)^2} = 1$ choosing the center of the earth as the center of the coordinate system

EXERCISE 18-6

1. $C(0, 0), V(\pm 4, 0), F(\pm 5, 0)$, endpts. $(0, \pm 3)$, asym. $y = \pm\dfrac{3}{4}x$

3. $C(0, 0), V(0, \pm 6), F(0, \pm\sqrt{37})$, endpts. $(\pm 1, 0)$, asym. $y = \pm 6x$
5. $C(2, -3), V(2 \pm 2, -3), F(2 \pm \sqrt{29}, -3)$, endpts. $(2, -3 \pm 5)$, asym.

$y = -3 \pm \left(\dfrac{5}{2}\right)(x - 2)$ **7.** $C(0, 0), V(\pm\sqrt{3}, 0), F(\pm\sqrt{6}, 0)$, endpts.

$(0, \pm\sqrt{3})$, asym. $y = \pm x$ **9.** $C(2, -1)$, $V(2 \pm 4, -1)$, $F(2 \pm 5, -1)$,

endpts. $(2, -1 \pm 3)$, asym. $y = -1 \pm \dfrac{3}{4}(x - 2)$ **11.** $\dfrac{y^2}{4} - \dfrac{x^2}{5} = 1$

13. $\dfrac{x^2}{16} - y^2 = 1$ **15.** $\dfrac{(y-2)^2}{4} - \dfrac{(x+4)^2}{12} = 1$

17. $\dfrac{(x+2)^2}{4} - \dfrac{4(y-4)^2}{9} = 1$ **19.** $\dfrac{(y-4)^2}{4} - \dfrac{x^2}{5} = 1$

21. $x = 0, y = 0$ **23.** $x = 0, y = 0$ **25.** $x = 0, y = 0$

27.

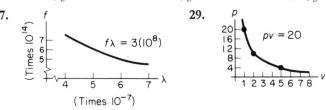

$f\lambda = 3(10^8)$

(Times 10^{-7})

29.

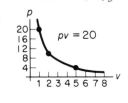

$pv = 20$

EXERCISE 18-7

1. circle   **3.** parabola   **5.** parabola   **7.** hyperbola   **9.** ellipse
11. hyperbola   **13.** ellipse   **15.** ellipse   **17.** parabola
19. hyperbola

EXERCISE 18-8

1.

$2y = x + 2$

3.

$y = \dfrac{1}{x}$

5. $y = -x^2 + 5x - 6$

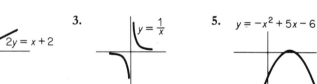

7.

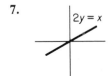

$2y = x$

9. $(x+1)^2 + (y+2)^2 = 4$

-1

-2

11. $x = \dfrac{1-t}{t}, y = 1 - t$ **13.** $x = \dfrac{2t}{1+t^3}, y = \dfrac{2t^2}{1+t^3}$

15. $x = \dfrac{1-t}{1+2t}, y = \dfrac{t(1-t)}{1+2t}$

17. $x^2 + y^2 = r^2$, the equation of a circle   **19.** $t = \dfrac{v_0 \sin \alpha}{g}$

21. $\dfrac{x^2}{a^2} + \dfrac{y^2}{b^2} = 1$

REVIEW PROBLEMS—CHAPTER 18

1. circle, $C(2, -3)$, radius $= 3$ **2.** hyperbola, $C(-1, 0)$, $V(-1, \pm 2)$,
$F(-1, \pm 2\sqrt{2})$, asymptotes: $y = \pm(x + 1)$ **3.** hyperbola, $C(-1, 0)$,

$V(-1 \pm 2, 0)$, $F(-1 \pm 2\sqrt{5}, 0)$, $y = \pm 2(x + 1)$ **4.** parabola,

$V\left(-\dfrac{5}{2}, 1\right)$, $F(-2, 1)$ **5.** hyperbola, $C(0, 0)$ **6.** ellipse, $C(2, -1)$,

$V(2 \pm 3, -1)$, $F(2 \pm \sqrt{5}, -1)$ **7.** parabola, $V(0, 2)$, $F(3, 2)$

8. No locus   **9.** ellipse, $C(-5, 2)$, $V(-5, 2 \pm 2\sqrt{2})$, $F(-5, 2 \pm 2)$

10. circle, $C(0, -2)$, radius $= 6$ **11.** parabola, $V(3, 0)$, $F\left(3, -\dfrac{1}{4}\right)$

12. ellipse, $C(0, 0)$, $V(\pm 4, 0)$, $F(\pm 2\sqrt{3}, 0)$ **13.** parabola, $V(0, 0)$,

$F\left(0, \dfrac{1}{12}\right)$ **14.** hyperbola, $C(-2, 6)$, $V(-2, 6 \pm 1)$, $F(-2, 6 \pm \sqrt{2})$,

$y = 6 \pm (x + 2)$ **15.** No locus   **16.** $4y = x^2 - 10x + 24$

17. $x = \dfrac{3t}{2 - t^3}$, $y = \dfrac{3t^2}{2 - t^3}$ **18.** (a) ellipse   (b) hyperbola

(c) parabola   (d) parabola   (e) circle

EXERCISE 19-1

1. $3, 6, 9, 12$ **3.** $1, 3, 5, 7$ **5.** $\dfrac{1}{2}, \dfrac{2}{3}, \dfrac{3}{4}, \dfrac{4}{5}$ **7.** $0, \dfrac{1}{3}, \dfrac{1}{2}, \dfrac{3}{5}$

9. $\dfrac{1}{2}, \dfrac{1}{2}, \dfrac{3}{8}, \dfrac{1}{4}$ **11.** $1, 0, -1, 0$ **13.** $\dfrac{e}{2}, \dfrac{e^2}{2}, \dfrac{e^3}{2}, \dfrac{e^4}{2}$

15. $1, -4, 9, -16$ **17.** $4n$ **19.** $2(n + 1)$ **21.** $\left(\dfrac{1}{3}\right)^{n-1}$

23. $\dfrac{(-1)^{n+1}}{n + 1}$

EXERCISE 19-2

1. arith., -71 **3.** geom., $-\dfrac{3}{16}$ **5.** geom., 243 **7.** arith., $\dfrac{11}{3}$

9. geom., $\dfrac{1}{128}$ **11.** neither   **13.** arith., 4 **15.** geom., $\dfrac{729}{64}$

17. 142 **19.** $\dfrac{16384}{243}$ **21.** 6 **23.** 7 ft   **25.** $6561°C$

EXERCISE 19-3

1. $2 + 3 + 4$ **3.** $-1 + 0 + 3$ **5.** $5 - 10 + 17$ **7.** $2 + 4 + 6$

9. $\dfrac{1}{2} + \dfrac{1}{4} + \dfrac{1}{8}$ **11.** 110 **13.** -360 **15.** -108 **17.** 620

19. 1092 **21.** -126 **23.** $.3333333333$ **25.** $62(1 + \sqrt{2})$

27. $a_1 = 10, n = 9; a_1 = -\dfrac{39}{4}, n = 88$ **29.** $S_{16} = 528, a_{16} = 48$

31. $n = 23, a_n = 36$ **33.** $a_1 = 64, a_6 = 2$ **35.** 8

37. $\dfrac{121}{243}$ or $-\dfrac{61}{243}$ **39.** $4.65 **41.** 385 **43.** 10,100

45. 2304 ft **47.** $\dfrac{81}{256}$

EXERCISE 19-4

1. div **3.** 0 **5.** $\dfrac{11}{4}$ **7.** 0 **9.** 0 **11.** $\dfrac{3}{2}$ **13.** div

15. 0 **17.** 0

EXERCISE 19-5

1. 60 **3.** 10 **5.** 1 **7.** 2 **9.** $\dfrac{27}{50}$ **11.** 6 **13.** 18

15. $-\dfrac{8}{3}$ **17.** $\dfrac{50}{27}$ **19.** $\dfrac{21}{4}$ **21.** $\dfrac{8}{33}$ **23.** $\dfrac{3209}{999}$ **25.** $\dfrac{2111}{99000}$

27. 12 ft **29.** 100 in. **31.** 50 ft

EXERCISE 19-6

1. $x^5 + 15x^4 + 90x^3 + 270x^2 + 405x + 243$

3. $a^5 - \dfrac{5a^4}{b} + \dfrac{10a^3}{b^2} - \dfrac{10a^2}{b^3} + \dfrac{5a}{b^4} - \dfrac{1}{b^5}$

5. $x^6 - 12x^5y + 60x^4y^2 - 160x^3y^3 + 240x^2y^4 - 192xy^5 + 64y^6$

7. $x^6 + 3x^5y + \dfrac{15x^4y^2}{4} + \dfrac{5x^3y^3}{2} + \dfrac{15x^2y^4}{16} + \dfrac{3xy^5}{16} + \dfrac{y^6}{64}$

9. $x^7 + \dfrac{7x^6}{2} + \dfrac{21x^5}{4} + \dfrac{35x^4}{8} + \dfrac{35x^3}{16} + \dfrac{21x^2}{32} + \dfrac{7x}{64} + \dfrac{1}{128}$

11. $16x^4 - 160x^3 + 600x^2 - 1000x + 625$

13. $1 + \dfrac{6x}{y^2} + \dfrac{15x^2}{y^4} + \dfrac{20x^3}{y^6} + \dfrac{15x^4}{y^8} + \dfrac{6x^5}{y^{10}} + \dfrac{x^6}{y^{12}}$

15. $\dfrac{x^7}{y^7} + \dfrac{7x^5}{y^5} + \dfrac{21x^3}{y^3} + \dfrac{35x}{y} + \dfrac{35y}{x} + \dfrac{21y^3}{x^3} + \dfrac{7y^5}{x^5} + \dfrac{y^7}{x^7}$

17. $x^{15} + 15x^{14}y + 105x^{13}y^2 + 455x^{12}y^3$

19. $2^{21}x^{21} - (21)2^{20}x^{20}b^2 + (210)2^{19}x^{19}b^4 - (1330)2^{18}x^{18}b^6$

21. $x^{16} - \dfrac{16x^{15}}{y} + \dfrac{120x^{14}}{y^2} - \dfrac{560x^{13}}{y^3}$

23. $x^{14} - 28x^{13}w^2y + 364x^{12}w^4y^2 - 2912x^{11}w^6y^3$

25. $a^{40} + 20a^{38}b^2 + 190a^{36}b^4 + 1140a^{34}b^6$

27. $1 - 2x + 3x^2 - 4x^3$

29. $\dfrac{1}{x^4} - \dfrac{4y}{x^5} + \dfrac{10y^2}{x^6} - \dfrac{20y^3}{x^7}$ **31.** $1 - \dfrac{2x}{3} + \dfrac{5x^2}{9} - \dfrac{40x^3}{81}$

33. $\dfrac{\sqrt{2a}}{2a} + \dfrac{3c\sqrt{2a}}{8a^2} + \dfrac{27c^2\sqrt{2a}}{64a^3} + \dfrac{135c^3\sqrt{2a}}{256a^4}$

35. $3003x^{10}y^5$ **37.** $-15504x^5y^{15}$ **39.** $10{,}264{,}320x^8y^4$ **41.** $\dfrac{6435y^8}{x^{14}}$

43. $59136x^6y^6$ **45.** $184756a^{20}b^{20}$ **47.** $-286x^{10}y^3$
49. $-(12285)2^{12}y^{12}x^6$ **51.** 7.071   **53.** 7.810   **55.** 3.072
57. 2.991   **59.** .923

REVIEW PROBLEMS—CHAPTER 19

1. 1200   **2.** 11   **3.** $\dfrac{23}{99}$ **4.** $-\dfrac{112b^4}{243a^{10/3}}$ **5.** $-\dfrac{4}{11}$

6. $\dfrac{35b^8}{4a^{9/2}}$ **7.** $924x^{12}$ **8.** $\dfrac{255}{64}$ **9.** $\dfrac{3}{2}$ **10.** 4.987

11. $\dfrac{4}{5}$ **12.** $\dfrac{1}{2^{n-1}}$ **13.** 2^{15} **14.** -1 **15.** (b) and (c)

16. $\dfrac{4}{3}$ **17.** $\dfrac{1}{\sqrt[8]{165}}$ **18.** $\dfrac{243}{128}$ ft   **19.** 56 ft

EXERCISE A-6

1. 6   **2.** 7   **3.** 10   **4.** 9.1   **5.** 6.75   **6.** 9.62   **7.** 49.8
8. 340   **9.** 47.0   **10.** .0826   **11.** 3220   **12.** .836   **13.** 9.87
14. 3.08

EXERCISE A-7

1. 15   **2.** 15.77   **3.** 3530   **4.** 42.1   **5.** .001322   **6.** 1737
7. 9.98   **8.** 1341   **9.** 244   **10.** 57.1   **11.** .1621   **12.** .201
13. 170.5   **14.** 5620   **15.** 6890   **16.** 2870

EXERCISE A-8

1. 2.32   **2.** 165.2   **3.** .0767   **4.** 106.1   **5.** .000713   **6.** 77.5
7. 1861   **8.** 26.3   **9.** 1.154   **10.** .0419   **11.** .936   **12.** 1.535
13. 616   **14.** .0298   **15.** 4.96   **16.** .332

EXERCISE A-9

1. 36.7   **2.** 8.35   **3.** .0000632   **4.** 3400   **5.** .00357
6. 13,970   **7.** 1586   **8.** .0223   **9.** .01311   **10.** 2.35   **11.** .0414
12. .0977

EXERCISE A-10

1. 25.2   **2.** 9.27   **3.** .1202   **4.** 123.0   **5.** 11.93   **6.** 45900
7. 5.07   **8.** 54.9   **9.** 449.5, 817.5, 992.1, 1435, 1598, 2095, 2328,
2533, 3000

EXERCISE A-12

1. .0625, .00385, 1.389, 15.39, .0575, .0541, .01490
2. .00253, .000559, 6.21   3. 2,160   4. 74.0, 10.97
5. 199.5, 8.55   6. See answers to Problems 2, 3, 4.
7. 2.74, .364, .392, 2.56, 12.54, .0797

EXERCISE A-13

1. 2.83	2. 3.46	3. 4.12	4. 9.43	5. 2.98	6. 29.8
7. .943	8. 85.3	9. .252	10. .00797	11. 252	12. 316
13. 231	14. .279	15. 5720			

EXERCISE A-14

1. 24.2	2. .416	3. 8.54	4. .0698	5. 4.43	6. 1.176
7. 32.8	8. 398	9. 43.7	10. 29.4	11. 64.2	12. 11.41
13. 109.1	14. .0602	15. 1.525×10^5	16. 1.589		

EXERCISE A-15

| 1. 2.06 | 2. 3.11 | 3. 9.00 | 4. 9.47 | 5. 19.69 | 6. .1969 |
| 7. .424 | 8. .914 | 9. 44.7 | 10. .855 | 11. 909 | 12. 2.15 |

EXERCISE A-17

2. a. .5 b. .616 c. .0581 d. 1 e. .999 f. .0276 g. .253 h. .381 i. .204
j. .783   3. a. .866 b. .788 c. .998 d. 0 e. .0349 f. 1.00 g. .968
h. .925 i. .979 j. .623   4. a. 30° b. 61° c. 22° d. 5.74° e. .86°
f. 38.3° g. 3.55° h. 1.775° i. 66.9°   5. a. 60° b. 29° c. 68° d. 84.26°
e. 89.14° f. 51.7° g. 86.45° h. 88.22° i. 23.1°

EXERCISE A-18

1. 0.142, 0.515, 1.907, 0.0177, 3.55, 19.1, 1.09,
   7.03, 1.94, 0.524, 56.40, 0.282, 0.0524, 0.918
2. a. 13.50° b. 38.15° c. 42.60° d. 28.37° e. 3.38° f. 4.69° g. 23.36°
h. 2.465° i. 0.855° j. 20.5° k. 74.95° l. 77.91° m. 86.63° n. 45.85°
o. 50.95°   3. a. 76.5° b. 51.85° c. 47.40° d. 61.63° e. 86.6° f. 85.3°
g. 66.6° h. 87.54° i. 89.145° j. 69.5° k. 15.05° l. 12.09° m. 3.37°
n. 44.15° o. 39.05°   4. a. 28.55° b. 24.09° c. 63.4° d. 51.7° e. 50.2°
f. 83.14°

EXERCISE A-19

1. a. .0247 b. .01454 c. .0436 d. .0466   2. a. 1.044° b. 2.65°
c. 4.96°   3. a. .0627 b. .000627 c. .627 d. 6.27   4. a. 1.696°

b. 16.96°   c. .01696°   **5.** a. 15.92   b. 7.59   c. .00000548   d. 50.6
6. .0597, .0597, 16.75   **7.** .000977, .000977, 1023   **8.** .00436, 229, .00436

EXERCISE A-20

1. 6.21, 3.91, 0.693, 0.3365, 0.0421   **2.** −6.21, −3.91, −.693, −.3365, −.0953, −.01975   **3.** a. 4.33   b. 2.03   c. 2.22   d. −.1744   e. −1.93 f. −.1132   g. −.0954   h. .358   i. .0421   **5.** 1.386, .25, 1.1488, .8705, 1.01396, .98623, 1.001386, .998614

EXERCISE A-21

1. a. 20.1   b. .0498   c. 1.492   d. .670   e. 1.03562   f. .9656   g. 3.827 h. .2613   i. .0854   j. 34.8   k. .9740   l. 1.0890   **2.** a. 8.33   b. 0.1200 c. 1.2361   d. 0.8090   e. 1.02143   f. .9790   **3.** a. 54.6   b. 3640   c. 1.537 d. 1.0216   e. .0334   f. .8496   g. .9817   h. .00203

index